Agrobiodiversity Conservation: Securing the Diversity of Crop Wild Relatives and Landraces

Agrobiodiversity Conservation: Securing the Diversity of Crop Wild Relatives and Landraces

Edited by

Nigel Maxted

School of Biosciences, The University of Birmingham, UK

M. Ehsan Dulloo

Bioversity International, Italy

Brian V. Ford-Lloyd

School of Biosciences, The University of Birmingham, UK

Lothar Frese

Julius Kühn-Institut, Federal Research Centre for Cultivated Plants Institute for Breeding Research on Agricultural Crops, Germany

José M. Iriondo

Departamento de Biología Y Geología, Universidad Rey Juan Carlos, Spain

Miguel A.A. Pinheiro de Carvalho

ISOPlexis Gene Bank, Campus da Penteada, Portugal

www.cabi.org

CABI is a trading name of CAB International

CABI
Nosworthy Way
Wallingford
Oxfordshire OX10 8DE
UK

Tel: +44 (0)1491 832111
Fax: +44 (0)1491 833508
E-mail: cabi@cabi.org
Website: www.cabi.org

CABI
875 Massachusetts Avenue
7th Floor
Cambridge, MA 02139
USA

Tel: +1 617 395 4056
Fax: +1 617 354 6875
E-mail: cabi-nao@cabi.org

A catalogue record for this book is available from the British Library, London, UK.

Library of Congress Cataloging-in-Publication Data

Crop wild relative conservation and use / edited by N. Maxted ... [et al.].
p. cm.
ISBN 978-1-84593-099-8 (alk. paper)--ISBN 978-1-84593-307-4 (ebook)
1. Crops--Germplasm resources. 2. Germplasm resources, Plant. 3. Genetic resources conservation. I. Maxted, Nigel. II. Title.

SB123.3.C768 2007
333.95′34--dc22

2007017714

ISBN-13: 978 1 84593 851 2

Commissioning editor: Sarah Hulbert
Editorial assistant: Alexandra Lainsbury
Production editor: Simon Hill

Typeset by SPi, Pondicherry, India.
Printed and bound by CPI Group (UK) Ltd, Croydon, CR0 4YY.

Contents

Preface

The advent of the Convention on Biological Diversity (CBD) and the FAO International Treaty on Plant Genetic Resources for Food and Agriculture (ITPGRFA) has refocused national and international biodiversity conservation on to *in situ* activities and for socio-economic species this has placed a much greater emphasis on on-farm and genetic reserve conservation. Prior to this, agrobiodiversity conservation had been primarily *ex situ* (largely gene bank), but by 1992, when the United Nations Conference on the Environment and Development meeting resulted in the Convention on Biological Diversity, with the exception of crop wild relatives (CWR), the vast majority of crop diversity, had already been effectively conserved *ex situ*. Although there have been significant advances in both *in situ* and *ex situ* agrobiodiversity conservation since the 1960s, there has remained a lack of good illustrative examples of *in situ* CWR and landrace (LR) conservation, particularly where both *in situ* and *ex situ* approaches have been applied in an efficient complementary manner. Slowly but steadily in recent years a methodological base has been developed that facilitates, through conservation action, the practical implementation of on-farm and genetic reserve protocols. However, even today too often protected areas are designated solely to conserve mega-fauna or ecosystem, and agrobiodiversity conservation is ignored.

In Europe the establishment of CWR and LR *in situ* conservation has been led by the European Cooperative Programme for Plant Genetic Resources – *In Situ* and On-Farm Network (www.ecpgr.cgiar.org/networks/iin_situi_and_on_farm.html). The Network first spawned the EC FP5 funded PGR Forum project (http://pgrforum.org/), which made significant *in situ* conservation methodological advances, and then the EC GenRes funded AEGRO project (http://aegro.jki.bund.de), which developed further the methodologies for both CWR and LR *in situ* conservation, and a conference entitled 'Towards the establishment of genetic reserves for crop wild relatives and landraces in Europe' was held at the University of Madeira, Funchal, Portugal between 13 and 17 September 2010 – with the aim of making the advances achieved by AEGRO available to the various stakeholder communities including users of agrobiodiversity and conservationists. This book represents the Conference Proceedings including the presentations given in Funchal, together with some invited papers by those unable to attend. The conference and this volume are aimed at providing exemplars for CWR and LR conservation and use with practical methods, protocols and initiatives being described. Specifically, the objectives of the book are synonymous with those of the conference:

- Development of *in situ* management (ISM) work plans based on the genetic reserve and on-farm concepts;
- Identification of sites suited to organize genetic reserves and on-farm conservation;
- Promotion of discussion and establishment of European genetic reserve and on-farm networks, to ensure European *in situ* crop agrobiodiversity conservation;
- Provision of generic data quality standards for *in situ* CWR and LR diversity conservation, ensuring integration with existing projects;
- Provision of improved linkage between *in situ* and *ex situ* CWR and LR diversity conservation;
- Dissemination of AEGRO products to the European and global PGR community, and discussion of their wider application and continued use.

Although the conference and the volume specifically focused on methodological application in Europe, the *in situ* methods described are generic and are transferable across the globe. Therefore the text is designed to be of use to conservation professionals, geneticists, ecologists, agriculturalists, agro- and biodiversity policy makers and postgraduate students in developed and developing countries throughout the world.

'Towards the establishment of genetic reserves for crop wild relatives and landraces in Europe' conference participants.

Contributors

Afonin, A.A., Saint Petersburg State University, 10 Liniya VO, 33, Saint Petersburg, Russia. Tel: 812 465 0751, Fax: 812 328 7146, Email: acer737@gmail.com

Akparov, Z.I., Institute of Genetic Resources, Azerbaijan National Academy of Sciences, 155 Azadliq Ave, 1106, Baku, Azerbaijan. Tel: (994-12) 5629171, Fax: (994-12) 4499221, Email: akparov@yahoo.com

Ambrose, M., John Innes Centre, Norwich Research Park, Norwich, UK. Tel: (+44) (0)1603 450 630, Email: mike.ambrose@bbsrc.ac.uk

Argento, S., Dipartimento di OrtoFloroArboricoltura e Tecnologie Agroalimentari (DOFATA), University of Catania, Italy. Tel: +39 095 234307, Fax: (+39) 095 234329, Email: sargento@unict.it

Arnaud, E., Bioversity International, Via dei Tre Denari 472/a, 00057 Maccarese, Rome, Italy. Tel: (+39) 06 61181 204, Fax: (+39) 06 61979 661, Email: e.arnaud@cgiar.org

Aronsson, M., Swedish Species information Center, Box 7007, 750 07, Uppsala, Sweden. Tel: (+46) 18673414, Fax: (+46) 18673480, Email: Mora.Aronsson@ArtData.slu.se

Asdal, Å., Norwegian Genetic Resource Centre, The Norwegian Forest and Landscape Institute, PO Box 115, N – 1431, Norway. Tel: (+47) 40622920, Fax: (+47) 37044278, Email: asmund.asdal@skogoglandskap.no

Avagyan, A., Armenian State Agrarian University, Teryan Str. 74, 0009, Erevan, Armenia. Tel: (+374) 93 415037, Fax: (+374) 10 541800, Email: alvinaav@hotmail.com

Barata, A.M., Banco Português de Germoplasma Vegetal, Instituto Nacional de Recursos Biológicos, I.P. (INRB, I.P.), Quinta S. José, S. Pedro de Merelim, 4700-859 Braga, Portugal. Tel: 253 300 962, Fax: 253 300 961, Email: ana.barata@inrb.pt

Barbagiovanni Miracolo, I., Area for resource protection, farm control and production quality, ARSIAL (Regional Agency for Development and Innovation of agriculture in Lazio), via Rodolfo Lanciani, 38, 00162 Roma, Italy. Tel: (+39) 068 6272450, Fax: (+39) 068 6273270, Email: biodiver@arsial.it

Barocco, R., Dipartimento Biologia Applicata, Università degli Studi di Perugia, Perugia, Italy. Tel: (+39) 075 585 62 18, Fax: (+39) 075 585 62 24, Email: barocco@unipg.it

Bartha, B., ProSpecieRara, Pfrundweg 14, 5000, Aarau, Switzerland.

Begemann F., Federal Office for Agriculture and Food (BLE), 513- Information and Coordination Centre for Biological Diversity, Bonn, Germany. Tel: (+49) 228 99 6845 3240, Email: Frank.Begemann@ble.de

Benediková, D., Research Institute of Plant Production – Genebank, Slovak Agricultural Research Centre, Bratislavská cesta 122, 921 68, Piešt'any, Slovakia, Tel: (+421-33) 7722311/12/26/27, Fax: (+421-33) 7726306, Email: benedikova@vurv.sk

Berishvili, T., Biological Farming Association 'Elkana', 16th Nakvety, 3rd Delisi Street, 0177, Tbilisi, Georgia. Tel: (+995-32) 328321, Fax: (+995-32) 536487, Email: crop@elkana.org.ge

Bettencourt, E., Genetic Resources, Ecophysiology and Plant Breeding Unit, Instituto Nacional de Recursos Biológicos, I.P. (INRB, I.P.), Quinta do Marquês, 2784-505 Oeiras, Portugal (currently on leave of absence). Tel: (+39) 06 61181 204, Fax: (+39) 0661979661, Email: eliseu.bettencourt@gmail.com

Bilz, M., IUCN Species Programme, Red List Unit, IUCN, 219c Huntingdon Road, Cambridge, CB3 0DL, UK. Tel: (+44) 1223 277 966, Fax: (+44) 1223 277 845, Email: melanie.bilz@iucn.org

Bjørn, G.K., Department of Horticulture, Aarhus University, Kirstinebjergvej 10, 5792 Aarslev, Denmark. Fax: (+45) 89993490, Email: GitteK.Bjorn@agrsci.dk

Bocci, R., Associazione Italiana Agricoltura Biologica, via Piave 14, Roma, Italy, Tel: (+39) 328 3876663, Fax: (+39) 064 5437469, Email: r.bocci@aiab.it

Boller, B., Agroscope Reckenholz-Tänikon, Research Station ART, Reckenholzstrasse 191, CH-8046 Zürich, Switzerland. Tel: (+41) (0)44 377 73 63, Fax: (+41) (0)44 377 72 01, Email: beat.boller@art.admin.ch

Bonneuil, C., Centre Koyré de Recherche en Histoire des Sciences et des Techniques, Museum National d'Histoire Naturelle (UMR 8560, CNRS-MNHN-Ehess-CSI), 75005 Paris, France. Tel: (+33) (0)1 40 79 80 01, Fax: (+33) (0)1 73 73 35 64, Email: bonneuil@damesme.cnrs.fr

Branca, F., Dipartimento di OrtoFloroArboricoltura e Tecnologie Agroalimentari (DOFATA), University of Catania, Italy, Tel: (+39) 095 234307, Fax: (+39) 095 234329, Email: fbranca@unict.it

Bullinska-Radomska, Z., Plant Breeding and Acclimatization Institute, Poland. Tel: (+48) 227 334652, Fax: (+48) 227 334649, Email: z.bulinska@ihar.edu.pl

Carnide, V., Centre of Genomics and Biotechnology, Institute of Biotechnology and Bioengineering (CGB/IBB), University of Trás-os-Montes and Alto Douro, 5001-801 Vila Real, Portugal. Tel: (+351) 259 350 530, Fax: (+351) 259 350 572, Email: vcarnide@utad.pt

Carter, S., Sgoil Lionacleit, Isle of Benbecula, Western Isles, HS7 5PJ Scotland. Tel: (+44) (0)1870 603690, Email: sgoillionacleit@gnes.net

Ceccarelli, S., ICARDA – BIGM, PO Box 5466 Aleppo, Syrian Arab Republic. Tel: (+963) 21 6221590, Fax: (+963) 21 6221591, Email: s.ceccarelli@cgiar.org

Chable, V., SAD-Paysage, INRA, 65, rue de Saint-Brieuc, 35042 Rennes, France. Tel: (+33) 608 279 679, Email: chable@rennes.inra.fr

Cop, J., Agronomy Department, Biotechnical Faculty, University of Ljubljana, Jamnikarjeva 101 - PO Box 2995, 1001, Ljubljana, Slovenia. Tel: (+386) 1 3203274, Fax: (+386) 1 4231088, Email: jure.cop@bf.uni-lj.si

Costanza, M.T., Area for resource protection, farm control and production quality, ARSIAL (Regional Agency for Development and Innovation of Agriculture in Lazio), via Rodolfo Lanciani, 38, 00162 Roma, Italy. Tel: (+39) 068 6272450, Fax: (+39) 068 6273270, Email: biodiver@arsial.it

Coupland, G., Department of Plant Developmental Biology, Max Planck Institute for Plant Breeding Research (MPIPZ), Cologne, Germany. Tel: (+49) (0)39482 5 571, Email: coupland@ipk-gatersleben.de

Curtis, T., Davitt Road, Bray, Co. Wicklow, Ireland. Tel: (+353) 87 7631315, Email: tomcurtis@eircom.net

Dantas, J.C., Sabores do Minho, Grijó, 4970-315 Rio Frio, Arcos de Valdevez, Portugal. Tel: (+351) 926134396, (+351) 258531064, (+351) 258523871, Website: www.saboresdominho.com

Daugstad, K., Norwegian Institute for Agricultural and Environmental Research, 2940 Heggenes, Norway. Tel: (+47) 406 22 927, Fax: (+47) 61 34 06 65, Email: kristin.daugstad@bioforsk.no

Demeulenaere, E., UMR7206 Eco-anthropologie & Ethnobiologie, MNHN Département Hommes, Natures, Sociétés – CP 135 – 57 rue Cuvier 75231 PARIS Cedex 05, France. Tel: (+33) (0)1 40 79 34 28, Email: elise.demeulenaere@gmail.com

Dempewolf, H., Global Crop Diversity Trust, c/o Food and Agriculture Organization (FAO) of the United Nations, Via delle Terme di Caracalla, 00153 Rome, Italy. Tel: (+39) 06 570 53324, Fax: (+39) 06 570 55634, Email: Hannes.Dempewolf@croptrust.org

Dias, S., Bioversity International, Via dei Tre Denari 472/a, 00057 Maccarese, Rome, Italy. Tel: (+39) 06 61181 204, Fax: (+39) 0661979661, Email: s.dias@cgiar.org

Dordas, C., Laboratory of Agronomy, School of Agriculture, Aristotle University of Thessaloniki, 541 24 Thessaloniki. Tel: (+30) 2310 998602, Fax: (+30) 2310 998634, Email: chdordas@agro.auth.gr

Duarte, M.C., Instituto de Investigação Científica Tropical, Trav. Conde da Ribeira, 9, 1300-142, Lisboa, Portugal. Tel: (+351) 213616340, Email: mcduarte@iict.pt

Dulloo, M.E., Bioversity International, Via dei Tre Denari 472/a, 00057 Maccarese, Rome, Italy. Tel: (+39) 06 61181 204, Fax: (+39) 0661979661, Email: m.dulloo@cgiar.org. Currently at Food and Agriculture Organization of the United Nations, Via delle Terme di Caracalla, 00153 Rome, Italy. Email: ehsan.dulloo@fao.org

Dzmitryeva, S., V.F. Kuprevich Institute for Experimental Botany of the National Academy of Sciences of Belarus, Akademicheskaya Str. 27, 220027, Minsk, Belarus. Tel: (+375) 17 2842015, Fax: (+375) 2841853, Email: karyology@biobel.bas-net.by

Dzyubenko, E., N.I. Vavilov All-Russian Institute for Plant Industry, 42, Bolshaya Morskaya St., Saint Petersburg, Russia. Tel: 812 315 5093, Fax: 812 571 8762, Email: n.dzyubenko@vir.nw.ru

Dzyubenko, N., N.I. Vavilov All-Russian Institute for Plant Industry, 42, Bolshaya Morskaya St., Saint Petersburg, Russia. Tel: 812 315 5093, Fax: 812 571 8762, Email: elena.dzyubenko@gmail.com

Eastwood, R.J., Millennium Seed Bank, Seed Conservation Department, Royal Botanic Gardens, Kew, Wakehurst Place, Ardingly, West Sussex RH17 6TN, UK. Tel: (+44) (0)1444 894197, Fax: (+44) (0)1444 894110, Email: Ruth.Eastwood@kew.org

Efken, J., Institute of Market Analysis and Agricultural Trade Policy (MA), Johann Heinrich von Thünen-Institute (vTI), Federal Research Institute for Rural Areas, Forestry and Fisheries, Bundesallee 50, D-38116 Braunschweig, Germany. Tel: (+49) 531 596 5307, Email: josef.efken@vti.bund.de

Enders, M., Julius Kühn-Institut, Institute for Breeding Research on Agricultural Crops, Quedlinburg, Germany. Tel: (+49) 3946 47 701, Fax: (+49) 3946 47 255, Email: matthias.enders@jki.bund.de

Engels, J.M.M., Bioversity International, Via dei Tre Denari 472/a, 00057 Maccarese-Fiumicino, Rome, Italy. Tel: (+39) 066118251, Fax: (+39) 0661979661.

Fasoula, D., Agricultural Research Institute, PO Box 22016, 1516 Nicosia, Cyprus. Tel: (+357) 22403124, Fax: (+357) 22316770, Email: dfasoula@arinet.ari.gov.cy

Ferant, N., Slovenian Institute for Hop Research and Brewing, Cesta Žalskega tabora 2, 3310, Žalec, Slovenia. Tel: (+386) 3 7121600, Fax: (+386) 3 7121620, Email: natasa.ferant@guest.arnes.si

Fontinha, S., ISOPlexis Genebank Bank, Biotechnology and Genetic Resources Group, University of Madeira, 9000-390 Funchal, Portugal. Tel: (+351) 291 705 386, Fax: (+351) 291 705 249, Email: quercus@uma.pt

Ford-Lloyd, B.V., School of Biosciences, University of Birmingham, Edgbaston, Birmingham, B15 2TT, UK. Tel: (+44) 121 414 5571, Fax: (+44) 121 414 5925, Email: b.ford-lloyd@bham.ac.uk

Freitas, G., ISOPlexis Genebank Bank, Biotechnology and Genetic Resources Group, University of Madeira, 9000-390 Funchal, Portugal. Tel: (+351) 291 705 386, Fax: (+351) 291 705 249, Email: quercus@uma.pt

Frese, L., Julius Kühn-Institut, Institute for Breeding Research on Agricultural Crops, Quedlinburg, Germany. Tel: (+49) 3946 47 701, Fax: (+49) 3946 47 255, Email: lothar.frese@jki.bund.de

Freudenthaler, P., Austrian Agency for Health and Food Safety, Wieningerstrasse 8, 4020, Linz, Austria. Tel: (+43) 50 555 41200, Fax: (+43) 50 555 41119, Email: paul.freudenthaler@ages.at

Friedel, S., Genebank Department, Leibniz Institute of Plant Genetics and Crop Plant Research (IPK), Gatersleben, Germany. Tel: +49 (0)39482 5 571, Email: friedel@ipk-gatersleben.de

Garzia, J.H., Area for resource protection, farm control and production quality, ARSIAL (Regional Agency for Development and Innovation of Agriculture in Lazio), via Rodolfo Lanciani, 38, 00162 Roma, Italy. Tel: (+39) 068 6272450, Fax: (+39) 068 6273270, Email: biodiver@arsial.it

Germeier, C.U., Julius Kühn Institute, Federal Research Centre for Cultivated Plants (JKI), Institute for Breeding Research on Agricultural Crops, Quedlinburg, Germany. Tel: (+49) 3946 47 701, Fax: (+49) 3946 47 255, Email: christoph.germeier@jki.bund.de

Goldringer, I., UMR de Génétique Végétale, Ferme du Moulon, 91190 Gif-sur-Yvette, France. Tel: (+33) 1 69 33 23 46, Fax: (+33) 1 69 33 23 40, Email: isa@moulon.inra.fr

Graner, A., Genebank Department, Leibniz Institute of Plant Genetics and Crop Plant Research (IPK), Gatersleben, Germany. Tel: (+49) (0)39482 5-571, Email: graner@ipk-gatersleben.de

Green, N., Science and Advice for Scottish Agriculture, Roddinglaw Rd, Edinburgh, EH12 9FJ, Scotland. Tel: (+44) (0)131 244 8853, Fax: (+44) (0)131 244 8940, Email: niall.green@sasa.gsi.gov.uk

Greene, S.L., USDA, Agricultural Research Service, 24106 N. Bunn Road, Prosser, WA 99352, USA. Tel: 509-786-9265, Fax: 509-786-9370, Email: stephanie.greene@ars.usda.gov

Guarino, L., Global Crop Diversity Trust, c/o Food and Agriculture Organization (FAO) of the United Nations, Via delle Terme di Caracalla, 00153 Rome, Italy. Tel: (+39) 06 570 53324, Fax: (+39) 06 570 55634, Email: luigi.guarino@croptrust.org

Hadas, R., Israel Plant Gene Bank, ARO Volcani Center, POB 6 Bet-Dagan 50250, Israel. Tel: (+972) 3 9683897, Fax: (+972) 3 9683895, Email: rihadas@volcani.agri.gov.il

Harrer, S., Federal Office for Agriculture and Food (BLE), 513- Information and Coordination Centre for Biological Diversity, Bonn, Germany. Tel: (+49) 228 99 6845 3240), Email: Siegfried.Harrer@ble.de

Heinonen, M., MTT AgriFood Research Finland, Biotechnology and Food Research/Genetics Research, ET-building, FI-31600 Jokioinen, Finland. Tel: (+358) 019768154, Email: maarit.heinonen@mtt.fi

Höhne, C., Institute for Breeding Research on Agricultural Crops, Julius Kühn-Institut, Federal Research Centre for Cultivated Plants (JKI), Erwin-Baur-Str. 27, D-06484 Quedlinburg, Germany. Tel: (+49) 3946 47702, Email: Claire.Höhne@jki.bund.de

Holly, L., Central Agricultural Office, Directorate of Plant Production and Horticulture, Research Centre for Agrobotany, Külsömezö 15, 2766, Tápiószele, Hungary. Tel: (+36 53) 380069, Fax: (+36 53) 380072, Email: lholly@agrobot.rcat.hu

Holubec, V., Crop Research Institute, Dept. of Gene Bank, Drnovska 507, 161 06 Praha 6 Ruzyne, Czech Republic. Tel: (+420) 233 022 497, Fax: (+420) 233 022 286, Email: holubec@vurv.cz

Hübner, S., Genebank Department, Leibniz Institute of Plant Genetics and Crop Plant Research (IPK), Gatersleben, Germany. Tel: (+49) (0)39482 5-571, Email: hübner@ipk-gatersleben.de

Hunter, D., Bioversity International, Via dei Tre Denari 472/a, 00057 Maccarese, Rome, Italy. Tel: (+39) 06 61181 204, Fax: (+39) 0661979661, Email: D.hunter@cgiar.org

Ibraliu, A., Agricultural University of Tirana, Faculty of Agriculture, Department of Crop Production, Kamez, Tirane, Albania. Tel: (+355) 4 266897, Fax: (+355) 4 227804/353874, Email: alzhama@yahoo.com

Iriondo, J.M., Área de Biodiversidad y Conservación, Depto. Biología y Geología, Universidad Rey Juan Carlos, c/ Tulipán s/n, 28933 Móstoles (Madrid), Spain. Tel: (+34) 91 488 8144, Fax: (+34) 91 664 7490, Email: jose.iriondo@urjc.es

Ivanovska, S., Dept. of Genetics & Plant Breeding, Faculty for Agricultural Sciences and Food, University of Skopje, Bul Aleksandar Makedonski, bb 91000, Skopje, Macedonia FYR. Tel: (+389 2) 3115277, Fax: (+389 2) 3134310, Email: s.ivanovska@zf.ukim.edu.mk

Jinjikhadze, T., Institute of Botany, Cultural Flora Department, Kodjori road #1, 0131, Tbilisi, Georgia. Tel: (+995 93) 256367 / (+995 32) 227326, Fax: (+995 32) 595386 / 001153, Email: tamrikoj@yahoo.com

Kamari, G., Dept. of Biology, University of Patras, 265 00, Patras, Greece. Tel: (+30) 2610997276, Fax: (+30) 2610997275, Email: kamari@upatras.gr

Karpavičienė, B., Institute of Botany of Nature Research Centre, Žaliųjų Ežerų g. 49, LT-08406 Vilnius, Lithuania. Fax: (+370) 5 2729950, Email: birute.karpavičienė@botanika.lt

Kastler, G., Réseau Semences Paysannes, Cazalens, 81600 BRENS, France. Tel: 05 63 41 72 86, Email: guy.kastler@wanadoo.fr

Katsiotis, A., Agricultural University of Athens, Plant Breeding & Biometry Lab., 118 55 Athens, Greece. Tel: (+30) 210 5294634, Email: katsioti@aua.gr

Keilwagen, J., Genebank Department, Leibniz Institute of Plant Genetics and Crop Plant Research (IPK), Gatersleben, Germany. Tel: (+49) (0)39482 5 571, Email: keilwagen@ipk-gatersleben.de

Kell, S.P., School of Biosciences, University of Birmingham, Edgbaston, Birmingham, B15 2TT, UK. Tel: (+44) 121 414 5571, Fax: (+44) 121 414 5925, Email: s.p.kell@bham.ac.uk

Khoury, C., Global Crop Diversity Trust, c/o Food and Agriculture Organization (FAO) of the United Nations, Via delle Terme di Caracalla, 00153 Rome, Italy. Tel: (+39) 06 570 53324, Fax: (+39) 06 570 55634, Email: colin.khoury@croptrust.org

Kik, C., Centre for Genetic Resources, the Netherlands, PO Box 16, 6700 AA, Wageningen, the Netherlands. Tel: (+31 317) 480861, Fax: (+31 317) 423110, Email: chris.kik@wur.nl

Kilian, B., Genebank Department, Leibniz Institute of Plant Genetics and Crop Plant Research (IPK), Gatersleben, Germany. Tel: (+49) (0)39482 5 571, Email: kilian@ipk-gatersleben.de

Knüpffer, H., Genebank Department, Leibniz Institute of Plant Genetics and Crop Plant Research (IPK), Gatersleben, Germany. Tel: (+49) (0)39482 5 571, Email: knupffer@ipk-gatersleben.de

Koehler, C., Swiss Commission for the Conservation of Cultivated Plants, Domaine de Changins, CP 1012, CH-1260 Nyon 1, Switzerland. Tel: (+41) (0)22 363 47 02, Fax: (+41) (0)22 362 13 25, Email: Christopher.koehler@art.admin.ch

Konovalov, F.A., Genebank Department, Leibniz Institute of Plant Genetics and Crop Plant Research (IPK), Gatersleben, Germany. Tel: (+49) (0)39482 5 571, Email: konovalov@ipk-gatersleben.de

Koop, L., Bioversity International, Via dei Tre Denari 472/a, 00057 Maccarese-Fiumicino, Rome, Italy. Tel: (+39) 066118251, Fax: (+39) 0661979661, Email: l.koop@cgiar.org

Korpelainen, H., Dept of Applied biology, University of Helsinki, PO Box 27 (Latokartanonkaari 7), FIN-00014, Helsinki, Finland. Tel: (+358 9) 191 58383, Fax: (+358 9) 191 58727, Email: helena.korpelainen@helsinki.fi

Kristiansen, K., Department of Horticulture, Aarhus University, Kirstinebjergvej 10, 5792 Aarslev, Denmark.

Kyratzis, A., Agricultural Research Institute, Ministry of Agriculture, Natural Resources & Environment, PO Box 2016, 1516, Nicosia, Cyprus. Tel: (+357) 22 403131, Fax: (+357) 22 316770, Email: a.kyratzis@arinet.ari.gov.cy

Labokas, J., Institute of Botany of Nature Research Centre, Žaliųjų; Ežerų g. 49, LT-08406 Vilnius, Lithuania. Fax: (+370) 5 2729950, Email: juozas.labokas@botanika.lt

Ladizinsky, G., Hebrew University of Jerusalem, Faculty of Agriculture, Rehovot, 76 100 Israel. Tel: (+972) 8 948 9253, Email: ladizins@agri.huji.ac.il

Lara-Romero, C., Departamento Biología y Geología, Universidad Rey Juan Carlos, c/ Tulipán s/n, E-28933 Móstoles, Madrid, Spain. Tel: (+34) 914 888 144, Fax: (+34) 916 647 490, Email: carlos.lara@urjc.es

Lelli, L., Area for resource protection, farm control and production quality, ARSIAL (Regional Agency for Development and Innovation of Agriculture in Lazio), via Rodolfo Lanciani, 38, 00162 Roma, Italy. Tel: (+39) 0686272450, Fax: (+39) 0686273270, Email: biodiver@arsial.it

Lopes, V.R., Banco Português de Germoplasma Vegetal, Instituto Nacional de Recursos Biológicos, I.P. (INRB, I.P.), Quinta S. José, S. Pedro de Merelim, 4700-859 Braga, Portugal. Tel: 253 300 962, Fax: 253 300 961, Email: violeta.lopes@inrb.pt

Louwaars, N., Centre for Genetic Resources, Postbus 16, 6700AA, Wageningen, the Netherlands. Tel: 0317480854, Email: niels.louwaars@wur.nl

Ložienė, K., Institute of Botany of Nature Research Centre, Žaliųjų Ežerų g. 49, LT-08406 Vilnius, Lithuania. Fax: (+370) 5 2729950, Email: kristina.ložienė@botanika.lt

Lu, Bao-Rong, Ministry of Education Key Laboratory for Biodiversity and Ecological Engineering, The Department of Ecology and Evolutionary Biology, School of Life Science, Fudan University, Shanghai 200433, China. Tel: (+86) 21 65643668, Email: brlu@fudan.edu.cn

MacPherson, N., Sgoil Lionacleit, Isle of Benbecula, Western Isles, HS7 5PJ, Scotland. Tel: (+44) (0)1870 603690, Email: sgoillionacleit@gnes.net

Maggioni, L., Bioversity International, Via dei Tre Denari 472/a, 00057 Maccarese-Fiumicino, Rome, Italy. Tel: (+39) 066118251, Fax: (+39) 0661979661.

Magos Brehm, J., Universidade de Lisboa, Museu Nacional de História Natural, Jardim Botânico, R. Escola Politécnica 58, 1250-102 Lisboa, Portugal; Universidade de Lisboa, Faculdade de Ciências, Centro de Biologia Ambiental, Campo Grande C2, 1749-016 Lisboa, Portugal. Tel: (+351) 213921882, Email: jbrehm@museus.ul.pt

Maloupa, E., Agricultural Research Center of Northern Greece, PO Box 60458, 57001, Thermi, Greece. Tel: (+30) 2310 471 613, Fax: (+30) 2310 478907, Email: maloupa@bbgk.gr

Martinez, J.J.R., Universidad Miguel Hernandez, Carretera de Beniel km 3.2, 03312, Orihuela, Alicante, Spain. Tel: (+34) 966 749600/01, Fax: (+34) 966 749619, Email: juanj.ruiz@umh.es

Martins-Loução, M.A., Jardim Botânico, Museu Nacional de História Natural, Universidade de Lisboa, Rua da Escola Politécnica 58, 1250-102 Lisboa, Portugal. Tel: (+351) 213921882, Email: maloucao@reitoria.ul.pt

Matos, M., Centre of Genomics and Biotechnology, Institute of Biotechnology and Bioengineering (CGB/IBB), University of Trás-os-Montes and Alto Douro, 5001-801 Vila Real, Portugal. Tel: (+351) 259 350 530, Fax: (+351) 259 350 572, Email: mmatos@utad.pt

Maxted, N., School of Biosciences, University of Birmingham, Edgbaston, Birmingham, B15 2TT, UK. Tel: (+44) 121 414 5571, Fax: (+44) 121 414 5925, Email: nigel.maxted@dial.pipex.com

Mendes-Moreira, P., Departamento de Ciências Agronómicas, Escola Superior Agrária de Coimbra, Portugal. Tel: 239 802 940, Fax: 239 802 979, Email: pmoreira@esac.pt

Milla, R., Área de Biodiversidad y Conservación, Universidad Rey Juan Carlos. c/ Tulipán s/n, E-28933 Móstoles, Madrid, Spain. Tel: (+34) 91 488 8517, Fax: (+34) 91 664 7490, Email: ruben.milla@gmail.com

Miranda, J., ARDAL – Associação Regional de Desenvolvimento do Alto Lima, Praça do Município, 4974-003 Arcos de Valdevez, Portugal. Tel: (+351) 258 520 503, Fax: (+351) 258 520 503, Email: jorgemiranda@ardal.pt

Müller, J.V., Millennium Seed Bank, Seed Conservation Department, Royal Botanic Gardens, Kew, Wakehurst Place, Ardingly, West Sussex RH17 6TN, UK. Tel: (+44) (0)1444 894197, Fax: (+44) (0)1444 894110, Email: j.mueller@kew.org

Munro, N., Garden Organic, Ryton Organic Gardens, Coventry, Warwickshire CV8 3LG, UK. Tel: (+44) 024 7630 3517, Fax: (+44) 024 7663 9229, Email: nmunro@gardenorganic.org.uk

Musayev, M., Genetic Resources Institute, Azerbaijan National Academy of Sciences, 155, Azadliq Ave., AZ1106, Baku, Azerbaijan. Tel: (+994) 12 4424907, Fax: (+994) 12 4499221, Email: m_musayev4@yahoo.com

Nachtigall, M., Julius Kühn-Institut, Institute for Breeding Research on Agricultural Crops, Quedlinburg, Germany. Tel: (+49) 3946 47 701, Fax: (+49) 3946 47 255, Email: gerlinde.nachtigall@jki.bund.de

Nardi, P., Area for resource protection, farm control and production quality, ARSIAL (Regional Agency for Development and Innovation of Agriculture in Lazio), via Rodolfo Lanciani, 38, 00162 Roma, Italy. Tel: (+39) 0686272450, Fax: (+39) 0686273270, Email: biodiver@arsial.it

Negri, V., Dipartimento Biologia Applicata, Università degli Studi di Perugia, Perugia, Italy. Tel: (+39) 075 585 62 18, Fax: (+39) 075 585 62 24, Email: vnegri@unipg.it

Neumann, K., Genebank Department, Leibniz Institute of Plant Genetics and Crop Plant Research (IPK), Gatersleben, Germany. Tel: (+49) (0)39482 5-571, Email: neumann@ipk-gatersleben.de

Nóbrega, H., ISOPlexis Genebank Bank, Biotechnology and Genetic Resources Group, University of Madeira, 9000-390 Funchal, Portugal. Tel: (+351) 291 705 386, Fax: (+351) 291 705 249, Email: quercus@uma.pt

Özkan, H., Department of Field Crops, Faculty of Agriculture, University of Cukurova, Adana, Turkey. Tel: (+90) 456 511 8669, Fax: (+90) 456 511 8679, Email: hozkan@mail.cu.edu.tr

Pacicco, L., Dipartimento Biologia Applicata, Università degli Studi di Perugia, Perugia, Italy. Tel: (+39) 075 585 62 18, Fax: (+39) 075 585 62 24, Email: pacicco@unipg.it

Paoletti, S., Area for resource protection, farm control and production quality, ARSIAL (Regional Agency for Development and Innovation of Agriculture in Lazio), via Rodolfo Lanciani, 38, 00162 Roma, Italy. Tel: (+39) 0686272450, Fax: (+39) 0686273270, Email: biodiver@arsial.it

Parra-Quijano, M., Departamento de Biología Vegetal, Universidad Politécnica de Madrid, 28040 Madrid, Spain. Tel: (+34) 913365462, Fax: (+34) 913365656, Email: mauricio.parra@agrobiodiversidad.org

Pasam, R.K., Genebank Department, Leibniz Institute of Plant Genetics and Crop Plant Research (IPK), Gatersleben, Germany. Tel: +49 (0)39482 5-571, Email: pasam@ipk-gatersleben.de

Pego, S.E., Fundação Bomfim. Rua da Boavista, 152-154, 4700-416 Braga, Portugal. Tel: 253 412 412, Fax: 253 216 236.

Pinheiro de Carvalho, M.Â.A., ISOPlexis Genebank Bank, Biotechnology and Genetic Resources Group, University of Madeira, 9000-390 Funchal, Portugal. Tel: (+351) 291 705 386, Fax: (+351) 291 705 249, Email: quercus@uma.pt

Pinto-Carnide, O., Centre of Genomics and Biotechnology, Institute of Biotechnology and Bioengineering (CGB/IBB), University of Trás-os-Montes and Alto Douro, 5001-801 Vila Real, Portugal. Tel: (+351) 259 350 530, Fax: (+351) 259 350 572, Email: ocarnide@utad.pt

Porfiri, O., Area for resource protection, farm control and production quality, ARSIAL (Regional Agency for Development and Innovation of Agriculture in Lazio), via Rodolfo Lanciani, 38, 00162 Roma, Italy (Consultant for ARSIAL). Email: ori@mercurio.it

Preston, J.M., School of Biosciences, University of Birmingham, Edgbaston, Birmingham, B15 2TT, UK. Tel: (+44) 0121 414 2975, Fax: (+44) 0121 123456789, Email: jxp707@bham.ac.uk

Radun, M., Genetic Resources Institute, University of Banjaluka, Bulevar vojvode Petra Bojovica 1a, 78000, Banjaluka, Republika Srpska, Bosnia and Herzegovina. Tel: (+387) 51330988, Email: marinabl14@yahoo.com

Radušienė, J., Institute of Botany of Nature Research Centre, Žaliųjų Ežerų g. 49, LT-08406 Vilnius, Lithuania. Fax: (+370) 5 2729950, Email: jolita.radušienė@botanika.lt

Ralli, P., Greek Gene Bank, Agricultural Research Centre of Northern Greece, National Agricultural Research Foundation, 570 01 Thermi-Thessaloniki, Laboratory of Agronomy, School of Agriculture, Aristotle University of Thessaloniki, 541 24 Thessaloniki, Greece. Tel: (+30) 2310 471110, Fax: (+30) 2310471209, Email: rallip@agro.auth.gr

Rea, R., Area for resource protection, farm control and production quality, ARSIAL (Regional Agency for Development and Innovation of Agriculture in Lazio), via Rodolfo Lanciani, 38, 00162 Roma, Italy. Tel: (+39) 0686272450, Fax: (+39) 0686273270, Email: biodiver@arsial.it

Reichling A., Am Zainhammer 4, Eberswalde, Germany. Tel: (+49) 3334387666, Email: areichling@web.de

Reis, A., Banco Português de Germoplasma Vegetal, Instituto Nacional de Recursos Biológicos, I.P. (INRB, I.P.), Quinta S. José, S. Pedro de Merelim, 4700-859 Braga, Portugal. Tel: 253 300 962, Fax: 253 300 961, Email: aida.reis@inrb.pt

Rocha, F., Banco Português de Germoplasma Vegetal, Instituto Nacional de Recursos Biológicos, I.P. (INRB, I.P.), Quinta S. José, S. Pedro de Merelim, 4700-859 Braga, Portugal. Tel: 253 300 962, Fax: 253 300 961, Email: filomena.rocha@inrb.pt

Sandru, D., Banca de Resurse Genetice Vegetale Suceava, Bulevardul 1 Decembrie 1918 nr. 17, 720237, Suceava, Romania. Tel: (+40 230) 521016, Fax: (+40 230) 521016, Email: genebank@suceava.astral.ro

Sarikyan, K., Scientific Center of Vegetable and Industrial Crops, Ministry of Agriculture, Ararat Marz, Darakert Community, 0808, Ararat region, Armenia. Tel: (+374 94) 867030, Email: Karuine_Sarikyan@mail.ru

Schierscher-Viret, B., Biodiversity Conservation Information Systems, BCIS.CH, Le Ravin, CH-1268 Begnins, Switzerland. Tel: (+41) 795804351, Email: beate.schierscher@bcis.ch

Scholten, M., Crops and Soil System Research Group, Scottish Agricultural College, West Mains Road, Edinburgh, EH9 3JG, Scotland. Tel: (+44) 774 6671984, Email: maria_scholten@hotmail.com

Shaaf, S., Genebank Department, Leibniz Institute of Plant Genetics and Crop Plant Research (IPK), Gatersleben, Germany. Tel: (+49) (0)39482 5 571, Email: shaaf@ipk-gatersleben.de

Sharma, R., Genebank Department, Leibniz Institute of Plant Genetics and Crop Plant Research (IPK), Gatersleben, Germany. Tel: (+49) (0)39482 5 571, Email: sharma@ipk-gatersleben.de

Sherman, R., Garden Organic, Ryton Organic Gardens, Coventry, Warwickshire CV8 3LG, UK. Tel: (+44) 024 7630 3517, Fax: (+44) 024 7663 9229, Email: rsherman@hdra.org.uk

Smekalova, T., Agrobotany Dept, N.I.Vavilov All-Russian Institute of Plant Industry, 42/44, Bolshaja Morskaya Street, 190000, St Petersburg, Russia. Tel: (+7 812) 4142234, 3144848, Fax: (+7 812) 5718762, Email: t.smekalova@vir.nw.ru

Spataro, G., Istituto Nazionale di Ricerca per gli Alimenti e la Nutrizione (INRAN) Attività di gestione ex Ente Nazionale Sementi Elette, c/o Ministero delle Politiche Agricole Alimentari e Forestali, via XX Settembre 20, 00187 Roma, Italy. Tel: (+39) 06 51494548, Fax: (+39) 06 51494550, Email: spataro@inran.it

Spoor, B., Crops and Soil System Research Group, SAC, West Mains Road, Edinburgh, EH9 3JG, Scotland. Tel: (+44) (0)131 535 4000, Email: bill.spoor@sac.ac.uk

Stehno, Z., Genebank Department, Crop Research Institute, Drnovská 507, 161 06, Praha 6 – Ruzyne, Czech Republic. Tel: (+420) 233022364, Fax: (+420) 233022286, Email: stehno@vurv.cz

Stoilova, T., Institute of Plant Genetic Resources, 4122, Sadovo distr. Plovdiv, Bulgaria. Tel: (+359) 32 629026, Fax: (+359) 32 629026, Email: tz_st@abv.bg

Strajeru, S., Banca de Resurse Genetice Vegetale Suceava, Bulevardul 1 Decembrie 1918 nr. 17, 720237, Suceava, Romania. Tel: (+40) 230 521016, Fax: (+40) 230 521016, Email: genebank@suceava.astral.ro

Svalheim, E., Norwegian Institute for Agricultural and Environmental Research, Division Kvithamar, N-7500 Stjørdal, Norway. Tel: (+47) 45210350, Fax: (+47) 37044278, Email: ellen.svalheim@bioforsk.no

Šveistytė, L., Plant Gene Bank, Stoties g. 2, LT-58343 Akademija, Kėdainių r., Lithuania. Fax: (+370) 5 2374722, Email: laima.sveistyte@botanika.lt

Tan, A., Department of Plant Genetic Resources, Aegean Agricultural Research Institute, P.K. 9 Menemen, 35661, Izmir, Turkey. Tel: (+90 232) 8461331, Fax: (+90 232) 8461107, Email: ayfer_tan@yahoo.com

Tanca, M., Area for resource protection, farm control and production quality, ARSIAL (Regional Agency for Development and Innovation of Agriculture in Lazio), via Rodolfo Lanciani, 38, 00162 Roma, Italy. Tel: (+39) 0686272450, Fax: (+39) 0686273270, Email: biodiver@arsial.it

Taviani, P., Area for resource protection, farm control and production quality, ARSIAL (Regional Agency for Development and Innovation of Agriculture in Lazio), via Rodolfo Lanciani, 38, 00162 Roma, Italy. Tel: (+39) 0686272450, Fax: (+39) 0686273270, Email: biodiver@arsial.it

Teeling, C., School of Biosciences, University of Birmingham, Edgbaston, Birmingham, B15 2TT, UK. Tel: (+44) 121 4142975, Email: cxt755@bham.ac.uk

Thomas, M., UMR de Génétique Végétale, Ferme du Moulon, 91190 Gif-sur-Yvette, France. Tel: (+33) 1 69 33 23 46, Fax: (+33) 1 69 33 23 40, Email: thomas@moulon.inra.fr

Torres, E., Departamento de Biología Vegetal, Universidad Politécnica de Madrid, 28040 Madrid, Spain. Tel: (+34) 913365462, Fax: (+34) 913365656, Email: etorres@agricolas.upm.es

Tribulato, A., Dipartimento di OrtoFloroArboricoltura e Tecnologie Agroalimentari (DOFATA), University of Catania, Italy. Tel: (+39) 095 234307, Fax: (+39) 095 234329, Email: atribulato@unict.it

van Hintum, Th., Centre for Genetic Resources, the Netherlands (CGN), Wageningen University and Research Centre, Wageningen, the Netherlands. Tel: (+31 317) 480913, Fax: (+31 317) 423110, Email: theo.vanhintum@wur.nl

Venanzoni, R., Dipartimento Biologia Applicata, Università degli Studi di Perugia, Perugia, Italy. Tel: (+39) 075 585 62 18, Fax: (+39) 075 585 62 24, Email: venanzoni@unipg.it

Veronesi, F., Dipartimento Biologia Applicata, Università degli Studi di Perugia, Perugia, Italy. Tel: (+39) 075 585 62 18, Fax: (+39) 075 585 62 24, Email: veronesi@unipg.it

Veteläinen, M., MTT Agrifood Research Finland, Biotechnology and Food Research, Genetic diversity, ET-building, FIN-31600, Jokioinen, Finland. Tel: (+358) 3 41883683, Fax: (+358) 3 41883618, Email: merja.vetelainen@mtt.fi

Vögel, R., State Agency for Environment, Health and Consumers Protection Brandenburg, Eberswalde, Germany. Tel: (+49) 33342728, Fax: (+49) 3334662710, Email: rudolf.voegel@lugv.brandenburg.de

von Korff, M., Department of Plant Developmental Biology, Max Planck Institute for Plant Breeding Research (MPIPZ), Cologne, Germany. Tel: (+49) (0)39482 5 571, Email: vonkorff@ipk-gatersleben.de

Vorosvary, G., Central Agricultural Office, Directorate of Plant Production and Horticulture, Research Centre for Agrobotany, Külsömezö 15, 2766, Tápiószele, Hungary. Tel: (+36 53) 380070, Fax: (+36 53) 380072, Email: gaborv@agrobot.rcat.hu

Weißgerber, W., Genebank Department, Leibniz Institute of Plant Genetics and Crop Plant Research (IPK), Gatersleben, Germany. Tel: (+49) (0)39482 5 571, Email: Weißgerber@ipk-gatersleben.de

Acknowledgements

We acknowledge the financial support of the EC-funded AEGRO project ('An integrated European *in situ* management work plan: implementing genetic reserves and on-farm concepts', project number AGRI GENRES 057, through the EC DG AGRI council regulation 870/2004) and ECPGR (European Cooperative Programme for Plant Genetic Resources) for help to facilitate the conference 'Towards the establishment of genetic reserves for crop wild relatives and landraces in Europe', from which these proceedings are drawn. Specifically, the authors of Chapters 1–10, 16 and 31 are grateful for the co-funding of the EC-funded AEGRO project. We also acknowledge co-funding by Fundação para a Ciência e a Tecnologia, through the Fund for the Support of Scientific Community.

1 Genetic Reserve Conservation of European Crop Wild Relative and Landrace Diversity

L. Frese, G.K. Bjørn, F. Branca, B.V. Ford-Lloyd, C.U. Germeier, J.M. Iriondo, A. Katsiotis, S.P. Kell, N. Maxted, V. Negri and M.Â.A. Pinheiro de Carvalho

1.1 Introduction

Genetic variation used in variety breeding originates to a large extent from the ancestral species of the crops and is being managed by breeders in breeding pools. Genetic bottlenecks caused right at the beginning of the domestication process or caused by breeding at a later stage of the domestication process (e.g. Kilian *et al.*, 2006) necessitates the introgression of novel genetic variation from the crop's gene pool into the breeding pools. Crop wild relatives (CWR) and landraces (LR) are part of this gene pool, which is tapped by breeders if need arises. Without this flow of novel genes from wild species and landraces into the breeding pools the adaptability of crops to changing environmental conditions and user demands would be very difficult to maintain today and in the future. The demand for traits and trait variation can be forecasted by plant breeders for the next 10–20 years while a prognosis of the future needs of agriculture is not possible beyond this time horizon. As long as the full range of interspecific variation within wild species related with a crop species continues to exist and evolve, breeding, and therefore food security, is hedged against accidental or inbuilt losses of genetic variation within their breeding pools and will be able to cope with future challenges. There is, however, strong evidence that a significant number of CWR species are on the brink of extinction (Kell *et al.*, Chapter 28, this volume) and therefore may no longer serve as an effective backup gene pool for plant breeding.

Within the agricultural sector it is the responsibility of plant breeding research to judge the risks for the plant breeding and seed trade sector arising out of the loss of CWR and LR diversity, which form a significant part of the plant genetic resources for food and agriculture (PGRFA). In the field of PGRFA protection the interests of the PGRFA conservation community coincide to a large extent with the interests of the community of conservation biologists as approximately three-quarters of all Euro-Mediterranean plants species have socio-economic value (Kell *et al.*, 2008). Many of them play a very significant role as gene donors (e.g. Maxted and Kell, 2009; McGrath *et al.*, 2010). There is, however, a slight but significant difference in the conservation approaches of both communities.

While the agricultural sector accentuates the usefulness of PGRFA, the nature conservation sector considers habitat and species protection as an end in itself and has called since the 1870s for the establishment of *in situ* conservation programmes. The conservation biologists stress the value of genetic wildlife conservation which recognizes the long-term needs – 'need for continued evolution within natural environments, and the need for conservation of wild biota as potentially useful resource for the future' (Jain, 1975) – and launched the genetic reserve concept in the 1970s. With the coming into force of the Convention on Biological Diversity the *in situ* conservation was raised to the level of an international strategy which is supplemented by the *ex situ* conservation of genetic resources in gene banks. Maxted *et al.* (1997a) proposed a methodology to underpin the 'genetic reserve conservation technique', which has subsequently been significantly enhanced (Iriondo *et al.*, 2008; Maxted *et al.*, 2008).

As the need for traits and trait variation in agriculture cannot be precisely foreseen, it is reasonable to focus on the conservation of evolutionary processes and genetic variation as such. Genetic variation is a necessary precondition for continued evolution of species (Namkoong *et al.*, 1996). The genetic system plays a key role in these adaptation processes. The genetic system is defined by all mechanisms that serve the organization (e.g. chromosomes, genes), reproduction, combination, and distribution over space and time of genetic information. The maintenance of the adaptability of a species requires that: (i) the mechanisms of its genetic system are intact; (ii) the external conditions under which the mechanisms function are realized; and (iii) the genetic variation required for alteration of these mechanisms is available (Gregorius, 2001). The management of the genetic variation contained in such a dynamic system by means of a genetically representative sample of occurrences is the fundamental aim of *in situ* management.

The major gap in CWR and LR conservation is the lack of good practical worked examples of their conservation. Slowly but steadily in recent years the science has begun to be developed into action that applies both *in situ* and *ex situ* techniques in a complementary manner (Maxted *et al.*, 1997b; Iriondo *et al.*, 2008; Veteläinen *et al.*, 2009). The advent of the CBD refocused national and international biodiversity conservation on to *in situ* activities and for socio-economic species this placed a much greater emphasis on to CWR conservation. Prior to this, the focus of agrobiodiversity conservation had been almost exclusively *ex situ* and largely focused on gene bank facilities, and socio-economic species in protected areas was largely ignored as they were primarily designated for mega-fauna or ecosystem conservation. Methodologies and conservation techniques such as the genetic reserve conservation technique are required to realize the *in situ* conservation strategy as called for by Art. 5.1 (d) of the ITPGRFA. The project 'An Integrated European In Situ Management Work Plan: Implementing Genetic Reserves and On Farm Concepts' (AEGRO, 2011) aimed to enhance methodologies and techniques, to test them and reveal the constraints impairing the applications of the genetic reserve conservation technique in practice.

1.2 Enhancement of the *In Situ* Conservation Strategy

A strategy consists of a *concept* (i.e. an idea of the goals to be achieved), a *management plan*, and a *recognized position of a team* in the expert community as well as a recognizable type of argumentation which is called a '*behaviour pattern*' in management science (Wikipedia, 2011). The PGR Forum research project (EC Framework 5 Programme for Research, Key action 2.2.3 Assessing and conserving biodiversity, 2003–2005), coordinated by the University of Birmingham, elaborated the perspectives for *in situ* conservation of CWR as well as concepts and methods. A part of the PGR Forum project team continued to collaborate within AEGRO, both of which arose out of partnership initiated within the ECPGR *In situ* and On-farm Conservation Network Coordinating Group. In the case of AEGRO, members of the ECPGR working groups on *Avena*, *Beta*, *Brassica* and *Prunus* also participated. Although the composition of the project groups changed, all partners share a common perspective and organize their work on a commonly agreed management plan, thus forming a team with a recognizable position in the scientific community and share a common behaviour pattern. This is an important and indispensable precondition if a long-term strategy, as the *in situ* conservation of PGRFA clearly is, is to be implemented based on volatile project funds (see Maxted *et al.*, Chapter 43, this volume). A strategy cannot be implemented without a concept and plan, a methodology and tools, (human) resources and (financial) means.

1.3 Concept and Plan

There are several approaches to achieving the systematic conservation of CWR and LR diversity, but three distinct approaches may be characterized as the individual, national and global approach (Maxted and Kell, 2009). The global or monographic approach focuses on the conservation of priority crop gene pools and has been mostly taken into consideration in AEGRO. Diversity is conserved systematically via a global *in situ* network of genetic reserves, with *ex situ* backup in national and international collections. Priorities for the global approach are established and applied worldwide, irrespective of political boundaries and are likely to mean that genetic reserves are primarily established in sites selected on scientific grounds.

1.4 Methodology and Tools

Each approach to the conservation of priority crop gene pools aims to conclude with the incorporation of CWR conservation within existing protected areas, which involves the amendment of existing management plans and results in genetic variation of a species being actively maintained within the protected areas. There are four basic steps in the methodology developed for the monographic approach: (i) taxon delineation; (ii) selection of target taxa; (iii) ecogeographic diversity analysis; and (iv) selection of target sites (Maxted *et al.*, 2008). After a test of the methodology using the model crops oat (*Avena*), beet (*Beta*/*Patellifolia*), crop brassicas (*Brassica*) and cherry (*Prunus*) and further development, the methodology was made available in the public domain (see http://aegro.jki.bund.de/) as part of the 'CWR *In Situ* Strategy Helpdesk'. The helpdesk is provided in first instance as a guide and information facility for national PGRFA programme leaders involved in the development of a CWR *in situ* conservation strategy (Kell *et al.*, Chapter 2, this volume). For the identification of CWR genetic reserve sites for a target crop gene pool, precise instructions are provided. The end point of the step-wise procedure is the identification of 'ideal' CWR genetic reserve sites. The next step beyond the methodology for identification of genetic reserves is to make recommendations for site and population management and to promote the political and legal decisions that need to be taken to establish the genetic reserves as a discrete component in the protected area (Kell *et al.*, Chapter 2, this volume).

The four-step-methodology has been applied and candidate genetic reserve sites have been identified for each model crop gene pool by the crop experts. A list of recommended genetic reserve sites is provided by GenResIS (2011).

The helpdesk proved to be a valuable decision support system. Informed decisions for the selection of genetic (GR) sites within the distribution area of a species require high quality distribution data on individual occurrences and information on the ecogeographic patterns of genetic diversity of that species. High quality distribution data were not available to the AEGRO project but had to be generated for the model crop gene pools by downloading, compiling, supplementing and harmonizing data from various information sources (Germeier *et al.*, Chapter 31, this volume). The lack of readily available high quality distribution data is a significant constraint to a wider application of the genetic reserve conservation technique. The quality of distribution data required for the execution of the monographic approach can be improved on the long run if the data acquisition at the national level and the data flow within the

EU-27/Europe would be organized as described by Vögel and Reichling (Chapter 30, this volume) and Harrer *et al.* (Chapter 29, this volume).

Information on the structure of genetic diversity available in literature is fragmented and difficult to assess as the information was generated for purposes other than making decisions on the choice of GR sites. There is an increasing number of genetic markers produced by the plant breeding research sector suited to measure differentiation between occurrences of species (e.g. McGrath *et al.*, 2010) and the costs for performing genetic analysis with genetic markers are decreasing. Genetic markers are useful tools for describing genetic processes within and between occurrences of a species such as the spatial and temporal distribution of genetic information. They can assist in selecting 'ideal' GR sites or help to decide on the location and shape of a GR, but unless they are not linked to genes under natural selection the data are to be interpreted prudently (González-Martínez *et al.*, 2006).

A systematic methodology for selecting and ranking genetic reserve sites for the creation of a network at the European level, described by Parra-Quijano *et al.* (Chapter 3, this volume), may mitigate the lack of information on the structure of genetic diversity. The ecogeographic approach described by Parra-Quijano *et al.* (Chapter 3, this volume) is based on a Geographic Information System established for *Beta* gathering information to the European level on administrative divisions, protected areas, bioclimatic, geophysical and edaphic information, and use of the territory (CORINE, 2010) as well as occurrence of *Beta* species. A number of target genetic reserve sites were identified using these non-genetic criteria. The selected sites are to be visited to confirm the presence of the target species and to assess the suitability of the protected areas to hold a genetic reserve according to management quality criteria outlined by Iriondo *et al.* (Chapter 10, this volume).

The systematic genetic reserve selection process follows safety criteria, as sites are located within protected areas, efficiency criteria, as they include areas with the greatest number of populations of the selected species, and representativeness criteria, covering ecogeographic units where the species are commonly found, as well as those where the species are marginally found. It is assumed that by conserving the greatest ecogeographic variability of the species it will be possible to gather the greatest genetic diversity of adaptive importance and, possibly, the most interesting allelic variation in the genes of interest for plant breeding. The ecogeographic approach can be considered the first step of the genetic analysis of a target species as it reduces the number of objects to those assumed to contribute to the conservation of genetic diversity the most and thereby reduces the amount of analytical work associated with genetic investigations (Maxted *et al.*, 1995; Dulloo *et al.*, 2008).

Independently from the ecogeographic approach focusing on CWR as detailed by Parra-Quijano *et al.* (Chapter 3, this volume), Negri *et al.* (Chapter 16, this volume) developed a similar concept for LR conservation. Six areas in central Italy were identified showing the highest concentration of landraces within the investigated region. These areas also contain CWR of the four AEGRO model crops and were nominated as the Most Appropriate Areas (MAA) for PGRFA conservation where LR and CWR diversity can be managed jointly.

1.5 Resources and Means

The essential resource within the context of a decision support system is the human resource, i.e. the experts' knowledge, motivation and engagement. The strategy was developed and tested by members of the ECPGR *In situ* and On-farm Conservation Network, its working groups and a few experts from ECPGR crop-specific working groups. The wider application of the techniques for *in situ* and on-farm management of PGRFA within Europe will require the engagement of the whole ECPGR community and will depend on the formalization of a process similar to AEGIS, the European Genetic Resources Integrated System (AEGIS, 2011). Within Europe a rich and diverse expertise in the field of PGRFA management exists. These human resources can be mobilized. For transforming *in situ* and on-farm management research and implementation projects into a durable programme, means are required which are less volatile than project funds.

It is not a matter of lacking European vision, expertise, knowledge, tools and agencies which hamper progress in *in situ* and on-farm management of plant genetic resources, it is rather a lack of sufficient cooperation between experts, agencies and ministries at all levels. The cooperation can be improved if the countries in Europe would agree on a common strategy for *in situ*/on-farm management. This would release synergies and promote the sustainable management and use of landraces and crop wild relatives native to Europe.

1.6 Raising Professional Awareness

As stated above the major gap in CWR and LR science is good practical worked examples of their conservation and use. In this context, it is clear that the AEGRO project significantly developed further the methodologies for both CWR and LR *in situ* conservation. As it is important that these advances are made available to potential users and other stakeholders, AEGRO project partners, in collaboration with the European Cooperative Programme for Plant Genetic Resources *In situ* and On-farm Network, organized a dissemination conference, 'Towards the establishment of genetic reserves for crop wild relatives and landraces in Europe', held at the University of Madeira, Funchal, Portugal between 13 and 17 September 2010. The conference and this volume are meant to provide a comprehensive set of CWR and LR conservation and use exemplars of practical methods, protocols and initiatives, in fact a 'toolkit' for future European CWR and LR conservation and use. Although the conference and the volume specifically focused on methodological application in Europe, the *in situ* methods described are generic and are transferable globally. The objectives of the conference and this text are, therefore, to:

- Develop *in situ* management workplans based on the genetic reserve and on-farm concepts.
- Methodologically identify sites suited to organize genetic reserves and on-farm conservation.
- Perform GIS analysis of genetic reserve and on-farm sites as a means of targeting locations for *in situ* conservation establishment.
- Discuss and establish a European genetic reserve and on-farm network to ensure European *in situ* CWR and LR diversity conservation.
- Provide generic data quality standards for *in situ* CWR and LR diversity conservation, ensuring integration with existing projects.
- Provide improved linkage between *in situ* and *ex situ* CWR and LR diversity conservation.
- Promote the use of *in situ* conserved CWR and LR diversity.

As such, this text aims to be a dissemination tool for AEGRO products to the European and global PGR community, while at the same time, promote discussion of their wider application and continued use.

References

AEGIS (2011) *AEGIS – A European Genebank Integrated System*. Available at: http://aegis.cgiar.org/home.html/ (accessed 21 April 2011).

AEGRO (2011) *An Integrated European In Situ Management Work Plan: Implementing Genetic Reserves and On Farm Concepts*. Available at: http://aegro.jki.bund.de/ (accessed 21 April 2011).

CORINE (2010) *Land cover 2000, version 2009*. Available at: www.eea.europa.eu/publications/COR0-landcover (accessed 21 April 2011).

Dulloo, M.E., Labokas, J., Iriondo, J.M., Maxted, N., Lane, A., Laguna, E., Jarvis, A. and Kell, S.P. (2008) Genetic reserve location and design. In: Iriondo, J.M., Maxted, N. and Dulloo, E. (eds) *Plant Genetic Population Management*. CAB International, Wallingford, UK, pp. 23–64.

GenResIS (2011) Genetic Reserve Information System. Available at: www.agrobiodiversidad.org/aegro (accessed 21 April 2011).

González-Martínez, S.C., Krutovsky, K.V. and Neale, D.B. (2006) Forest-tree population genomics and adaptive evolution. *New Phytologist* 170(2), 227–238.

Gregorius, H.-R. (2001) Sustainable treatment of resources: the genetic basis. In: Müller Starck, G. and Schubert, R. (eds) *Genetic Response of Forest Systems to Changing Environmental Conditions*, Vol. 70, pp. 203–222.

Iriondo, J.M., Dulloo, E. and Maxted, N. (eds) (2008) *Conserving Plant Genetic Diversity in Protected Areas: Population Management of Crop Wild Relatives*. CAB International, Wallingford, UK.

Jain, S.K. (1975) Genetic reserves. In: Frankel, O.H. and Hawkes, J.G. (eds.), *Crop genetic resources for today and tomorrow*. Cambridge University Press, Cambridge, pp. 379–396.

Kell, S.P., Knüpffer, H., Jury, S.L., Ford-Lloyd, B.V. and Maxted, N. (2008) Crops and wild relatives of the Euro-Mediterranean Region: making and using a conservation catalogue. In: Maxted, N., Ford-Lloyd, B.V., Kell, S.P., Iriondo, J.M., Dulloo, M.E. and Turok, J. (eds) *Crop Wild Relative Conservation and Use*. CAB International, Wallingford, UK, pp. 69–109.

Kilian, B., Özkan, H., Kohl, J., von Haeseler, A., Barale, F., Deusch, O., Brandolini, A., Yucel, C., Martin, W. and Salamini, F. (2006) Haplotype structure at seven barley genes: relevance to gene pool bottlenecks, phylogeny of ear type and site of barley domestication. *Molecular Genetics and Genomics* 276, 230–241.

Maxted, N. and Kell, S. P. (2009) *Establishment of a global network for the* in situ *conservation of crop wild relatives: status and needs*. FAO Commission on Genetic Resources for Food and Agriculture, Rome, 266 pp.

Maxted, N., Van Slageren, M.W. and Rihan, J. (1995) Ecogeographic surveys. In: Guarino, L., Ramanatha Rao, V. and Reid, R. (eds) *Collecting Plant Genetic Diversity: Technical Guidelines*. CAB International, Wallingford, UK, pp. 255–286.

Maxted, N., Hawkes, J.G., Ford-Lloyd, B.V. and Williams, J.T. (1997a) A practical model for *in situ* genetic conservation. In: Maxted, N., Ford-Lloyd, B.V. and Hawkes, J.G. (eds) *Plant Genetic Conservation: the* in situ *Approach*. Chapman & Hall, London, pp. 545–592.

Maxted, N., Ford-Lloyd, B.V. and Hawkes, J.G. (1997b) *Plant genetic conservation: the* in situ *approach*. Chapman & Hall, London.

Maxted, N., Dulloo, E., Ford-Lloyd, B.V., Iriondo, J. and Jarvis, A. (2008) Genetic gap analysis: a tool for more effective genetic conservation assessment. *Diversity and Distributions* 14, 1018–1030.

McGrath, J.M., Panella, L. and Frese, L. (2010) Wild species in beet improvement. In: Kole, C. (ed.) *Wealth of Wild Species: Role in Plant Genome Elucidation and Improvement*. Springer, Berlin (in press).

Namkoong, G., Boyle, T., Gregorius, H.R., Joly, H., Savolainen, O., Ratnam, W. and Young, A. (1996) Criteria and Indicators for Impacts of Forest Management on Genetic Variation. Testing Criteria and Indicators for Sustainable Management of Forests. *CIFOR Working Paper No. 10*.

Negri, V., Barocco, R., Pacicco, L., Veronesi, F. and Venanzoni, R. (2011) Landraces in Europe: an approach towards identifying landrace rich areas as a priority for protection. In: Maxted, N., Dulloo, M.E., Ford-Lloyd, B.V., Frese, L., Iriondo, J.M. and Pinheiro de Carvalho, M.A.A. (eds) *Agrobiodiversity Conservation: Securing the Diversity of Crop Wild Relatives and Landraces*. CAB International, Wallingford, UK.

Veteläinen, M., Negri, V. and Maxted, N. (2009) *European Landraces: On-farm conservation, Management and Use*. Bioversity Technical Bulletin 15. Bioversity International, Rome, Italy, pp. 1–359.

Wikipedia (2011) Unternehmensstrategie. Available at: http://de.wikipedia.org/wiki/Unternehmensstrategie#cite (accessed 21 April 2011).

2 *In Situ* Conservation of Crop Wild Relatives: A Strategy for Identifying Priority Genetic Reserve Sites

S.P. Kell, N. Maxted, L. Frese and J.M. Iriondo

2.1 Introduction

Crop wild relatives (CWR) are species closely related to crops and are defined by their potential ability to contribute beneficial traits for crop improvement (Maxted *et al.*, 2006). They have been used increasingly in plant breeding since the early 20th century and have provided vital genetic diversity for crop improvement – for example, to confer resistance to pests and diseases, improve tolerance to environmental conditions such as extreme temperatures, drought and flooding and to improve nutrition, flavour, colour, texture and handling qualities (Maxted and Kell, 2009). In monetary terms, CWR have contributed significantly to the agricultural and horticultural industries, and to the world economy (Maxted *et al.*, 2008a; Maxted and Kell, 2009).

Today, agricultural production is challenged by climate change. The International Panel on Climate Change (IPCC, 2007) estimates that by 2100, maize and wheat yields will be reduced by 40% at low latitudes, while in China, rice yields will decrease by up to 30% unless climate change mitigation is undertaken. Breeders will therefore have to provide varieties able to cope with the impacts of changing growing conditions. Due to the breadth of genetic diversity inherent in CWR populations, which are adapted to a wide range of environmental conditions, they are likely to be needed more than ever before to maintain the adaptability of crops. Thus, CWR are a critical component of plant genetic resources for food and agriculture (PGRFA) and are vital for future food security; however, despite their recognized value, they have historically received relatively little systematic conservation attention.

There are two primary techniques for CWR conservation: *in situ*, primarily in natural habitats managed as genetic reserves (GRs) (Box 2.1) and *ex situ* as seed in gene banks. Historically, CWR conservation has focused almost entirely on *ex situ* collection and storage, but it can be argued that *in situ* conservation is more appropriate because the genetic diversity inherent in and between wild CWR populations is constantly changing in response to their environment; therefore, CWR populations are a component of natural ecosystems that cannot effectively just be maintained *ex situ* (Maxted *et al.*, 2008a).

A number of recent initiatives have raised the profile of CWR and put them on the international conservation agenda. However, conservationists and policy makers are faced with the difficult challenge of how to conserve the large number of CWR species and the genetic diversity that they

Box 2.1 The Genetic Reserve Concept

A genetic reserve is defined as 'the location, management and monitoring of genetic diversity in natural populations within defined areas designated for long-term conservation' (Maxted *et al.*, 1997). The concept combines *in situ* conservation with active management and a long-term approach. The rationale for this type of conservation is that it is: (i) applicable to all plant species; (ii) allows for continued evolution; and (iii) allows for multiple-taxon conservation. Moreover, it conserves the genetic diversity of the target taxon in a dynamic way, as well as its habitat and all existing biotic and abiotic interactions (including humans).

Several approaches to GR conservation can be identified, each with different aims and strategies, depending on the approach (see Maxted and Kell, 2009). For example, the aims for CWR GRs in Europe are to conserve genetic diversity in the widest range of priority CWR taxa at the European scale; therefore, the aim is to design a network of reserves that adequately and efficiently maintains the genetic diversity of the target taxa. When we talk about 'adequately' maintaining the genetic diversity of target taxa, we mean conserving a good representation of the genetic diversity of adaptive and agricultural value present in such taxa. Similarly, by 'efficiently' we mean to obtain this goal using the minimum number of GRs.

Coordination with *ex situ* holdings and crop databases is an important part of the genetic reserve concept. *Ex situ* seed banks can be a relevant component in the functioning of GRs as they provide a back-up of genetic diversity in case any catastrophe should occur. Furthermore, they facilitate information exchange, access for breeding and other research, and promote use.

contain. If a broad definition of a CWR is used (i.e. all the species in the same genus as a cultivated plant), there are more than 16,000 crop and CWR species in the territories of the EU Member States – 13,875 of these are native and at least 2665 are endemic (Kell *et al.*, 2008). CWR are under threat from habitat loss, agricultural intensification, over-collection, climate change and lack of conservation attention, yet only 9% of PGR accessions in European gene bank collections are CWR (Dias *et al.*, Chapter 33, this volume), while most wild populations *in situ* are not actively monitored and managed in or outside protected areas (PAs) (Maxted *et al.*, 2008b). There is therefore a real challenge to the nature conservation and PGRFA sectors to conserve these valuable resources.

The establishment of GRs for CWR is a priority in order to maintain a broad range of genetic diversity within and between populations; however, with a large number of species to conserve, a systematic approach to the identification of GR sites is needed to maximize resource use. This chapter provides a generic methodology that can be used to prioritize taxa on the basis of their potential use for crop improvement and relative threat status, gather the necessary data to undertake diversity and gap analysis for target taxa, and select the most appropriate CWR GR sites. It is built on those proposed by Maxted *et al.* (2008c), Maxted and Kell (2009) and Maxted *et al.* (in prep.), which address floristic and monographic approaches to CWR conservation (Box 2.2). Although it may be necessary to adjust parts of the methodology according to the specific biological, ecological and geographical attributes of individual crop complexes, it provides a generic framework for the conservation of any crop gene pool.

2.2 Methodology for Identifying CWR Genetic Reserve Sites for a Target Crop Gene Pool

In this section, a summary of the methodology for identifying CWR GR sites for a target crop gene pool is provided and illustrated with a case study for the gene pool of cultivated beets. For more detailed step by step guidance on implementing the methodology, including a list of data sources, the reader is referred to the 'CWR *In Situ* Strategy Helpdesk' (http://aegro.Jki.bund.de/aegro/index.PhP?id=188), which is provided as a

Box 2.2 Floristic and Monographic Approaches to CWR Conservation (Maxted *et al.*, 2011)

Floristic and monographic approaches relate to the breadth of coverage of the CWR conservation strategy. A floristic approach involves the development of a CWR conservation strategy for CWR diversity that occurs in a defined geographical area, which may be a sub-national area such as an administrative unit or protected area, a whole country, a supra-national region, or even the whole world. A monographic approach on the other hand is restricted to certain crop gene pools, but like the floristic approach may be carried out at any geographic scale.

The floristic approach is comprehensive because it attempts to encompass all CWR diversity that occurs within a geographical unit; however, while being comprehensive for the geographical unit, the full geographic range of an individual taxon may or may not be included, depending on whether it is endemic to that geographical unit. The monographic approach focuses on CWR diversity within target crop gene pools, which are usually identified on the basis of their perceived value for food security and/or economic stability. Both approaches will ultimately conclude with the systematic conservation of priority CWR diversity via a network of conservation sites and genetic reserves, with backup in *ex situ* collections.

Whether a floristic or monographic approach is taken is likely to depend on: (i) the quantity and quality of existing data; and (ii) the resources available to prepare the conservation strategy. The scope of the parent organization undertaking the conservation may also impact the approach; for example, an international cereal research institute is likely to focus monographically on cereal crops, while a national biodiversity institute is likely to adopt a more floristic approach. It is worth noting that if the goal is to maximize CWR diversity, it is likely that both approaches need to be combined.

guide and information facility for national programmes, research institutes, NGOs, PA managers, or individuals involved in the development of a CWR *in situ* conservation strategy.

There are four basic steps in the *in situ* methodology: (i) taxon delineation; (ii) selection of target taxa; (iii) diversity analysis; and (iv) selection of target sites. The end point of the methodology is the identification of 'ideal' CWR GR sites. The political and legal steps that need to be taken beyond this point to establish the GRs are not part of the methodology. The next step beyond the methodology for identification of GRs is to make recommendations for site and population management (see Maxted *et al.*, 2008d).

2.2.1 Step 1: taxon delineation

The starting point for a crop gene pool CWR conservation strategy is a list of target taxa; therefore, for the target crop gene pool it is necessary to:

1. Generate a list of taxa that occur in the crop gene pool. Although not all the taxa in the gene pool will necessarily be immediately included in the CWR conservation strategy, the complete list of taxa provides a reference point for future potential conservation actions of lower priority taxa.

2. Generate a list of taxa that occur within the defined geographic range of the conservation strategy (i.e. national, regional or global). These may be both native and introduced, but the conservation strategy is most likely to focus on native species.

To achieve these two steps, online information sources and/or literature (monographs, crop-specific studies etc.) need to be consulted (see Maxted and Guarino, 2003).

At this stage, it is necessary to adopt an accepted taxonomy to form the basis of the taxon list and the subsequent conservation strategy. The list of taxa should show the accepted taxon name and authority and list primary synonyms with authorities. This is important because different information systems use different accepted taxonomies; therefore, when searching for information on a specific taxon it could be possible to miss important information if synonymy is not taken into account.

Beet case study step 1

The beet gene pool consists of two genera – *Beta* and *Patellifolia* (Table 2.1). The genus *Beta* is divided into section *Beta* with three species and two subspecies, section *Corollinae*, which is composed of three base species and two hybrid species, as well as *B. nana* (the only species of the former section *Nanae*). The genus *Patellifolia* encompasses three species. Nine wild relatives of cultivated beets are native to Europe.

2.2.2 Step 2: selection of target taxa

In general, it is not practical to attempt to actively conserve all the taxa within the crop gene pool due to resource limitations; therefore, we need to prioritize and select taxa from the list that will be proposed for active conservation. Factors that can be used to ascribe 'value' and establish conservation priorities include (Maxted *et al.*, 1997):

- Current conservation status;
- Socio-economic use;
- Threat of genetic erosion;
- Genetic distinctiveness;
- Ecogeographic distribution;
- Biological importance;
- Cultural importance;
- Cost, feasibility and sustainability;
- Legislation;
- Ethical and aesthetic considerations;
- Priorities of the conservation agency.

For CWR, an initial, simple prioritization on the basis of socio-economic use of the associated crop (a step which will already have been taken in selecting the target crop gene pool) and relative threat has been proposed (e.g. Ford-Lloyd *et al.*, 2008; Magos Brehm *et al.*, 2008). In addition, Maxted and Kell (2009) proposed that within each crop gene pool, the closest wild relatives should be afforded higher conservation priority over the more distantly related species because these are the taxa that can more

Table 2.1. *Beta* working taxonomy and Gene Pool concept.[a]

Gene pool	Taxa
Primary	Section *Beta* Transhel
	B. vulgaris L. subsp. *vulgaris* (cultivated beets)
	Leaf Beet Group
	Garden Beet Group
	Fodder Beet Group
	Sugar Beet Group
	B. vulgaris L. subsp. *maritima* (L.) Arcang.*
	B. vulgaris L. subsp. *adanensis* (Pamuk.) Ford-Lloyd & Will.*
	B. macrocarpa Guss.*
	B. patula Ait.*
Secondary	Section *Corollinae* Ulbrich
	Base species:
	B. corolliflora Zosimovich
	B. macrorhiza Steven
	B. lomatogona Fisch & Meyer
	Hybrid species:
	B. intermedia Bunge
	B. trigyna Wald. & Kid.*
	B. nana Boiss. & Heldr.*
Tertiary	Genus *Patellifolia* Williams, Scott & Ford-Lloyd
	P. procumbens (Smith) A.J. Scott, Ford-Lloyd & J.T. Williams*
	P. webbiana (Moq.) A.J. Scott, Ford-Lloyd & J.T. Williams*
	P. patellaris (Moq.) A.J. Scott, Ford-Lloyd & J.T. Williams*

[a]Synonyms are not shown in this table but are recorded in an associated database.
*Wild relative native to Europe

easily be used in crop improvement using conventional breeding methods. However, the literature on the taxa within the target crop gene pool should be thoroughly searched to check for cases where a more distantly related taxon has been highlighted as a gene donor (or potential gene donor), and these taxa should also be afforded conservation priority. Of these prioritized taxa, those in most urgent need of conservation action (i.e. those with a very limited geographic range, often rare or endemic taxa, and/or known to be under threat) are given precedence.

This methodology therefore primarily targets the taxa that are most closely related to the crop species (or that have shown promise in crop improvement programmes) and that are threatened or have restricted distribution ranges. However, ideally, national and regional *in situ* networks of CWR GRs should in the long term be expanded to ensure that all taxa of potential importance for crop improvement are actively conserved. In particular, selected populations of the closest wild relatives that are widespread and common should be actively conserved throughout their range, ensuring that populations representing the extremes of the range (both geographically and topographically) are conserved. Individual populations of these taxa may harbour important genes adapted to particular environmental conditions – genes that may confer important traits to improve crops in the future. Populations of these taxa that already occur within PAs should also be monitored. In many cases, if a floristic approach is taken, it is possible to establish a reserve that conserves multiple CWR taxa, which, when possible, has obvious advantages.

There are two stages to the selection of target taxa: (i) creation of a level 1 prioritized list based on actual or potential use as gene donors; and (ii) creation of a level 2 prioritized list based on threat and/or distribution. In this methodology, the two steps are presented sequentially (i.e. the level 2 prioritized list is based on the level 1 prioritized list). The advantage of this approach is that in cases where there is limited information on the distribution of the taxa and/or for gene pools containing a very large number of taxa, the level 1 prioritization narrows the list of taxa down to those that are likely to be most important as gene donors for crop improvement and further information is only sought for that list of taxa.

The disadvantage of this approach is that some of the more distantly related taxa in the gene pool that are threatened or have restricted distributions may be missed in the conservation planning process. Therefore, in cases where a gene pool contains a relatively small number of taxa or where distribution data are readily available for all the taxa (e.g. in the case of the beet gene pool), it is desirable to undertake the prioritization in the reverse order by collating threat and distribution data on all taxa in the gene pool first, then applying the second level of prioritization based on potential use as gene donors. Using this approach, more distantly related taxa that are threatened or have restricted distributions can be highlighted as a conservation priority on that one criterion, and even though they may still not be given the highest level of priority for immediate conservation action, they may be promoted as candidates for conservation at a later date. Furthermore, if it is not immediately possible to put in place *in situ* conservation measures for these taxa, they can be earmarked for collection and storage in *ex situ* collections.

To organize the list of taxa within the crop gene pool according to their degree of relationship to the crop, a literature search should be carried out on the crop complex. Taxa should be organized into a table showing primary, secondary or tertiary wild relatives using one of three methods:

1. Where genetic information is available and taxa have been classified using the Gene Pool (GP) concept (Harlan and de Wet, 1971), organize the taxa into the table listing those in GP1B as primary wild relatives, those in GP2 as secondary wild relatives and those in GP3 as tertiary wild relatives.
2. Where genetic information is not available, if possible, substitute the Gene Pool concept with the Taxon Group (TG) concept

(Maxted *et al.*, 2006), which provides a proxy for taxon genetic relatedness. Organize the taxa into the table listing those in TG1b as primary wild relatives, those in TG2 as secondary wild relatives, and those in TG3 and TG4 as tertiary wild relatives.

3. For crop genera that have not been classified using the GP concept and not subclassified into sections and subgenera, the available information on genetic and/or taxonomic distance must be analysed to make reasoned assumptions about the most closely related taxa.

Whichever system is used, it is important to ensure that references are provided to substantiate the assumptions made about taxon relatedness.

In general, the primary and secondary wild relatives are selected as a priority for conservation action, but tertiary wild relatives that have been highlighted as gene donors or potential gene donors should also be added to the priority list. As carried out under Step 1, taxa in the priority list that occur within the geographical area of the conservation strategy are then tagged for further action.

To select taxa on the basis of relative threat and/or distribution (either the entire gene pool or the priority taxon list based on use potential): (i) consult the IUCN Red List of Threatened Species and national or regional Red Lists or carry out a literature search which may reveal important information about the threat status of a taxon; and (ii) compare the geographical range of the taxa. At this stage, a degree of objectivity is required, since there is no clear dividing line between a taxon with a limited range and one with a distribution that is deemed to enable 'classification' of the taxon as one not in immediate need of conservation action, unless very detailed information is already available about genetic erosion of the taxa. However, where the range of a taxon is known, the methodology proposed by Ford-Lloyd *et al.* (2008, 2009) can be used as a guide when establishing taxon conservation priorities at regional level (e.g. across Europe). Generally speaking, taxa that are known to be endemic to a country or subnational unit or those that occur in only a few countries or subnational units are more likely to be under threat at regional level. Similarly, at national or subnational level, available information must be gathered on the range of the taxa in order to establish which are most likely to be threatened by their limited distribution range.

Step 2 results in a reduced list of taxa that have been selected on the basis of their value as gene donors and relative threat. This list of target taxa now forms the basis for immediate conservation planning for the crop gene pool.

Beet case study step 2

The taxa in the gene pool were organized according to their degree of relationship to cultivated beets (Table 2.1). All wild species in the beet gene pool are either known as potential donors of useful genes or have already been used in crop enhancement programmes; therefore, all taxa are considered as a priority for conservation action on the basis of their potential use value. Considering relative threat, a recent initiative to carry out regional Red List assessments of a selection of European CWR (see Kell *et al.*, Chapter 28, this volume) highlighted five wild relatives of beet as a priority on the basis of their threat status: *B. patula* and *Patellifolia webbiana* (Critically Endangered), *B. macrocarpa* (Endangered), *B. vulgaris* subsp. *adanensis* and *B. nana* (Vulnerable). The remaining four taxa native to Europe were assessed as Data Deficient (*B. trigyna*) and Least Concern (*B. vulgaris* subsp. *maritima*, *P. patellaris* and *P. procumbens*).

It is important to note that the selection of target taxa on the basis of relative threat (whether based on Red List assessments or relative distribution) is likely to vary depending on the geographical scope of the conservation strategy. For example, in the case of beet, at European level, *P. patellaris* and *P. procumbens* are not immediate priorities for conservation action due to their relatively widespread distribution. However, if the scope of the conservation strategy is national, these taxa may be targeted as a priority for conservation action; for example, in Portugal, they are both a priority due to the fact that only a few subpopulations occur.

2.2.3 Step 3: diversity analysis

Once the priority list of CWR species has been identified (Step 2), the next step is to collate the available ecogeographic information to assist in further formulation of the CWR conservation strategy. This involves the collation and analysis of geographic, ecological, environmental and genetic data. These data are predictive and aid the location of the CWR taxonomic (inter-taxa) and genetic (intra-taxon) diversity that can then be targeted for conservation. As the goal is to maximize conserved genetic diversity, information on the partitioning of genetic diversity across the ecogeographic distributions of the target taxa is useful in identifying sites or combinations of sites of maximum diversity. However, even with rapidly decreasing costs of analysing genetic diversity, this information may be extremely limited; in which case, analysis of ecological and environmental data associated with the sites at which the populations occur can be used as a proxy for genetic diversity. The culmination of the diversity analysis should be a set of areas with high concentrations of the priority CWR species and populations of CWR taxa containing or thought to contain complementary and/or unique genetic diversity.

Geographic data are of two types – coordinate and descriptive. Ideally, coordinate data should be used for accuracy (however, even coordinate data can sometimes be misleading, depending on the accuracy and quality of the original data). Descriptive data can be converted to coordinate data by consulting gazetteers. At this stage in the analysis, issues of data quality have to be taken into account and steps may need to be taken to improve the accuracy of the distribution data to remove any erroneous entries. For example, it has been suggested that only population occurrences with geographic coordinates that have two decimal digits or more are used in the analysis. Another limitation is that the availability of occurrence data may be very heterogeneous across the range of the target taxon – this needs to be taken into account when making decisions on the selection of target sites (Step 4). Where distribution data are too sketchy or otherwise incomplete or inaccurate, it may be necessary to recommend that a detailed ecogeographic survey is undertaken before further analysis.

Genetic diversity analysis is only possible where the necessary information already exists or where resources permit the generation of novel genetic diversity information. There are two types of genetic diversity information of interest for the establishment of GRs and for backup in *ex situ* collections: intra-population and inter-population diversity. The precise method of generating genetic diversity information is taxon-specific. Decisions regarding the type of genetic analysis to undertake can be based on existing studies of related taxa or taxa sharing similar biological attributes. Literature searches can be undertaken to obtain this information, as well as consulting specialist databases and taxon experts.

Ecological and environmental data associated with the target taxa can be of two types: actual (i.e. data directly linked to a taxon) or secondary (i.e. data indirectly linked to a taxon via the attributes of the site in which it is found). Actual ecological and environmental data can be sourced by obtaining characterization and evaluation data associated with *ex situ* accessions, and/or by consulting the available literature on the target taxon – for example, there may be published or grey literature as a result of ecological studies of the taxon or of associated taxa that occur in the same habitats – or by collecting fresh data in the field. Secondary data are obtained by gathering data associated with known locations of a taxon (e.g. climate, soil type, geological substrate, habitat type, altitudinal range and land use). Some of these data are readily available in the form of Geographical Information System (GIS) files, which are overlaid with the distribution data, and from which inferences can be made about the ecological preferences of a taxon.

The data collated are analysed to build detailed taxon ecogeographic profiles. A GIS program such as ArcGIS can be used to create distribution maps overlaid with ecological, environmental and genetic data, and locate complementary GR locations (i.e. those that represent the widest range of ecogeographic diversity of the target taxa as possible). The analysis may be simple to complex, depending on availability of data, expertise, time and resources. The data should also be imported into an appropriate information management system from which standard taxon data sheets can be extracted to form the basis of GR proposals and management plans.

Complementarity analysis may also be undertaken. This aims to maximize taxonomic diversity conservation in the minimum number of sites and may be useful when dealing with gene pools containing a large number of taxa or for multiple gene pools. The GIS program DIVA GIS (see Hijmans *et al.*, 2001) is useful for undertaking complementarity analysis and is available for download free of charge.

Beet case study step 3

Diversity analysis of the beet gene pool was carried out in two stages. First, a review and compilation of the available geographic, ecological, environmental and genetic information for each of the target taxa was carried out. At this stage, the emphasis was on the use of genetic data to establish the ecogeographic pattern of genetic diversity. Second, a detailed ecogeographic diversity analysis of the target taxa using GIS was undertaken (see Parra-Quijano *et al.*, Chapter 3, this volume). This part of the analysis involved the compilation of information on factors related to abiotic adaptation upon a GIS background containing environmental variables (climatic, edaphic and geophysical). Using this method, each potential site for the establishment of a GR was environmentally characterized to aid the final selection of target sites.

2.2.4 Step 4: selection of target sites

In some cases, the range of the target taxon will define the precise site or sites where active *in situ* conservation is needed. Obviously, for a taxon that is known only to occur at one location and is considered a high priority as a potential gene donor, then that single location must be targeted for reserve establishment. Where the geographic range of the target taxon is broader, sites should be selected that represent the widest range of ecogeographic characteristics as possible.

Once the target taxon distribution has been identified and mapped, and diversity analysis undertaken (Step 3), PA overlays are used to ascertain whether the target taxon populations occur within the boundaries of existing PAs. CWR, like any other group of wild plant species, are located both within and outside existing PAs; however, the most efficient approach in the first instance (to avoid the purchase and establishment of new sites) is to establish CWR GRs within existing PAs (Maxted *et al.*, 2007). Therefore, the most appropriate PAs (e.g. national parks and heritage sites) within which to locate GRs should be identified.

GIS analysis using PA shapefiles provides an indication of which PAs contain populations of the target taxa. In addition, this method can be used to predict which PAs contain high concentrations of CWR diversity. To be certain that the populations do exist within the PA(s), it is necessary to confirm their presence before GR establishment is recommended. This information is not always easy to obtain; however, if the taxon expert is not certain of its presence at the site, it may be possible to contact the agency responsible for the management of the PA to see if they have an inventory of taxa available or whether it is possible for site staff to confirm the presence of the taxon. If possible, ground truthing by visiting the site(s) personally should be undertaken. This is of course subject to available time and resources.

Where target taxon populations are found to already occur within existing PAs, these populations should be prioritized for inclusion in the CWR GR network on the basis that they have already been afforded some degree of protection, even if only by default. However, it is important to stress that even though a target taxon population may occur within the boundaries of a PA, this does not automatically mean that the population is actively conserved. On the contrary, few PAs are established to conserve specific target taxa, and those that have tend to focus on animal conservation. To conserve the range of genetic diversity inherent in CWR populations, active site management and monitoring is needed (see Iriondo *et al.*, Chapter 10, this volume) – some PAs do not even have management plans, and those that do are often limited by financial resources and lack of capacity to put the plan into practice.

In cases where a few to several PAs are found to contain populations of a target taxon, results of the diversity analysis can be used to select sites that best represent the ecogeographic diversity within the target taxon. A further consideration for the selection of PAs is the option for multiple taxa GRs. Analysis of all target taxa within the crop gene pool (and preferably across several crop gene pools) may reveal that some PAs contain populations of more than one taxon. In terms of expediency of resource use, multi-taxa reserves have obvious advantages over those that only contain a population of one taxon.

Where target taxon populations do not already occur within existing PAs, these populations should also be prioritized for inclusion in the CWR GR network on the basis that they have not already been afforded any degree of protection; especially for rare or threatened species. Obviously, justifying the need for and actually establishing new PAs will involve a significant initial injection of time and resources. Nomination of GRs at the target locations may of course be hindered by a range of socio-political factors, such as legal issues, land use conflicts, issues of land ownership, or lack of local support. Therefore, if possible a range of alternative sites should be recommended and ranked according to their suitability based on ecogeographic considerations.

The main criterion for allocating priorities to sites is the conservation of the maximum genetic diversity possible. When assigning priorities for a particular target taxon, the ecogeographic analysis will form the basis of the priority ranking of sites. When the aim is to conserve multiple taxa within the same sites, a balance has to be met between prioritizing those sites that contain the greatest taxonomic diversity and those that contain less taxonomic diversity, but more genetic diversity specific to particular target taxa. Other factors to take into account when assigning priority ranking to selected sites include: land use, potential development pressures (e.g. sites closer to towns and cities may be less secure), presence of invasive species (particularly on islands), level and quality of site management, legal status, potential conflict with existing site management aims and social unrest. A thorough assessment of all factors, both scientific and socio-political, must be made and considered when selecting the ideal sites.

The potential effects of climate change on populations of the target taxa also need to be taken into account. Considerations include the particular vulnerability of populations in coastal and high altitude areas, whether there is sufficient intra-population genetic diversity and reproductive success in populations to allow adaptation to new conditions, and whether small, fragmented populations with little migration will be able to colonize new sites (Veteläinen *et al.*, 2007). In the absence of detailed studies on individual target taxa, it will not be possible to predict exactly where sites need to be established because: (i) we will not know whether populations of a taxon will have the ability to adapt to new conditions at current sites; (ii) we will not know whether populations will have the ability to migrate to new sites; and (iii) if migration occurs, how quickly it will take place and in what direction. However, greater emphasis on habitat protection to prevent and reduce habitat fragmentation and the establishment of corridors between habitat patches to facilitate range shifts of mobile species is likely to be important for many CWR taxa (Jarvis *et al.*, 2008).

Beet case study step 4

The selection of target GR sites for the beet gene pool involved collaboration between a taxon expert with good knowledge of the European populations and a GIS expert who carried out the detailed ecogeographic diversity analysis. Initially, candidate sites for the target taxa were identified by the taxon expert using genetic distance and genetic diversity data, as well as geographic data. Sites were recommended for immediate action if: (i) the occurrence was known to be distributed in a Natura 2000 protected site; and/or (ii) the occurrence was known to represent a unique or specific fraction of the taxon's genetic diversity. As one of main objectives of the *in situ* management strategy is the maintenance of the highest possible amount of genetic diversity of a gene pool in nature, which is prerequisite for the evolutionary processes generating novel genetic variation, ten candidate GR sites for *B. vulgaris* subsp. *maritima* were also proposed. Although the taxon was not included in the prioritized list, these sites are known to harbour genetic variation for traits useful for plant breeding. As this taxon is widely distributed there is no immediate need for active management; however, as little is known about the geographic distribution pattern of traits useful for plant breeding, the establishment of GRs should be seen as a precautionary measure to secure these materials.

In parallel, a systematic protocol (using the ecogeographic information obtained in Step 3 and species occurrence data) was developed by the GIS expert to assist the taxon expert in the selection and ranking of GR sites for the creation of a network at the European level. This protocol is described in detail by Parra-Quijano *et al.* (Chapter 3, this volume). The approach is based on the generation of an Ecogeographical Land Characterization (ELC) map that identifies different ecogeographical units that are likely to promote local adaptation in the target species populations. It maximizes the ecogeographical representation of the selected sites that fall within PAs and positively informs other criteria such as the occurrence of other taxa of the same genus at the site, generating a selected number of potential sites. This information assisted the taxon expert to produce a list of 28 candidate sites distributed in seven European countries.

Geographic information of selected sites provided by the taxon expert was verified by the GIS expert by importing coordinates of the target taxon populations into a GIS. These data were overlaid with the geographic coordinates of the PAs proposed for the establishment of GRs by the taxon expert. Thus, the preliminary list of PAs containing proposed GRs were obtained and a map with these areas was developed. Subsequently, the information provided by the taxon expert concerning the list of proposed PAs was checked. When inconsistencies were found between the location of the populations of the target taxa and the location of the proposed sites, alternative PAs were suggested to the expert for consideration. When no PAs could be found where suitable populations of a particular target taxon occurred, the taxon expert provided the geographic coordinates of the target population of the target taxon where a GR could be established. Thus, a final list of locations (mainly within PAs) was identified where GRs could potentially be established (Table 2.2) and a final map was obtained.

After this stage, further information relating to the sites and populations was gathered to aid the documentation and verification of the selected sites. This involved the collation of habitat types, land use and conservation status of the sites, as well as information on the status of the populations of the target taxa at the sites.

2.3 Conclusion

Crop wild relatives contain a wide pool of genetic diversity that is important to maintain for its use in plant breeding for crop improvement. The highest priority CWR for food security are not adequately conserved,

Table 2.2. Target genetic reserve sites for the beet gene pool.

Target taxon(a)	Country	Site name	Site code	Designation
Beta macrocarpa	Portugal	Ria Formosa – Castro Marim	PTCON0013	Natura 2000
	Spain	Amagro	ES7010011	Natura 2000
	Spain	Archipielago Chinijo	ES7010045	Natura 2000
	Spain	Cabo de Gata – Níjar	ES0000046	Natura 2000
	Spain	Costa del Norte de Fuerteventura	ES0000348	Natura 2000
	Spain	Interian	ES7020081	Natura 2000
	Spain	Lomo del Carretón	ES7020037	Natura 2000
	Spain	Salinas de Santa Pola	ES0000120	Natura 2000
B. macrocarpa, *Patellifolia procumbens*	Spain	Anaga	ES0000109	Natura 2000
B. nana	Greece	Notoanatolikos Parnassos – Ethnikos Drymos Parnassou - Dasos Tithoreas	GR2450005	Natura 2000
	Greece	Oros Gkiona	GR2450002	Natura 2000
	Greece	Oros Olympos	GR1250001	Natura 2000
B. patula, *P. procumbens*	Portugal	Parque Natural da Madeira – Ponta de São Lourenço	PTMAD0003	Natura 2000
B. vulgaris subsp. *adanensis*	Greece	Arkoi, Lepsoi, Agathonisi Kai Vrachonisides	GR4210010	Natura 2000
	Greece	Rodos: Profitis Ilias – Epta Piges – Petaloudes	GR4210006	Natura 2000
B. vulgaris subsp. *maritima*	Denmark	Gisseløre, Houget	DNK349790	CDDA
	Germany	Helgoland mit Helgoländer Felssockel	DE1813391	Natura 2000
	Germany	Küstenstreifen west – Und Nordfehmarn	DE1532391	Natura 2000
	Italy	Delta del Po: Tratto Terminale e Delta Veneto	IT3270017	Natura 2000
	Portugal	Ria de Aveiro	PTZPE0004	Natura 2000
	France	Estuaires et Littoral Picards (Baies de Somme et d'Authie)	FR2200346	Natura 2000
P. procumbens	Portugal	Laurisilva da Madeira	PTMAD0001	Natura 2000
	Spain	Cuenca de Benchijigua – Guarimiar	ES7020107	Natura 2000
	Spain	El Hierro	ES0000103	Natura 2000
	Spain	Laderas de Enchereda	ES7020101	Natura 2000
P. webbiana	Spain	Area maritima de la Isleta	ES7010016	Natura 2000

either *in situ* or *ex situ*, and we cannot rely only on *ex situ* conservation of these resources as it does not maintain the evolutionary process of adaptation found in wild populations – this can only be achieved by managing *in situ* populations in CWR genetic reserves. The value of CWR for food security and the need for their conservation has recently been placed firmly on the international conservation agenda, but due to the large number of taxa that exist, coupled with limited resources, a means of setting priorities for their conservation is needed. Further, as CWR have largely been neglected by the conservation community, we face the challenge of identifying where and how to conserve them *in situ*.

In this chapter, we have proposed a methodology for the identification of GR sites for a target crop gene pool, which addresses the need to conserve the maximum range of genetic diversity in the highest priority taxa in terms of their known or potential value for crop improvement and relative threat. This approach has already been applied at a global scale for a number of globally important crop gene pools (see Maxted and Kell, 2009) and for four crop gene pools important for food security and economic stability in Europe – oats (*Avena* spp.), beet (*Beta* spp.), brassicas (*Brassica* spp.) and cherry (*Prunus* spp.). By applying the methodology across a range of different crop types, including cereals, leafy vegetables, root crops and fruit trees, it has been possible to: (i) reveal different perspectives on its application by several experts; (ii) investigate ways in which the application of the individual steps may differ between different crop groups; (iii) scrutinize the methodology to confirm its applicability to a range of crop groups; and (iv) refine the methodology to ensure that it is widely applicable to any crop gene pool and easily understood by all those involved in CWR *in situ* conservation strategy planning.

We have found that the model can be widely used; however, it is clear that its application will necessarily be slightly adapted according to the different crop gene pools to which it is applied. For example, the means of selecting target taxa varies from one gene pool to another, depending on:

- The number of species in the gene pool (e.g. *Brassica* is a large genus compared to *Avena*, *Beta* and *Patellifolia*).
- The number of crops in the gene pool (e.g. *Brassica* contains several crops).
- Knowledge of the genetic relationship between taxa (e.g. there is better knowledge for *Avena*, *Beta* and *Patellifolia* than for *Brassica*).
- Knowledge of the breeding potential of species (e.g. there is better knowledge for *Avena*, *Beta* and *Patellifolia* than for *Brassica*).

Further, the means of selecting target sites varies from one gene pool to another, depending on: (i) existing knowledge of intra- and inter-specific genetic diversity of target taxa; and (ii) existing knowledge of localities of target species (e.g. population size, threats and suitability of the site to establish a GR). In addition, the application of the methodology highlighted the difficulty of dealing with taxonomic data (i.e. different nomenclature in different information systems) and occurrence data (e.g. lack of coordinates, problems of data quality, and evenness of data quality across Europe). However, these are challenges that we face in conservation planning in general and are not specific to this model.

To conclude, a logical and systematic framework for CWR conservation is needed that is applicable to any country or region and to any crop gene pool. This may involve both the floristic and monographic approaches, but in order to conserve the maximum range of genetic diversity in the highest priority crop gene pools for global, regional and local food security, a crop gene pool approach is needed that can be applied in tandem with the floristic approach at national level. The methodology presented in this chapter can now be applied to develop conservation strategies for more priority crop gene pools with the aim of eventually ensuring that the genetic diversity that we may rely on in years to come is secured in a network of national, regional and global CWR genetic reserves.

References

Ford-Lloyd, B.V., Maxted, N. and Kell, S.P. (2008) Establishing conservation priorities for crop wild relatives. In: Maxted, N., Ford-Lloyd, B.V., Kell, S.P., Iriondo, J., Dulloo, E. and Turok, J. (eds) *Crop Wild Relative Conservation and Use*. CAB International, Wallingford, UK, pp. 110–119.

Ford-Lloyd, B.V., Maxted, N. and Kell, S. (2009) Prioritization of wild *Beta* species for conservation: the PGR Forum experience. In: Frese, L. and Maggioni, L. (compilers) *Report of a Working Group on* Beta *and World* Beta *Network, Third Joint Meeting, 8–11 March 2006, Puerto de la Cruz, Tenerife, Spain*, pp. 27–30.

Harlan, J. and de Wet, J. (1971) Towards a rational classification of cultivated plants. *Taxon* 20, 509–517.

Hijmans, R.J., Guarino, L., Cruz, M. and Rojas, E. (2001) Computer tools for spatial analysis of plant genetic resources data: 1. DIVA-GIS. *Plant Genetic Resources Newsletter* 127, 15–19.

IPCC (2007) *Fourth Assessment Report Climate Change 2007: Synthesis Report*. Intergovernmental Panel on Climate Change, Geneva, Switzerland.

Jarvis, A., Lane, A. and Hijmans, R.J. (2008) The effect of climate change on crop wild relatives. *Agriculture, Ecosystems and Environment* 126(1–2), 13–23.

Kell, S.P., Knüpffer, H., Jury, S.L., Ford-Lloyd, B.V. and Maxted, N. (2008) Crops and wild relatives of the Euro-Mediterranean region: making and using a conservation catalogue. In: Maxted, N., Ford-Lloyd, B.V., Kell, S.P., Iriondo, J., Dulloo, E. and Turok, J. (eds) *Crop Wild Relative Conservation and Use*. CAB International, Wallingford, UK, pp. 69–109.

Magos Brehm, J., Maxted, N., Ford-Lloyd, B.V. and Martins-Loução, M.A. (2008) National inventories of crop wild relatives and wild harvested plants: case-study for Portugal. *Genetic Resources and Crop Evolution* 55, 779–796.

Maxted, N. and Guarino, L. (2003) Planning plant genetic conservation. In: Smith, R.D., Dickie, J.B., Linington, S.H., Pritchard, H.W. and Probert, R.J. (eds) *Seed Conservation: Turning Science into Practice*. Royal Botanic Gardens, Kew, London, pp. 37–78.

Maxted, N. and Kell, S. (2009) *Establishment of a Network for the* In Situ *Conservation of Crop Wild Relatives: Status and Needs*. Commission on Genetic Resources for Food and Agriculture. Food and Agriculture Organization of the United Nations, Rome, 211 pp.

Maxted, N., Hawkes, J.G., Ford-Lloyd, B.V. and Williams, J.T. (1997) A practical model for *in situ* genetic conservation. In: Maxted, N., Ford-Lloyd, B.V. and Hawkes, J.G. (eds) *Plant Genetic Conservation: the* In Situ *Approach*. Chapman and Hall, London, pp. 339–367.

Maxted, N., Ford-Lloyd, B.V., Jury, S.L., Kell, S.P. and Scholten, M.A. (2006) Towards a definition of a crop wild relative. *Biodiversity and Conservation* 15(8), 2673–2685.

Maxted, N., Scholten, M.A., Codd, R. and Ford-Lloyd, B.V. (2007) Creation and use of a national inventory of crop wild relatives. *Biological Conservation* 140, 142–159.

Maxted, N., Ford-Lloyd, B.V. and Kell, S.P. (2008a) Crop wild relatives: establishing the context. In: Maxted, N., Ford-Lloyd, B.V., Kell, S.P., Iriondo, J., Dulloo, E. and Turok, J. (eds) *Crop Wild Relative Conservation and Use*. CAB International, Wallingford, UK, pp. 3–30.

Maxted, N., Iriondo, J., Dulloo, E. and Lane, A. (2008b) Introduction: the integration of PGR conservation with protected area management. In: Iriondo, J.M., Maxted, N. and Dulloo, E. (eds) *Plant Genetic Population Management*. CAB International, Wallingford, UK, pp. 1–22.

Maxted, N., Dulloo, E., Ford-Lloyd, B.V., Iriondo, J. and Jarvis, A. (2008c) Genetic gap analysis: a tool for more effective genetic conservation assessment. *Diversity and Distributions* 14, 1018–1030.

Maxted, N., Iriondo, J., De Hond, L., Dulloo, E., Lefèvre, F., Asdal, A., Kell, S.P. and Guarino, L. (2008d) Genetic Reserve Management. In: Iriondo, J.M., Maxted, N. and Dulloo, E. (eds) *Plant Genetic Population Management*. CAB International, Wallingford, UK, pp. 65–87.

Maxted, N., Kell, S. and Magos Brehm, J. (2011) *Options to Promote Food Security: On-farm Management and* In Situ *Conservation of Plant Genetic Resources for Food and Agriculture*. Food and Agriculture Organization of the United Nations, Rome.

Maxted, N., Kell, S.P., Magos Brehm, J. and Ford-Lloyd, B.V. (in prep.) Systematic strategies for crop wild relative conservation.

Veteläinen, M., Helgadóttir, Á. and Weibull, J. (compilers) (2007) *Climatic change and genetic resources in northern Europe. Report of a Workshop, 18–19 September 2006, Rovaniemi, Finland*. Bioversity International, Rome. Available at: www.ecpgr.cgiar.org/publications/publication/issue/climatic_change_and_genetic_resources_in_northern_europe.html (accessed 9 May 2011).

3 Spatial and Ecogeographic Approaches for Selecting Genetic Reserves in Europe

M. Parra-Quijano, J.M. Iriondo, L. Frese and E. Torres

3.1 Genetic Reserve Site Selection for Crop Wild Relatives

Most crop wild relative (CWR) species are distributed across diverse habitats and geographic locations (Hawkes *et al.*, 1997). Since CWR have identical biological and ecological characteristics to other wild species, CWR require similar strategies to those used in the *in situ* conservation of threatened wild species. One of these strategies involves species conservation in protected areas.

In CWR, a particular approach to *in situ* conservation was defined through the establishment of 'genetic reserves', ideally a network of connected areas sustaining at least the minimum of genetic variation required for the adaptation of the species to changing environmental conditions through natural selection (Jain, 1975). While *ex situ* conservation aims at the maintenance of a genetic snapshot of the species' diversity taken at a specific time and place, the genetic reserve conservation technique focuses on the maintenance of evolutionary processes. Namkoong *et al.* (1996) defined four processes (drift, selection, mating system, migration) and four genetic indicators required to judge the sustainability of these evolutionary processes. They further stressed that sufficient and adequate genetic variation is the indispensable requirement of evolution. The optimal location of these genetic reserves may or may not coincide with the location of currently established protected areas. Maxted (2003) pointed out that priorities used to establish protected areas do not necessarily have to coincide with CWR species conservation requirements. However, the creation of new protected areas for the establishment of genetic reserves for CWR may be in most cases an unrealistic scenario, at least in Europe, where the protected areas contained by the Natura 2000 network and the nationally designated protected areas cover over 20% of the territory. Therefore, a more realistic approach for the conservation of CWR species in Europe is to determine which of the currently existing protected areas could be used for the establishment of genetic reserves. Maxted *et al.* (2008a, b) and Dulloo *et al.* (2008) offer a detailed explanation about genetic reserves nature and operation.

3.2 Representativeness on *In Situ* Conservation

A recurrent objective in gene banks is to preserve representative samples of the genetic diversity of the target species, as it is impossible in practice to conserve all their alleles or genotypes. Representativeness can be approached at three different scales in gene banks: population, species and group of species. In the first approach, the objective is to conserve the germplasm of a particular population through an appropriate sampling of individuals. At the single-species scale, the goal is to represent the natural genetic diversity of a species in the gene bank in the most reliable way through an appropriate sampling of the populations. Finally, when a group of species is considered, efforts on *ex situ* conservation tend to balance the diversity representation of all single-species included in the group and the priority awarded to each species from a conservation perspective.

Representativeness in *in situ* conservation must accomplish similar objectives to those expressed for the *ex situ* approach. However, the genetic characterization required to assess the genetic representativeness of a group of species, single-species or several populations can be unaffordable due to the potentially large number of samples to be analysed. Ecogeographic representativeness, as proposed for *ex situ* conservation (Parra-Quijano *et al.*, 2008), can be used as a surrogate of genetic representativeness for *in situ* conservation, with similar advantages such as: low cost, easy-to-use, appropriate for large numbers (species, populations), and assisted by powerful geographic information system (GIS) tools. In this approach it is assumed that ecogeographic difference is highly correlated with genetic difference. If this is the case, the selection of a set of areas using the ecogeographic approach would also maximize the genetic diversity within the whole set of occurrences of a species growing within these areas. This approach is not only rational and pragmatic. Selecting components of the genetic reserve network using the surrogate measure reduces the set of occurrences to a manageable number, which in turn facilitates species-specific genetic analyses required to verify or refuse the initial hypothesis. One could argue that the amount of genetic variation present in the whole set of recommended occurrences and the genetic differences between them would be higher than in an equal number of occurrences sampled at a random if the aforementioned correlation is high enough.

In representativeness studies, the ecogeographic approach attempts to include all environmental adaptation features, with special attention on species/populations adapted to extreme or marginal environments. Nevertheless, the ecogeographic representativeness approach is only applicable to the group of species or single-species scales, because the resolution of ecogeographic variables is not accurate enough to discriminate within-population environmental variability, at least in populations whose occupancy area is below 5 km diameter. Furthermore, local adaptation at the within-population scale may not necessarily be widespread among species. This fact does not represent a handicap since *in situ* conservation programmes for crop and CWR species are usually focused on large areas and several target species, where the intra-population scale becomes less meaningful.

Ecogeographic representativeness for *in situ* conservation of plant genetic resources can be used to select genetic reserves in both on-farm and protected areas. In this chapter, we will explore strategies for designing a network of genetic reserves in protected areas based on the concept of ecogeographic representativeness and combining subjective and objective criteria.

Following the principle of representativeness for *in situ* conservation, the location of genetic reserves should be selected to include as much genetic diversity as possible. Using ecogeographic data as a surrogate of genetic characterization, a network of genetic reserves can be selected within the current network of protected areas to conserve a particular species.

3.3 Ecogeographic Maps

Ecogeographic land characterization (ELC) maps have been proposed as a suitable technique to assess the ecogeographic representativeness of germplasm preserved in *ex situ* collections (Parra-Quijano *et al.*, 2008). This type of map reflects as many categories as environmental adaptive scenarios occur over a particular workspace. Adaptive scenarios group environments based on bioclimatic, geophysical and edaphic characteristics. In a recent study, a generalist ecogeographic map was evaluated in terms of plant adaptation for eight species with good fit in some species and poor in others (Parra-Quijano *et al.*, 2011). The two main conclusions of this study were that ecogeographic maps are most useful when specifically developed for single-species or related species groups (for example related legumes or grass species) and that this type of map can be applied both to CWR and landrace germplasm.

3.4 Stepwise Process to Select Genetic Reserves for Crop Wild Relatives

A stepwise methodology to support genetic reserve selection process is introduced. Each step of the methodology is supported by GIS tools and a summarized schematic structure can be visualized in Fig. 3.1. Along the process, subjective and objective criteria are included.

3.4.1 Subjective criteria: expert knowledge

Expert knowledge has been used to determine what, how and where to conserve biodiversity in protected areas strategies (Cowling *et al.*, 2003). In a parallel way, more objective algorithms and software have been developed with similar aims (Margules and Pressey, 2000). Despite the subsequent controversy about which approach would be more fitted to select appropriate conservation areas, Cowling *et al.* (2003) agree that both are more complementary than opposite approaches. Following this argument, we propose to include expert knowledge criteria in two steps of the process: (i) selection of ecogeographic variables that are considered to be most directly related to plant adaptation;

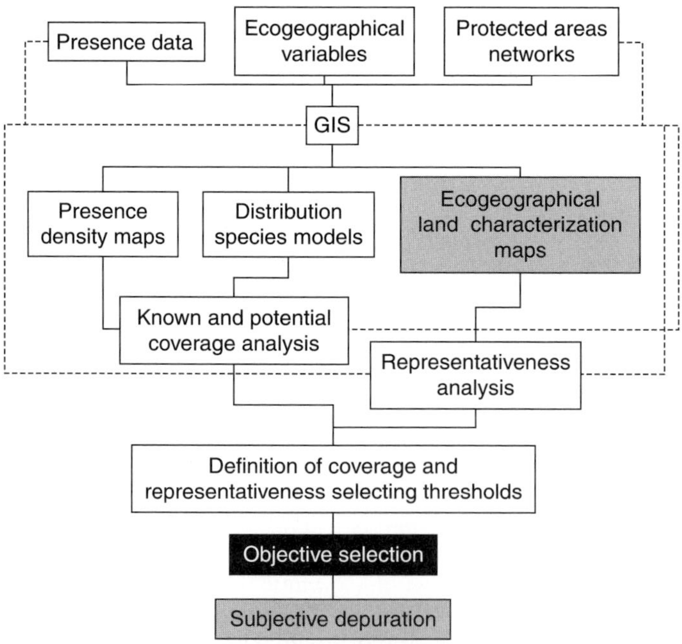

Fig. 3.1. Structure of a genetic reserve selection process. The steps of the process indicated in white squares mainly use objective criteria, while the steps indicated in grey and black squares include subjective criteria. In grey squares the subjective component is from expert knowledge and in the black square essentially involves researcher decisions.

and (ii) final process of depuration of pre-selected candidate locations for the establishment of genetic reserves.

There are other researcher interventions in the process, but they are more related to definition of statistical thresholds or selection of modelling algorithms than to biological or agronomical knowledge of the species.

3.4.2 Objective criteria: ecogeographic approach

The objective component of the proposed methodology can be understood as a series of data management and statistical processes to obtain enough information to select the most appropriate conservation areas for CWR. The methodologies to be applied are a decision of the researcher, but the results for themselves define which areas would be appropriate using objective criteria. Some statistical and data management methodologies require definition of thresholds by the researcher that reduce objectiveness. However, highly objective methodologies have been gradually developed and applied through the development of multivariate statistics, modelling algorithms and data mining research areas (Moilanen *et al.*, 2009). In any case there is always a human bias inside any objective process, for instance the composition of presence data or the initial ecogeographical dataset.

3.4.3 Defining potential sites for genetic reserves

The selection process presented here is based on the idea of establishing genetic reserves in already existing protected areas, supported by minimum legal and administrative standards, and it is focused on a target group of species (i.e. priority congeneric species of a particular crop). For illustration purposes, some data and images of the genetic reserve selection process for *Beta* species developed in the context of the AEGRO project will be shown.

The first step is to compile presence data from a CWR target species. At this level, the georeference quality of presence data is a critical issue because the final result is highly influenced by this factor. Georeference quality affects species distribution models (Soberón and Peterson, 2004) and the ecogeographic representativeness analysis. For this reason, high quality georeferenced presence data should be selected in this step. We propose to evaluate the georeference quality based on three criteria: (i) precision of geographic coordinates (also considering the date of presence data) and locality descriptions (if available); (ii) occurrence in sites suitable for plants; and (iii) concordance between the locations described by passport data and the locations derived from geographic coordinates.

A second step is to identify areas with high density of species presence (see Fig. 3.2a). Genetic reserves should be located in protected areas with a large number of matches with target species presence data (expressed as 'high presence density areas') to ensure two aspects: (i) the real occurrence of the species within the area; and (ii) to conserve the greatest possible number of populations of the target species.

Several algorithms, such as the kernel density estimator (Silverman, 1986), can be applied to obtain estimates of density. A threshold should then be established to create a map of high presence density areas.

The third step involves the use of the map of current protected areas (Fig. 3.2b), which is overlapped on the high presence density areas map. As a result, the protected areas that match with high presence density areas are pre-selected.

At this time, an ecogeographic land characterization (ELC) map covering the workspace should be created (Fig. 3.3a). Details about the creation and validation of ELC maps can be found in Parra-Quijano *et al.* (2008, 2011). A critical step in the process of generating this map is the selection of the ecogeographic variables that are

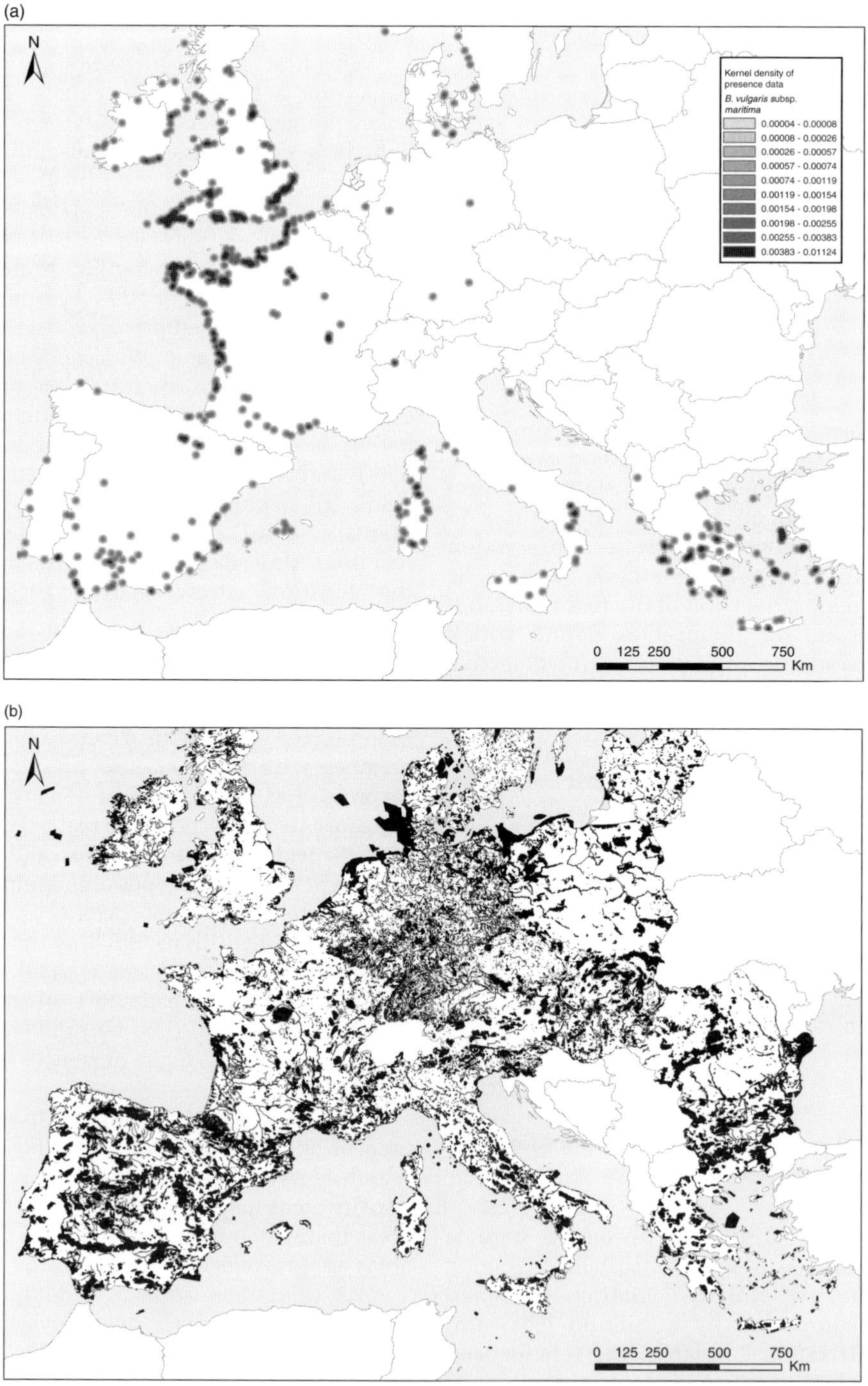

Fig. 3.2. Examples of the second and third steps followed in the case of *Beta* species (*Beta vulgaris* subsp. *maritima*, *Beta vulgaris* subsp. *adanensis*, *Beta macrocarpa* and *Beta nana*) genetic reserve selection process in the AEGRO project. (a) Presence density (Kernel) map for *B. vulgaris* subsp. *maritima*. (b) Map of Natura 2000 network of protected areas.

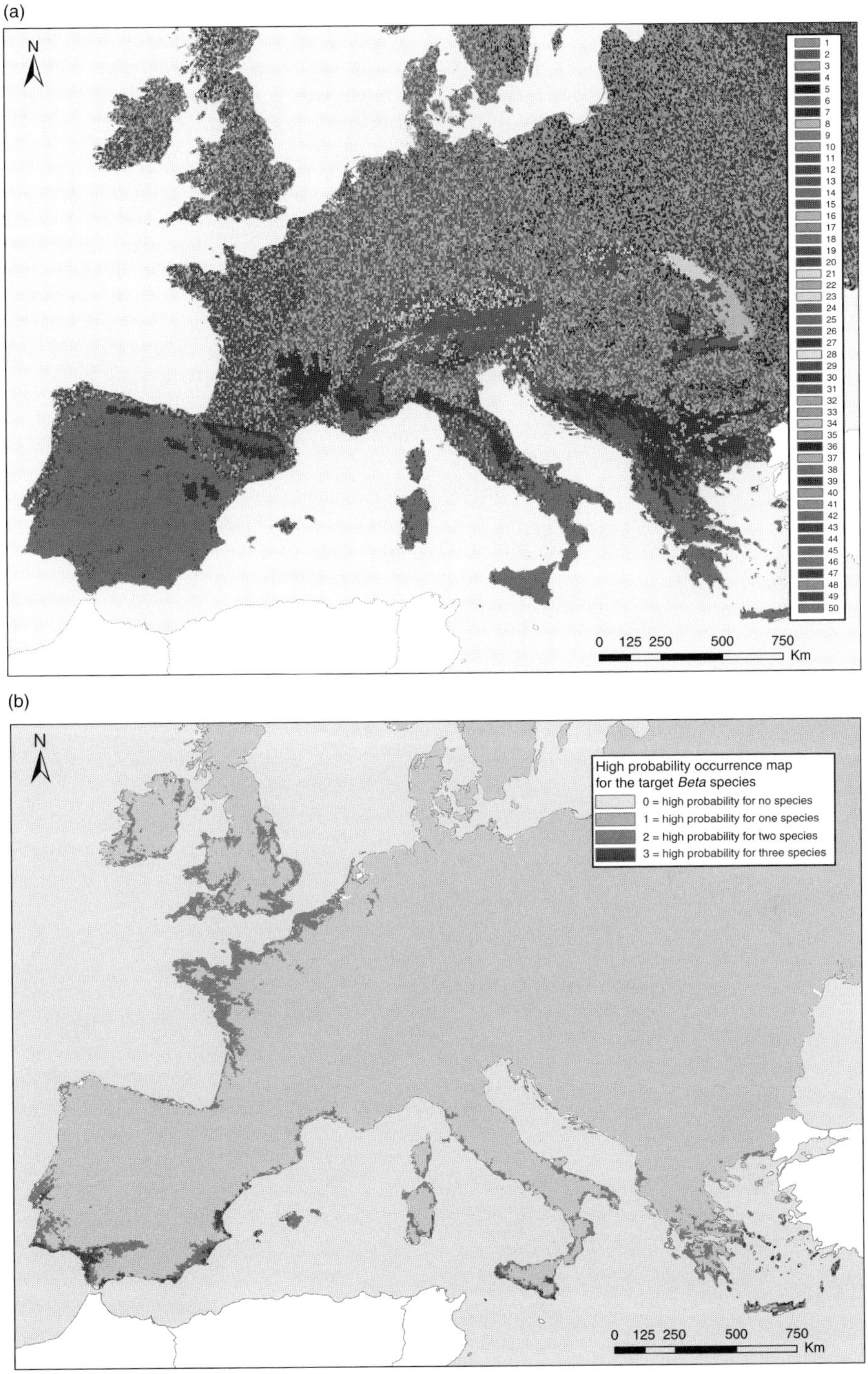

Fig. 3.3. Additional maps included in the third step of the process. (a) Ecogeographic land characterization map designed for *Beta* species with 50 ecogeographic categories. (b) Potential species richness map for three *Beta* species.

most relevant for the target species in terms of environmental adaptation. We propose the joint use of subjective and objective criteria for the selection of these variables, based on expert's surveys (subjective) and statistical tests that assess the contribution of these ecogeographic variables to the goodness of fit of predictive distribution models (objective).

Species distribution models (SDM) should then be applied to each target species to determine high probability of occurrence areas (single species) and potential species richness (summarizing binarized maps of single-species high probability of occurrence, see Fig. 3.3b). There are several modelling techniques and software developed although for only-presence data Maxent (Phillips *et al.*, 2006) or GARP (Stockwell and Peters, 1999) prevail.

At the fourth step the ecogeographic category of each presence data (point) is extracted from the ELC map and an ecogeographic category distribution is obtained. Each category is then classified according to the species' presence frequency (Parra-Quijano *et al.*, 2011). Categories with low frequency values are prioritized because they represent the limit of the species' range and extreme environmental conditions where many interesting genetic configurations can take place. Finally, the species presence data belonging to priority ecogeographic categories and the potential species richness map are matched against the pre-selected protected areas. Based on expert knowledge, some decision criteria are fixed and the final list of genetic reserves is selected (Fig. 3.4). These criteria should satisfy two conditions:

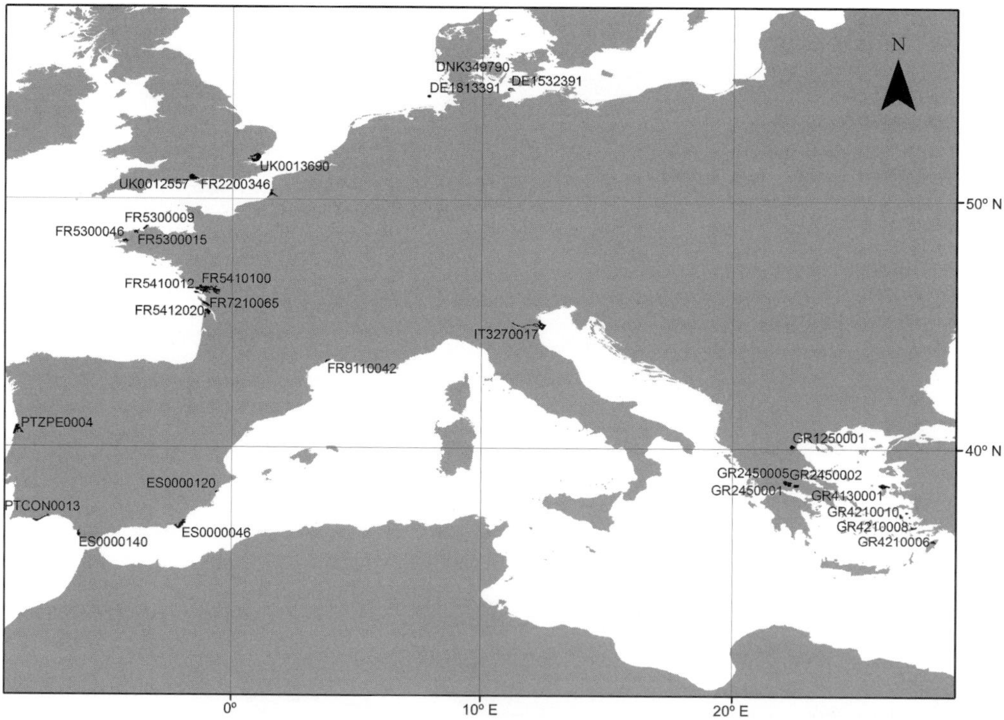

Fig. 3.4. Map of the locations of the genetic reserves generated by the ecogeographic approach for priority *Beta* species in continental Europe and the Mediterranean islands, upon which expert criteria were applied for final selection.

1. To include as much as possible all ecogeographic categories for the species with special attention to those classified as 'less frequent' (representativeness criterion);
2. To prioritize locations which contain high potential species richness areas (where several target species are being considered at the time).

This criterion tends to maximize the efficiency through the protection of the highest number of target species populations in the lowest number of genetic reserves.

Criteria and thresholds can be relaxed or adjusted to the real scenario, taking into account expert knowledge on particularly interesting traits in certain populations and/or the specific needs of target taxa conservation. Species presence data in nearby surrounding areas may also be integrated in the analysis as this may imply the existence of additional populations within the protected area that are not properly recorded.

3.5 Evaluation of the Selected Genetic Reserves

Once genetic reserves are selected, it is very important to assess the coverage of populations from target and non-target CWR species in quantitative and qualitative terms. An *in situ* survey is essential to ratify presence data obtained from databases and assess the precise location and demographic status of the populations. Some statistical and ecogeographic evaluators can be applied to determine the conservation performance of the proposed genetic reserve network. Since introduction of subjective criteria may provide different results, evaluation of possible alternatives is advisable. As a result of this, some of the candidate locations for the establishment of a genetic reserve may be finally dropped off and substituted by others that get a better *in situ* evaluation.

References

Cowling, R.M., Pressey, R.L., Sims-Castley, R., le Roux, A., Baard, E., Burgers, C.J. and Palmer, G. (2003) The expert or the algorithm? – comparison of priority conservation areas in the Cape Floristic Region identified by park managers and reserve selection software. *Biological Conservation* 112, 147–167.

Dulloo, M.E., Labokas, J., Iriondo, J.M., Maxted, N., Lane, A., Laguna, E., Jarvis, A. and Kell, S.P. (2008) Genetic reserve location and design. In: Iriondo, J.M., Maxted, N. and Dulloo, E. (eds) *Conserving Plant Genetic Diversity in Protected Areas. Population Management of Crop Wild Relatives*. CAB International, Wallingford, UK, pp. 23–64.

Hawkes, J.G., Maxted, N. and Zohary, D. (1997) Reserve design. In: Maxted, N., Ford-Lloyd, B.V. and Hawkes, J.G. (eds) *Plant Genetic Conservation. The* in situ *Approach*. Chapman and Hall, London, pp. 132–143.

Jain, S.K. (1975) Genetic reserves. In: Frankel, O.H. and Hawkes, J.G. (eds) *Crop Genetic Resources for Today and Tomorrow*. Cambridge University Press, Cambridge, pp. 379–396.

Margules, C.R. and Pressey, R.L. (2000) Systematic conservation planning. *Nature* 405, 243–253.

Maxted, N. (2003) Conserving the genetic resources of crop wild relatives in European protected areas. *Biological Conservation* 113, 411–417.

Maxted, N., Iriondo, J.M., De Hond, L., Dulloo, E., Lefèvre, F., Asdal, A., Kell, S.P. and Guarino, L. (2008a) Genetic reserve management. In: Iriondo, J.M., Maxted, N. and Dulloo, M.E. (eds) *Conserving Plant Genetic Diversity in Protected Areas. Population Management of Crop Wild Relatives*. CAB International, Wallingford, UK, pp. 65–87.

Maxted, N., Iriondo, J.M., Dulloo, E. and Lane, A. (2008b) Introduction: The integration of PGR conservation with protected area management. In: Iriondo, J.M., Maxted, N. and Dulloo, E. (eds) *Conserving Plant Genetic Diversity in Protected Areas. Population Management of Crop Wild Relatives*. CAB International, Wallingford, UK, pp. 1–22.

Moilanen, A., Wilson, K.A. and Possingham, H.P. (2009) *Spatial Conservation Prioritization*. Oxford University Press, New York.

Namkoong, G., Boyle, T., Gregorius, H.-R., Joly, H., Savolainen, O., Ratnam, W. and Young, A. (1996) *Testing Criteria and Indicators for Assessing the Sustainability of Forest Management: Genetic Criteria and Indicators*. Working Paper No. 10. Center for International Forestry Research, Jakarta, Indonesia.

Parra-Quijano, M., Draper, D., Torres, E. and Iriondo, J.M. (2008) Ecogeographical representativeness in crop wild relative *ex situ* collections. In: Maxted, N., Ford-Lloyd, B.V., Kell, S.P., Iriondo, J.M., Dulloo, M.E. and Turok, J. (eds) *Crop Wild Relative Conservation and Use*. CAB International, Wallingford, UK, pp. 249–273.

Parra-Quijano, M., Iriondo, J.M. and Torres, E. (2011) Ecogeographical land characterization maps as a tool for assessing plant adaptation and their implications in agrobiodiversity studies. *Genetics Resources and Crop Evolution*, doi 10.1007/s10722-011-9676-7.

Phillips, S.J., Anderson, R.P. and Schapire, R.E. (2006) Maximum entropy modeling of species geographic distributions. *Ecological Modelling* 190, 231–259.

Silverman, B.W. (1986) *Density Estimation for Statistics and Data Analysis*. Chapman and Hall, London.

Soberón, J. and Peterson, A.T. (2004) Biodiversity informatics: managing and applying primary biodiversity data. *Philosophical Transactions of the Royal Society of London, Series B, Biological Sciences* 359, 689–698.

Stockwell, D.R.B. and Peters, D.P. (1999) The GARP modelling system: problems and solutions to automated spatial prediction. *International Journal of Geographic Information Systems* 13, 143–158.

4 The Challenges of Modelling Species Distribution: A Case Study of Wild Cherry (*Prunus avium* L.) in Europe

C. Teeling, N. Maxted and B.V. Ford-Lloyd

4.1 Introduction

Wild cherry (*Prunus avium* L.) is a crop wild relative species, native to most of Europe. The aim of this study is to consider this species' distribution at a continental scale with the aim of suggesting sites for genetic reserves. *Prunus avium* can be found across the temperate zones of Europe with a maximum altitude of about 1900 m reported in France, however, in Britain it is essentially a lowland species (Russell, 2003) and in central Europe it is generally found at lower altitudes (Gomory, 2004). In Andalusia, Spain, Vivero *et al.* (2001) found scattered populations of wild cherry over several mountains at elevations from 1300 to 1800 m.

Reproduction in *P. avium* occurs sexually through insect-mediated pollination and subsequent seed dispersal by animals (Mariette *et al.*, 1997), especially birds such as pigeons, thrushes, starlings and jays, and by small mammals (Russell, 2003). Wild cherry's asexual reproduction is a complementary adaptational strategy via suckering; it does not reproduce prolifically from seeds (Holtken and Gregorius, 2006) and, when it does, dispersal is somewhat limited (Vaughan *et al.*, 2007) despite its importance in gene flow. It is a typical pioneer plant favouring plenty of light and little competition (Bratber and Vestrheim, 2004); it is a colonizing species of early forest successional stages due to its capacity for rapid vegetative propagation (Gomory, 2004) and when large gaps appear, such as after disturbance. Wild cherry is tolerant of a wide range of environmental conditions, for example, it can be found growing in dry sites, not out of preference, but because it cannot be out-competed there by species such as European beech (*Fagus sylvatica* L.) (Stojecová and Kupka, 2009). It also demonstrates a guerrilla strategy (Vaughan *et al.*, 2007), having widely spaced ramets (individuals within a genetically identical population), allowing it to spread to more favourable areas. *Prunus avium* has a long history of cultivation across its native range for fruit, timber and ornament; as a noble hardwood, it is still valued today for panelling and cabinet making (Diaz *et al.*, 2007). It is also the progenitor of much of the sweet cherry eaten today and is the primary rootstock in use for sweet cherry production (Mariette *et al.*, 2007).

Wild cherry is commonly found in field margins and is a valuable component of mixed temperate broadleaf woodland. Conservation of the temperate woodland habitat in which it is found, largely depends upon the priorities of the country in which conservation is planned. Small and heavily managed landscapes, such as the British Isles, have only a few natural populations remaining (despite archaeological records that date back as far as 1000 BC from wood remains in Ireland) in the form of ancient woodland and hedgerow individuals, and non-indigenous trees of old and new varieties (for fruit and timber). The distributions of this species in the European countries included in this study are not well known; wild cherry has a scattered distribution over a wide range (Russell, 2003) but is considered rare in some countries (Stojecová and Kupka, 2009). In Andalusia, Spain, it is found in wet ravines of mountain ranges, where the effects of the summer drought are lessened. A combination of already long, dry summers and increasing temperatures may further restrict the range of the species (Vivero *et al.*, 2001). Further, survival of populations such as those in southern Spain – relict with declining populations – depends largely upon the overall conservation planning for the broader area. For instance, the best stand is considered to be on the northern slopes of the Sierra Nevada National Park; a very well-protected, mixed old broadleaf forest with other threatened species (Vivero *et al.*, 2001).

The aim of our study was to map the distribution of wild cherry using existing information about sites of occurrence and interpolate those data to identify other potential sites within the same extent (Europe), in order to select the most suitable and diverse sites for genetic reserves. The most freely available data were in the form of historical records, usually gathered from online repositories or herbaria. The species' known distribution was mapped with these occurrence data and then incorporated into a species distribution model (SDM) for further analysis. This chapter will describe the methods, using this study for context, and discuss common problems that should be considered when employing SDM techniques more generally.

4.1.1 Species distribution modelling

SDM is predicated upon the niche concept, whereby species' geographic distributions are modelled, usually because of a lack of detail and availability of data, with regard to large-scale climatic and topographic variables. Many modelling methods are intuitive and easy to use and produce good results (Franklin, 2009), but the conclusions drawn from studies are very much dependent on the questions being asked (Elith and Leathwick, 2009b). In the case of the *P. avium* data, the most important issue was validation; the need to refine and sort good quality data from those available, then select from the remaining records to provide a reliable indication of potential sites for the conservation of this species.

Spatial scale is an important element of modelling and the aims of the study usually influence the extent under consideration (Elith and Leathwick, 2009a). Spatial grain ideally reflects the properties of the data, including the predictor variables and grid cell size, and also spatial accuracy and precision of the species records (Elith and Leathwick, 2009b). Spatial grain has an influence on how definitively the results can be applied; for instance, using the grid-based survey method of the Botanical Society of the British Isles (BSBI), the individual is recorded at the centre of the grid square regardless of where it actually occurs in sites of 2 or 10 km^2. This does not provide precise information about the location of the individual, nor does it give any information about true absence. For the current study, a broad scale was employed because a large part of Europe comprised the study area and the native range of this species.

4.1.2 Sources of bias

In the present study, the site locations of many accessions within the *Prunus* database are imprecise and known to be incomplete; some of the records are thought to be spurious, therefore a subset of more accurate records (for example, very recent accessions with full passport information) was produced with a sampling method specifically chosen to reduce spatial bias. Applying a sub-sampling method can lessen the effect of geopolitical boundaries and reduce the impact of historical collecting bias (see Fig. 4.1), for example,

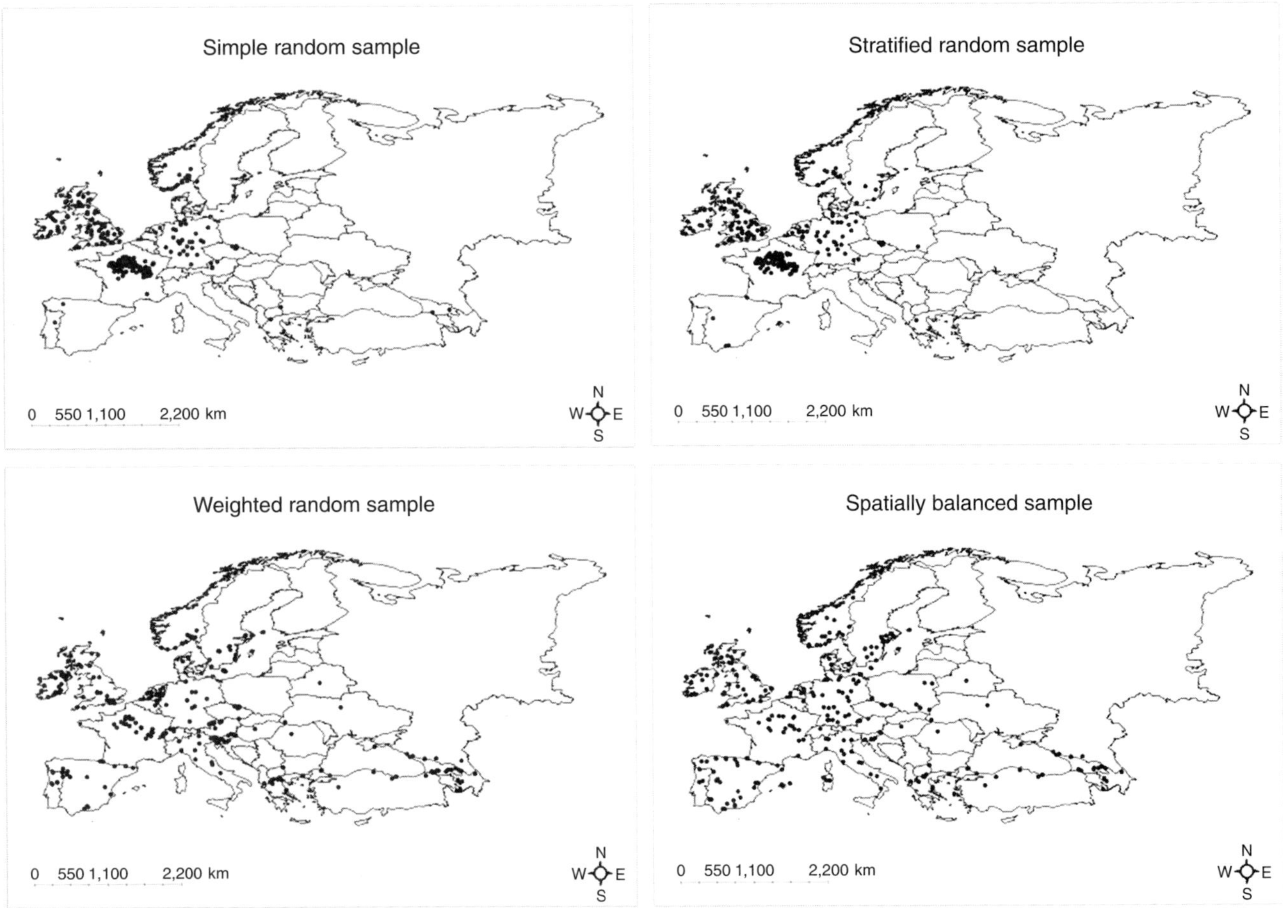

Fig. 4.1. Maps of sampled distributions using four types of sampling method to select approximately 300 points: simple random (from all records since 1960), stratified random (according to anthropogenic biome), weighted random (weight applied based on preponderance of records from country) and spatially balanced (using R script and a random sample based on probability weight).

the preponderance of recorded occurrences in UK and Germany, thus providing an adequate sample and ensuring that the records give as accurate a picture of the optimum environmental space for the species as possible.

There are several sources of bias and error that can arise when using a SDM; they will occur to some degree in all cases so potential errors must be considered in any subsequent analysis yet, providing the limitations are recognized, they can be acknowledged as an inherent part of the process. Examples of the types of potential error include: incorrect identification of species or change of classification; inaccurate spatial referencing of samples; and date recorded, and therefore current status of individuals/populations. Bias due to sampling is also a factor of modelling, for example, a preponderance of easily accessible locations sampled, such as along roads and rivers and near towns or biological stations. Historical trends can also influence sample bias; in the *Prunus* study, western European countries with a long history of biological recording (Britain, France, Germany) accounted for 80% of the total database records. When using records from historical collections, the modeller should remember that data are not generally collected with the purpose of determining the distributional limits of a species; rather, sampling for museum collections tends to be biased towards rare and previously unknown species. It has, however, been shown that small sample size can degrade model performance more than sample bias (Loiselle *et al.*, 2008) and, in some cases, biased sampling may lead to non-representative sampling of the available environmental conditions, although this is not necessarily the case.

4.2 Materials and Methods

The present study used the Maxent modelling algorithm, which provides a best approximation of an unknown distribution because it draws on everything that is known without assuming anything that is not known (Franklin, 2009). It has relevant settings for conservation and ecological applications, and can be used with presence-only data, which is particularly common when using biological records (Phillips and Elith, 2010).

Relevant variables for inclusion in the analysis are available at a variety of resolutions, and spatial scale should be maintained where possible for variables as well as species occurrences. During this study, a GIS database of environmental layers was collated and used to generate input predictor variables that influence distribution of a species. Environmental variables can be in the form of categorical or continuous data and usually comprise climate and habitat variables (Philips *et al.*, 2006). A principal component analysis (PCA) was used in the wild cherry study, because from the initial data gathering, more than 90 variable layers were found, although not all were usable or useful. These variables included topography, land use change, soil type, precipitation and maximum daily temperature. Variable selection and availability are important factors when studying niche breadth, as is scale, as organisms are differentially affected at different scales. The data are stored and formatted as layers in ArcGIS v10 (ESRI, 2011), but this software is also crucial for visualizing model results and carrying out additional processing of model output (Pearson, 2008).

A spatial database of the selected variable layers was collated and point data for each variable extracted and incorporated into the model. This number of occurrence records (300) split randomly into training and testing (70/30%) was chosen because, although not as sensitive to sample size as some other modelling algorithms (Wisz, 2008), Maxent has been found to be of value when used with very small numbers of records – as few as five localities (Pearson, 2008). Kadmon *et al.* (2003) found that, although average accuracy generally increases with increase in observations, there is hardly any effect on accuracy with more than 75 presence records in a climate envelope model, even though it is expected that accuracy should increase until a maximum accuracy potential is met (Hernandez *et al.*, 2006).

During modelling, known species distributions are mapped, then the model algorithm is applied and calibrated and the predicted distribution, made using predictor variables presented as grids or sampled data, is produced. These are then tested using alternative predictor variables – those incorporating the most variation from the PCA – and variations of sample sizes and settings for validation. In the *Prunus* study, the model algorithm was applied and calibrated using mostly the settings applied as standard within the Maxent interface (Phillips and Dudik, 2008). In fitting the model, constraints are set that enable the prediction to reflect patterns in the samples, and then the model is selected that maximizes entropy (chooses most uniform or spread out samples). Maxent is able to model complex relationships between the species and environment (Elith and Leathwick, 2009a).

4.3 Results and Discussion

The outcomes of this study are two-fold. First, sites can be selected as genetic reserves for *P. avium* and given priority because they have the optimum combination of suitable variables affecting the species' distribution, then potential distribution of this species can be modelled with a view to predicting sites for introduction, where other factors are favourable for the long-term conservation of this (and potentially incorporating other) species, e.g. as influenced by climate or land use change. Second, the predictor variables influencing the distribution of *P. avium* can be elucidated, apropos of the samples selected above.

In interpreting output, the choices made during modelling are important to consider; results can have different interpretations and conclusions resulting from inappropriate or incorrect assumptions. For example, the area under the receiver operating-characteristic curve (AUC) represents the model's ability to discriminate between presence and absence, a value of 1 denotes high accuracy, whereas 0.5 or lower indicates that the prediction is no better than chance. It is the most widely used measure of SDM performance, but is affected by the choice of area to include from outside the known distribution of the species (Termansen *et al.*, 2006). Due to the uncertainty of absences (a species may not be absent just because it is not easily detectable), false absences occur more often than false presences (Lobo *et al.*, 2007) and relying on a model with such errors can lead to underestimation of the importance of variables, and therefore the exclusion of 'good' habitats from management planning (Gu and Swihart, 2004).

SDM provides an approach that can help extend the usefulness of direct observation and improve our interpretation and understanding of species distributions. In addition to final suggestions made for reserve site selection, an important outcome of this study is the recognition of the need for data quality checking, and the importance of acknowledging the limitations of the chosen variables and field data, and their spatial attributes. Recent methods formally estimate probability of presence but need statistical tests to measure whether the model is well calibrated, which can in turn indicate how predictions could be improved and show potential hidden problems (Phillips and Elith, 2010).

Even with a complete map of a species' current distribution and where the location of every individual in an area is known, predictions of how a species' distribution might be altered in the face of environmental change – such as global warming or land cover change – may still not be definitive. The complete range of suitable environmental conditions for the species (the fundamental ecological niche) is different to the realized niche, which is the narrower 'space' that the species actually occupies, and is most adapted to, due to the interaction of other factors such as competition. SDMs are underpinned by niche theory and the environmental predictors included, and the quality and resolution of the data held within will affect accuracy and outcome.

SDMs are important because they are driven by the need to forecast the impact of management actions or environmental change on patterns of biodiversity from local to global scales. Any suggestions made for genetic reserves have to be considered in the context of the origin of the initial sample data and its potential bias, through to the choices made and thresholds selected. The resulting prediction is always preliminary until it has been confirmed (Wisz *et al.*, 2008), and it is at this stage where an expert opinion can be used. This is sometimes seen to be subjective and to have the potential to introduce more error and bias, such as when experts are called in to fill gaps in knowledge when it is problematic to make a management decision or wait for more data. However, expert advice can be in the form of data to support the predictive models to achieve the end of informing conservation planning objectives and strategies, and this is often the most accessible and cost-effective way to obtain current ecological information and often invaluable knowledge (Drew *et al.*, 2011).

References

Bratber, E. and Vestrheim, S. (2004) Sweet cherries in Norway. In: Dille, L.L. (ed.) *Nordic Gene Resources 2004*. Nordic Gene Bank, Alnarp, Sweden, pp. 1–23.

Diaz, R., Zas, R. and Fernandez-Lopez, J. (2007) Genetic variation of *Prunus avium* in susceptibility to cherry leaf spot (*Blumeriella jaapii*) in spatially heterogeneous infected seed orchards. *Annals of Forest Science* 64, 21–30.

Drew, C.A., Wiersma, Y.F. and Huettmann, F. (eds) (2011) *Predictive Species and Habitat Modeling in Landscape Ecology*. Springer, New York.

Elith, J. and Leathwick, J.R. (2009a) *Conservation prioritisation using species distribution models*. In: Moilanen, A., Wilson, K.A. and Possingham, H. (eds) *Spatial Conservation Prioritization: Quantitative Methods and Computational Tools*. Oxford University Press, Oxford, pp. 70–93.

Elith, J. and Leathwick, J.R. (2009b) Species distribution models: ecological explanation and prediction across space and time. *Annual Review of Ecology, Evolution and Systematics* 40, 677–697.

ESRI (Environmental Systems Research Institute) (2011) *Overview. What is GIS?* Available at: www.esri.com/what-is-gis/index.html (accessed 20 August 2010).

Franklin, J. (2009) *Mapping Species Distributions: Spatial Inference and Prediction*. Cambridge University Press, Cambridge.

Gomory, D. (2004) Mutual links of demographic and genetic processes in a wild cherry population during the colonization. *Biologia* 59(4), 493–500.

Gu, W. and Swihart, R.K. (2004) Absent or undetected? Effects of non-detection of species occurrence on wildlife–habitat models. *Biological Conservation* 116, 195–203.

Hernandez, P.A., Graham, C.H., Master, L.L. and Albert, D.L. (2006) The effect of sample size and species characteristics on performance of different species distribution modeling methods. *Ecography* 29, 773–785.

Holtken, A.M. and Gregorius, H.R. (2006) Detecting local establishment strategies of wild cherry (*Prunus avium*). *BMC Ecology* 6(13), 13 pp.

Kadmon, R., Farber, O. and Danin, A. (2003) A systematic analysis of factors affecting the performance of climatic envelope models. *Ecological Applications* 13(3), 853–867.

Lobo, J.M., Baselga, A., Hortal, J., Jiménez-Valverde, A. and Gómez, J.F. (2007) How does the knowledge about the spatial distribution of Iberian dung beetle species accumulate over time? *Diversity and Distributions* 13(6), 772–780.

Loiselle, B.A., Jørgensen, P.M., Consiglio, T., Jimenez, I., Blake, J.G., Lohmann, L.G. and Montiel, O.M. (2008) Predicting species distributions from herbarium collections: does climate bias in collection sampling influence model outcomes? *Journal of Biogeography* 35, 105–116.

Mariette, S., Lefranc, M., Legrand, P., Taneyhill, D., Frascaria-Lacoste, N. and Machon, N. (1997) Genetic variability in wild cherry populations in France. Effects of colonizing processes. *Theoretical and Applied Genetics* 94, 904–908.

Mariette, S., Balsemin, E., Stoeckel, S., Tavaud, M., Le Boler, H., Santi, F. and Verger, M. (2007) Parental participation in progeny and effective population sizes in experimental seed orchards of wild cherry *Prunus avium* L. (Batsch). *Annals of Forest Science* 64, 553–559.

Pearson, R.G. (2008) *Species' distribution modelling for conservation educators and practitioners*. Synthesis. American Museum of Natural History. Available at: http://ncep.amnh.org (accessed 30 January 2010).
Philips, S.J., Anderson, R.P. and Schapire, R.E. (2006) Maximum entropy modeling of species geographical distributions. *Ecological Modelling* 190, 231–259.
Phillips, S.J. and Dudík, M. (2008) Modeling of species distributions with Maxent: new extensions and a comprehensive evaluation. *Ecography* 31, 161–175.
Phillips, S.J. and Elith, J. (2010) POC plots: calibrating species distribution models with presence-only data. *Ecology* 91(8), 2476–2484.
Russell, K. (2003) *EUFORGEN Technical Guidelines for Genetic Conservation and Use for Wild Cherry (*Prunus avium*)*. International Plant Genetics Resources Institute, Rome.
Stojecová, R. and Kupka, I. (2009) Growth of wild cherry (*Prunus avium* L.) in a mixture with other species in a demonstration forest. *Journal of Forest Science* 55(9), 264–269.
Termansen, M., McClean, C.J. and Preston, C.D. (2006) The use of genetic algorithms and Bayesian classification to model species distributions. *Ecological Modelling* 192, 410–424.
Vaughan, S.P., Cottrell, J.E., Moodley, D.J., Connolly, T. and Russell, K. (2007) Distribution and fine-scale spatial-genetic structure in British wild cherry (*Prunus avium*). *Heredity* 98, 274–283.
Vivero, J.L., Hernandez-Bermejo, J.E. and Prados Ligero, J. (2001) Conservation strategies and management guidelines for wild *Prunus* genetic resources in Andalusia, Spain. *Genetic Resources and Crop Evolution* 48(5), 533–546.
Wisz, M.S., Hijmans, R.J., Li, J., Peterson, A.T., Graham, C.H. and Guisan, A. (2008) Effects of sample size on the performance of species distribution models. *Diversity and Distributions* 14, 763–773.

5 Towards the Establishment of a Genetic Reserve for *Beta patula* Aiton

M.Â.A. Pinheiro de Carvalho, H. Nóbrega, G. Freitas, S. Fontinha and L. Frese

5.1 Introduction

The distribution area of *Beta* species extends from Turkey and adjacent countries to the Macaronesian archipelagos as well as from Morocco to south Norway. *Patellifolia* species mainly occur on the Canary Islands and Madeira, in the southern part of the Iberian peninsula and Morocco (Frese, 1990). *Beta* and *Patellifolia* species are related to cultivated beets and have been widely used by breeders for introgression of resistance genes into the sugarbeet crop (Frese, 2002). Borges *et al.* (2008) confirmed the presence of four crop wild relatives (CWR) of cultivated beets on Madeira: *Beta patula* Aiton, *B. vulgaris* subsp. *maritima*, *Patellifolia procumbens* (C.Sm. ex Hornem) A.J. Scott, B.V. Ford-Lloyd & J.T. Williams and *Patellifolia patellaris* (Moq.) A.J. Scott, B.V. Ford-Lloyd & J.T. Williams. *Beta patula* and *B. vulgaris* subsp. *maritima* are components of the primary gene pool, while the genus *Patellifolia* belongs to the tertiary genepool of beet crops (Frese, 2011). Luterbacher *et al.* (2004), confirmed by Schliephake *et al.* (2010), detected variation for Beet Mild Yellowing Virus (BMYV) resistance in *B. patula*, which may be used to enhance resistance in cultivated *Beta*.

Beta patula is a rare species, endemic to Madeira, and a genetic resource for plant breeding. Under good growing conditions, plants develop stems of up to 30 cm in length, branching and spreading or ascending freely from the base, with linear to lanceolate leaves, narrowed at the base. The cymes are composed of seven flowers on average (Fig. 5.1) (Press, 1994). The species is self-pollinating but out-crossing is easily possible. The multigerm fruits contain the highest seed number of all *Beta* species (Letschert, 1993), but the total average single plant seed yield is the lowest of all *Beta* section *Beta* species (Frese, 2011). In the past *B. patula* occurrences were found on Ponta de São Lourenço, the Desertas Islands and Porto Santo, only (Aiton, 1789; Ulbrich, 1934; Coons, 1954; Borges *et al.*, 2008). Ponta de São Lourenço is a peninsula in the extreme east of Madeira island ending in two small islets: Ilhéu do Desembarcadouro (ID) and Ilhéu Farol. The Desertas Islands is a complex of the three small islands Ilhéu do Chão (IC), Deserta Grande and Bugioh located 18 km south-east of Madeira. According to the current state of knowledge, *B. patula* can only be found on ID and IC where it can grow sympatrically with *B. vulgaris* subsp. *maritima*. Because of its endemicity and narrow geographical distribution, *B. patula* is considered one of the 100 most endangered species of Macaronesia (Martin *et al.*, 2008), justifying its inclusion in the list of species under protection of the Habitats Directive of

Fig. 5.1. *Beta patula* Aiton, adult plants.

the Natura 2000 protected area network (Anonymous, 1992). Recently, the IUCN has included *B. patula* in the European CWR Red List (see Kell *et al.*, Chapter 28, this volume).

The step-wise methodology for the identification of genetic reserves sites for a target crop gene pool (see Kell *et al.*, Chapter 2, this volume) was applied in the AEGRO Crop Case Study *Beta* (Frese *et al.*, in preparation). The results of this study suggest the need for establishing a genetic reserve for *B. patula*. This species is therefore a good model for testing and developing elements of the genetic reserve conservation technique in practice. The present chapter describes the ecogeographic analysis and provides the demographic baseline, while Frese *et al.* (Chapter 6, this volume) focuses on the genetic diversity analysis of the species and draws conclusions.

5.2 Ecogeographical Analysis

Field surveys were conducted between April 2007 and July 2010 to identify sites potentially suited for the establishment of a genetic reserve. During the first surveys, data on the distribution and abundance of the four species were recorded. The surveys in 2007/2008 specifically aimed to:

- Establish a demographic baseline;
- Describe the geological, edaphic and topographic features of the distribution area;
- Define sampling areas and establish sampling plots; and
- Sample leaf probes required for the analysis of genetic diversity.

In subsequent surveys the abundance of *B. patula* and the location of plant groups were recorded to get more insight into the population dynamics of the species and to understand which factors could promote or threaten the species.

The grassland habitat type (Code 45 IUCN) is the dominant habitat of ID and IC. Information on mean annual precipitation, temperature, relative humidity, thermal amplitude and solar irradiation were compiled into a cartographic database held by the Meteorological Institute, Funchal (Anonymous, 2007). The geophysical description included topography, soil and land use.

Georeferenced data on altitude, aspect and slope, and soil type were recorded using a Global Positioning System and military maps. Physical and chemical soil properties were determined as described by Pinheiro de Carvalho *et al.* (2003).

The main factors determining the ecogeographic diversity of the islets are described below.

5.2.1 Geophysical description

The geographic centre coordinates of ID are 32° 44′ 01″ N and 016° 40′ 17″ W. The islet is 1.95 km long with a width of 0.43 km at its broadest part and 0.06 km at its narrowest part. Altitude ranges from 0 to 104 m asl. IC is located approximately 25 km south-west of Madeira. It is the smallest of the Desertas Islands, located at 32° 35′ 07″ N and 016° 32′ 30″ W and has an area of approximately 0.5 km^2. It has a table-like shape rising over 80 m asl.

5.2.2 Area geology

The geology is characterized by basaltic pyroclastic deposits as a result of intense marine erosion, particularly along the north-eastern coast (Goodfriend *et al.*, 1996; Klügel *et al.*, 2009). The Desertas Island complex and Ponta de São Lourenço are part of a continuous landmass which has existed for about 18,000 years and share a common geological origin and history (Carvalho and Brandão, 1991).

5.2.3 Topography

In spite of their common geological origin, ID and IC show a distinct landscape. Ponta de São Lourenço is a straight and deeply eroded peninsula of Madeira. Therefore, the relief of ID is diverse. Some natural barriers, uneven terrain, rocky formations and soil units can partially impede the distribution of beet CWRs. On the contrary, the landscape of IC is completely flat without relief barriers that could limit the expansion of wild beet populations. Based on this information and field observations, ID was divided in six areas for fieldwork and plant distribution analysis (Pinheiro de Carvalho *et al.*, 2010).

5.2.4 Bioclimatic description

Both islets have a similar ecology. Annual thermo-pluviometric variation is shown in Fig. 5.2. *Beta patula* is distributed in an area with an average annual temperature of 19.5°C (minimum and maximum temperatures of 17°C and 24°C, respectively), annual precipitation of less than 800 mm (estimated average monthly rainfall less than 100 mm) occurring mostly between October and December. The summer is extremely dry, although air humidity is rather high (60–70%) due to the strong influence of the ocean. Mean daily solar radiation is 4500 Wh/m^2 day with minimum values of 2500 Wh/m^2/day in January and maximum values of 6500 Wh/m^2/day in July. Saline spray water is transported by oceanic winds to inland areas. Thus, the climate of both sites can be classified as upper inframediterranean, mesothermic (Koppen classification) and semi-arid (Thornthwaite classification) (Machado, 1970).

5.2.5 Edaphic conditions

The soil at both sites can be described as loam-clay and rocky, poor in organic matter, low in moisture content and high in salinity. ID can be roughly divided into three different soil units: haplic calcisoils, eutric accident soils and eutric rocky soils (FAO, 1988). Soil samples were taken during field work and 23 soil parameters were analysed to establish edaphic conditions within *B. patula* patches and to assess possible threats of sites where species' plant number appeared to decline.

These results show that the soil varies from alkaline ($n = 2$ samples) to acid ($n = 7$), the majority has a pH close to neutral ($n = 8$).

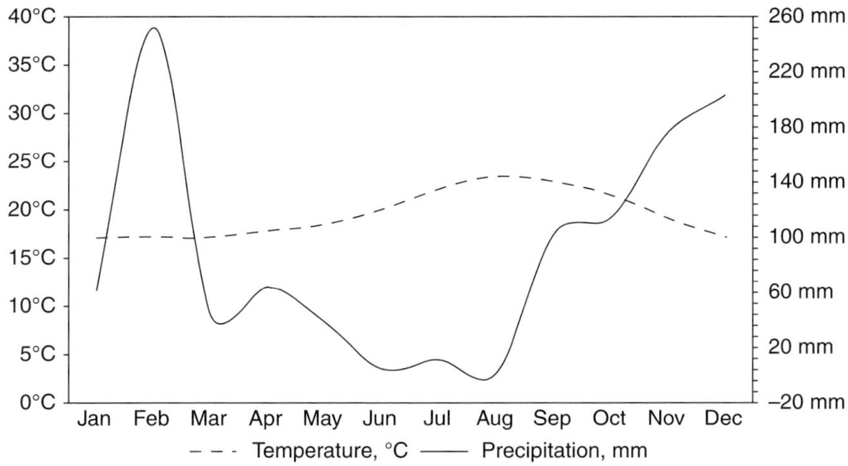

Fig. 5.2. Establishment of *B. patula* ecogeographic and bioclimatic conditions. The plot shows the annual thermo-pluviometric variation on Desembarcadouro islet.

Table 5.1. Edaphic conditions of ID and IC islets. Major soil parameters determined for randomly collected and mixed soil samples collected in *B. patula* areas with different vegetation cover. IDA1 to IDA6 are the six sample areas delimited in ID. IC1 to IC3 are the sample areas in IC. BP1 to BP4 are the *B. patula* spots/patches in ID.

Sampling area	pH	OM (%)	N (ppm)	P (ppm)	K (meq/100 g)	F ass (ppm)	Mg (meq/100 g)	C* (mS/cm)
IDA1	7.5	4.6	125.0	1237.0	4.8	1440.0	11.9	5.5
IDA2	7.2	3.7	5.0	137.0	5.4	1488.0	15.3	6.7
IDA3	5.9	4.4	125.0	2198.0	6.2	1680.0	16.3	13.2
IDA4	5.9	6.6	130.0	1924.0	5.8	1728.0	12.5	5.4
IDA5	7.1	4.6	140.0	3023.0	7.6	1632.0	12.9	4.6
IDA6	6.9	4.0	145.0	1278.0	6.5	1584.0	14.1	4.4
IC1	7.7	2.9	45.0	4534.0	10.2	1488.0	10.9	4.5
IC2	6.3	3.2	30.0	3710.0	6.0	1632.0	8.2	1.2
IC3	5.5	7.5	20.0	989.0	6.0	3840.0	9.0	1.2
BP1	5.6	4.1	155.0	234.0	7.5	1536.0	16.2	3.6
BP2	7.3	4.6	175.0	289.0	5.4	1531.0	14.4	4.1
BP3	7.4	3.0	10.0	453.0	0.9	1728.0	14.4	7.0
BP4	6.6	2.4	50.0	371.0	6.1	1536.0	10.5	6.3
ES1	7.7	0.8	150.0	343.0	3.8	1608.0	15.5	69.0
ES2	5.8	2.2	135.0	41.0	6.0	1584.0	14.9	61.1
AS1	6.0	6.0	145.0	1319.0	6.0	1680.0	11.2	2.3
AS2	6.6	2.1	140.0	398.0	6.5	1728.0	16.1	2.3

ES, eroded soil plots; AS, plots with alien species; OM, organic matter; N, nitrate content; P, phosphorus; K, assimilable potassium; F, assimilable fluor; Mg, magnesium; C, soil conductivity; *parameter with statistically significant differences detected among areas, using one-way ANOVA ($P<0.05$).

Organic matter content varies considerably, but is on average less than the average value obtained for non-cultivated soils of Madeira (4.1%) (Ganança *et al.*, 2007). The results for the eight most important soil parameters are summarized in Table 5.1. These parameters have been selected because they have more weight in the PCA separation of sampled areas, according to their edaphic conditions.

5.3 Species Distribution and Population Baseline

The distribution pattern of the wild beet species was established during survey missions on both islets. On ID plant findings were recorded along four linear parallel transects stretching across the islet from east to west, and the data were used to establish the boundaries of plant patches of *B. patula*, *B. vulgaris* subsp. *maritima* and *P. procumbens* within the total distribution area. The boundaries were mapped on a scale of 1:25,000. This procedure was used to calculate the CWRs' distribution area with the assistance of Remote Sensing Images from DRIGOT (Regional Office for Geographical Information and Territorial Planning).

5.3.1 *Beta patula* distribution

Plant surveys were used to establish the *B. patula* distribution pattern on both islets. The total distribution area (TDA) occupied by *B. patula* is 81,979.9 m^2 on ID distributed through six sites (Pinheiro de Carvalho *et al.*, 2010), while on IC it is only about 27,500 m^2, with all beet CWRs occurring in a single site.

5.3.2 Census of the wild beet species

Patch boundaries and two parallel transects crossing over the majority of the *B. patula* spots detected during the first survey were used to establish 50 square meter plots at 40–50 m intervals lengthwise. These plots were used to perform the census and collect plant material. Each plot was marked with a coloured steel rod and georeferenced to allow for future monitoring of the species. Within these plots, leaf samples were collected for the genetic diversity analysis (see Frese *et al.*, Chapter 6, this volume). Seed samples were collected from sites likely to represent the ecogeographic diversity of the islets and are included in the collection of ISOPlexis Genebank as a backup collection.

A total of 541 adult *B. patula* plants were counted in 27 of the 50 plots on ID, while 132 plants were counted on IC. These plant counts were used to estimate population sizes and the demographic baseline for *B. patula* (Pinheiro de Carvalho *et al.*, 2010).

5.3.3 Establishing *B. patula* population sizes

The main purpose of establishing the demographic baseline for *B. patula* populations is the determination of population sizes. Population size was extrapolated from census data using the formula:

$$P = (NP \times SDA \times SR) \quad (5.1)$$

where *P* is the population size determined for a site of species occurence, *NP* the number of plants counted in all plots of a sampling site, *SDA* the species site distribution area, *SR* the sampling ratio calculated by dividing the surveyed plots area by *SDA*.

Subsequently, species population sizes in each islet (*PS*) were determined by:

$$PS = \sum_{i=1}^{n} P_i \quad (5.2)$$

where PS is the population size determined as the sum of the population size established for each site of species occurrence, with *n* varying from 1 to 6 in ID and being equal to 1 in IC.

According to our calculations, species population sizes on ID range between 2730 and 4620 individuals, while the values for the IC population vary between 924 and 1840 individuals (Table 5.2).

5.3.4 Anthropogenic land use and species threat assessment

Although *B. patula* occurs in the most remote areas of the Madeira Archipelago, its distribution and population size could have been influenced by human activities. Agricultural and livestock activities were reported for the peninsula of Ponta de São

Table 5.2. Demographic baseline and population sizes of *B. patula* on Desembarcadouro (ID) and Chão (IC) islets. The minimium and maximum population sizes have been established applying the formula (Eqn 5.2) and varying the values of sampling ratio (SR). The SR ratio for the minimum population size is calculated using the number of plots where *B. patula* was detected, and the SR was estimated with all plots set up in the sampling site for the maximum population size (see Pinheiro de Carvalho *et al.*, 2010).

Species	Islet	TDA (m^2)	NP	MiS	MaS
B. patula	ID	81,979.9	541	2,730	4,620
	IC	27,500.0	132	924	1,840

MaS, maximum population size; MiS, minimum population size; NP, total number of plants counted in the census species; TDA, species total distribution area.

Lourenço and IC until the end of the first half of last century (Silva and Meneses, 1984). Historically, human pressure on the *B. patula* habitat is low due to the difficult physical access to the islets and their status as Natura 2000 protected sites. Nevertheless, a threat assessment was conducted to ascertain the viability of the species. The assessment of *B. patula* was based on the analysis of four major threat categories, namely habitat loss, alien species occurrence, soil erosion and population fragmentation. These threats were assessed through the observation of alien species present in the vegetation cover and the land surface covered by them (Paulo *et al.*, 2008), and the survey of eroded and unaffected plant canopy areas. These observations seem to show the existence of relative high pressure of introduced species in some of *B. patula* and other beet CWR growth patches, and a decrease of their capability to occupy soil eroded areas, loss of edaphic conditions and soil ground, or compete for growth surface in areas where *Mesembryanthenum crystallinum* and other alien species are present. However, these observations need to be confirmed by a timeline series of surveys, because of dramatical fluctuations in the size of *B. patula* patches.

5.4 Discussion

The application of the genetic reserve conservation technique requires knowledge of the ecogeographic diversity of the target species' habitat as well as the biology, distribution and abundance of the species. Most of this kind of information was unavailable for *B. patula* until Pinheiro de Carvalho *et al.* (2010) determined the exact distribution of this species and three additional wild beet species on Madeira.

Ecogeographic information allows us to understand which factors are likely to influence the species' genetic diversity. Furthermore, the data would allow us to select ecogeographically distinct sites within the total distribution area. These sites could form a network of genetic reserves that covers a large fraction of the genetic diversity present in *B. patula*.

The edaphic conditions of the habitat were assessed using 23 soil parameters. The main sampling areas (ID and IC) are separated (Fig. 5.3) according to the results of a multivariate analysis of all parameters. A separation between eroded and unaffected areas on ID was also observed. Salinity and soil conductivity seem to be the main factors determining the distribution of *B. patula*. These factors also appear to be associated with soil erosion and acidification resulting from seagull guano concentrated in certain locations.

On ID, the species showed a fragmented distribution. The small distances between patches and the absence of effective geographic barriers that could effectively hamper the gene flow leads us to assume that *B. patula* plants form a single unique population and that the fragmented distribution is probably caused by site-specific environmental factors promoting or restricting the presence of plants. By contrast, *B. patula* forms a non-fragmented plant

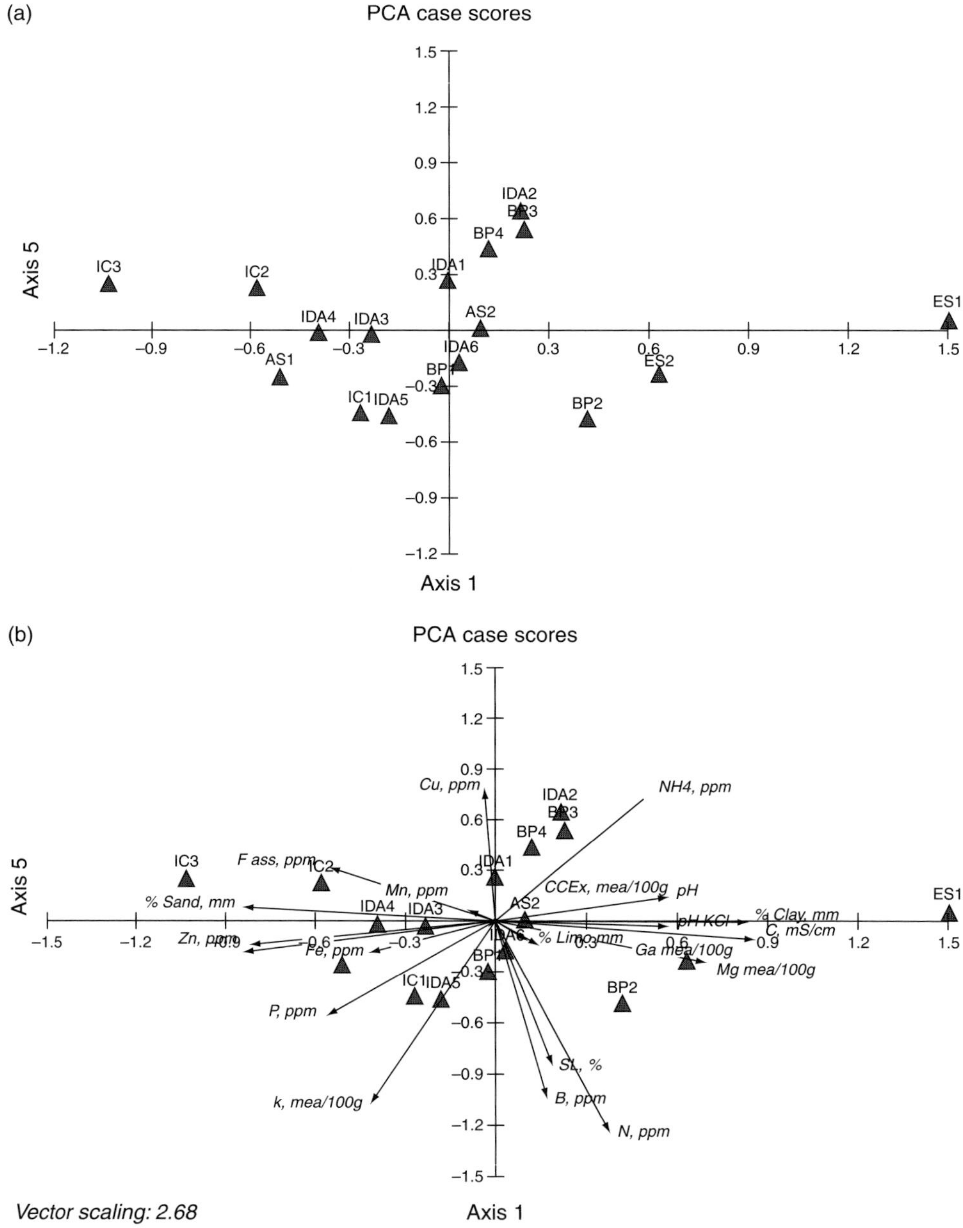

Fig. 5.3. Assessment and discrimination of edaphic conditions of *B. patula* sites on Desembarcadouro (ID) and Chão (IC) islets, as well as between *B. patula* distribution (BP) and eroded (ES) and alien species (AS) areas on Desembarcadouro. (a) The principal component analysis (PCA) explains 71.7% of all variability across five axes, with an eigenvalue of 5.2 in axis 1 and 2.1 in axis 5. (b) Biplots show the contribution of parameter vectors to the cases separation along the axes.

group on IC. This plant group seems to be an autochthonous population isolated by distance from the population growing on ID.

By definition the genetic reserves conservation technique requires active population management. To enable the impact of management measures, a demographic and genetic baseline monitoring should be established for monitoring. Early estimations of a population size (of less than 50 adult specimens (Carvalho *et al.*, 2008) date back to investigations resulting in the classification of Ponta de São Lourenço as a protected area of Natural Park of Madeira under the Natura 2000 network (Anonymous, 1992).

Protection measures undertaken by Natural Park of Madeira (NPM) staff have increased *B. patula* population sizes by 10.8 and 2.6 times on ID and IC, respectively, according to the census performed during this study. Data extrapolation (Table 5.2) indicates the existence of an even higher number of plants, i.e. an average of 3675 and 1386 plants for ID and IC, respectively. However, these sizes are only approximations that need to be confirmed by time-series surveys. Lehmkhul (1984) showed that depending on the survey area and other environmental factors, the number of individuals in a population can be two to three times higher than census data, but effective population size (N_e) can be smaller than actual population size (N) or even census data (Iriondo *et al.*, 2008). Effective population size needs to be established to ensure that the criteria of a minimum viable population (MVP) are met by *B. patula*. Our observations show that the sizes of the ID population suffer strong fluctuations in time, with the number of plants declining or even disappearing in more than 60% of the surveyed plots. This trend is common to the majority of annual and biannual plant species (Iriondo *et al.*, 2008).

Well developed seed banks play an important role for the survival of populations under harsh environmental conditions. Seed banks contribute to the establishment of new plant patches on sites not known before. This issue can be a matter of further research. At the moment we can only hypothesize that *B. patula* populations do not reach the critical MVP size, which underpins the need for active management of its intraspecific genetic diversity.

Acknowledgements

We are grateful for the assistance given by the staff of the Natural Park of Madeira during the project.

References

Aiton, W. (1789) *Hortus kewensis*, Catalogue of plants cultivated in Royal Botanical Garden of Kew, London, Vol. 1, 315.

Anonymous (1992) Habitats Directive (92/43/CEE Annex I).

Anonymous (2007) Atlas do Ambiente. Ministério das Cidades, Ordenamento do Território e Ambiente. APA, Agência Portuguesa do Ambiente. Available at: www.iambiente.pt/atlas/est/index.jsp?zona=madeira&grupo=madeira.m0&tema=madeira.m0 (accessed 28 May 2009).

Borges, P.A.V., Abreu, C., Aguiar, A.M.F., Carvalho, P., Jardim, R., Melo, I., Oliveira, P., Sérgio, C., Serrano, A.R.M. and Vieira, P. (eds) (2008) A list of the terrestrial fungi, flora and fauna of Madeira and Selvagens archipelagos, Funchal and Angra do Heroísmo. Direcção Regional do Ambiente da Madeira and Universidade dos Açores.

Carvalho, A.M.G. de and Brandão, J.M. (1991) Geologia do Arquipélago da Madeira. Museu Nacional de História Natural, Lisboa, 170 pp.

Carvalho, J.A., Fernandes, F. and Jardim, R. (2008) *Beta patula* Aiton. In: Martin, J.L., Arechavaleta, M., Borges, P.A.V. and Faria, B. (eds) *Top 100. Las 100 especies amenazadas prioritárias de gestión en la region euopea biogeográfica de la Macaronesia*. Consejeria de Medio Ambiente y Ordenación Territorial, Gobierno de Canarias, pp. 218–219.

Coons, G.H. (1954) The wild species of *Beta*. *Proceedings of the American Society of Sugar Beet Technology* 8, 142–147.

FAO/UNESCO (1988) Soil Map of the World. Revised Legend. *World Soil Resources, Report 60*, Food and Agriculture Organization of the United Nations, Rome.

Frese, L. (2002) Combining static and dynamic management of PGR: a case study of *Beta* genetic resources. In: Engels, J.M.M., Rao, V.R., Brown, A.H.D. and Jackson, M.T. (eds) *Managing Plant Genetic Diversity*. CAB International, Wallingford, UK, pp. 133–148.

Frese, L. (2011) Conservation and access to sugarbeet germplasm. *Sugar Technology* 12(3–4), 207–219.

Frese, L., Meijer, E. de and Letschert, J.P.W. (1990) New wild beet genetic resources from Portugal and Spain. *Zuckerind* 115(11), 950–955.

Ganança, J.F.T., Abreu, I., Sousa, N.F., Paz, R.F., Caldeira, P., Santos, T.M.M., Costa, G., Slaski, J.J. and Pinheiro de Carvalho, M.A.A. (2007) Soil conditions and evolution of aluminium resistance among cultivated and wild plant species on the Island of Madeira. *Plant Soil Environment* 53(6), 239–246.

Goodfriend, G.A., Cameron, R.A.D., Cook, L.M., Courty, M.-A., Fedoroff, N., Livett, E. and Tallis, J. (1996) The quaternary sequence of Madeira: stratigraphy, chronology, and paleoenvironmental interpretation. *Palaeogeography, Palaeoclimatology, Palaeoecology* 120, 195–234.

Iriondo, J.M., Ford-Lloyd, B., De Hond, L., Kell, S.P., Lefevre, F., Korpelainen, H. and Lanes, A. (2008) Plant population monitoring methodologies for the *in situ* genetic conservation of CWR. In: Iriondo, J.M., Maxted, N. and Dulloo, M.E. (eds) *Conserving Plant Genetic Diversity in Protected Areas*. CAB International, Wallingford, UK, pp. 99–120.

Klügel, A., Schwarz, S., Bogaard, Pvd., Hoernle, K.A., Wohlgemuth-Ueberwasser, C.C. and Köster, J.J. (2009) Structure and evolution of the volcanic rift zone at São Lourenço peninsula, Madeira. *Bulletin of Volcanology* 71, 671–685, doi:10.1007/s00445-008-0253-7.

Lehmkhul, J.F. (1984) Determining size and dispersion of minimum viable populations for land management planning and species conservation. *Environmental Management* 8(2), 167–176.

Letschert, J.P.W. (1993) *Beta* section Beta: biogeographical patterns of variation and taxonomy. *Wageningen Agricultural University Papers*, Wageningen, the Netherlands, 93-1.

Luterbacher, M., Asher, M., Deambrogio, E., Biancardi, E., Stevenato, P. and Frese, L. (2004) Sources of resistance to diseases of sugar beet in related *Beta* germplasm: I. Foliar diseases. *Euphytica* 139, 105–121.

Machado, M. (1970) *Balanço Hídrico e Clima da Ilha da Madeira*. Instituto Nacional Meteorologia e Geofísica, Lisboa.

Martin, J.L., Arechavaleta, M., Borges, P.A.V. and Faria, B. (2008) *Top 100. Las 100 especies amenazadas prioritárias de gestión en la region euopea biogeográfica de la Macaronesia*. Consejeria de Medio Ambiente y Ordenación Territorial, Gobierno de Canarias, Santa Cruz, Tenerife, Spain.

Paulo, J., Pires, R. and Fontinha, S. (2008) *Plantas da Ponta de São Lourenço*. Edição do Serviço do Parque Natural da Madeira, Funchal, Portugal. ISBN 978-989-95497-1-5.

Pinheiro de Carvalho, M.A.A., Slaski, J.J., dos Santos, T.M.M., Ganança, F.T., Abreu, I., Taylor, G.J., Clemente Vieira, M.R., Popova, T.N. and Franco, E. (2003) Identification of aluminium resistant genotypes among madeiran regional wheats. *Communications on Soil Sciences and Plant Analysis* 34(19–20), 2973–2985.

Pinheiro de Carvalho, M.A.A., Nóbrega, H., Frese, L., Freitas, G., Abreu, U., Costa, G. and Fontinha, S. (2010) Distribution and abundance of *Beta patula* Aiton and other crop wild relatives of cultivated beets on Madeira. *Journal für Kulturpflanzen* 62(10), 357–366.

Press, J.R. (1994) Chenopodiaceae. In: Press, M.J. and Short, M.J. (eds) *Flora of Madeira*. HMSO, London, pp. 68–74.

Schliephake, E., Habekuß, A. and Frese, L. (2010) Resistance of *Beta patula* Ait. to Beet mild yellowing virus. Abstract Book of '*CWRs & Landraces in Europe: Towards the establishment of genetic reserves for crop wild relatives and landraces in Europe*'. Meeting of the ECPGR In situ and On-farm Conservation Network and the EU project AGRI GENRES 057 – AEGRO, 13–16 September 2010. Funchal, Portugal, p. 91.

Silva, A.F. and Meneses, C.A. (1984) *Elucidário madeirense*, vol.1. SRTC – DRAC, Funchal, 431 pp.

Ulbrich, E. (1934) Trib. 3. Betoideae-Beteae. In: Engler, A. and Harms, H. (eds) *Die natürlichen Pflanzenfamilien nebst ihren Gattungen und wichtigsten Arten, insbesondere den Nutzpflanzen. Angiospermae: Reihe Centrospermae*. Verlag von Wilhelm Engelmann, Leipzig, Germany, Band 16c, pp. 457–480.

6 *Beta patula* Aiton: Genetic Diversity Analysis

L. Frese, M. Nachtigall, M. Enders and M.Â.A. Pinheiro de Carvalho

6.1 Introduction

Beta patula Ait. is known as a rare species with a very limited distribution area. Except for a small dataset published by Letschert (1993), neither the exact distribution of the species nor any demographic or genetic data were available previous to the work performed by Pinheiro de Carvalho *et al.* (2010) in the context of the AEGRO project (AEGRO, 2010). The analysis of the intraspecific genetic diversity was initiated to provide the nature conservation agency responsible for the protection of *B. patula* with information required for the establishment and management of a genetic reserve. The justification for the establishment of a genetic reserve for *B. patula* is detailed in Chapter 5, this volume, by Pinheiro de Carvalho *et al.* When the planning of a genetic reserve has been initiated, a set of populations representing most of the genetic diversity of the target taxon is to be identified and located. For this purpose the geographic distribution pattern of genetic diversity within a species can be assessed using genetic traits as recommended in step 3 of the methodology for identifying CWR genetic reserve sites for a target crop gene pool (see Kell *et al.*, Chapter 2, this volume).

At the end of the planning process a network of genetic reserves for the target species can be installed and populations managed actively (Maxted *et al.*, 1997). Genetic reserve management plans include the monitoring of the populations' response to management interventions. For any monitoring actions a baseline needs to be established. The demographic baseline of *B. patula* is described by Pinheiro de Carvalho *et al.* (Chapter 5, this volume), while this chapter adds the genetic baseline.

6.2 Genetic Diversity Analysis

Leaves of *B. patula* and *B. vulgaris* subsp. *maritima* (L.) Arcang. were sampled in seven areas on the two small islets of Ilhéu do Desembarcadouro (ID) and Ilhéu Chão (IC) off Madeira, Portugal (Pinheiro de Carvalho *et al.*, 2010). A total of 242 individual plants of *B. patula* and 20 *B. vulgaris* subsp. *maritima* plants sympatrically growing on two of the ID areas were used for genetic analysis. Six sampling areas were located on ID and the seventh on IC.

A large set of simple sequence repeat (SSR) markers has been published for *B. vulgaris* (e.g. McGrath *et al.*, 2007) facilitating the performance of a *B. patula* genetic diversity analysis. Approximately 500 mg fresh leaf material per plant was harvested on both islets, processed according to a modified method after Chase and Hills (1991) and sent to the Julius Kühn-Institut (Germany) where total genomic DNA was extracted using a modified mini-preparation procedure as described by Dorokhov and Klocke (1997). Then a PCR was carried out using 25 SSR markers (Table 6.1) developed for *B. vulgaris* subsp. *vulgaris* and subsp. *maritima*. The amplicons were separated and detected on an automatic dual laser fluorescence DNA-sequencer (LI-COR). The resulting fragment sizes were stored in a database designed by Enders (2010). All computations and calculations were based upon these data.

To visualize the genetic structure and the dependencies between geographic distribution and genetic diversity, a factorial analysis was carried out using the software package Darwin (Perrier and Jacquemound-Collet, 2006). The results are shown in Fig. 6.1.

A clear separation of *B. vulgaris* subsp. *maritima* (ID3/ID4) from *B. patula* (BPIC, BP1 to BP6) as well as a separation between the occurrences of *B. patula* on the two islets (BPIC versus BP1 to BP6) was detected. The genetic differences between plant groups sampled in the six areas on Ilhéu do Desembarcadouro proved to be smaller, indicating a closer relationship between these plant groups. This higher similarity may be due to gene flow between the plant groups on the ID islet. This finding was

Table 6.1. SSR markers used to study genetic diversity in *B. patula* (BP) and *B. vulgaris* subsp. *maritima* (BVM). The number of alleles per locus, PIC-values, observed (Ho) and expected (He) heterozygosities are presented.

Marker locus	Linkage group	Reference	No. of alleles	PIC value	H_o	H_e
2KWS	2	5	9	0.8041	0.2366	0.8263
Bmb1[$]	–	2	4	0.5530	0.0191	0.6266
Bmb3*	–	2	0	0.0000	0.0000	0.0000
Bmb4	–	2	7	0.7373	0.1183	0.7704
Bmb6[$]	–	2	4	0.4876	0.0496	0.5718
BQ487642	6	1	7	0.7784	0.0267	0.8035
BQ583448*	1	1	0	0.0000	0.0000	0.0000
BQ584037[$]	2	1	3	0.2103	0.0916	0.2219
BQ588629	1	1	9	0.6324	0.3168	0.6684
DX580252*	3	1	0	0.0000	0.0000	0.0000
DX581113	9	1	5	0.5304	0.0191	0.5816
FDSB1007	8	4	4	0.4210	0.0687	0.4839
FDSB1027	3	4	5	0.4342	0.1221	0.4679
FDSB1033[+]	9	4	3	0.1197	0.0573	0.1234
FDSB1250	7	4	3	0.3234	0.0000	0.4006
FDSB1300	2	4	5	0.6905	0.0344	0.7380
GTT1	6	1	4	0.4439	0.0076	0.4922
SB04	5	3	5	0.4174	0.0649	0.4829
SB06	4	3	4	0.4876	0.0229	0.5437
SB07	4	3	6	0.7172	0.1298	0.7526
SB11[$]	–	3	4	0.3171	0.0115	0.3399
SB13	3	3	5	0.4300	0.0916	0.4634
SB15	5	3	6	0.5184	0.0153	0.5475
USDA29	8	1	5	0.3742	0.0611	0.4242
USDA3	7	1	2	0.2675	0.0000	0.3182
Average	–	–	4.5	–	0.0626	0.466

The most informative markers are in bold.
*did not amplify, [+]only amplified in BVM, [$]only amplified in BP.
Reference: [1]McGrath *et al.*, 2007; [2]Cureton *et al.*, 2002; [3]Richards *et al.*, 2004; [4]Laurent *et al.*, 2007; [5]kindly provided for research purposes by KWS Saat AG, Einbeck, Germany.

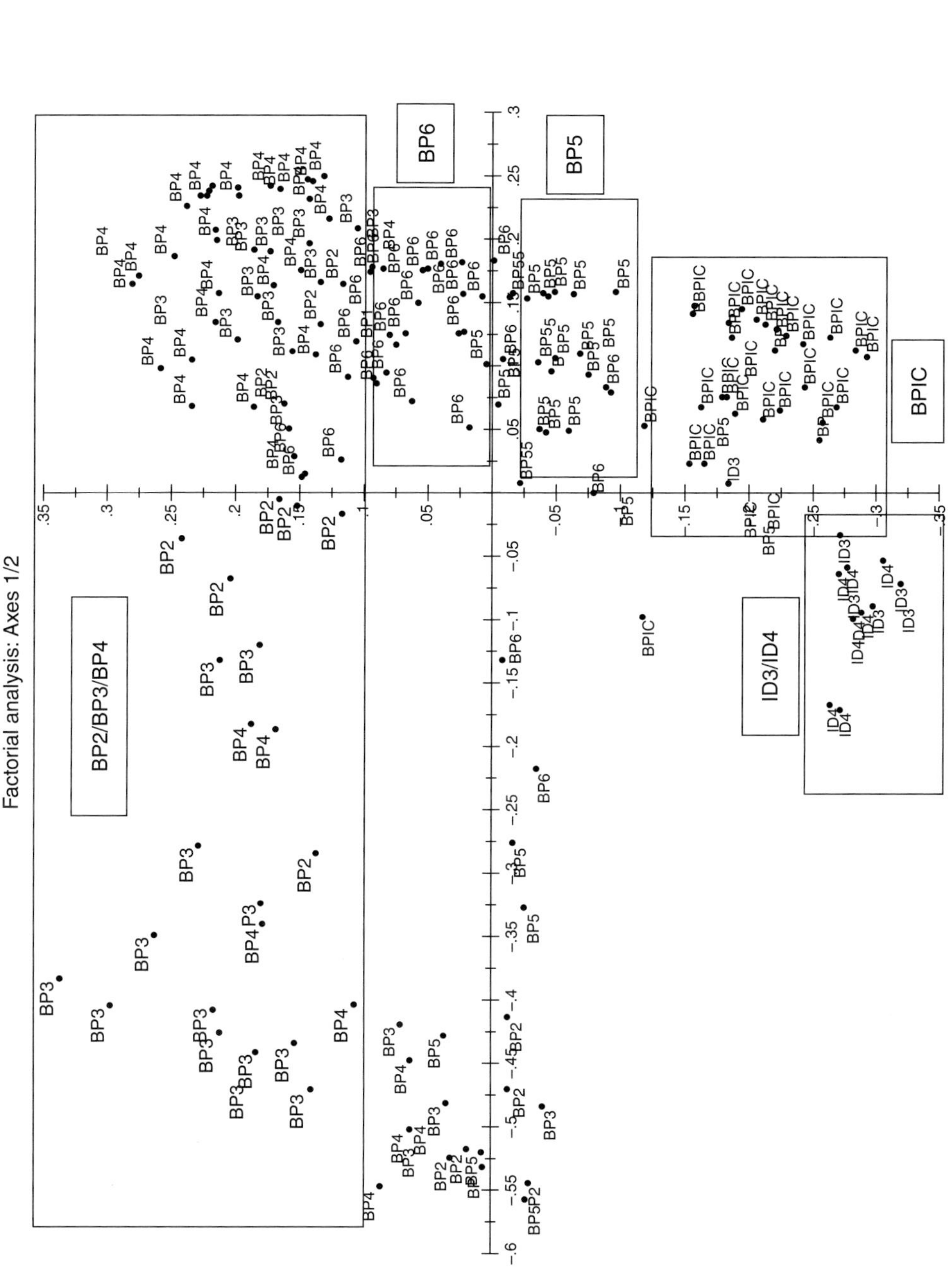

Fig. 6.1. Genetic structure of *B. patula* (BPIC, BP1 to BP6) and *B. vulgaris* subsp. *maritima* (ID3, ID4). BP1 is represented by only three individuals and does not form a distinct group. Axes one and two describe 42.48% of the total variance.

supported by a population structure analysis carried out with the software package Geneland (Guillot *et al.*, 2005) using Markov chain Monte Carlo (MCMC) methods. This software divided the sampled plants into different groups. All *B. vulgaris* subsp. *maritima* plants (ID3/ID4) formed one group, while all plants from the sampling area BPIC were separated into two distinct groups. The *B. patula* plants from the ID islet were arranged into six groups. The first three groups were comprised of plants from only one sampling area each with no plants from a different area except for a maximum of one migrant. The other three groups contained plants from up to four different proximate sampling areas.

All polymorphic loci were individually tested for Hardy-Weinberg-Equilibrium (HWE) to ascertain the applicability of F-statistics (Wright, 1932). Significant deviations from HWE were not detected. Furthermore, the PIC (Polymorphic Information Content) value as well as observed heterozygosity (H_o) and expected heterozygosity (H_e) were calculated. A large difference was observed between the latter two (averaged H_o = 0.0626 versus averaged H_e = 0.466). The results are shown in Table 6.1.

To understand the reasons for this deviation, two F-statistic parameters (F_{is}, the Inbreeding coefficient and F_{st}, the Fixation index) (Wright, 1932) were computed using FStat Software (Goudet, 2001). The results indicate a moderate deficiency in heterozygote frequencies, which is caused by geographic isolation (average F_{st} = 0.329). Inbreeding (average F_{is} = 0.826) proved to be the major cause of heterozygote frequency reduction. As expected, a negative F_{is} value (average F_{is} = −0.162) was computed for *B. vulgaris* subsp. *maritima* plants growing within the same area. The high inbreeding of *B. patula* is most likely the result of a higher rate of self-pollination as compared to strictly allogamous *B. vulgaris* subsp. *maritima*. These findings underpin Letschert (1993), who asserted that *B. patula* is a self-compatible species.

In order to identify the locations containing the main fraction of genetic diversity, the allele numbers at marker loci per sampling area were determined. As expected, a lower number of alleles were detected in the peripheral areas than in the central sampling areas, except for the westernmost sampling plot 'BP6', which had slightly higher allele numbers.

Two conclusions can be drawn from the genetic analysis:

1. The genetic reserve ought to include both islets.
2. The results presented in this study suggest that there is no gene flow between *B. patula* and *B. vulgaris* subsp. *maritima* species on Ilhéu do Desembarcadouro. As *P. procumbens*, a species of the tertiary *Beta* gene pool, also grows in this area which is part of the Natura 2000 protected area network (Pinheiro de Carvalho *et al.*, 2010), three species could be managed within the same genetic reserve site.

6.3 Suggestions and Elements for a Genetic Monitoring Plan

The genetic data indicate that the genetic reserve and consequently genetic monitoring should include all sites where the threatened species occurs, although the Ilhéu Chão is difficult to access making active management very impractical. If for capacity reasons active management is to be restricted, priority should be given to sampling areas 'BP3' to 'BP6', which represent the major fraction of genetic diversity of the species as measured with SSR markers.

The Royal Society (Anonymous, 2003) stressed that any project aiming to measure diversity in long-term series can only be of durable nature if it satisfies the requirements of interested parties. Once these interests have been identified, a framework for actions can be defined and agreed upon. In the case of *B. patula* the interested parties are: (i) the species conservation sector represented by the Madeiraean authorities who are interested in the maintenance of a rare and threatened species; and (ii) the plant genetic resources for food and agriculture conservation sector represented by the

ECPGR working group on *Beta*, which is interested in the maintenance of a genetic resource for breeding. The common interest of both groups is readily defined. Both must be interested in quickly detecting any negative impacts leading to reduced viability of the species within the genetic reserve. Hence, a project like the present one matches the conditions stated by the Royal Society.

The development of a complete monitoring design as provided by Iriondo *et al.* (2008) is beyond the scope of this chapter as is the development of a genetic monitoring schema. Genetic monitoring aims at *indicating the state of endangerment of a genetic resource* (Gregorius and Degen, 2007) and this objective requires more than just counting alleles of SSR marker loci. However, some elements of a monitoring plan can be described based on the *B. patula* diversity study. These elements are the following:

- *Genetic diversity analysis* – within this pilot study the work flow ranging from sampling of leaf probes in the habitat to SSR marker analysis has been organized and tested. The experiences can be shared with the ECPGR working groups and others interested in the subject.
- *Genetic diversity data management* – for the recording of georeferenced single plant data, a data acquisition tool has been developed and tested (see Germeier *et al.*, Chapter 31, this volume). A database was modelled and implemented to support genetic monitoring by storing and post-processing all laboratory and field data in a well documented and structured way (Enders, 2010). The information system keeps the data available for the analysis of time series.
- *Genetic diversity trend reports* – the first set of results has been documented and discussed with the representatives responsible for the site management and will be considered in the site management plan of the Madeira nature conservation authorities.

Having conducted a species census on both islets and a genetic diversity analysis, a time frame for monitoring is to be suggested. *Beta patula* has continued to exist from the time of its first description by Aiton (1789) until today. However, its survival in the past is no guarantee for its survival in the near future. In particular, if climate change reduces the distribution area of *B. patula* during the next five decades in a similar dimension as forecasted by Jarvis *et al.* (2008) for *Arachis*, *Solanum* and *Vigna*, actions need to be undertaken in time to safeguard the species. As there are no indications that the climate policy will effectively control climate warming, it is reasonable to organize monitoring actions. These actions, embedded in a framework allowing feedback and adaptation to needs, should at first comprise a regular annual survey of both islets by the local nature protection service, and demographic monitoring at 5-year intervals, i.e. in the year 2013, 5 years after the first census and sampling. It should also be kept in mind that *B. patula* is a small-sized species and every leaf sample taken in the natural habitat is a disadvantage to the species. The *genetic monitoring* action should therefore only be triggered if the results from the surveys and the demographic monitoring indicate significant and alarming negative population trends.

In the meantime, components of the work flow can be improved to reduce error sources and costs of SSR marker analysis. These improvements concern two aspects:

1. A genetic reserve management plan will be elaborated with a detailed prescription of the data recording and sampling procedures to reduce handling errors. Data recorded in the natural habitat and data produced in the laboratory should be kept in a single information system. Ideally, the genetic reserve manager as well as the laboratory producing the SSR marker data should upload their standardized data on to a web-based information system whose operation is guaranteed for the next few decades.

2. In the case of ID and IC as potential genetic reserve areas for *B. patula*, the cost/information ratio may be reduced by approximately 40% in the future simply by restricting genetic diversity analysis to the most informative SSR markers (Table 6.1). Marker loci with the highest factor scores as derived from the factor analyses are considered the most informative in this context (for details see Enders, 2010). Private alleles, i.e. alleles only found in one sampling area, have been identified using GDA Software (Lewis and Zaykin, 2001). By observing the appearance, spreading or disappearance of private alleles within the different sampling areas and/or in the whole population, changes in the pattern of genetic diversity could be detected with minimal input in terms of laboratory work.

References

AEGRO (2010) An Integrated European *In Situ* Management Work Plan: Implementing Genetic Reserves and On Farm Concepts (AEGRO). Available at: http://aegro.jki.bund.de (accessed 15 December 2010).

Aiton, W. (1789) *Hortus kewensis*. Catalogue of plants cultivated in Royal Botanical Garden of Kew, London, Vol. 1, 315.

Anonymous (2003) *Measuring Biodiversity for Conservation*. The Royal Society, Policy document 11/03. Royal Society, London.

Chase, M.W. and Hills, H.H. (1991) Silica gel: an ideal material for field preservation of leaf samples for DNA studies. *Taxon* 40, 215–220.

Cureton, A.N., Burns, J.M., Ford-Lloyd, B. and Newbury, H.J. (2002) Development of simple sequence repeat (SSR) markers for the assessment of gene flow between sea beet (*Beta vulgaris* ssp. *maritima*) populations. *Molecular Ecology Notes* 2(4), 402–403.

Dorokhov, D.B. and Klocke, E. (1997) A rapid and economic technique for RAPD analysis of plant genomes. *Russian Journal of Genetics* 33, 358–365.

Enders, M. (2010) Entwicklung und Anwendung molekularer und informatorischer Werkzeuge zum genetischen Monitoring bei Wildrüben. Diplomarbeit im Fach Bioinformatik. Martin-Luther-Universität Halle-Wittenberg, Naturwissenschaftliche Fakultät III, Institut für Informatik.

Goudet, J. (2001) FSTAT, a program to estimate and test gene diversities and fixation indices (version 2.9.3). Available at: www2.unil.ch/popgen/softwares/fstat.htm (accessed 4 January 2011).

Gregorius, H.-R. and Degen, B. (2007) Monitoring genetischer Ressourcen – Prinzipien und Methoden. In: Begemann, F., Schröder, S., Wenkel, K.-O. and Weigel, H.-J. (eds) *Monitoring und Indikatoren der Agrobiodiversität*. Tagungsband eines Symposiums am 7. und 8. November 2006 in Königswinter. Agrobiodiversität, Schriftenreihe des Informations- und Koordinationszentrums für Biologische Vielfalt, Band 27, pp. 39–65.

Guillot, G., Mortier, F. and Estoup, A. (2005) Geneland: a computer package for landscape genetics. *Molecular Ecology Notes* 5, 712–715.

Iriondo, J.M., Ford-Lloyd, B., De Hond, L., Kell, S.P., Lefèvre, F., Korpelainen, H. and Lane, A. (2008) Plant population monitoring methodologies for the *in situ* genetic conservation of CWR. In: Iriondo, J.M., Maxted, N. and Dulloo, M.E. (eds) *Conserving Plant Genetic Diversity in Protected Areas*. CAB International, Wallingford, UK, pp. 88–123.

Jarvis, A., Lane, A. and Hijmans, R.J. (2008) The effect of climate change on crop wild relatives. *Agriculture Ecosystems & Environment* 126, 13–23.

Laurent, V., Devaux, P., Thiel, T., Viard, F., Mielordt, S., Touzet, P. and Quillet, M.C. (2007) Comparative effectiveness of sugar beet microsatellite markers isolated from genomic libraries and GenBank ESTs to map the sugar beet genome. *Theoretical and Applied Genetics* 115(6), 793–805.

Letschert, J.P.W. (1993) *Beta* section Beta, biogeographical patterns of variation and taxonomy. *Wageningen Agricultural University Papers* 1(1), 32.

Lewis, P.O. and Zaykin, D. (2001) Genetic Data Analysis: computer program for the analysis of allelic data, Version 1.0 (d16c). Available at: http://lewis.eeb.uconn.edu/lewishome/software.html (accessed 21 October 2010).

Maxted, N., Ford-Lloyd, B.V. and Hawkes, J.G. (1997) Complementary conservation strategies. In: Maxted, N., Ford-Lloyd, B.V. and Hawkes, J.G. (eds) *Plant Genetic Conservation: the In Situ Approach*. Chapman & Hall, London, pp. 20–55.

McGrath, J.M., Trebbi, D., Fenwick, A., Pannella, L., Schulz, B., Laurent, V., Barnes, S. and Murray, S.C. (2007) An open source first-generation molecular genetic map from a sugarbeet x table beet cross and its extension to physical mapping. *Crop Science* 47(Suppl.)1, 27–44.

Perrier, X. and Jacquemound-Collet, J.P. (2006) DARwin software. Available at: http://darwin.cirad.fr/darwin (accessed 21 October 2010).

Pinheiro de Carvalho, M.Â.A., Nóbrega, H., Frese, L., Freitas, G., Abreu, U., Costa, G. and Fontinha, S. (2010) Distribution and abundance of *Beta patula* Aiton and other crop wild relatives of cultivated beets on Madeira. *Journal für Kulturpflanzen* 62(10), 357–366.

Richards, C.M., Brownson, M., Mitchell, S.E., Kresovich, S. and Panella, L. (2004) Polymorphic microsatellite markers for inferring diversity in wild and domesticated sugar beet (*Beta vulgaris*). *Molecular Ecology Notes* 4(2), 243–245.

Wright, S. (1932) The roles of mutation, inbreeding, crossbreeding and selection in evolution. *Proceedings of the VI International Congress of Genetics* 1, 356–366.

7 Assessing Genetic Reserves in Sicily (Italy): The *Brassica* Wild Relatives Case Study

F. Branca, S. Argento and A. Tribulato

7.1 The Genus *Brassica* and its Wild Relatives

The *Brassica* genus, belonging to the Brassicaceae, represents one of the most complex genera used in the vegetable industry. As many as 320 taxa (species, varieties, con-varieties, subspecies and cultigen groups) are included in the genus and a number of them are widely used as important sources of vegetables, condiments, and both edible and industrial oils. The ancestors of modern cultivated Brassicas have been used in the past in human diet, and they are still harvested in the wild in some countries.

The elucidation of the origin of crops of *B. oleracea* L. is still an important research task. Studies of DNA polymorphisms suggested a monophyletic origin (Song *et al.*, 1988), whereas for other authors the domestication of several cultigen groups of *B. oleracea* is polyphyletic and is strictly related to several wild *Brassica* ancestors that represent a common complex gene pool (Snogerup *et al.*, 1990). The latter assumed that a number of crops, widespread from the European Atlantic coast to the North Sea, originated from *B. oleracea* subsp. *oleracea*, while recent studies using molecular genetic marker evidenced a strong divergence among Mediterranean *Brassica* wild species and the *B. oleracea* wild type occurrences of northern European Atlantic coasts. It was concluded that the domestication of *B. oleracea* cultigen groups occurred along the Greek-speaking area of the Mediterranean basin (Mitchell, 1976; Maggioni *et al.*, 2010). The origin of cauliflower and broccoli from *B. oleracea* group seems to be located in the mentioned area and is linked to *B. cretica*, *B. incana*, *B. insularis*, *B. macrocarpa*, *B. montana*, *B. rupestris* and *B. villosa* as wild relatives (Gòmez-Campo and Gustafsson, 1991). These species are perennial with a woody stem, up to 1.5 m tall, large leaves and high glucosinolate content; they are diploid ($2n=18$), often self-incompatible, with high tendency to mutations, and freely cross-pollinate among them and with *B. oleracea* crops grown in the surroundings.

Brassica wild relatives have been extensively used for more than a century to increase resistance/tolerance to biotic and abiotic stress in several cultivars. Recently, high antioxidant properties were found in some Sicilian broccoli landraces. The use of Sicilian broccoli landraces and wild relative species in breeding allowed obtaining new cultivars with high levels of glucosinolate (Mithen *et al.*, 2003). Sicilian occurrences of crop wild relatives (CWR) were further investigated within the work package 'Case Study *Brassica*' of the targeted action AGRI GEN RES 057 (AEGRO, 2011).

7.2 Description of Sicilian *Brassica* Wild Species ($n = 9$)

Brassica incana: C genome diploid. This species is spread along the Tyrrhenian, Ionian and Adriatic coasts of Italy and along Balcanian coasts. In Sicily occurrences are present mainly along the east and north-east coasts. The species is diffused in limestone cliffs, rocky slopes with maquis, disturbed vegetation of roadsides and townships, from sea level up to at least 900 m where grazing and fire represent the main threats. For this species 16 sites were found along the north-east coast and one in the south-east. The different occurrences of *B. incana* are mainly spread in protected areas.

Brassica macrocarpa: C genome diploid. This species is endemic to the Egadi Islands, west Sicily, where it can be encountered in limestone cliffs and rocky slopes (Fig. 7.1) near the sea and where grazing and fire represent the main threats. This species grows in two sites, one in Favignana and the other one in Marettimo, of which the latter is a legally protected area. Both occurrences of *B. macrocarpa* are diffused in the marine protected area of Egadi Islands.

Brassica rupestris: C genome diploid. This species is widespread along the north coast of Sicily, mainly in western areas, in limestone or rarely sandstone cliffs, usually with northern exposure or otherwise in comparatively shaded positions, from sea level to 1100 m. Grazing and fire represent the main threats. For this species 21 sites were identified, 19 along the north-west coast of Sicily, one in the south-east and one in the north-east. Several occurrences of *B. rupestris* are spread in protected areas (Madonie and Nebrodi regional parks).

Brassica villosa: C genome diploid. This species is widespread in Sicily in the central-west inland areas and along the north-west coast, in limestone or rarely sandstone cliffs, usually with northern exposure or otherwise in comparatively shaded positions, from sea level to 1000 m

Fig. 7.1. *Brassica macrocarpa* in its natural environment in Favignana.

where grazing and fire are the main threats. For this species 27 sites were identified along the north-west coasts and in the central-west inland areas of Sicily. The occurrences in these sites belong to the subspecies *bivoniana*, *drepanensis*, *tinei* and *villosa* and showed morphological differences which may allow classifying them as species. Also for this species, several occurrences grow in protected areas (Madonie and Nebrodi regional parks).

In some cases, along the north-west coast, both of the latter species, *B. rupestris* and *B. villosa*, coexist in the same areas and quite likely cross-pollinate among them and with *B. oleracea* landraces widespread in home gardens and in peri-urban vegetable farms of some villages.

7.3 Characterization of Wild *Brassica*

The case study aimed at characterizing Sicilian *Brassica* wild relatives on the basis of bio-morphological and genetic traits. To this purpose, seeds of accessions of *Brassica incana*, *B. macrocarpa*, *B. rupestris* and *B. villosa* were sown and 40 plants per accession were transplanted in a cold greenhouse at the experimental agriculture farm of the University of Catania in autumn 2008 (Fig. 7.2). One-year-old plants were characterized on the basis of the main IBPGR and UPOV descriptors (IBPGR, 1990; UPOV, 1988) at the reproductive stage. In order to check for putative similarities in different regions, two occurrences of *B. rupestris* from Sicily (Roccella Valdemone and Ragusa) and two from Calabria (Stilo and Pazzano) were characterized and compared (Table 7.1).

Brassica macrocarpa and one occurrence of *B. villosa* flowered during autumn 2009 while the other species started flowering during winter 2010. The reproductive phase began more than 1 year after transplanting. The traits of stem branching, leaf hairiness, leaf and petiole anthocyanin content, shape and size showed high variation (Table 7.2). For the occurrences of *B. rupestris* a high variation regarding stem branching, leaf shape and, in a more general extent, the plant habitus was observed.

Fig. 7.2. Characterization of *Brassica* CWRs at the experimental fields of Catania University.

Table 7.1. Wild *Brassica* occurrences collected and studied by DOFATA and their origin.

Accession number	Species	Origin	Geographical coordinates
UNICT 3253	*B. macrocarpa*	Favignana	N 37°55′59″ E 12°90′00″
UNICT 3270	*B. rupestris*	Stilo	N 38°28′51″ E 16°27′53″
UNICT 3410	*B. rupestris*	Pazzano	N 38°28′12″ E 16°27′33″
UNICT 3405	*B. rupestris*	Roccella Valdemone	N 37°56′02″ E 15°00′55″
UNICT 3458	*B. rupestris*	Ragusa	N 36°55′32″ E 14°44′24″
UNICT 3512	*B. incana*	Augusta	N 37°18′33″ E 15°07′25″
UNICT 3944	*B. villosa*	Marianopoli	N 37°36′53″ E 13°56′27″
UNICT 4158	*B. incana*	Sortino	N 37°08′28″ E 15°02′32″

Brassica incana showed a wide variation with respect to colour and size of the petiole, leaf colour and plant shape. Clear differences with respect to the leaf hairiness, a trait present in *B. incana* and *B. villosa* and not in *B. macrocarpa* and in *B. rupestris* were detected.

A hierarchical cluster analysis was carried out and a dendrogram generated using as unit the squared Euclidean distance and the complete linkage furthest neighbour method (Fig. 7.3). The characterized occurrences were classified into three main groups. Group A represented all occurrences of *B. rupestris*, Group B represented all occurrences belonging to *B. incana* and *B. villosa*, and Group C contained *B. macrocarpa*. The latter is distinct from the other species and occurrences on the basis of the used descriptors, confirming previous taxonomy studies (Raimondo *et al.*, 1991).

Young leaves collected from 20 plants per accession were used for molecular characterization. To achieve this goal we analysed BoAP1 locus, with the aim to detect simple sequence repeat (SSR) polymorphism and to observe the variation that occurs in *Brassica* for this nucleotide site, which in *B. oleracea* is highly correlated with phylogenetic patterns (King, 2003). This locus was studied because it expresses one of the MADS-box genes involved in reproductive processes and in particular in modification of inflorescence shape and size.

Regarding molecular characterization, on the basis of the BoAP1 SSR locus high variation of alleles among the different wild *Brassica* occurrences was observed. In fact, it was found that 6 out of the 13 alleles identified for the studied nucleotide were rarely locally frequent while one was the most frequent in all occurrences with a percentage as high as 100% for *B. incana* and *B. macrocarpa*. For *B. macrocarpa* a high presence of one allele with a frequency of about 92% was found.

The allele variability for BoAP1 nucleotide was higher in the wild *Brassica* occurrences studied than in some cauliflower and broccoli landraces and F1 hybrid used as control.

7.4 Prioritization for *Brassica* Genetic Reserve Assessment

Based on the results presented in the previous chapters, species and their occurrences were prioritized for *in situ* conservation and sites for the establishment of genetic reserves were suggested. Among the species, highest priority was given to *B. macrocarpa* Guss for being an endemism and for its distinct profile among the species studied on the basis of biomorphological and molecular marker traits variation. The assessment, as CR (Critically Endangered) in the European Commission Environment red list of threatened species, is in progress. *Brassica macrocarpa* is being assessed as CR because its extent of occurrence is less than 100 km², its area of occupancy is less than 10 km², it is severely fragmented as it occurs in only two isolated occurrences and there is a continuing decline in the extent and quality of its habitat due to grazing pressure, reforestation, construction

Table 7.2. Characterization of *Brassica* wild occurrences on the basis of the IBPGR and UPOV descriptors.

	B. rupestris				*B. incana*		*B. macrocarpa*	*B. villosa*
Descriptor	Pazzano	Roccella Valdemone	Stilo	Ragusa	Sortino	Augusta	Favignana	Marianopoli
Stem length (cm)	33.1	32.8	47.6	30.4	28.0	31.5	39.1	56.1
Stem width (cm)	1.8	2.1	2.1	2.3	1.9	2.2	1.8	2.0
Stem length/ diameter ratio	18.9	16.0	22.4	12.9	15.1	14.2	21.6	27.8
Emergency date (d)	30.0	27.0	29.0	49.0	–32.0	27.0	–13.0	70.0
Plant height (cm)	57.1	60.5	70.0	59.7	50.9	67.5	65.3	77.9
Plant diameter (cm)	82.3	83.0	74.1	66.2	81.5	95.7	86.1	87.5
Plant shape	3.0	3.0	5.0	1.0	3.0	5.0	3.0	5.0
Leaf anthocyanin colour	9.0	9.0	5.8	8.5	1.0	2.6	2.1	1.0
Leaf colour	2.0	4.0	3.0	2.9	2.0	2.0	3.0	2.0
Leaf blade shape	2.0	5.0	5.0	2.0	2.0	2.0	2.0	2.0
Leaf blade length (cm)	26.2	28.6	26.6	27.4	36.8	61.5	26.3	40.8
Leaf blade width (cm)	20.1	19.9	19.4	17.2	22.2	27.5	20.6	25.4
Leaf shape	0.8	0.7	0.7	0.6	0.6	0.4	0.8	0.6
Petiole attitude	3.9	3.4	3.5	2.7	3.8	3.5	3.0	3.6
Leaf lamina attitude	5.8	5.7	3.8	6.2	5.5	5.8	5.7	6.6
Leaf blade density of curling	1.0	1.0	1.0	1.0	1.0	1.0	1.0	1.0
Leaf blade blistering	5.0	3.4	3.3	3.0	3.5	2.8	4.3	3.9
Leaf lobes	2.0	1.9	2.0	2.0	2.0	1.7	2.0	2.0
Petiole length (cm)	22.9	23.1	20.1	19.6	9.4	0.0	24.4	15.3
Petiole width (cm)	0.8	0.9	0.8	1.1	0.5	0.0	0.9	0.7
Petiole enlargement	3.0	3.1	3.0	3.0	1.6	0.0	3.0	2.3
Petiole and/or mid-vein colour	4.0	3.6	2.7	2.4	0.5	0.0	2.0	0.7
Branching	0.0	0.2	2.8	0.2	0.0	0.0	7.0	0.0
Heading habit	0.0	0.0	0.0	0.0	0.0	0.0	0.0	0.0
Plant leaves (*n*)	18.7	20.7	34.4	18.9	15.5	13.5	50.6	16.5
Leaf scars (*n*)	8.9	5.0	8.9	11.6	4.4	4.2	7.4	4.1
Leaf blade: folding in cross sections	3.5	4.6	3.9	3.7	–	4.1	3.4	3.7
Hairiness	0.0	0.0	0.0	0.0	7.0	7.0	0.0	5.0

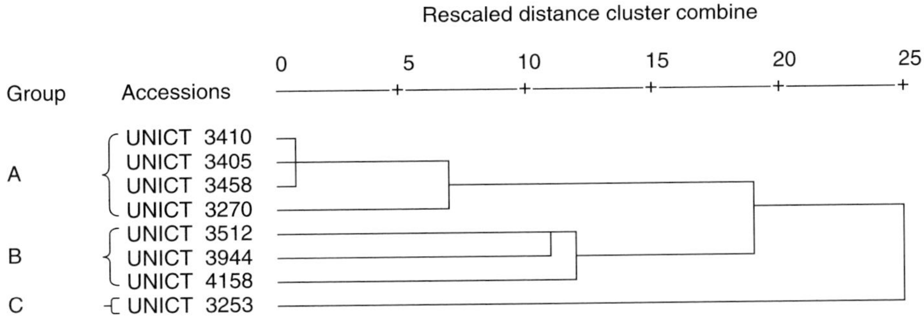

Fig. 7.3. Dendrogram based upon characterization data using complete linkage and further neighbour method.

of holiday resorts and associated improvements to the road infrastructure, as well as recreational activities and fires.

Snogerup *et al.* (1990) recorded three occurrences ranging in size between 101 and approximately 1000 individuals. According to recent inspections of the sites the occurrence on Favignana comprises 750 individuals. On Marettimo it can be found at three localities and the occurrence comprises less than 500 individuals. One occurrence on the island of Levanzo is now extinct but the remaining two occurrences were reported to be stable in the period from 2004 to 2005 (CEC, 2009) and are currently stable though at risk of fragmentation. Grazing pressure is causing a decrease in the number of new recruitments to these sites.

Moreover, *B. macrocarpa* is listed as a priority species in Annex II of the EU Habitats Directive (CEC, 1992) and in Annex I of the Convention on the conservation of European wildlife and natural habitats (Council of Europe, 1979). At national level it has been assessed as Critically Endangered (Pignatti *et al.*, 2001) and was reported by the CEC (2009) to have poor prospects of surviving, thus the species is likely to struggle unless habitat conditions change.

The species is interesting due to the high glucosinolate content in leaves (Branca *et al.*, 2002) and the effectiveness of its dry leaves tilled into the soil to control root knot nematodes (*Meloidogyne* spp.) in tomato (Branca *et al.*, 2008). In addition, *B. macrocarpa* has been used in breeding programmes with the aim to increase aliphatic glucosinolate in broccoli lines with the aim to hydrolyse high amounts of the isothiocyanate sulforaphane, a compound involved in cancer protection in mammals (Qiu *et al.*, 2008).

A number of European research groups focused attention on the protection of *Brassica* CWRs. Seed samples are conserved *ex situ* in a number of gene banks and germplasm holdings as documented in the ECPGR *Brassica* Database (BRAS-EDB, 2010). Considering all facts, DOFATA proposed therefore the establishment of a genetic reserve for *B. macrocarpa* aiming at the conservation of the whole variability of the species distributed on both islands. The strategies for proposing and managing the genetic reserve of Egadi Island for *in situ* conservation of *B. macrocarpa* are jointly elaborated by DOFATA and the Centro Universitario per la Tutela e la Gestione degli Ambienti Naturali ed Agricoli (CUTGANA – University of Catania). The latter already manages seven nature reserves in Sicily. The other *Brassica* CWR, *B. incana*, *B. rupestris* and *B. villosa*, already frequently occur in protected areas facilitating their protection in collaboration with the management staff of the Sicilian Department of Forestry and Nebrodi, Madonie and Alcantara park administration boards.

7.5 Conclusions

The activities supported by the AEGRO project allowed the identification of several sites of *Brassica* CWR occurrences, the collection of germplasm for *ex situ* conservation, the characterization of the accessions on the basis of the IBPGR and UPOV descriptors list and the collection of leaf samples for molecular genetic marker analysis. The large variation observed among the analysed accessions confirmed the importance of the Sicilian material of the primary gene pool of *Brassica* ($n=9$), endorsing the need for promoting their protection by the establishment of genetic reserves.

References

AEGRO (2011) An Integrated European *In Situ* Management Work Plan: Implementing Genetic Reserves and On Farm Concepts (AEGRO). Available at: http://aegro.jki.bund.de (accessed 28 April 2011).

Branca, F., Li, G., Goyal, S. and Quiros, C. (2002) Survey of aliphatic glucosinolates in Sicilian wild and cultivated Brassicaceae. *Phytochemistry* 59, 717–724.

Branca, F., Argento, S. and Tribulato, A. (2008) Work in progress on characterization and exploitation of *Brassica macrocarpa* Guss. *Brassica 2008 5th ISHS International Symposium on Brassicas and the 16th Crucifer Genetics Workshop*. Magnor Hansen, Lillehammer, Norway, 170.

BRAS-EDB (2010) The ECPGR Brassica Database. Available at: http://documents.plant.wur.nl/cgn/pgr/brasedb/ (accessed 20 October 2010).

CEC (1992) European Commission Habitats Directive. Council directive 92/43/EEC of 21 May 1992 on the conservation of natural habitats and of wild fauna and flora. European Commission, Brussels.

CEC (2009) *Composite Report on the Conservation Status of Habitat Types and Species as required under Article 17 of the Habitats Directive*. Report from the Commission to the Council and the European Parliament, Brussels.

Council of Europe (1979) Convention on the conservation of European wildlife and natural habitats. *European Treaty Series, No. 104*, Council of Europe, Bern, Switzerland.

Gómez-Campo, C. and Gustafsson, M. (1991) Germplasm of wild *n*=9 Mediterranean species of Brassica. *Botanika Chronika* 10, 429–434.

IBPGR (1990) *Descriptors for* Brassica *and* Raphanus. International Board for Plant Genetic Resources, Rome.

King, G.J. (2003) Using molecular allelic variation to understand domestication processes and conserve diversity in Brassica crops. *Acta Horticulturae* 598, 181–186.

Maggioni, L., von Bothmer, R., Poulsen, G. and Branca, F. (2010) Origin and domestication of cole crops (*Brassica oleracea* L.): linguistic and literary considerations. *Economic Botany* 64(2), 109–123.

Mitchell, N.D. (1976) The Status of *Brassica oleracea* L. subsp. *oleracea* (Wild Cabbage) in the British Isles. *Watsonia* 11, 97–103.

Mithen, R., Faulkner, K., Magrath, R., Rose, P., Willianson, G. and Marquez, J. (2003) Development of isothiocianate-enriched broccoli and its enhanced ability to induce phase 2 detoxification in mammalian cells. *Theoretical and Applied Genetics* 106, 727–734.

Pignatti, S., Menegoni, P. and Giacanelli, V. (2001) *Liste Rosse e Blu della Flora Italiana*. ANPA, Rome.

Qiu, D., Gao, M., Ellison, C., Li, G., Huang, F., Haggard, J.E., Negre-Zakharov, F., Branca, F. and Quiros, C. (2008) QTL analysis of isothiocianate content in a *Brassica* interspecific hybrid. *XVI International Conference on Plant and Animal Genome*, San Diego, 173.

Raimondo, F.M., Mazzola, P. and Ottonello, D. (1991) On the taxonomy and distribution of *Brassica* sect. Brassica (Cruciferae) in Sicily. *Flora Mediterranea* 1, 64–86.

Snogerup, S., Gustafsson, M. and von Bothmer, R. (1990) *Brassica* sect. Brassica (Brassicaceae). I. Taxonomy and variation. *Willdenowia* 19, 271–365.

Song, K., Osborn, T.C. and Williams, P.H. (1988) Brassica taxonomy based on nuclear restriction fragment length polymorphism (RFLPs). 2. Preliminary analysis of subspecies within *B. rapa* (syn. *campestris*) and *B. oleracea*. *Theoretical and Applied Genetics* 76, 593–600.

UPOV (1988) Guidelines for the conduct of test for distinctness, homogeneity and stability: Turnip, Turnip rape (*Brassica rapa* L. emend. Metzg.). International Union for the protection of New Varieties of Plants, Geneva, 34 pp.

8 The AEGRO *Brassica* Case Study

K. Kristiansen and G.K. Bjørn

8.1 Introduction

Soupizet (2002) compiled the information on the biology and ecology of *Brassica insularis* Moris., *B. montana* Pourret and *B. oleracea* L. as well as on the geographic distribution, genetic variability and the conservation status of these species in France. She concluded that the French populations incorporate an important part of the genetic diversity of these species in the global context and that actions should be initiated ensuring the conservation of this important germplasm *in situ*. Occurrences of the species were identified within protected areas, in particular for *B. insularis*, which is a protected species in France. Soupizet proposed concrete management actions required to improve the *in situ* conservation status of that species. The list of species and sites suggested by Soupizet is to be considered an early and first component of an EU-wide *Brassica* genetic reserve network.

The aim of the AEGRO *Brassica* Case Study was to further contribute to the development of this network. The case study focused on relatives of *B. oleracea*. The four-step methodology developed by AEGRO and made available online at the CWR *In Situ* Strategy Helpdesk was used (AEGRO, 2010; Kell *et al.*, Chapter 2, this volume) to select target species and to identify genetic reserve sites. The four steps are: (i) delineation of taxon within the crop gene pool; (ii) selection of target taxa according to conservation status, potential use, threat and distribution area; (iii) analysis of ecogeographic diversity by collating geographic, ecological and genetic data; and (iv) selection of target sites by investigating whether target taxa occur within already established protected areas. The case study is divided into a European and a Sicilian part; the latter is published by Branca *et al.* (Chapter 7, this volume).

8.2 The Crop *Brassica*

Brassica crops are grown worldwide as vegetables and for the production of oils for food, fodder and industrial uses. The genus *Brassica* contains the major crop species *B. juncea* (Indian mustard), *B. napus* (rapeseed), *B. oleracea* (cabbage) and *B. rapa* (Chinese cabbage, turnip) as well as the minor economic important species *B. carinata* (Abyssinian mustard) and *B. nigra* (black mustard) (FitzJohn *et al.*, 2007). According to the triangle of U (U, 1935), *B. nigra*, *B. oleracea* and *B. rapa* are the ancestral parents of the other three mentioned *Brassica* crops.

Brassica oleracea and *B. rapa* are closely related whereas *B. nigra* is more distant to the two other species (Warwick and Sauder, 2005). Nine further species (*B. cretica*, *B. hilarionis*, *B. incana*, *B. insularis*, *B. macrocarpa*, *B. montana*, *B. rupestris*, *B. tournefortii* and *B. villosa*) are or have been collected locally for direct use (CWRIS-PGR Forum, 2010).

Brassica is included in the tribe Brassiceae, which contains approximately 50 genera and 240 species. Based on chloroplast restriction site polymorphisms and nuclear ribosomal internal transcribed spacer as well as chloroplast *trn*L intron sequences, Warwick and Sauder (2005) concluded that the genus *Brassica* is of polyphyletic origin.

8.3 Intergeneric and Interspecific Hybridizations

After interspecific hybridizations between six *Brassica* crops with 21 wild *Brassica* species only *B. balearica* did not produce hybrids (Table 8.1) (FitzJohn *et al.*, 2007). Intergeneric hybrids with *Brassica* crops have been obtained with species of *Capsella*, *Coincya*, *Crambe*, *Diplotaxis*, *Enarthrocarpus*, *Eruca*, *Erucastrum*, *Hirschfeldia*, *Moricandia*, *Orychophragmus*, *Physaria*, *Raphanus*, *Rapistrum*, *Rorippa*, *Sinapidendron* and *Sinapis* (FitzJohn *et al.*, 2007; Warwick *et al.*, 2009) (Table 8.1). In some cases *in vitro* culture of immature embryos were necessary to obtain hybrids, and somatic hybrids with *Arabidopsis*, *Armoracia*, *Barbarea*, *Camelina*, *Descurainia*, *Matthiola* and *Trachystoma* have been produced by protoplast fusion (Warwick *et al.*, 2009). It has been possible to make successful crosses between species and genera with different chromosome numbers, and as a large proportion of combinations have successfully yielded offspring even under natural conditions, the gene pool of the crop *Brassica* is very large. Backcrosses or production F_2 populations have been possible in a number of the interspecific and intergeneric hybrids, thus many *Brassica* species seem to be within the primary gene pool of at least one of the *Brassica* crops, and also other Brassicaceae genera should be included in the primary gene pool (Table 8.1). Of the genera successfully hybridized with *Brassica* crops the species of *Capsella*, *Crambe*, *Diplotaxis*, *Eruca*, *Erucastrum*, *Raphanus*, *Rorippa* and *Sinapis* are distributed in Europe. These genera should also be taken into consideration when multi-species genetic reserves are established within the network of Natura 2000 protected areas.

8.4 Target Species and Genetic Reserves

Brassica cretica Lam. ssp. *cretica* and ssp. *laconica* Gustafsson & Snogerup, *B. incana* Ten., *B. insularis* Moris and *B. montana* Pourr. were selected as target species in the case study. They are largely self-incompatible, suffrutescent perennials, which can grow to a height of 1.5 m. They grow in the Mediterranean area with climates characterized by hot and dry summers and winter rain and they are most often found as chasmophytes in limestone substrates, often in fissures of cliffs and rocks (Snogerup *et al.*, 1990). For several of the target species relative large genetic differences among populations have been found, which is probably caused by isolation of populations by distance (Hurtrez-Boussès, 1996; Rao *et al.*, 2002; Widen *et al.*, 2002; Edh *et al.*, 2007; Noël *et al.*, 2010). Thus, to protect a large part of the genetic variation several populations of each species need to be included in genetic reserves.

The descriptions of the target species given in the next sections are taken from Warwick *et al.* (2009) while the locations of occurrences are from Snogerup *et al.* (1990), GBIF (2010), CWRIS-AEGRO PLIS (2010) and EUNIS (2010).

8.4.1 *Brassica cretica*

Brassica cretica ssp. *cretica* grows on limestone cliffs, gorges and flat pavements of small non-grazed ravines at altitudes up

Table 8.1. Successful interspecific and intergeneric hybridizations with *Brassica* crops. Summarized from FitzJohns *et al.* (2007) and Warwick *et al.* (2009).

Crop	Interspecific hybrids	Intergeneric hybrids
B. carinata (2n=34)	*B. fruticulosa, B. juncea,* * *B. maurorum, B. napus,* * *B. nigra,* * *B. oleracea,* * *B. rapa, B. tournefortii*	*Diplotaxis assurgens, D. tenuisiliqua, D. virgata, Enarthrocarpus lyratus,* * *Erucastrum abyssinicum, E. gallicum, Orychophragmus violaceus,* * *Raphanus sativus, Sinapis alba, S. arvensis, S. pubescens*
B. juncea (2n=38)	*B. carinata,* * *B. gravinae,* * *B. maurorum,* * *B. napus,* * *B. nigra,* * *B. oleracea, B. oxyrrhina,* * *B. rapa,* * *B. tournefortii**	*Crambe hispanica, Diplotaxis catholica, D. erucoides, D. muralis, D. siettiana, D. siifolia, D. tenuifolia, D. virgata,* * *Enarthrocarpus lyratus,* * *Eruca vesicaria, Erucastrum abyssinicum, E. virgatum, Moricandia arvensis, Orychophragmus violaceus,* * *Raphanus raphanistrum, R. sativus, Sinapidendron frutescens, Sinapis alba, S. arvensis, S. pubescens**
B. napus (2n=38)	*B. bourgeaui,* * *B. carinata,* * *B. cretica,* * *B. fruticulosa, B. gravinae,* * *B. juncea,* * *B. maurorum,* * *B. montana,* * *B. nigra,* * *B. oleracea,* * *B. rapa,* * *B. tournefortii*	*Capsella bursa-pastoris, Diplotaxis catholica, D. erucoides,* * *D. harra,* * *D. muralis,* * *D. siifolia, D. tenuifolia, D. virgata, Enarthrocarpus lyratus,* * *Eruca vesicaria, Erucastrum gallicum,* * *Hirschfeldia incana,* * *Moricandia arvensis,* * *Orychophragmus violaceus,* * *Physaria fendleri, Raphanus raphanistrum,* * *R. sativus,* * *Rapistrum rugosum, Rorippa islandica,* * *Sinapis alba,* * *S. arvensis,* * *S. pubescens**
B. nigra (2n=16)	*B. barrelieri, B. carinata,* * *B. fruticulosa, B. juncea,* * *B. maurorum, B. napus,* * *B. oleracea,* * *B. oxyrrhina, B. rapa,* * *B. spinescens, B. tournefortii*	*Coincya monensis, Diplotaxis erucoides, D. siettiana, D. tenuifolia, Erucastrum cardaminoides, E. varium, E. virgatum, Hirschfeldia incana, Moricandia arvensis, Orychophragmus violaceus,* * *Raphanus sativus,* * *Sinapis alba,* * *S. arvensis*
B. oleracea (2n=18)	*B. bourgeaui, B. carinata,* * *B. cretica,* * *B. incana,* * *B. insularis,* * *B. juncea, B. macrocarpa,* * *B. maurorum, B. montana,* * *B. napus,* * *B. nigra, B. oxyrrhina, B. rapa,* * *B. rupestris,* * *B. tournefortii, B. villosa**	*Coincya monensis, Diplotaxis erucoides, D. muralis, D. tenuifolia, Enarthrocarpus lyratus,* * *Eruca vesicaria, Erucastrum abissinicum, Hirschfeldia incana, Moricandia arvensis,* * *M. nitens, Orychophragmus violaceus, Raphanus sativus,* * *Sinapis alba, S. arvensis, S. pubescens*
B. rapa (2n=20)	*B. barrelieri, B. bourgeaui,* * *B. carinata,* * *B. cretica,* * *B. fruticulosa, B. gravinae, B. incana, B. insularis, B. juncea,* * *B. macrocarpa, B. maurorum, B. montana,* * *B. napus,* * *B. nigra,* * *B. oleracea,* * *B. oxyrrhina, B. spinescens, B. tournefortii*	*Capsella bursa-pastoris, Diplotaxis catholica, D. erucoides, D. muralis,* * *D. siettiana, D. siifolia, D. tenuifolia,* * *D. tenuisiliqua, D. virgata, Enarthrocarpus lyratus,* * *Eruca vesicaria,* * *Erucastrum abyssinicum, E. canariense, E. cardaminoides, E. gallicum,* * *E. leucanthum, Hirschfeldia incana, Moricandia arvensis, Ochryphragmus violaceus,* * *Raphanus sativus,* * *Sinapis alba, S. arvensis*

*Combinations with successful production of either backcrosses or F_2 populations.

to 700 m. The main distribution area is Crete with localities along the entire island as well as the northern part of Peloponnisos (Snogerup *et al.*, 1990). On Crete the localities of Topolia, Gonies and Moni Kapsa represent a large part of the genetic variation (Rao *et al.*, 2002) and could be selected as genetic reserves. The Gonies population is not part of an already established protected area, but further populations could be protected by including e.g. the Samaria Gorge as a genetic reserve. On northern Peloponnisos relatively large populations of *B. cretica* ssp. *cretica* have been reported from Akrokorinthos and Xilokerisa (Snogerup *et al.*, 1990) and these localities should be investigated in order to preserve this genetic variation. However, these localities are not within protected areas. *Brassica cretica* ssp. *laconica* grows in open, almost vertical limestone cliffs, from 150 to 650 m on southern and western Peloponnisos and is not found in true maritime locations. Most of the populations of this subspecies are found in inaccessible cliffs, gorges and ravines. Snogerup *et al.* (1990) report several populations with more than 1000 plants and regarded this taxon as not being threatened. However, these records are more than 20 years old; thus their conservation status should be investigated.

Widen and co-workers (e.g. Rao *et al.*, 2002; Widen *et al.*, 2002; Edh *et al.*, 2007) have published a number of papers regarding population genetics, evolution and diversification of *B. cretica* during the last 10 years. Nuclear and chloroplast microsatellite markers revealed exceptionally high levels of population differentiation and relatively little within-population diversity on Crete (Edh *et al.*, 2007).

8.4.2 *Brassica incana*

Brassica incana Ten. grows on limestone cliffs, on rocky slopes with maquis and in areas with disturbed vegetation. It occurs at altitudes up to 900 m and has been found in coastal areas of southern and central Italy, eastern Sicily as well as in coastal areas of the Croatian islands around Curzola and on Corfu (Snogerup *et al.*, 1990).

For *B. incana* only a few reports regarding its genetic diversity have been published. Ellis *et al.* (1999, 2000) found some resistance against cabbage root fly (*Delia radicum*) and cabbage aphid (*Brevicoryne brassicae*) within the species.

Outside Sicily, the Sorrento peninsula of Italy should be included as a genetic reserve as several populations are present and they were already under severe threat in late 1980s (Snogerup *et al.*, 1990). Further genetic reserves for *B. incana* should be established on Curzola and Corfu as these populations are isolated from the Sicilian and Italian ones.

8.4.3 *Brassica insularis*

Brassica insularis Moris grows mainly on limestone cliffs in open and sunny positions. It is found on Corsica, Sardinia and at the Tunisian coast, mainly at altitudes below 500 m, however on Corsica populations were recorded growing at altitudes up to 1200 m (Snogerup *et al.*, 1990). The populations on Corsica have been monitored rather intensively (Noël *et al.*, 2010), where the species is restricted to nine cliff areas with population sizes up to 5000 individuals and with marked differences among populations (Hurtrez-Boussès, 1996). Petit *et al.* (2001) report a minor gene flow between the populations. On Corsica the populations at Teghime/Monti Rossi, Inzecca and Punta Corbaghiola (Soupizet, 2002) could be first choices as genetic reserves. On Sardinia genetic reserves could be established at Capo Caccia and the islets of Isola Rossa and Isole de Cavoli as they had quite large populations (Snogerup *et al.*, 1990).

Mithen and Lewis (1988) reported a hypersensitive resistance in *B. insularis* towards the fungal pathogen, *Leptosphaeria maculens*, causing stem canker in oilseed rape.

8.4.4 *Brassica montana*

Brassica montana Pourr. grows in limestone cliffs and rocks, in maquis and disturbed ground. It occurs mainly in coastal areas stretching from north-west Italy to the Gerona coast of north-east Spain. In Alpi Apuanae, Italy it grows up to an altitude of 1000m (Snogerup *et al.*, 1990). *Brassica montana* risks extinction and/or introgression from garden grown *Brassica* along the Italian and French Riviera (Snogerup *et al.*, 1990).

Genetic studies in *B. montana* are very scarce. Cauwet *et al.* (1992) cited by Soupizet (2002) studied genetic diversity using polymorphic LAP, AcPh and Est enzyme loci. The within-population diversity of 17% of the total variation was low as compared to the between-population diversity. A good correlation between genetic distance and geographic distance was observed. Data on phenotypic variation were published by Perez-Garcia (2005) who found differences between Spanish populations in germination percentages (0–86%) and mean germination time (3.6–6.1 days).

Perez-Garcia collected seeds at seven places along the Gerona coast of north-east Spain. The Spanish components of genetic diversity of the species could be secured by establishing genetic reserves at these locations. In France, Mount Faron just north of Toulon and the Mercantour National Park between Moulinet and Sospel previously had large populations and within these regions genetic reserves may be established. In Italy, Capo di Noli and Parco Alpi Apuanae may qualify as genetic reserves.

8.4.5 *Further Brassica species*

Brassica balearica from Mallorca, *B. bourgeaui* from the Canary Islands, *B. cadmea* from central Greece, *B. gravinae* from central and southern Italy, *B. hilarionis* from northern Cyprus, *B. nivalis* from Mt Pirin Planina in Bulgaria and Mt Olympus in Greece, *B. procumbens* (Poir.) from Corsica and Giglio Island in Italy and *B. tyrrhena* from Sardinia (Giotta *et al.*, 2002) have also rather restricted distribution areas and are also to be considered when establishing multi-species genetic reserves. Not all of these species are included in the *B. oleracea* cytodeme.

References

AEGRO (2010) CWR *in situ* strategy helpdesk – a step-wise methodology for identifying CWR genetic reserve sites for a target crop gene pool. Available at: http://aegro.jki.bund.de (accessed 15 December 2010).

CWRIS-AEGRO PLIS (2010) Population level information system. Available at: http://aegro.bafz.de/aegroprod_brassica/home.seam (accessed 15 December 2010).

CWRIS-PGR Forum (2010) Crop Wild Relative Information System. Available at: www.pgrforum.org/cwris/cwris.asp (accessed 15 December 2010).

Edh, K., Widen, B. and Ceplitis, A. (2007) Nuclear and chloroplast microsatellites reveal extreme population differentiation and limited gene flow in the Aegean endemic *Brassica cretica* (Brassicaceae). *Molecular Ecology* 16, 4972–4983.

Ellis, P.R., Pink, D.A.C., Barber, N.E. and Mead, A. (1999) Identification of high levels of resistance to the cabbage root fly, *Delia radicum*, in wild *Brassica* species. *Euphytica* 110, 207–214.

Ellis, P.R., Kift, N.B., Pink, D.A.C., Lynn, J.J. and Tatchell, G.M. (2000) Variation in resistance to the cabbage aphid (*Brevicoryne brassicae*) between and within wild and cultivated *Brassica* species. *Genetic Resources and Crop Evolution* 47, 395–401.

EUNIS (2010) European Nature Information System. Available at: http://eunis.eea.europa.eu/ (accessed 15 December 2010).

FitzJohn, R.G., Armstrong, T.T., Newstrom-Lloyd, L.E., Wilton, A.D. and Cochrane, M. (2007) Hybridisation within *Brassica* and allied genera: evaluation of potential for transgene escape. *Euphytica* 158, 209–230.

GBIF (2010) Global Biodiversity Information Facility. Available at: http://data.gbif.org/welcome.htm (accessed 15 December 2010).

Giotta, C., Piccitto, M. and Arrigoni, P.V. (2002) Un nuovo endemismo della Sardegna: *Brassica tyrrhena* sp. nov. (Brassicaceae). A new endemic species from Sardinia: *Brassica tyrrhena* sp. nov. (*Brassicaceae*). *Webbia* 57, 1–5.

Hurtrez-Boussès, S. (1996) Genetic differentiation among natural populations of the rare Corsican endemic *Brassica insularis* Moris: Implications for conservation guideline. *Biological Conservation* 76, 25–30.

Mithen, R.F. and Lewis, B.G. (1988) Resistance to *Leptosphaeria maculens*. I. Hybrids of *Brassica oleracea* and *B. insularis*. *Journal of Phytopathology* 123, 253–258.

Noël, F., Maurice, S., Mignot, A., Glemin, S., Carbonell, D., Justy, F., Guyot, I., Olivieri, I. and Petit, C. (2010) Interaction of climate, demography and genetics: a ten year study of *Brassica insularis*, a narrow endemic Mediterranean species. *Conservation Genetics* 11, 509–536.

Perez-Garcia, F. (2005) Seed germination of different populations of wild (*n*=9) *Brassica montana* and *B. oleracea*. *Spanish Journal Agricultural Research* 3, 331–334.

Petit, C., Freville, H., Mignot, A., Colas, B., Riba, M., Imbert, E., Hurtrez-Boussès, S., Virevaire, M. and Olivieri, I. (2001) Gene flow and local adaptation in two endemic plant species. *Biological Conservation* 100, 21–34.

Rao, G.-Y., Andersson, S. and Widen, B. (2002) Flower and cotyledon asymmetry in *Brassica cretica*: genetic variation and relationships with fitness. *Evolution* 56, 690–698.

Snogerup, S., Gustafsson, M. and von Bothmer, R. (1990) *Brassica* sect. Brassica (Brassicaceae) I. Taxonomy and variation. *Willdenowia* 19, 271–365.

Soupizet, F. (2002) Gestion in situ de la diversité génétique des espèces sauvages apparentées aux espèces cultivées. Études des *Brassica* L., *Beta* L., *Olea* L. *Les Études du bureau des resources génétiques (BRG). No. 2*. ISBN 2-908447-21-5.

U, N. (1935) Genome analysis in *Brassica* with special reference to the experimental formation of *B. napus* and peculiar mode of fertilization. *Japanese Journal of Botany* 7, 389–452.

Warwick, S.I., Francis, A. and Gugel, R.K. (2009) Guide to wild germplasm of *Brassica* and allied crops (tribe Brassiceae, Brassicaceae) 3rd edn. Available at: www.brassica.info/info/publications/guide-wild-germplasm.php (accessed 15 December 2010).

Warwick, S.U. and Sauder, C.A. (2005) Phylogeny of tribe Brassiceae (Brassicaceae) based on chloroplast restriction site polymorphism and nuclear ribosomal internal transcribed spacer and chloroplast trnL intron sequences. *Canadian Journal Botany* 83, 467–483.

Widen, B., Andersson, S., Rao, G.-Y. and Widen, M. (2002) Population divergence of genetic (co)variance matrices in a subdivided plant species, *Brassica cretica*. *Journal Evolutionary Biology* 15, 961–970.

9 Surveying and Conserving European *Avena* Species Diversity

A. Katsiotis and G. Ladizinsky

9.1 Introduction

The genus *Avena* consists of diploid ($2n = 2x = 14$), tetraploid ($2n = 4x = 28$) and hexaploid ($2n = 6x = 42$) species, with the basic chromosome number of seven ($x = 7$). All species are annuals and self-pollinated, with the exception of *Avena macrostachya* Bal. ex Coss. Et Dur., which is a perennial, cross-pollinated, autotetraploid. Although some diploid (*Avena strigosa* Schreb.) and tetraploid (*Avena abyssinica* Hochst.) species are rarely cultivated, the commonly cultivated species are *Avena sativa* L. and *Avena byzantina* C. Koch. Most *Avena* species are wild and weedy, and are distributed mainly around the Mediterranean Sea and the Canary Islands. During the last 20 years new species have been described, such as diploid *A. atlantica* (Baum and Fedak, 1985a), and tetraploids *A. agadiriana* (Baum and Fedak, 1985b) and *A. insularis* (Ladizinsky, 1998).

The genus highest diversity is found in northern Africa (Tunisia, Algeria and Morocco), southern Spain and Sicily (Italy). All major biological species are found in Morocco except *A. canariensis*, *A. ventricosa*, *A. insularis* and *A. macrostachya*. Species *A. agadiriana*, *A. atlantica* and *A. magna* are found only in Morocco. *Avena canariensis* is unique to the Canary Islands, *A. damascena* to Syria and Morocco and *A. insularis* to Tunisia and Sicily (Italy). *Avena murphyi*, *A. longiglumis* and *A. prostrata* are found in restricted areas in southern Spain, and *A. ventricosa* is found in Cyprus (Leggett, personal communication), within Europe. A large number of *Avena* accessions throughout the world are held in gene banks, but some of the species are well underrepresented.

The EURISCO database contains 34,307 *Avena* accessions but only 369 (about 1%) of them are listed as wild (EURISCO, 2011), thus there is a need to identify and prioritize those species for *in situ* conservation in order to maintain their genetic variability. Prioritization of species for *in situ* conservation can only be accomplished if genomic and species relationships are known. During the last years molecular techniques have been used to unravel these relationships (Drossou *et al.*, 2004; Nikoloudakis and Katsiotis, 2008; Nikoloudakis *et al.*, 2008). Another factor considered when prioritizing species is their distribution. Thus, taking into consideration all of the above factors, *Avena* species that have been recognized to be in need of *in situ* conservation in Europe are as follows.

1. *Avena ventricosa* ($2n=2x$), present in Baku (Azerbaijan), Oran (Algeria) and Cyprus. This species belongs to the tertiary gene pool, and until recently its distribution in Cyprus was not known.
2. *Avena insularis* ($2n=4x$), a recently discovered species which appears to be the tetraploid progenitor of all hexaploid types, including the common oat (Ladizinsky, 1998), belonging to the oat secondary gene pool. *Avena insularis* is present in two areas in southern Sicily: north of Gela and west of Catalgirone. In these areas *A. insularis* is restricted to uncultivated heavy soil.
3. *Avena murphyi* ($2n=4x$), a member of the common oat secondary gene pool, present at the southern tip of Spain, west of Tarifa, where it is restricted to uncultivated heavy soil. The natural habitat of *A. murphyi* was traditionally used as pasture land (see Fig. 9.1), but in the last 30 years or so this soil has been turned into arable land and the area of this species is rapidly declining.

The present chapter provides an ecogeographic analysis, a demographic baseline and a genetic diversity analysis for the three prioritized *Avena* species.

9.2 Species Distribution

The three *Avena* species presented here have a very restricted distribution within Europe. *Avena ventricosa* is present only in the island of Cyprus, *A. insularis* only in Sicily, and *A. murphyi* at the south-western tip of Spain. Field surveys were conducted in May 2009 for *A. ventricosa* in Cyprus, and in May 2010 for *A. insularis* and *A. murphyi* in Sicily and Andalusia, respectively.

9.2.1 *Avena ventricosa*

Avena ventricosa is native to Algeria, Cyprus, Iraq, Saudi Arabia and Azerbaijan

Fig. 9.1. Typical *Avena murphyi* habitat in the area of Bolonia, Andalusia, Spain.

(Baum, 1977). According to *Flora of Cyprus* (Meikle, 1977) *A. ventricosa* has been recorded to be present in at least four sites: near Kampos (recorded in 1928), at Athalassa Farm and near the Agricultural Research Institute (ARI) (recorded in 1967 by Merton), between Nissou and Stavrovouni (recorded in 1933) and near Nicosia airport (recorded in 1951 by Merton). All of the above mentioned areas were visited during the May 2009 survey to record the presence/absence of the species. No *A. ventricosa* populations were found in the area of Kampos, or even in close proximity to this area, thus a misidentification of the species is possible. Between Kampos and Nicosia large *A. ventricosa* populations were recorded, in some cases coexisting with *A. sterilis* and/or *A. hirtula*. *Avena ventricosa* was also present just outside the Cypriot Agricultural Research Institute (ARI), although a new pine forest was planted during the last 10–15 years, and at Athalassa Farm, a National Park. New sites where *A. ventricosa* was present included Mt Macheras and the Salt lake (Alykes), close to Larnaka airport.

9.2.2 *Avena insularis*

This species was first discovered by Ladizinsky in 1996 (Ladizinsky, 1998) in Sicily, around Lake Comunelli and at Mt Bubonia, and few years later in Tunisia, at Temime and Bargou regions (Ladizinsky and Jellen, 2003). In May 2010 Sicily was visited to further collect, survey the species populations and search for new sites where *A. insularis* might be present. In at least four sites around Lake Comunelli *A. insularis* populations were found. One of the sites had a pure, large *A. insularis* population (with no *A. sterilis* present), while at least two more populations were found in a fenced area by the Regione Siciliana, Assessorato Agricolotura e Foreste, planted with pine trees. Another population, mixed with *A. sterilis*, was found at Mt Giase, also fenced by the same agency, where eucalyptus and pines trees were planted. A new site for *A. insularis* was found close to Borgo Franchetto. In 2011 eight more *A. insularis* populations have been mapped in Sicily by Ladizinsky. In areas investigated north of Mt Etna no *A. insularis* plants were found. Thus, it seems that this species is present south of the Catania-Palermo highway and east of Agrigento.

9.2.3 *Avena murphyi*

Avena murphyi is native to a restricted area in southern Spain between Tarifa and Vejer de la Frontera (Baum, 1977). During 2007, a collection mission for *A. murphyi* and *A. prostrata* in Andalusia was funded by the ECPGR (ECPGR, 2011). In 2010 the same areas were visited to survey changes in the *A. murphyi* populations. During the last mission at least two fields, close to the road going to Bolonia, were found with pure dense populations of *A. murphyi*, while in neighbouring fields no *A. murphyi* plants were present (Fig. 9.1). New sites with small *A. murphyi* populations were located close to the cities of Atlantera and Barbate. Another site with this species present was north of Alcalá de los Gazules, a place that was indicated by the Consejeria de Medio Ambiente of the Junta de Andalusia in Seville. This is an interesting site because it is not the typical *A. murphyi* habitat.

9.3 Ecogeographic Analysis

The common approach that was adopted for all three field surveys included gathering geological and soil maps (if available), obtaining recent climatic data through national agencies, and identifying protected areas under either NATURA 2000 (EEA, 2011) or national/regional laws.

9.3.1 Geology and edaphic conditions

Avena ventricosa

Avena ventricosa is mostly found in the central-east part of Cyprus south of Nicosia and west of Nicosia towards the area of Linou, while it is absent from the south-west part of Cyprus. According to the geological maps *A. ventricosa* is present in areas that include upper and lower pillow lavas and basal groups, alluvium sedimentary formations and fanglomerate, terrace deposits. Soil types included calcaric-lithic leptosols, calcaric-leptic regosols, calcaric-rendzic leptosols and eutric-lithic leptosols.

Avena insularis

Avena insularis was initially found around Lake Comunelli, in southern Sicily (Ladizinsky, 1998). The soil types where *A. insularis* is found are regosols, brown soils and/or vertic brown soils. This type of soil covers about 13.4% of Sicily, making up a large part of the alluvial crumbling clayey hill area, found extensively in the provinces of Agrigento and Caltanissetta. The bedrock is a combination of sandstone, clay and conglomerate rocks.

Avena murphyi

Avena murphyi has a very restrictive range of habitat north-west of Tarifa, where heavy alluvial soils are found.

9.3.2 Bioclimatic description

Cyprus has a typical Mediterranean climate. Summers (mid-May to mid-September) are hot and dry, winters (November to mid-March) are rainy, and spring and autumn are short seasons of rapid change in weather conditions. The day length ranges from 9.8 h in December to 14.5 h in June. About 60% of the annual precipitation falls during December–February, while rainfall in the warmer months contributes little or nothing to water resources and agriculture. Summer thunderstorms contribute less than 5% to the annual precipitation. Annual rainfall in the south-west is on average 450 mm, increasing to 1100 mm at the top of the central Troodos massif and decreasing steadily northwards and eastwards to between 300 and 350 mm in the central plain and the flat south-eastern parts of the island. The average temperature on the central plain is 10°C in January (with a minimum of 5°C) and 29°C during July–August (with a maximum of 36°C). *Avena ventricosa* is found east of the Troodos massif.

Sicily also has a Mediterranean climate, with dry summers and wet winters. Average maximum temperature in the area of Agrigento during July–August is 33°C, while the average lowest during February is about 7°C. The average annual precipitation in the same area is about 450 mm most of which (94%) occurs during the months September to April. Average sunlight in the month of December is 4.0 h/day, reaching 10.3 h/day in July.

Avena murphyi has a very restricted distribution in the area north-west of Tarifa. This area has an Atlantic coastal oceanic Mediterranean climate. The average annual precipitation is close to 600 mm, most of it between September and May (97%). Total sunshine hours in December are 142, reaching a high of 307 h in July. Average low temperature in January is 11°C and average high is 24°C in August.

9.3.3 Ecogeographic diversity analysis

For the ecogeographic diversity analysis an updated version of the European Avena Database (EADB, 2011) was used, containing 23,889 accessions (duplicates excluded), which is the representation of 34 *Avena* taxa in 81 countries. However, a number of obstacles were encountered, such as non-consistent taxonomy, misspellings of species names, wrong coordinates and above all the poor representation of the prioritized species in *ex situ* collections. In the EURISCO database, which contains information from the EADB, there are 16 accessions of *A. murphyi*, seven accessions of *A. insularis*, 21 accessions of *A. ventricosa* (19 of which have been obtained through the collection organized by the AEGRO project in collaboration with the Avena Working Group of ECPGR), most of them rarely include coordinates. Furthermore, identification of *Avena* species is difficult and within species genetic variation can be revealed only by using molecular markers.

9.4 Species Population Baseline

Different approaches have been used for the three *Avena* species. Original source population sizes were not possible to be defined for all species, since populations cover large areas that are not clearly delineated, densities in most cases are low, and individuals are hard to be distinguished and identified macroscopically.

As described earlier, *A. ventricosa* is widespread in Cyprus, and in most cases coexists with one or more of the following species, *A. sterilis*, *A. hirtula*, *A. barbata* and *A. eriantha*, making it hard to estimate population sizes (the main difference between *A. ventricosa* and the other species is its long and sharp callus at the bottom of the dispersal unit as well as the uneven length of the glumes). Fourteen sites were surveyed and 16 populations were sampled; in all cases, except one, *A. ventricosa* plants had a low density. In one site a large *A. ventricosa* population was present as a pure stand with few scattered *A. hirtula* and *A. sterilis* plants, having 6–10 plants/m^2.

Avena insularis is morphologically very similar to *A. sterilis*, but can be distinguished mainly by its oblong disarticulation scar. In all areas surveyed *A. insularis* coexisted with *A. sterilis*, an aggressive noxious weed, hindering the estimation of population sizes. From the seven sites surveyed, *A. insularis* was present in five of them and only in two of them had good population stands, one of which was in an area protected by the Regione Siciliana, Assessorato Agricolotura e Foreste.

Finally, *A. murphyi* has a restricted geographical distribution in south-west Spain. The Consejeria de Medio Ambiente of the Junta de Andalusia in Sevilla has detailed geographic information system (GIS) maps and observations regarding population sizes of *A. murphyi*. However, in at least one case a large population (>1000 plants) has been identified and recorded by the agency north of the town Alcalá de los Gazules, but the actual population size was found to be much smaller, since only scattered plants were found, not exceeding 50. In total 17 sites were surveyed, 15 of which contained *A. murphyi* plants, in most cases coexisting with *A. sterilis*. The majority of those sites were roadsides and field edges, with very few plants (not more than ten). In three areas that were well defined (two agricultural fields and one roadside area) it was possible to estimate population sizes. At the roadside area, which was close to 500 m^2, there were ~250 plants. In one field, with a total surface area of 24,500 m^2, more than 10,000 plants were estimated, and in another close-by field with a total surface of 12,300 m^2, about 3000 plants. It is noteworthy that the surrounding fields had very few *A. murphyi* plants, indicating primarily differences in field management.

9.5 Genetic Diversity Analysis

Identification of *Avena* species is difficult and genetic variation within species using morphological characteristics is not possible. Thus, use of molecular markers is necessary to reveal the genetic diversity within a species. Different techniques have been used in the past including RAPDs, RFLPs, AFLPs, SSRs (Alicchio *et al.*, 1995; Nocelli *et al.*, 1999; Li *et al.*, 2000; Drossou *et al.*, 2004), and sequence analysis of the internal transcribed spacers (Rodionov *et al.*, 2005; Nikoloudakis *et al.*, 2008) to study genetic diversity in *Avena* species. In the present study we used RAPDs to reveal the genetic diversity within and among populations of a species. Ten random RAPD primers that have been previously evaluated and generated polymorphism have been used according to Drossou *et al.*'s (2004) protocol. In total, 34 individual plants originating from 14 sites and 16 populations (ranging from one to four plants per population) were used for *A. ventricosa*, 19 individual plants from five sites (ranging from three to five plants per population) for *A. insularis* and 27 individual plants from 15 different sites (ranging from one to five plants per population) for *A. murphyi*. All plants were grown from seeds except *A. murphyi*, where leaves were collected from the field, and DNA was extracted. PCR reactions produced 132 reproducible polymorphic bands for *A. ventricosa*, 112 for

A. insularis and 125 for *A. murphyi*. Genetic similarities were calculated using the Dice (1945) similarity coefficient and cluster analysis was performed using the UPGMA method. The cophenetic correlation coefficients calculated, indicating the goodness of fit of a set of data to the clustering method used, was for all three analyses performed >0.90. Genetic similarities among entries within *A. ventricosa* ranged from 0.72 to 0.99, for *A. insularis* from 0.80 to 0.95 and for *A. murphyi* from 0.75 to 0.96. For all three species no clustering according to their geographic origin of the samples was observed.

9.6 Species Threat Assessment

Taking into consideration all of the above information the three *Avena* species were found to have different levels of threat assessments. In general, *A. ventricosa* is still present in areas where it was recorded more than 75 years ago, such as a number of sites between Nissou and Stavrovouni. Nicosia airport is currently considered a buffer zone, so it was not visited. However, taking into consideration the current observations and the fact that no physical disturbance (cultivation, urbanization etc.) for the last 35 years has happened in the area, it is safe to assume that *A. ventricosa* populations are present. Also *A. ventricosa* populations are still nowadays present in areas around ARI and the Athalassa Park (first recorded in 1967) are not at risk, especially since these areas are under monitoring of ARI and the National Forest Department, respectively. Furthermore, most of the surveyed areas in Cyprus are within Natura 2000 (site CY2000002 Alykos Potamos-Agios Sozomenos and site CY6000002 Alykes Larnakas) or National Forest Law (Athalassa Farm) protected sites, thus providing the environment for longlasting *in situ* presence of the species with proper management and monitoring guidelines.

Avena insularis is a species that was discovered and described in Sicily only 15 years ago. This was the third time that the area was visited and the first time to assess dispersal and population sizes of the species. Except the areas where the species was discovered and is still present, new sites have been recorded for the first time. Thus the information regarding the actual population dispersal of the species cannot be considered complete. *Avena insularis* plants (coexisting with *Lygeum spartum* and *Hedysarum coronarium*) are found exclusively on uncultivated land, which had either never been cultivated, or had long been abandoned, explaining why *A. insularis* is so rare in that region, which is almost totally under cultivation. Two of the populations recorded were present in areas fenced by the Regione Siciliana, Assessorato Agricolotura e Foreste, without however, at the moment, providing conservation techniques and management protocols for the species. These sites are not considered protected areas but rather restricted activity areas, where reforestation takes place in order to control soil erosion, by planting eucalyptus and, more recently, pine trees. Recommendations for providing at least some niches for *A. insularis* populations within the fenced areas are a necessity for securing *in situ* conservation of the species.

During the last decade at least two collecting missions of *A. murphyi* germplasm in south-west of Spain have been funded. According to Pedro Garcia, who participated in both of them, and Gideon Ladizinsky who discovered the species (Ladizinsky, 1971), both of whom participated in the 2010 survey, there is a rapid decline of the species presence in the area. In the year 2007 no *A. murphyi* plants were found in areas where samples have being previously collected (Pedro Garcia, personal communication). Nowadays, the native species is mainly found as isolated plants on roadsides and field edges. Only three sites were found in the 2010 survey to contain reasonable size populations, the first one being a small roadside area which was also visited during 2007 and where *A. murphyi* plants coexist with *A. sterilis*. No differences were observed in species percentage coverage of the area compared to the 2007 visit. The other two sites were pastureland fields (including *Phalaris tuberosa*, *Hordeum bulbosum*, *Poa pratensis*, *Lolium perenne* and *Hedysarium coronarium*), in close proximity, one of which had a large *A. murphyi* population. Surrounding fields had no or very few scattered *A. murphyi* plants.

The reason for the presence of such a large population was the time of cow grazing. The farmer was bringing in the cows after the plants have shed their seeds on the soil, maintaining the seedbed. The area where *A. murphyi* thrives is mostly privately owned fields, part of which is under Natura 2000 (site ES0000337 Estrecho, including 41 bird, three mammal and one amphibian species under protection, but no plants). In order to conserve this species urgent action needs to be taken at the regional government level by informing farmers for proper management of their fields, in order to increase the number of *A. murphyi* seeds on the seedbed, and monitoring the population sizes.

Acknowledgements

We are grateful for the assistance given by Angelos Kyratzis, Agricultural Research Institute, Cyprus, Ferdinando Branca and Alessandro Tribulato, Università di Catania, Italy, and Pedro Garcia, Universidad de León, Spain during our field surveys.

References

Alicchio, R., Aranci, L. and Conte, L. (1995) Restriction fragment length polymorphism based phylogenetic analysis of *Avena* L. *Genome* 38, 1279–1284.

Baum, B.R. (1977) *Oats, wild and cultivated: a monograph of the genus* Avena *L. (Poaceae)*. Biosystematics Research Institute, Canada Department Of Agriculture, Research Branch, Ottawa, Ontario, Canada. Monograph no. 14. Supply and Services Canada, Ottawa.

Baum, B.R. and Fedak, G. (1985a) *Avena atlantica*, a new diploid species of the oat genus from Morocco. *Canadian Journal of Botany* 63, 1057–1060.

Baum, B.R. and Fedak, G. (1985b) A new tetraploid species of *Avena* discovered in Morocco. *Canadian Journal of Botany* 63, 1379–1385.

Dice, L.R. (1945) Measures of the amount of ecologic association between species. *Ecology* 26(3), 297–302.

Drossou, A., Katsiotis, A., Leggett, J.M., Loukas, M. and Tsakas, S. (2004) Genome and species relationships in genus *Avena* based on RAPD and AFLP molecular markers. *Theoretical and Applied Genetics* 109, 48–54.

EADB (2011) European Avena Database. Available at: http://eadb.jki.bund.de (accessed 10 May 2011).

ECPGR (2011) ECPGR Avena Working Group. Available at: www.ecpgr.cgiar.org/Workgroups/avena/Avena_CollTrip_Murcia_Andalusia_0607.pdf (accessed 10 May 2011).

EEA (2011) European Environment Agency. Available at: http://natura2000.eea.europa.eu (accessed 10 May 2011).

EURISCO (2011) European Search Catalogue. Available at: http://eurisco.ecpgr.org (accessed 10 May 2011).

Ladizinsky, G. (1971) *Avena murphyi* a new tetraploid species from southern Spain. *Israel Journal of Botany* 20, 297–301.

Ladizinsky, G. (1998) A new species of *Avena* from Sicily, possibly the tetraploid progenitor of hexaploid oats. *Genetic Resources and Crop Evolution* 45, 263–269.

Ladizinsky, G. and Jellen, E.N. (2003) Cytogenetic affinities between populations *Avena insularis* Ladizinsky from Sicily and Tunisia. *Genetic Resources and Crop Evolution* 50, 11–15.

Li, C.D., Rossnagel, B.G. and Scoles, G.J. (2000) The development of oat microsatellite markers and their use in identifying relationships among *Avena* species and oat cultivars. *Theoretical and Applied Genetics* 101, 1259–1268.

Meikle, R.D. (1977) *Flora of Cyprus*. The Bentham-Moxon Trust, Royal Botanic Gardens, Kew, UK.

Nikoloudakis, N. and Katsiotis, A. (2008) The origin of the C-genome and cytoplasm of *Avena* polyploids. *Theoretical and Applied Genetics* 117, 273–281.

Nikoloudakis, N., Skaracis, G. and Katsiotis, A. (2008) Evolutionary insights inferred by molecular analysis of the ITS1-5.8S-ITS2 & IGS *Avena* sp. sequences. *Molecular Phylogenetics and Evolution* 46, 102–115.

Nocelli, E., Giovannini, T., Bioni, M. and Alicchio, R. (1999) RFLP- and RAPD-based genetic relationships of seven diploid species of *Avena* with the A genome. *Genome* 42, 950–959.

Rodionov, A.V., Tyupa, N.B., Kim, E.S., Machs, E.M. and Loskutov, I.G. (2005) Genomic configuration of the autotetraploid oat species *Avena macrostachya* inferred from comparative analysis of ITS1 and ITS2 sequences: on the Oat karyotype evolution during the early events of the *Avena* species divergence. *Russian Journal of Genetics* 41, 518–528.

10 Quality Standards for Genetic Reserve Conservation of Crop Wild Relatives

J.M. Iriondo, N. Maxted, S.P. Kell, B.V. Ford-Lloyd, C. Lara-Romero, J. Labokas and J. Magos Brehm

10.1 Introduction

Crop wild relatives (CWR) are wild plant taxa closely related to species of direct socio-economic importance (e.g. food, ornamental, forestry, medicinal plants) (Kell *et al.*, 2003; Maxted *et al.*, 2006). CWR have been identified as a critical group vital for agriculture. They possess many desirable adaptive traits that can be bred into crops to address the impact of global climate change on agricultural production and changing market demands (Maxted *et al.*, 1997a, 2008a; Meilleur and Hodgkin, 2004; Heywood and Dulloo, 2006; Heywood *et al.*, 2007). However, the survival of these wild plants is increasingly threatened by unsustainable agroenvironmental management, habitat destruction and fragmentation, and climate change (Heywood *et al.*, 2007; Jarvis *et al.*, 2008; Maxted *et al.*, 2008a). Despite their socio-economic importance, the *ex situ* and *in situ* conservation of CWR have been largely overlooked (Maxted *et al.*, 1997b, 2008a; Heywood *et al.*, 2007). They represent only 5.6% of European *ex situ* gene bank holdings and only 6% of European CWR have been sampled (Maxted *et al.*, 2008a). Most CWR populations grow outside any form of systematic protection, and even when they are included in protected areas, they have not been subject to active management plans or included as specific objectives in conservation strategies (Maxted *et al.*, 1997b; Heywood *et al.*, 2007).

Genetic reserves (GRs) are areas designated for the active, long-term *in situ* conservation of wild populations where the primary consideration is to preserve its genetic diversity. Individual, national and global approaches to identifying GRs for CWR within protected areas have been developed (Maxted and Kell, 2009). In most cases the strategies for the *in situ* conservation of CWR involve the establishment of GRs within protected areas because it may lead to greater sustainability and efficacy of the GRs. However, it should be noted that this is not a necessary prerequisite and that *in situ* conservation of CWR is also possible (and may sometimes be necessary) outside protected area networks (Maxted *et al.*, 2008b).

The EU project 'An Integrated European In Situ Management Work Plan: Implementing Genetic Reserves and On Farm Concepts' (AEGRO, AGRI GENRES 057) focuses on the development and testing of *in situ* conservation work plans for CWR. One of the deliverables of the AEGRO project is the formulation of quality standards for GR conservation of CWR. The present chapter proposes a set of quality standards for the conservation of CWR taxa in GRs that include criteria for the establishment of GRs and management considerations to optimize GR efficacy.

The present standards have been developed taking into account the perspective of the manager of the GR. However, they may need to be complemented by the additional requirements that might be derived from the implementation of national, regional or global approaches to the *in situ* conservation of CWR (Maxted and Kell, 2009).

10.2 Rationale for a Set of Quality Standards for Genetic Reserves

The quality standards for genetic reserves are a tool for practitioners involved in the design of *in situ* conservation strategies for CWR, but also for protected area managers interested in the *in situ* conservation of CWR. They may be implemented in the last stage of the process of selecting the locations of a network of GRs (national, regional and global approaches) when multiple alternatives exist according to primary selection criteria (such as genetic diversity complementarity in single-species gene pool approaches or species richness in multi-species gene pool approaches).

As the establishment of GRs involves the implementation of active management including at some point, among other possible actions, demographic and genetic monitoring of the populations, the generation of a set of quality standards for GR conservation of CWR aims to ensure that conservation efforts are carried out following the most logical and efficient procedures that positively contribute to achieving the objectives. The conservation resources invested in the establishment of the GRs are then more likely to have long-term sustainability. The adoption of standards of good practices relies on the hypothesis that the projects or programmes that are executed this way achieve more rigour in the process of decision making, and more efficiency in the use of resources and in pursuing the objective of conservation (CMP, 2007; O'Neill, 2007).

In the formulation of standards we have set two levels of quality standards; namely, 'minimum' and 'optimal' quality standards. 'Minimum' quality standards assemble the baseline traits that would be required for any GR to operate and fulfil its conservation objectives, whereas 'optimal' quality standards include a more stringent set of requirements to be met by GRs in the longer term.

10.3 Objectives

The objectives that are pursued with the development of a set of quality standards for GR conservation of CWR are as follows.

1. To increase the efficiency of GR conservation of CWR, by optimizing the viability of the target populations, their genetic diversity, and environmental conditions that preserve their evolutionary potential.
2. To provide a tool for conservation managers to plan and execute conservation actions on CWR according to protocols that are considered 'good practices'.
3. To provide a learning tool for managers and conservationists where results can be shared and good practices improved.
4. To generate greater credibility in the institutions in charge of executing conservation and management actions.
5. To make the process of decision making more transparent.
6. To adopt the measures considered most appropriate by the CWR conservation community.
7. To facilitate the process of evaluating management efficacy.
8. To facilitate comparability and complementarity between sites managed to conserve CWR diversity.
9. To improve communications to society of the relevance of the efforts toward CWR conservation.

10.4 Process of Standard Elaboration

The present document is the result of a process of analysis and participation of researchers and practitioners of the CWR conservation community, led by the partners of the AEGRO project and involving the following stages.

1. Identification of the parameters and constraints that should be taken into account in GRs and in protected areas containing GRs in order to provide adequate *in situ* conservation of CWR.
2. Review of relevant literature concerning the establishment of quality standards in biodiversity conservation.
3. Preparation of a draft document.
4. Discussion of the draft document by the AEGRO project group.
5. Dissemination of the draft document to the CWR conservation community to obtain additional feedback.

10.5 Agreed Quality Standards for Genetic Reserve Conservation of CWR

The present quality standards are structured in three sections. The first section shows a set of quality standards that are meant to be applied in all GRs, whereas the two subsequent sections provide additional considerations to take into account when the GRs are established within the limits of already existing protected areas or as stand-alone GRs created in locations that had no previous protection.

10.5.1 Quality standards for genetic reserves

Location

MINIMUM STANDARDS

1. Located at sites that have been identified through a rigorous scientific process involving prioritization and careful site selection.
2. Located in a protected area network according to European, national or regional environmental law (e.g. Natura 2000 network or National Parks network).

OPTIMAL STANDARDS

1. Established in either natural or semi-natural environments.
2. Unless otherwise indicated by the specialists of the target CWR, isolated from exotic gene flow from crops related to target CWR and from exotic gene flow from congeneric wild plant species that could interbreed with target CWR. Threshold distances between target CWR populations and crops or other congeneric species should be determined depending on the pollination and seed dispersal strategies of each target CWR.
3. Easy to get to for monitoring and research purposes.

Spatial structure

MINIMUM STANDARDS

1. The polygon of the GR should be clearly defined and geo-referenced.
2. Sufficient extent to conserve the populations of the target taxon and its natural habitat and to maintain natural processes.

OPTIMAL STANDARDS

1. GRs consist of core and buffer zones and corridors, if necessary. Core zones are the central areas with stable habitat for the conservation of the target taxon. Buffer zones surround core zones and may be used for setting up *in situ* research experiments and for the establishment of infrastructure complementary to *in situ* conservation (e.g. glasshouse, seed bank or laboratory).

Target taxa

MINIMUM STANDARDS

1. The limits of the GR within the protected area are traced, taking into account as primary consideration the needs of target CWR populations.
2. GRs are designed to capture as much genetic diversity of each target taxon as possible, conserving at least the alleles that are common, widespread and localized *sensu* Marshal and Brown (1975).
3. A full survey that provides a demographic characterization of target CWR taxa has been made.

OPTIMAL STANDARDS

1. A full survey that provides a demographic characterization of other CWR taxa of conservation interest has been made.
2. GRs also maximize non-target species richness of CWR within its borders, thereby providing umbrella protection to other CWR.
3. A genetic characterization of the populations of the target CWR taxa has been made to provide a better understanding of the genetic structure within the GR and its contribution to the overall species genetic diversity.
4. Genetic diversity of the target CWR is complementary to that existing in other GRs where the same target taxon is conserved, rather than redundant.

Populations

MINIMUM STANDARDS

1. Population sizes are large enough to sustain long-term population viability and maintain evolutionary potential. Therefore, population sizes are larger than the demographic and genetic minimum viable population (MVP) estimates available for the target taxon.

OPTIMAL STANDARDS

1. GRs contain a sufficient number of target taxon populations or subpopulations to minimize extinction risk due to natural disasters such as drought or flooding.
2. Population age structure is balanced and significant recruitment takes place to avoid population bottlenecks that would decrease effective population size.
3. Necessary associated species are also present in sufficient numbers to underpin GR sustainability (e.g. pollinators, seed dispersers, keystone species, etc.).

Management

MINIMUM STANDARDS

1. The GR is recognized as such by the appropriate national environmental and/or agriculture agency.
2. Clearly defined, detailed, achievable and evaluable conservation objectives, including preservation of evolutionary potential of target species, are formulated.
3. It is verified that there are no contradictions or incompatibilities between the objectives of the GR and the objectives of management plans of the protected area of superior rank. If such conflicts arise then there are appropriate, pragmatic means of mitigating the contradictions or incompatibilities.
4. A management plan using participatory and evidence-based criteria is designed and implemented in which: (i) the target CWR taxa are clearly identified and located; (ii) a diagnostic of the conservation status of target CWR taxa based on demographic and genetic structure is made; (iii) problems, opportunities and trends are defined; (iv) social and ecological conditioning factors are identified; and (v) a detailed action plan that defines a baseline for time-series analysis is prepared and implemented.
5. Monitoring plans are designed and implemented. Demographic and genetic diversity indicators to evaluate the expected results of the actions are identified. A protocol to obtain the indicators is prepared and implemented at appropriate intervals to obtain a time-series analysis.
6. A framework for evaluating and reporting GR management effectiveness is established.
7. Necessary financial, technical and human resources are available to meet the start-up and ongoing costs to effectively implement management plans for target taxa or habitat restoration.
8. Institutional backing (local, national, regional) for this initiative is obtained through agreements, letters of support, etc.
9. Local social actors that have a direct or indirect relationship with the GR are identified. A strategy for involvement of the local community in GR conservation activities is formulated ensuring that both local people and other public stakeholders benefit from the establishment of the GR.
10. A clearly defined procedure to ensure and regulate the use of the genetic resources by breeders, researchers and other user

communities is defined. Accessibility to scientists for monitoring and collection of germplasm for research is explicitly stated.

OPTIMAL STANDARDS

1. A registry of actions and decisions is established to provide a systematic record of the decisions that are made, actions that are taken and results that are obtained. Periodic reports are written summarizing the development of the conservation plan and noting possible modifications. The establishment of the ISO 9000 family of standards for quality management systems would make this feedback process more rigorous.
2. Additional priority CWR taxa are identified and located. Monitoring plans are designed and implemented for additional priority CWR taxa present in the GR. Demographic and genetic structure of their populations using indicators defined for target taxa are measured.
3. Complementary *ex situ* conservation of the target taxa is effected through collaboration with a certified official public germplasm bank (ISO 9000 or alike) that operates within the territory.
4. A strategy of communication of results is prepared with previous identification of the target audience.
5. Sustainable utilization of conserved CWR is encouraged and links are encouraged between the GR and user community.
6. Clear institutional placement within formal network structures with other CWR GRs in the country, region or even globally.
7. The site is recognized at national or regional level as having particular CWR genetic diversity worthy of conservation.

10.5.2 Quality standards for the protected areas selected for the establishment of genetic reserves

Minimum standards

1. The protected area has a legal foundation that underpins long-term site stability with clearly defined conservation and protection objectives.
2. The governance of the protected area assumes a continuing commitment to the *in situ* conservation of target CWR taxa in the GR.
3. The management plan of the protected area acknowledges the existence of the GR and includes its maintenance among its objectives. Objectives and management of the GR are integrated with the mid- and long-term general management plans of the protected area.
4. An inventory of all CWR present in the protected area has been made.

Optimal standards

1. Clear institutional placement within the broader protected areas community networks in the country, region or even globally.

10.5.3 Stand-alone genetic reserve

In addition to the quality standards listed in section 10.5.1, GRs established outside existing protected areas should also meet minimum standards equivalent to those expressed in points 1 and 2 of section 10.5.2 for protected areas, i.e. an appropriate legal framework that warranties the long-term stability of the site and a continuing commitment to the *in situ* conservation of target CWR taxa in the GR.

10.6 Conclusions

In this chapter, we have proposed a set of quality standards for the *in situ* conservation of CWR taxa in GRs. The present standards may be a useful tool for protected area managers interested in conserving priority CWR in their protected areas (bottom-up approach), but also for all those involved in the last stage of the process of selecting locations for GRs when multiple alternatives exist according to primary selection criteria (top-down approach).

The CBD COP10, which identified targets for the new CBD Strategic Plan (2011–2020), specifically mentioned that the status of CWR should improve by 2020, and identified the establishment of specific protected areas for CWR as a means to achieve this conservation target (CBD, 2010). In this context, the CWR conservation community needs to further develop available methodologies for identifying and conserving CWR in protected areas. We believe that the establishment, application and subsequent revision of quality standards can be a helpful tool in this process. Thus, this set of quality standards is going to be tested on the candidate locations suggested for the establishment of GRs for priority species of *Avena*, *Beta*, *Brassica* and *Prunus* formulated in the AEGRO project, in order to obtain additional feedback.

References

CBD (2010) *Strategic Plan for Biodiversity 2011–2020*. Secretariat of the Convention on Biological Diversity, Montreal, Quebec, Canada.

CMP (2007) Open Standards for the practice of conservation, version 2.0. The Conservation Measures Partnership. Available at: www.conservationmeasures.org (accessed 2 April 2011).

Heywood, V. and Dulloo, E. (2006) *In situ* conservation of wild plant species, a critical global review of good practices. *IPGRI Technical Bulletin no 11*. IPGRI, Rome.

Heywood, V., Casas, A., Ford-Lloyd, B.V., Kell, S. and Maxted, N. (2007) Conservation and sustainable use of crop wild relatives. *Agriculture Ecosystems & Environment* 121, 245–255.

Jarvis, A., Lane, A. and Hijmans, R.J. (2008) The effect of climate change on crop wild relatives. *Agriculture Ecosystems & Environment* 126, 13–23.

Kell, S.P., Maxted, N. and Ford-Lloyd, B.V. (2003) Introducing PGR forum. *Crop Wild Relative* 1, 4–7.

Marshall, D.R. and Brown, A.H.D. (1975) Optimum sampling strategies in genetic conservation. In: Frankel, O.H. and Hawkes, J.D. (eds) *Crop Genetic Resources for Today and Tomorrow*. Cambridge University Press, Cambridge, pp. 53–80.

Maxted, N. and Kell, S. (2009) *Establishment of a Global Network for the* in situ *Conservation of Crop Wild Relatives: status and needs*. FAO Commission on Genetic Resources for Food and Agriculture, Rome.

Maxted, N., Ford-Lloyd, B.V. and Hawkes, J.G. (1997a) Complementary conservation strategies. In: Maxted, N., Ford-Lloyd, B.V. and Hawkes, J.G. (eds) *Plant Genetic Conservation: the* in situ *Approach*. Chapman & Hall, London, pp. 20–55.

Maxted, N., Hawkes, J.G., Ford-Lloyd, B.V. and Williams, J.T. (1997b) A Practical Model for *In Situ* Genetic Conservation. In: Maxted, N., Ford-Lloyd, B.V. and Hawkes, J.G. (eds) *Plant Genetic Conservation: the* in situ *Approach*. Chapman & Hall, London, pp. 545–592.

Maxted, N., Ford-Lloyd, B.V., Jury, S., Kell, S. and Scholten, M. (2006) Towards a definition of a crop wild relative. *Biodiversity and Conservation* 15, 2673–2685.

Maxted, N., Kell, S. and Ford-Lloyd, B. (2008a) Crop wild relative. Conservation and use: establishing the context. In: Maxted, N., Ford-Lloyd, B., Kell, S., Iriondo, J.M., Dulloo, E. and Turok, J. (eds) *Crop Wild Relative. Conservation and Use*. CAB International, Wallingford, UK, pp. 3–30.

Maxted, N., Iriondo, J.M., De Hond, L., Dulloo, E., Lefèvre, F., Asdal, A., Kell, S. and Guarino, L. (2008b) Genetic reserve management. In: Iriondo, J.M., Maxted, N. and Dulloo, E. (eds) *Conserving Plant Genetic Diversity in Protected Areas*. CAB International, Wallingford, UK, pp. 65–87.

Meilleur, B.A. and Hodgkin, T. (2004) *In situ* conservation of crop wild relatives: status and trends. *Biodiversity and Conservation* 13, 663–684.

O'Neill, E. (2007) Conservation audits: auditing the conservation process. Lessons Learned, 2003–2007. The Conservation Measures Partnership. Available at: www.conservationmeasures.org (accessed 2 April 2011).

11 Conservation of Biodiversity and Genetic Resources in Semi-Natural Grasslands in Norway

E. Svalheim and Å. Asdal

11.1 Introduction

The genetic resources of forage plants are considered to be among the most valuable indigenous genetic resources in Norway. A large diversity of genotypes of locally adapted plants has evolved over several hundreds of years of continuous farming under different climates, soil conditions and farming methods. At the same time, a large portion of these plants and genotypes are highly threatened, mainly due to the abandonment of farms and changes in farming methods.

Fodder plants such as grasses and clovers and ruminant domestic animals have, despite the cold climate and short summers, made it possible to farm and produce food in most parts of Norway. Farms have been established all over the country from south to north, from the coast to the interior as well as mountain farms above the treeline. As a result of different locations, climates, soil and farming methods a broad diversity of locally adapted genotypes has evolved over the centuries.

Unploughed and unfertilized grasslands used for grazing and hay mowing are considered to have a particularly rich biodiversity. A large number of plant species grow in such semi-natural fields and many of them depend on continuous low input farming for their survival. Important plant species that occur in such fields are timothy (*Phleum* sp.), fescues (*Festuca* sp.) and red and white clover (*Trifolium* sp.). About 30 other species of grasses and legumes occurring in unploughed low input fields have been identified for plant breeding purposes. In addition to forage plants a significant number of medicinal and aromatic plants have such fields as their main habitats. An even higher number of species can be defined as crop wild relatives (CWR). The fact that such fields also shelter red-listed plant species and rare insects and fungi increases their value further. It also strengthens the importance of cooperation between the agricultural and the environmental sectors.

Over the last 50 years, agriculture in Norway has, as in other European countries, experienced major changes. For economical reasons many farms in remote areas have been abandoned and the associated biodiversity found in these cultivated fields has been lost. Without management, the fields are quickly covered by trees and bushes. The alternative conversion to intensive cultivation with the use of fertilizers also threatens the semi-natural biodiversity.

For political reasons, several initiatives and projects have been carried out to safeguard the biodiversity of semi-natural grasslands in the cultural landscape.

The unfavourable trends have been known and caused concern for many years, and already during the 1990s the agricultural and environmental authorities carried out investigations and registered valuable and diversity rich fields of national (Category A), regional (Category B) and local (Category C) value.

At the overall political level, farmers, extension services and local authorities have been encouraged to maintain these fields using traditional methods. Governmental subsidies and grants have to some extent lowered the speed of this negative development. However, the unfavourable trends have continued and the results from research and mapping have been disappointing. Some reasons for the lack of results could be due to insufficient information and communication between the players, lack of knowledge about maintenance of traditional grasslands and actually also due to a fear of 'doing something wrong' in valuable fields.

11.2 'Our Semi-Natural Heritage' Project

In order to conserve biodiversity and especially the genetic resources in such semi-natural fields the project 'Our Semi-Natural Heritage' was established in 2006 by the Norwegian Institute for Agricultural and Environmental Research in cooperation with the Norwegian Genetic Resource Centre. The project aims at active and targeted management of the most valuable and species-rich localities of the cultural landscape. A brief presentation of the project is given in English at the website (Svalheim, 2006).

The project vision has been to ensure continued and traditional farming of a significant number of forage-producing fields in all parts of the country, covering a good cross-section of climatic and soil conditions and farming systems that can be found in Norway. The Category A and B fields from the survey of agricultural landscapes, carried out by the Norwegian Directorate for Nature Management, forms the practical basis for the work (Direktoratet for naturforvaltning, 2007). The next step was to clarify whether the owner and the farmer, where this is not the same person, have sufficient personal commitment and would, under certain conditions, be willing to ensure continued farming of such valuable fields. When these preconditions are met the next step has been to prepare a management plan for the field in question. Important background data for the plan is information on the farming history and methods. This includes the number and type of grazing animals, pasture periods and times for mowing and haymaking. Data about natural conditions have also been registered, including information about soils, climate, topography, present flora and habitat specifications (Bioforsk, 2010). The plan then concludes with guidelines on how the fields should be maintained in order to conserve biodiversity and genetic diversity in the field. This means often that current farming methods or the methods that were used earlier should be applied. The plan provides the farmer and owner of the field with a scientifically based description of the biological features of the field.

Long-term funding is essential for the conservation of localities, and the plans serve as a basis for applications for governmental grants for continued farming with traditional methods. It also serves as a legally binding document whereby the farmers receive governmental support for maintenance. Support for such traditional farming is sought from the general governmental grant schemes for promoting and conducting agricultural policy in Norway.

The plans also provide a baseline for monitoring how such fields change and develop. Monitoring the impacts of the implemented farming system has been carried out in some fields. Development and changes in the populations of certain plant species has been found to provide adequate status indicators (Bioforsk, 2010).

Good communications and cooperation with the regional agricultural and environmental regional authorities has been a definite prerequisite for the establishment of the project. Regional authorities hold the information about the farms and fields, and also the people involved, i.e. the owners, the farmers, the neighbours etc. They also administer national policies regionally and allocate government subsidies at the regional level.

Government subsidies are essential for the system to work and the project has facilitated access to such grants by adapting plans to overall policies and guidelines. Close cooperation is also essential with respect to regional and local planning of land use and also the provision of advice to farmers through the local extension services.

'Our Semi-Natural Heritage' was initially established in the two neighbouring counties of Aust-Agder and Vest-Agder in the far south of Norway. Good results have resulted in the extension of the project into other counties. At the end of 2010 management plans had been developed and implemented in 33 areas in five counties. The management plans state that the conserved fields are defined as *in situ* gene banks for forage grasses, legumes and CWR.

11.3 Action Plan for Hay Meadows

The concept was further strengthened in 2009 when the Norwegian Directorate for Nature Management launched the 'Action Plan for Hay Meadows'. This plan applies the experiences and methods from the 'Our Semi-Natural Heritage' project and establishes management plans, continued farming and monitoring at a national level.

Hay meadows used to be very common and widespread in Norway, but it is now estimated that the total area of scythed or mowed permanent grasslands in Norway is between only 500 and 2000 ha. The diversity and biology of the Norwegian hay meadows is not fully mapped and investigated, but it is quite clear that the botanical diversity of the cultivated meadow vegetation in Norway is unique in northern Europe (Direktoratet for naturforvaltning, 2009).

The Action Plan is a contribution to the Norwegian efforts in connection with the European agreements towards halting the loss of biodiversity. It aims at securing traditional farming in all nationally important (Category A) hay meadows and in most of the regionally important fields (Category B). In 2009 the national environmental database Naturbase (Direktoratet for naturforvaltning, 2007) had 241 Category A registered hay meadows and 612 in Category B. Already in 2010, 109 of these meadows were actively maintained using the 'Our Semi-Natural Heritage' model.

11.3.1 An example – Little Shoe Meadow

Little Shoe Meadow on the farm Eidsaa in Vest-Agder county illustrates several important aspects concerning the conservation of valuable extensively managed permanent grasslands (Fig. 11.1). The flora and fauna of the field has evolved through extensive hay mowing combined with late pasture by cattle until the 1940s. Since then, the field has been used for pasturing with different animals. The use of artificial fertilizer has always been limited (Svalheim, 2007). The meadow is especially known for its rich population of mountain arnica (*Arnica montana* L.). This species is a red-listed plant in Norway and is highly dependent on permanent grasslands. It is also a medicinal plant with significant importance and value. The population of *A. montana* in this field has been known by botanists since the 1960s. The meadow was also re-registered in 2003 and evaluated as a Category A locality with national importance. However, neither the landowner nor the farmer was aware of the plant and of other biodiversity values of the field before 2006 when information was provided through the project. When the persons involved were made aware of the situation, an atmosphere of cooperation was immediately established. A management plan has since been developed and implemented and the farmer has become eligible for grants from regional environmental schemes.

A monitoring programme has been established using *A. montana* as an indicator for measuring the effects of the farming system that has been implemented according to the management plan. The monitoring programme confirms so far that the recommended farming measures have resulted in an increase in the biodiversity values that the plan aims to conserve. Little Shoe Meadow is now considered as an *in situ* gene bank, especially for mountain arnica, but also for other species.

It has also been a very positive experience that as soon as information was made

Fig. 11.1. Monitoring of the population of *Arnica montana* L. is one of the applied indicators that are used in the Little Shoe Meadow. The owner Markus Haugland is participating in the counting of flowering stems in the field during the season. Photograph by Ellen Svalheim.

available and all involved were brought in contact with each other, it was quite easy to get overall commitment and compliance regarding the necessary steps to be taken. The Little Shoe Meadow has also gained national attention in the ministry, in the media and in local and regional forums.

11.3.2 Experience gained

The most important result from the project is that a widespread network of semi-natural fields maintained by traditional farming methods will constitute a national gene bank for *in situ* conservation of genetic resources, especially for forage crops. It is also noted that management plans made for one kind of field/habitat can serve as model plans towards similar objectives in other habitats. Some additional experiences can also be drawn. It is recognized that maintenance and conservation of semi-natural fields forms an arena for fruitful cooperation between the environmental and agricultural sectors. Further, the concept of conservation of biodiversity is more easily adopted by farmers when it also includes agricultural genetic resources (Svalheim, 2008).

References

Bioforsk (2010) Bioforsk home page. Available at: www.bioforsk.no/arvesolv (accessed October 2010).

Direktoratet for naturforvaltning (2007) *Kartlegging av naturtyper – verdisetting av biologisk mangfold. DN-håndbok 13, 2.* utgave 2006 (oppdatert 2007) and www.naturbase.no.

Direktoratet for naturforvaltning (2009) Handlingsplan for slåttemark [Action Plan for Hay meadows].

Svalheim, E. (2006) *Project 'Our semi-natural heritage' Bioforsk.* Available at: www.bioforsk.no/ikbViewer/page/prosjekt/tema?p_dimension_id=23259&p_menu_id=23270&p_sub_id=23260&p_dim2=23269 (accessed 2 March 2011).

Svalheim, E. (2007) Skjøtselsplan, Hauane og Lille skoenga, Eidså, Songdalen kommune, *Vest-Agder. Bioforsk rapport* 2 (215).

Svalheim, E. (2008) Biomangfoldet I kulturlandskapet – Arvesølvet på Agder. *Bioforsk rapport* 3 (47).

12 Effective Conservation of *Medicago* Crop Wild Relatives in Russia and Neighbouring Countries: Gap Analysis Points the Way Forward

S.L. Greene, A.A. Afonin, E. Dzyubenko and N. Dzyubenko

12.1 Introduction

The concept of gap analysis was introduced in the late 1980s as the conservation community explored more systematic, efficient strategies to conserve biodiversity than the traditional strategy that focused on the reactive rescue of endangered species (Scott *et al.*, 1993). Burley (1988) was among the first to describe the importance of identifying 'conservation gaps' by determining elements of biodiversity (e.g. habitats, species) not represented in existing protected areas. Knowledge of conservation gaps could then be used to guide further conservation efforts. Efficiency is gained by focusing on the conservation of habitats and their complement of species, rather than a single endangered species (Scott *et al.*, 1993). Conservation activities are less expensive, more effective and less socially contentious if species are protected before they become endangered (Jennings, 2000). The concept and benefits of gap analysis are relevant not only to wild species, but also economically important plant species and their close wild relatives. A major difference when applying gap analysis to conserve plant genetic resources (PGR) is that gaps are identified not only in protected areas, but also in *ex situ* germplasm collections. Recognizing the utility of gap analysis, the approach has been used to support PGR conservation activities in major and minor crop species and their wild relatives (e.g. Jarvis *et al.*, 2003; Maxted *et al.*, 2005, 2008b, 2011; Maxted and Kell, 2009; Upadhyaya *et al.*, 2009; Ramirez *et al.*, 2010; Ramirez-Villegas *et al.*, 2010). Methods of gap analysis for PGR are quite varied, depending on initial objectives, the availability of data, and the types of spatial analysis carried out. Maxted *et al.* (2008a) described a general approach.

Although gap analysis has been used in many food crops, the approach has been limited with forage crops. The objectives of this study were to focus on cultivated *Medicago* species and their crop wild relatives (CWR) that occur throughout the region of the Former Soviet Union (FSU). The area now includes the Russian Federation, the countries of Estonia, Latvia, Belarus and Ukraine in eastern Europe, Georgia, Armenia and Azerbaijan in the Caucasus region, and Uzbekistan, Turkmenistan, Tajikistan, Kyrgyzstan and Kazakhstan in south-central Asia. Our objectives were to carry out a gap analysis to: (i) determine if these taxa are adequately represented in *ex situ* germplasm collections; (ii) assess the extent to which these species are found in protected areas; and (iii) identify protected areas rich in species diversity that warrant further investigation to establish *in situ* reserves.

12.2 Cultivated *Medicago* Species and their CWR in the FSU region

The genus *Medicago* has over 60 species, and makes a significant contribution to the world's food supply by providing nutritious forage for livestock. As a legume, *Medicago* species have the added benefit of fixing atmospheric nitrogen and are widely used to improve soils. The most economically important crop is *Medicago sativa* L., a perennial tetraploid, outcrossing species commonly called lucerne or alfalfa. Sixteen annual *Medicago* species are cultivated (Wiersema and León, 1999). The genus is native to the Mediterranean basin and western Asia (Heyn, 1963). Primary centres of diversity include the Caucasus region, north-western Turkey and north-eastern Iran (Quiros and Bauchan, 1988).

Central Asia is considered an important centre of diversity for *M. sativa* (Ivanov, 1988). The infra-specific taxonomy of *M. sativa* is quite complex. Hybridization due to sympatric distribution, as well as the presence of unreduced gametes, lead to infrequent but successful crossing between diploid and tetraploid forms. Domestication has also blurred the line between cultivated and wild forms. Although the taxonomy of perennial *Medicago* species has been studied and reviewed by many scientists, this study uses the taxonomic classification of Small (2011), which is based on ploidy, hybridization, flower colour, fruit coiling and presence of glandular hairs on fruits. In this study we recognized 13 taxa of lucerne CWR in our study area (Table 12.1). Within the *M. sativa* complex, individuals within the same taxa can fall into the primary or tertiary gene pool, depending on ploidy level. CWRs in the primary lucerne gene pool include the tetraploid forms, all of which occurred in our study area. CWRs in

Table 12.1. Number of *ex situ* accessions of lucerne CWR taxa collected in the Russian Federation and surrounding NIS countries.

Crop wild relatives of lucerne (*Medicago sativa*)	Gene pool	Ecogeographic zone: Crimea	Caucasus	Mountain Central Asia	Steppe Belt	Eastern Siberia	Total
M. cancellata†	3	–	7	–	–	–	7
M. daghestanica	3	–	4	–	–	–	4
M. marina††	3	0	0	–	–	–	0
M. papillosa	2	–	3	–	–	–	3
M. rupestris††	2	0	–	–	–	–	0
M. sativa ssp. *varia*	1 or 3‡	11	41	14	242	0	308
M. sativa ssp. *caerulea*	3	–	53	–	48	–	101
M. sativa ssp. *falcata*	1 or 3‡	7	66	28	468	16	585
M. sativa ssp. *falcata* var. *viscosa*	1 or 3‡	8	–	–	–	–	8
M. sativa ssp. *glomerata*	1 or 3‡	1	43	–	1	–	45
M. sativa ssp. *sativa*	1	0	272	167	226	3	668
M. sativa ssp. *sativa* × *M. sativa* ssp. *glomerata*	1 or 3‡	1	20	–	–	–	21
M. saxatilis††	2	1	–	–	–	–	1
Total		29	509	209	985	19	1751

†Rare species (Red Book of the Russian Federation, 1988)
††Rare species (Red Book of Ukraine, 2009)
‡Tetraploid forms belong in gene pool 1, diploid forms belong in gene pool 3

the secondary gene pool include *M. papillosa* and *M. saxatilis*, which also occurred in our study area. CWRs in the tertiary pool include diploid forms within the *M. sativa* complex, including *M. sativa* ssp. *caerulea*. Since few accessions in *ex situ* collections have chromosome counts, we did not distinguish between diploid and tetraploid forms. Also in the tertiary gene pool are *M. cancellata*, *M. daghestanica*, *M. marina* and *M. rupestris*. *Medicago rupestris* occurs in a very restricted area in the Crimean mountains. It is interesting to note that of the five taxa that are not in the *M. sativa* complex, four were endemic to our study area (*M. cancellata*, *M. daghestanica*, *M. rupestris* and *M. saxatilis*) and four were listed as rare or vulnerable: *M. cancellata* (Red Book of the Russian Federation, 1988), *M. marina*, *M. rupestris* and *M. saxatilis* (Red Book of Ukraine, 2009).

The Mediterranean basin is the centre of diversity for the cultivated annual medics, however, the range of several important species extends northward into our study area. These species include *M. arabica*, *M. littoralis*, *M. lupulina*, *M. minima*, *M. orbicularis*, *M. polymorpha*, *M. rigidula*, *M. scutellata* and *M. truncatula*. With the exception of *M. littoralis* and *M. truncatula*, which readily cross with each other, CWRs are limited to the species themselves, and include forms that are cultivated, wild, and their hybrids.

12.3 Assessment of *ex situ* Conservation

12.3.1 Method used

Datasets listing *Medicago ex situ* accessions were obtained from VIR, the USDA National Plant Germplasm System (NPGS), the International Center for Agricultural Research in the Dryland Areas (ICARDA) and the South Australia Research and Development Institute (SARDI). The EURISCO (http://eurisco.ecpgr.org/static/index.html/) and GBIF (www.gbif.org) databases were also queried to identify *ex situ* accessions. Attempts were made to obtain National Inventory Databases of PGR collections identified by the Central Asia and Caucasus (CAC) Region (PGR) Inventory (Akimaliev, 2008). Data from the Azerbaijan gene bank were obtained from the EURISCO database. A database was assembled incorporating all accessions that had been collected in the wild and had latitude and longitude data or locality data sufficient to geo-reference. Accessions lacking taxonomic designation below the species level were not used. Taxonomic nomenclature among collections was standardized using Small's recent revision (Small, 2011). A single accession was retained when accessions were maintained by multiple institutes. The final dataset contained 2434 accessions. The dataset was mapped and compared with distribution maps of each of the taxa.

Species distribution maps were taken from the online *Interactive Agricultural Ecological Atlas of Russia and Neighboring Countries* (AgroAtlas) (Afonin *et al.*, 2009). Distribution maps were based on historic literature, occurrence data from herbarium specimens and germplasm collection sites, and recent field surveys carried out by forage scientists at VIR. Biological descriptions, photos and map metadata of each individual species can be found in the AgroAtlas (i.e. Dzyubenko and Dzyubenko, 2009). To summarize the *ex situ* representation of species across their distribution range, the study area was subdivided into five ecogeographic zones: Crimea, Caucasus, Steppe Belt, Eastern Siberia and Mountain Central Asia. *Ex situ* representation within the zones where the species occurred was tabulated. To obtain a more detailed understanding of where accessions had been collected and where gaps occur, we made species and observation richness maps of lucerne and annual medic CWRs using the *ex situ* collection dataset. DIVA-GIS version 7.3.0.1 (www.diva-gis.org) was used. The *ex situ* species richness maps were compared with species richness maps based upon distribution maps of lucerne and annual medic CWRs downloaded from the AgroAtlas. The species distribution maps from the AgroAtlas were converted into raster layers using IDRISI 2.3 (Clark University, USA). Pixels were assigned 1 if the species occurred and 0 if absent. The layers were summed to form the species richness map.

12.3.2 Results of *ex situ* gap analysis

The gap analysis suggested that *ex situ* representation of CWR species of lucerne was limited in the Crimea, Mountain Central Asia (with the exception of *M. sativa* ssp. *sativa*) and Eastern Siberia (Table 12.1). *Ex situ* representation of cultivated annual medic species reflected the Mediterranean origins of these species. Fewer species occurred in the more northern reaches of our study area and fewer accessions were represented in *ex situ* collections (Table 12.2).

Although the Crimea peninsula covers a relatively small area, it is rich in *Medicago* species diversity. Except for *M. sativa* ssp. *caerulea*, all CWR in the *M. sativa* complex occur there, yet were represented by only 29 accessions (Table 12.1). *Medicago rupestris* and *M. saxatilis* are in the secondary lucerne gene pool. Although they are endangered, there was only a single accession of *M. saxatilis* in our *ex situ* database. *Medicago marina*, a rare species in the tertiary gene pool, also occurs along the western coast of Crimea. Considering that the natural vegetation of the interior and coastal areas is rapidly disappearing due to agricultural and recreational development, further collecting of perennial *Medicago* species in the Crimea would be warranted. Most CWR species of annual medics are represented in the Crimea, except *M. truncatula* and *M. scutellata* (Table 12.2). A gap was identified in the interior of the peninsula (Fig. 12.1). A collection trip in 2008 focused on annual medic species in the Crimea, specifically *M. truncatula* (Greene, unpublished data, 2009). Although 120 accessions were collected, no samples of *M. scutellata* or *M. truncatula* were found. Review of herbarium specimens at the Nikitsky Botanical Garden Herbarium reflected the difficulties in identifying *M. truncatula* and its close associates. From our exploration, it is not clear whether *M. truncatula* ever occurred in Crimea, or has disappeared due to the extensive development of the Black Sea shoreline.

The Caucasus region had the richest occurrence of lucerne CWRs, and was also rich in annual medic CWRs. In our *ex situ* database 30% of the lucerne CWRs and 41% of annual medic CWRs were collected in the Caucasus area. Collecting trips in 2009 and 2010 in Armenia and Georgia, respectively, focused on finding *M. truncatula* (Greene, unpublished data 2010). Although the trips were successful, *M. truncatula* accessions

Table 12.2. Number of *ex situ* accessions of cultivated annual *Medicago* species collected in the Russian Federation and surrounding NIS countries.

Crop wild relatives of cultivated annual medic species	Gene pool	Ecogeographic zone: Crimea	Caucasus	Mountain Central Asia	Steppe Belt	Eastern Siberia	Total
M. arabica	1	14	26	–	–	–	40
M. littoralis	1	–	23	0	–	–	23
M. lupulina	1	8	40	10	96	0	154
M. minima	1	29	49	71	5	–	154
M. orbicularis	1	33	58	30	–	–	121
M. polymorpha	1	13	52	22	–	–	87
M. rigidula	1	25	33	40	–	–	98
M. scutellata	1	0	–	2	3	–	5
M. truncatula	1	–	–	1	–	–	1
Total		122	281	176	104	0	683

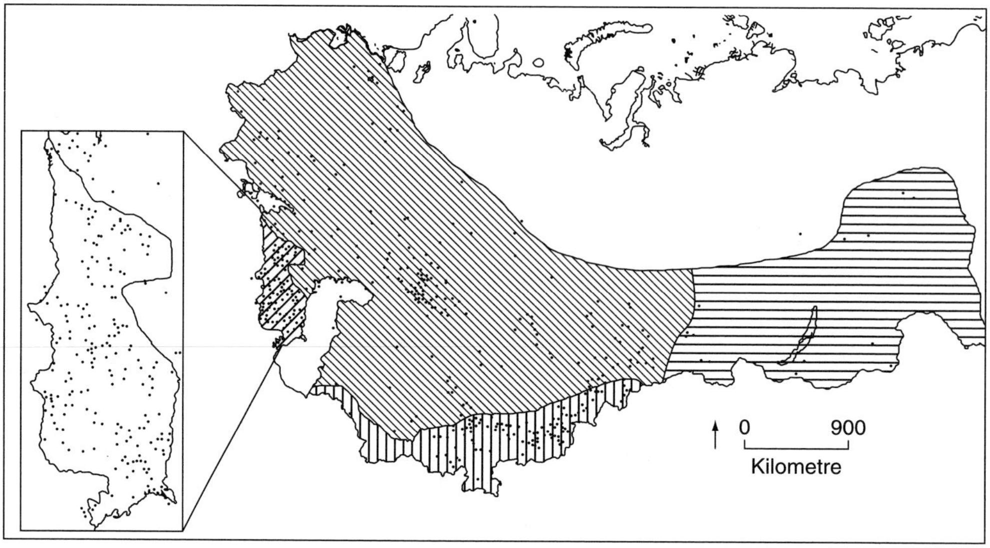

Fig. 12.1. Distribution of *ex situ* accessions of lucerne crop wild relatives. Ecogeographic zones are represented as follows: horizontal stripe, Eastern Siberia; vertical stripe, Mountain Central Asia; forward diagonal stripe, Steppe Belt; backward diagonal stripe, Caucasus.

were later re-identified as *M. rigidula* and *M. rigiduloides* (E. Small, Ottawa, 2010, personal communication). Further research is needed to determine if *M. truncatula* naturally occurs in the Caucasus. For both perennial and annual *Medicago* CWRs, there are still areas that have not been collected, such as northern Dagestan, Chechnya, Ingush, North Ossetia and Karachay-Cherkessia, probably due to political unrest. Areas that are stable and have limited representation include southern Azerbaijan, northern Georgia and southern Krasnodar.

In the Crimea and Caucasus areas, lucerne CWRs with glandular hair occur (*M. sativa* ssp. *glomerata*, hybrid forms between *M. sativa* ssp. *sativa* and *M. sativa* ssp. *glomerata*, and *M. sativa* ssp. *falcata* var. *viscosa*). This trait is considered an adaptation that conveys insect resistance (Small, 2011). In the USA, CWR introductions with glandular hairs have given rise to proprietary lucerne varieties that are resistant to potato leaf hopper (*Empoasca fabae*), a serious pest in the eastern USA (Shockley *et al.*, 2002). Because resistance is conveyed only in certain variants of the glandular hair trait, focus on collecting diversity within these taxa would be valuable given the proven economic value of utilizing this trait.

The Mountain Central Asia zone encompasses the Kopet Dag Mountains in the west and the Tian Shan Mountain system in the east. Species richness is limited in lucerne CWRs, however, the area is rich in annual medic diversity. Although international collecting trips during the 1990s were successful in obtaining wild forms of lucerne from the Tian Shan region, an *ex situ* gap occurs in the Kopet Dag Mountains (Fig. 12.1). There is also limited representation of annual medics from this region (Fig. 12.2). Examination of several herbarium specimens at VIR strongly supports the presence of *M. truncatula* in this area. A collecting trip to this region would be valuable to extend the range of *M. truncatula* diversity available to the genomics community. The Steppe Belt covered the largest area in our study. Over half of the lucerne CWR accessions in our database were collected from this zone, although gaps were identified in Estonia, Latvia and Belarus, and along the northern boundary of the Steppe Belt (Fig. 12.1). Eastern Siberia was the least represented in our

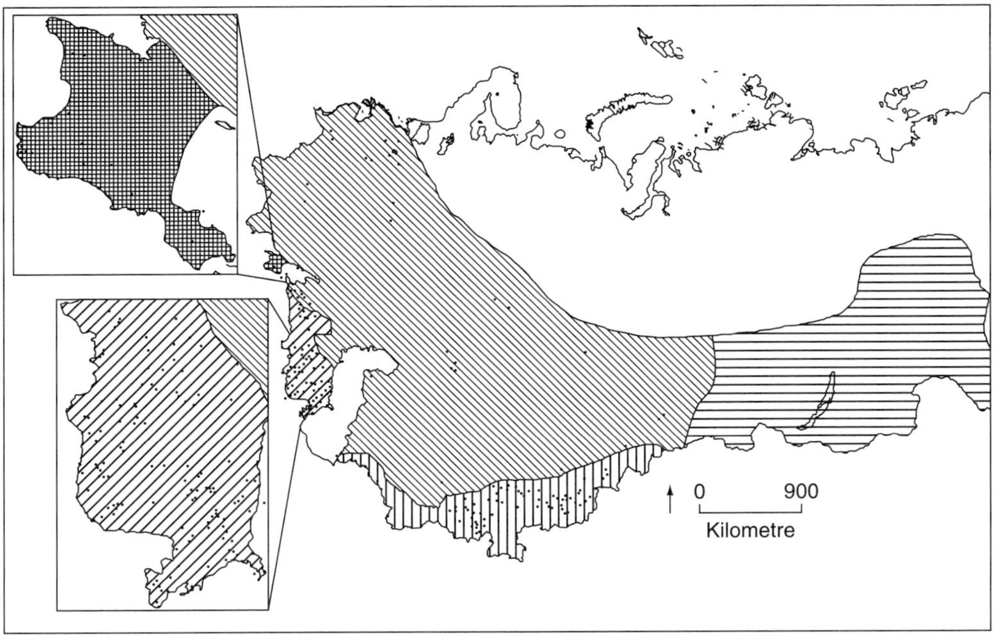

Fig. 12.2. Distribution of *ex situ* accessions of annual medic crop wild relatives. Ecogeographic zones are represented as follows: horizontal stripe, Eastern Siberia; vertical stripe, Mountain Central Asia; forward diagonal stripe, Steppe Belt; backward diagonal stripe, Caucasus; cross hatch, Crimea.

ex situ database (Tables 12.1, 12.2). CWR taxa of ssp. *sativa* and ssp. *falcata* have been collected in oblasts as far north as Sakha, and as far east as Sakhalin, but only 19 accessions represent this zone in *ex situ* collections (Fig. 12.1). Considering the extreme climate of eastern Siberia and the long-term cultivation and breeding of lucerne in Sakha, further collecting in this zone would be warranted to exploit lucerne germplasm adapted to extreme winter conditions. Only *M. lupulina*, *M. scutellata* and *M. minima* grow as far north as the Steppe Belt, and only *M. lupulina* occurs in Eastern Siberia. However, very few samples representing germplasm adapted to northern conditions have been collected (Fig. 12.2). Obtaining annual medics from these areas would be valuable to develop cold-hardy germplasm. Cold adapted annual medics might support more sustainable farming practices such as ley farming and cover cropping in more northern climates.

12.4 *In Situ* Assessment

In 2009, about 5% of the area once considered the Soviet Union had some form of land protection (IUCN and UNEP, 2009). Identifying protected areas (PA) where *Medicago* CWRs naturally occur is a first step for cost-effective *in situ* conservation of these important PGR. Polygons for protected areas were downloaded from the World Database on Protected Areas (WDPA) (IUCN and UNEP, 2009). Only PA classified as zapovedniks were included. Information on *Medicago* species that occurred in PA was acquired from a number of resources. Plant lists for various PA were found (www.biodat.ru/db/vid/index.htm). Crimean data were obtained from individual PAs (Marina Rudenko, Crimea, Ukraine, 2010, personal communication) and information on protected areas in middle Asia and Kazakhstan were obtained(http://iucnca.net/inforeserve1108/). Where species data were not available, *Medicago* distribution was compared to

PA distribution to estimate the number of targeted taxa that might occur in PA. Information was compiled in a spreadsheet that indicated presence/absence of each taxon in 109 PA. A map was developed that showed the location of the reviewed PA, and the percentage representation of *Medicago* species recorded by preserve managers or in literature, relative to the number of *Medicago* species that naturally occur in the area (Fig. 12.3). We determined that seven areas could capture the diversity and ecogeographic distribution of CWR species: the Crimea Peninsula, the north Caucasus region, the south Caucasus region, the Kopet-Dag region, western Kazakhstan, Altay/south-east Kazakhstan and the Yukutia region. Table 12.3 lists significant protected areas in these regions. Based on PA inventories, the majority of CWR species could be found in one or more of the following reserves: Dagestansky Zapovednik, Kavkazskiy Mab Biosphere Reserve, Orenburgsky Zapovednik, Krymskyi (Crimean) Nature Zapovednik, Karadagskiy Nature Zapovednik and Sunt Khasardag State Nature Reserve. None of the 109 reserves reviewed contained *M. marina*, *M. daghestanica*, *M. saxatilis*, *M. scutellata* or *M. rupestris*.

12.5 Conclusion

Gap analysis is an effective way to review and refine conservation strategies for CWRs. In this study we found that gaps exist in current *ex situ* efforts to conserve the CWRs of both lucerne and the cultivated annual medic species that occur in the area of the FSU. This region of the world is a centre of diversity for lucerne CWRs, and represents the northern boundary of many annual medic species. The gap analysis suggested *ex situ* collections have limited representation of perennial species in Crimea and eastern Siberia. The Caucasus region was well represented, but areas with little representation were identified, mainly areas with political unrest. In the Mountain Central Asia zone, representation was limited in Turkmenistan, an area where we can expect to find the easternmost distribution of *M. truncatula*, an important model species for genomic research, and currently unrepresented in *ex situ* collections. Surprisingly, there was a

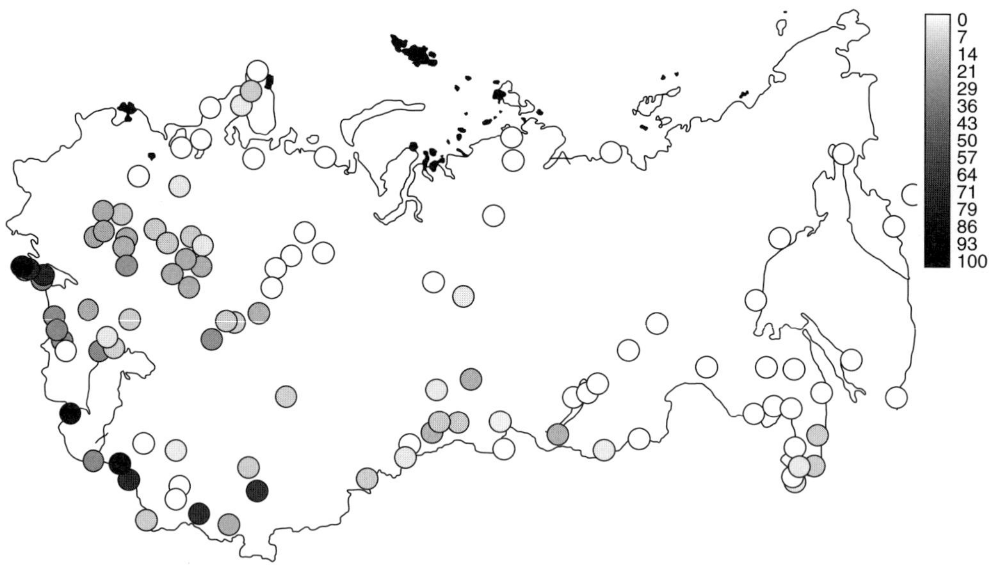

Fig. 12.3. Circles indicate protected areas that are at the reserve level of zapovedniks. Colour of circle corresponds to the percentage of targeted *Medicago* species listed in reserve flora inventories, relative to the number of species that naturally occur in the area.

Table 12.3. Established protected areas that have the potential of conserving important CWR species of *Medicago* (Data from www.protectedplanet.net).

Protected Area	Country	Latitude	Longitude	Size (km^2)	WDPA[†]	Target *in situ* area
Aksu Dzhabagly State Nature Reserve	Kazakhstan	42.3	70.6	750	1671	Altai/SW Kazakhstan
Borjomi Nature Reserve	Georgia	41.83	43.33	5300	1652	South Caucasus
Central Kolkheti Wetlands	Georgia	42.26	41.64	337	129931	South Caucasus
Dagestansky Zapovednik	Russia	44.57	46.77	190	15779	North Caucasus
Dilijan National Park	Armenia	41.71	44.84	280	1630	South Caucasus
Karadagskiy Nature Zapovednik	Ukraine	44.9	35.2	28	1754	Crimea
Karalars'kiy Regional Zakaznik	Ukraine	45.43	36.15	59	160988	Crimea
Kavkazskiy Mab Biosphere Reserve	Russia	43.86	40.26	2750	209778	North Caucasus
Kizil Agach State Nature Reserve	Azerbaijan	39.2	49	884	1983	South Caucasus
Kopetdag Preserve	Turkmenistan	37.86	58.05	497	167106	Kopet Dag
Krymskyi (Crimean) Nature Zapovednik	Ukraine	44.4	34	441	161100	Crimea
Lenskie Stolby Nature Park	Russia	61.77	129.72	4850	199988	Yukutia
Orenburgsky Zapovednik	Russia	51.22	56.71	216	68547	Western Kazakhstan
Ramit State Nature Reserve	Tadjikistan	38.8	69.3	161	1736	Altai/SW Kazakhstan
Russkiy Les Zakaznik	Russia	45.05	41.84	101	200842	North Caucasus
Sunt Khasardag State Nature Reserve	Turkmenistan	38.5	56.4	264	167124	Kopet Dag
Tamma Zakaznik	Russia	61.94	130.14	1772	19983	Yukutia
Turgayskiy State Nature Reserve	Kazakhstan	51.76	72.98	1830	17401	Western Kazakhstan

[†]WDPA – identifier number of protected area from World Protected Area Database

serious gap in *ex situ* collections for rare and/or vulnerable species, although they are in the secondary and tertiary gene pool of lucerne. The gap analysis helped us to identify seven geographic areas that, together, represent *Medicago* species and ecogeographic diversity of the study area. We identified a set of PA located in these areas, including six that reported among them the presence of the majority of our targeted *Medicago* species. This gap analysis has pointed out the next steps that need to be taken to ensure the *ex situ* and *in situ* conservation of important *Medicago* CWRs in the area of the FSU.

References

Afonin, A.N., Greene, S.L., Dzyubenko, N.I. and Frolov, A. (eds) (2009) Interactive Agricultural Ecological Atlas of Russia and Neighboring Countries: Economic Plants and their Diseases, Pests and Weeds – Online Database. Available at: www.agroatlas.ru (accessed 10 June 2010).

Akimaliev, J. (2008) Regional Strategy for the Conservation, Replenishment, and Use of Plant Genetic Resources for Food and Agriculture in Central Asia and the Caucasus for the period until 2015. Available at: www.croptrust.org/documents/regionalstrategies/centralasiaandcaucasus.pdf (accessed 5 November, 2010).

Burley, F.W. (1988) Monitoring biological diversity for setting priorities in conservation. In: Wilson, E.O. (ed.) *Biodiversity*. National Academy Press, Washington, DC, pp. 227–230.

Dzyubenko, N.I. and Dzyubenko, E.A. (2009) *Medicago daghestanica*. In: Afonin, A.N., Greene, S., Dzyubenko, N.I. and Frolov, A. (eds) Interactive Agricultural Ecological Atlas of Russia and Neighboring Countries. Economic Plants and their Diseases, Pests and Weeds – Online Database. Available at: www.agroatlas.ru/en/content/related/Medicago_daghestanica/ (accessed 23 December 2009).

Heyn, C.C. (1963) *The Annual Species of* Medicago. Magnes Press, Hebrew University, Jerusalem.

IUCN and UNEP (2009) The World Database on Protected Areas (WDPA). UNEP-WCMC, Cambridge. Available at: www.wdpa.org (accessed 2 January 2011).

Ivanov, A.I. (1988) *Alfalfa*. Amerind Publishing Company, New Delhi, India.

Jarvis, A., Ferguson, M., Williams, D., Guarino, L., Jones, P., Stalker, H., Valls, J., Pittman, R., Simpson, C. and Bramel, P. (2003) Biogeography of wild *Arachis*: assessing conservation status and setting future priorities. *Crop Science* 43, 1100–1108.

Jennings, M.D. (2000) Gap analysis: concepts, methods, and recent results. *Landscape Ecology* 15, 5–20.

Maxted, N., Mabuza-Dlamini, P., Moss, H., Padulosi, S., Jarvis, A. and Guarino, L. (2005) *An ecogeographic survey: African* Vigna. *Systematic and Ecogeographic Studies of Crop Genepools 10*. International Plant Genetic Resources Institute, Rome.

Maxted, N., Dulloo, E., Ford-Lloyd, B.V., Iriondo, J. and Jarvis, A. (2008a) Genetic gap analysis: a tool for more effective genetic conservation assessment. *Diversity and Distributions* 14, 1018–1030.

Maxted, N., White, K., Valkoun, J., Konopka, J. and Hargreaves, S. (2008b) Towards a conservation strategy for *Aegilops* species. *Plant Genetic Resource: Characterization and Utilization* 6, 126–141.

Maxted, N. and Kell, S.P. (2009) *Establishment of a global network for the in situ conservation of crop wild relatives: status and needs*. FAO Consultancy Report, FAO, Rome, pp. 1–265.

Maxted, N., Hargreaves, S., Kell, S.P., Amri, A., Street, K., Shehadeh, A., Piggin, J. and Konopka, J. (2011) Temperate forage and pulse legume genetic gap analysis. Bocconea (in press).

Quiros, C.F. and Bauchan, G.R. (1988) The Genus *Medicago* and the Origin of the *Medicago sativa* Complex. In: Hanson, A.A., Barnes, D.K. and Hill, R.R. (eds) *Alfalfa and Alfalfa Improvement*. Agronomy Monograph No. 29, American Society of Agronomy, Madison, Wisconsin, pp. 93–124.

Ramírez, J., Jarvis, A., Castañeda, N., van Etten, J. and Hijmans, R. (2010) Gap analysis of ex-situ collections of cropped species: landraces of major food crops and the Sorghum case study. Gap analysis of ex situ collections of cropped species, CGIAR System-wide Genetic Resources Programme Website. Available at: www.sgrp.cgiar.org (accessed 3 January 2011).

Ramírez-Villegas, J., Khoury, C., Jarvis, A., Debouck, D.G. and Guarino, L. (2010) A Gap Analysis Methodology for Collecting Crop Genepools: A Case Study with Phaseolus Beans. *PLoS ONE* 5(10): e13497. doi:10.1371/journal.pone.0013497.

Red Book of Ukraine (2009) Ukrainian Encyclopaedia, Kiev.

Red Data Book of the Russian Federation (1988) Online version of the Red Book of the Russian Federation, Volume 2, Plants. Available at: www.biodat.ru/db/rbp/ (accessed 2 September 2011).

Scott, J.M., Davis, F., Csuti B., Noss, R., Butterfield, B., Groves, C., Anderson, H., Caicco, S., D'Erchia, F., Edwards Jr, T., Ulliman, J. and Wright, G. (1993) Gap Analysis: A Geographic Approach to Protection of Biological Diversity. *Wildlife Monographs* 123, 3–41.

Shockley, F.W., Backus, E.A., Ellersieck, M.R., Johnson, D.W. and McCaslin, M. (2002) Glandular-haired alfalfa resistance to potato leafhopper (Homoptera: Cicadellidae) and hopperburn: development of resistance indices. *Journal of Economic Entomology* 95, 437–447.

Small, E. (2011) *Alfalfa and Relatives: Evolution and Classification of* Medicago. NRC Research Press/CAB International, UK.

Upadhyaya, H.D., Reddy, K.N., Irshad-Ahmed, M., Gowda, C.L.L. and Haussmann, B.I.G. (2009) Identification of geographical gaps in the pearl millet germplasm conserved at ICRISAT genebank from West and Central Africa. *Plant Genetic Resources: Characterization and Utilization* 8, 45–51. doi: 10.1017/S147926210999013X.

Wiersema, J. and León, B. (1999) *World Economic Plants: a standard reference*. CRC Press, Boca Raton, Florida.

13 Towards *In Situ* Conservation of Crop Wild Relatives in Lithuania

J. Labokas, B. Karpavičienė, L. Šveistytė, J. Radušienė and K. Ložienė

13.1 Legal Background

Two national laws make the legal background for the *in situ* conservation of plant genetic resources in Lithuania: the Law on National Plant Genetic Resources (LRS, 2001a) and the Law on Protected Areas (LRS, 2001b). The former provides that:

> for *in situ* conservation of national plant genetic resources, genetic reserves, gene conservation areas, seed collection stands shall be established or populations, groups or single trees shall be selected. *In situ* conservation of national plant genetic resources shall not be separated from the conservation of the natural habitats, maintenance of vital populations and regeneration through the creation of conditions favourable to their development.

The latter, the Law on Protected Areas, states that the genetic reserves are established 'for the protection of the populations of wild flora and fungi species having a genetic value' and 'The objective of establishment of genetic sites shall be to preserve resources of the genetic material required for activities. These sites shall be established in State-owned land for the preservation of seed forest stands and natural genetic resources of other species'. For the implementation of these statements some additional legal acts have been issued by the Ministry of Environment, including: (i) Regulations for Plant (Forest) Genetic Reserves; and (ii) Regulations for Seed Sites Attributed to the National Plant Genetic Resources. Based on these acts the plant (forest) genetic reserves and seed (genetic) sites are being established in different ecogeographic areas of the country.

13.2 Selection Criteria for *In Situ* Conservation of Crop Wild Relatives

Generally, crop wild relatives (CWR) are considered wild plant species, the progenitors or ancestors of the crops. As the term 'crop' is very broad in sense and can lead even to some misunderstanding, it is noteworthy to define the understanding of crop. In this chapter crop is considered any cultivated plant with existing cultivars and/or known cultivation at present or in the past, except for forest trees. This much narrowed definition, if compared to those of Harlan (1992) and Heywood (2008), also serves some pragmatic purposes, such as the prioritization of the target species. For the establishment of relationships between crops and their wild relatives we use the definition of a CWR proposed by Maxted *et al.* (2006), stating, *inter alia*, 'that this relationship is defined in terms of the CWR belonging to ... taxon groups 1 to 4 of the crop'. In brief, these are as follows: taxon group 1 (TG1) = taxa within the same species; TG2 = taxa within the same section or series; TG3 = taxa within the same subgenus; TG4 = taxa within the same genus.

Thus, the general criteria for genetic resource selection as stated in the Regulations for Seed Sites Attributed to the National Plant Genetic Resources include the evaluation of biological, ecological, genetic, social and economic value of the species. However, more concrete and even species-specific criteria are necessary to successfully define conservation objectives and implement conservation strategies. Developing specific selection criteria for the medicinal and aromatic plants (MAP) we use (i) species and (ii) population levels aiming to answer the following questions:

- What species could be considered as CWR of medicinal and aromatic plants in Lithuania?
- How many of them need conservation efforts?
- What priority levels are to be applied?

The answers to these questions may become clearer after the analysis of utilization of MAPs with emphasis on the dimensions of their cultivation in the country. This analysis provided the following groups of species:

1. Species utilized mainly through the cultivation of old and advanced cultivars, including locally released ones (peppermint, camomile, sage, hop, caraway, etc.).
2. Species which are not widely cultivated, but a few cultivars of which are known and used for cultivation (St John's wort, oregano, thyme, valerian, etc.).
3. Species being introduced into cultivation due to increased demand and other reasons.

One can observe that those three groups correspond principally to the different species domestication stages, from the oldest domesticated (group 1) to the most recently introduced and still undergoing introduction or domestication (group 3). Obviously, the species, which in historical times were first recognized as useful, were domesticated first. Undoubtedly, there are sufficiently large numbers of medicinal and aromatic plants that are not being introduced into cultivation at all. It also must be noted, that rare and endangered species deserve a special treatment in this context.

Selection criteria at population level include indicators of phenotypic and genetic diversity in the definition of target populations. In order to reduce costs of the selection, the phenotypic indicators were extensively used. A more detailed list of the criteria for the selection of medicinal and aromatic plant populations can be proposed, subdivided into two different groups: (i) intrinsic biological properties, such as the contents of bioactive substances, biological productivity, resistance to pests and diseases, winter hardiness, frost resistance, drought tolerance; and (ii) physical characteristics, such as size and distribution pattern of populations (subpopulations). This list only slightly differs from the generic ones by the indicator of bioactive substances. Nevertheless, each one of the criteria is applied in regard of the local conditions.

13.3 Implementation in the Field

The first fieldwork survey to select the seed (genetic) sites of medicinal and aromatic plants for *in situ* conservation was carried out in 2006. Due to practical considerations seed (genetic) sites were singled out based on the following criteria: number of target species and abundance of their resources, distinctive properties (phenotypic, genetic) of target species populations, position of sites with respect to ecogeographic conditions and clearly determined natural boundaries. The following workscheme was used:

1. Inventory of target species.
2. Description of plant community according to the Braun-Blanquet cover-abundance scale.
3. Establishment of coordinates and boundaries of the sites.

4. Calculation of site area, using GPS and web application (www.maps.lt).
5. Setting forth the motivation of a selection.
6. Mapping of a site.

A total of six seed (genetic) sites have been selected over the period of 2 years. All of them were established within existing protected areas: Regional Parks, with the exception of Leipalingis site. Total number of target species covered 33 species, total area covered 72.77 ha, mean size of a site was 12.12 ha with a size range of 0.4–30 ha. The sites were singled out based on a single prevailing species (marked ++ in Table 13.1), with the exception to the multispecies Dievenišk̇es site.

Among the 33 target species inventoried (Table 13.1), nine of them are wild relatives of well-known horticultural crops, namely: *Allium oleraceum*, *A. ursinum*, *Fragaria vesca*, *F. viridis*, *Malus sylvestris*, *Pyrus communis*, *Prunus spinosa*, *Ribes nigrum* and *Rubus caesius*. The remainder of the species can be classified into one of the three groups, based on the above mentioned grouping by cultivation:

1. CWR of species utilized mainly through cultivation: *Carum carvi*.
2. CWR of species not widely cultivated but having cultivars released: *Hypericum perforatum*, *Origanum vulgare*, *Thymus pulegioides*, *Valeriana officinalis*, *Vaccinium vitis-idaea*, *Sorbus aucuparia*, *Viburnum opulus*, *Corylus avellana* and *Arnica montana*.
3. CWR of species still being (potentially) introduced into cultivation: *Juniperus communis*, *Crataegus monogyna*, *Calluna vulgaris*, *Achillea millefolium*, *Primula veris*, *Rosa canina* and *Vaccinium uliginosum*. *Berberis vulgaris*, *Cornus sanguinea*, *Filipendula ulmaria*, *Gentiana cruciata*, *Lithospermum officinale*, *Potentilla erecta* and *Rhamnus cathartica* could also be considered within this group.

It must be noted that the above grouping applies with regard to the cultivation of the species as medicinal, aromatic or condiment (food) species. Meanwhile, the cultivation for the ornamental use of the same species would provide a different picture, as nearly all of them have ornamental cultivars released. In any case the above grouping is not strict. It only provides some tools for the prioritization of *in situ* conservation of medicinal and aromatic plants as well as related species. The group of the surveyed species which includes *Berberis vulgaris*, *Rhamnus cathartica*, *Potentilla erecta*, *Gentiana cruciata*, *Filipendula ulmaria* and *Cornus sanguinea* could be assigned to the lowest priority CWR or just wild collected species as their cultivation is not actually taking place. Rare species, such as *Arnica montana*, *Gentiana cruciata*, *Lithospermum officinale* and *Prunus spinosa* are to be treated with particular attention in order to safeguard their sparse resources, but this is beyond the objectives of the current paper.

The selection criteria targeted on population level were used for the justification of each site's selection. They were as follows: natural mutation (white lingonberry, Labanoras site) (Labokas *et al.*, 1997), within population genotypic diversity (common hawthorn, Dusia site) (Labokas, 2002), biological productivity (bear's garlic, Prienai site) (Karpavičienė, 2003), outstanding dimensions of population (hazelnut, Leipalingis site), optimal habitat of a rare species (blackthorn, Veliuona site).

So far the selection based on species occurrence has some priority due to the cheapest and simplest work. On the other hand, as the established sites already represent five vegetation classes, three types of habitats (forests, shrubs, meadows) and four geographic areas of Lithuania (South-Central, Southern, Eastern and South-Eastern) (Fig. 13.1) and the job is still ongoing in different parts of the country, a high probability exists that it covers the range of genetic diversity needed for the dynamic conservation of the target species. The only amendment required to the existing national legal acts is to define the crop wild relatives as such and therefore expand

Table 13.1. Target species inventoried in the six seed (genetic) sites.

#	Target species	1. Veliuona	2. Dusia	3. Labanoras	4. Leipalingis	5. Prienai	6. Dieveniškės
1	*Achillea millefolium* L.		+				
2	*Allium oleraceum* L.	+	+				
3	*Allium ursinum* L.					++	
4	*Arnica montana* L.						+
5	*Berberis vulgaris* L.	+					
6	*Calluna vulgaris* (L.) Hull						+
7	*Carum carvi* L.		+				
8	*Corylus avellana* L.				++	+	
9	*Cornus sanguinea* L.	+					
10	*Crataegus monogyna* Jacq.	+	++				
11	*Filipendula ulmaria* (L.) Maxim.					+	+
12	*Fragaria vesca* L.	+	+		+		
13	*Fragaria viridis* Weston	+	+				
14	*Gentiana cruciata* L.	+					
15	*Hypericum perforatum* L.		+				
16	*Juniperus communis* L.	+					+
17	*Lithospermum officinale* L.	+					
18	*Malus sylvestris* Mill.	+	+				
19	*Origanum vulgare* L.	+					
20	*Potentilla erecta* (L.) Raeusch						+
21	*Primula veris* L.	+	+				
22	*Prunus spinosa* L.	++					
23	*Pyrus communis* L.	+	+				
24	*Rhamnus cathartica* L.	+	+				
25	*Ribes nigrum* L.					+	
26	*Rosa canina* L.		+				
27	*Rubus caesius* L.	+	+				
28	*Sorbus aucuparia* L.					+	
29	*Thymus pulegioides* L.		+				
30	*Vaccinium uliginosum* L.						+
31	*Vaccinium vitis-idaea* L.			++			+
32	*Valeriana officinalis* L.						+
33	*Viburnum opulus* L.	+	+				
	Total number	17	15	1	2	5	9

++, prevailing (main) species.

Fig. 13.1. Established seed (genetic) sites (1–6) for medicinal and aromatic plant genetic resources conservation *in situ* and those pending confirmation (7–10). For the numbers of the established sites, see Table 13.1.

the list of target species correspondingly. Two priority tasks are needed to be implemented in the near future: creation of the national inventory of CWR; and development of management plans for the established seed (genetic) sites.

Acknowledgements

The work was carried out with the financial support of the Plant Gene Bank of the Ministry of Environment, Republic of Lithuania, contracts No. M-1 and No. M07-2.

References

Harlan, J.R. (1992) *Crops and Man*, 2nd edn. American Society of Agronomy and Crop Science Society of America, Madison, Wisconsin.

Heywood, V. (2008) Challenges of *in situ* conservation of crop wild relatives. *Turkish Journal of Botany* 32, 421–432.

Karpavičienė, B. (2003) Intensity of generative and vegetative reproduction of *Allium ursinum*. *Botanica Lithuanica* 9(1), 3–12.

Labokas, J. (2002) Properties of common hawthorn (*Crataegus monogyna* Jacq.) populations. *Biologija* 1, 59–61.

Labokas, J., Butkienė, Z., Butkus, V. and Stackevičienė, E. (1997) Distribution and intraspecific diversity of some berry plants in Lithuania. *Botanica Lithuanica* (Suppl.)1, 83–89.

LRS – Seimas of the Repubic of Lithuania (2001a) Republic of Lithuania Law on National Plant Genetic Resources. *Valstybės žinios* 90, 3144.

LRS – Seimas of the Republic of Lithuania (2001b) Republic of Lithuania Law on Protected Areas. *Valstybės žinios* 108, 3902.

Maxted, N., Ford-Lloyd, B.V., Jury, S.L., Kell, S.P. and Scholten, M.A. (2006) Towards a definition of a crop wild relative. *Biodiversity and Conservation* 15(8), 2673–2685.

14 *In Situ* Conservation of *Crocus cartwrightianus* in Cyclades and Crete

P. Ralli and C. Dordas

14.1 Introduction

Crocus cartwrightianus Herbert ($2n=2x=16$) is a rare endemic species and a wild relative of cultivated *Crocus sativus* L. ($2n=3x=24$) (Heywood and Zohary, 1995) found in southern Greece. The dried stigmas of *Crocus* have been used in the making of perfume, paint and medicine since prehistoric times as seen by wall paintings found on the island of Thera (Sarpaki, 2000) and Crete angiography and scripts (Figs 14.1 and 14.2). The centre of origin of the genus *Crocus* is not entirely certain, although Tammaro (1990) suggests that Asia Minor or the south-western Greek Aegean islands may be the probable area of origin. Results from a more recent study show that *C. sativus* was probably selected and domesticated in Crete during the Late Bronze Age (Negbi, 1999). Many authors suggest that *C. cartwrightianus* is the ancestor of *C. sativus* (Mathew, 1999; Grilli Caiola *et al.*, 2004), as flowering and other morphological traits in *C. cartwrightianus* show close similarities to *C. sativus*.

Crocus cartwrightianus is threatened with extinction in its natural environments. The decrease of populations is mostly considered to be a result of human activities (over-exploitation of its stigmas, which are used as a wild source of saffron, habitat destruction, etc.) and recent climatic and environmental changes. This results in the continuous reduction of its populations and loss of genetic diversity. Therefore, we studied the application of *in situ* conservation actions in the natural ecosystems of *C. cartwrightianus*. *In situ* conservation involves the location, designation, management and monitoring of target taxa in the location where they are found (Maxted *et al.*, 1997). This can be carried out in natural populations of wild species through the establishment of genetic reserves. Thus, genetic reserve conservation may be defined as the location, management and monitoring of genetic diversity in natural wild populations within defined areas designated for active, long-term conservation (Maxted *et al.*, 1997). This involves the conservation of large, significant parts of functioning ecosystems and the conservation of target species or groups of species. In this study we undertook some initial actions for the *in situ* conservation of *C. cartwrightianus* in the Cyclades islands and Crete. The main objective was to protect, manage and monitor selected populations in their natural habitats, so that natural evolutionary processes can be maintained, thus allowing new variation to be generated in the gene pool that will allow the species to adapt to gradual changes in environmental conditions (IPGRI, 2005).

Fig. 14.1. Fresco of saffron gatherers from the Bronze Age excavations in Akrotiri on the island of Thera.

Fig. 14.2. Red stigmas of *Crocus cartwrightianus*.

14.2 Materials and Methods

Crocus cartwrightianus is a perennial plant, adapted to overcome a dry dormant period in the form of an underground depressed-globose corm. It is an autumn-flowering species with purple, lilac or white perianth segments and three long bright red style branches (Mathew, 1999), the stigmas. It is vegetatively propagated with corms, which are the main reproductive system, although capsules with seeds are produced in autumn.

The study was conducted in the complex of the Cyclades islands (particularly Thera, Anafi and Christiana) and Crete to support the *in situ* conservation of *C. cartwrightianus*. Field surveys were carried out from 2007 to 2009 in Taksiarchis on the island of Thera. Populations of the target species were located and monitored. In 2009, additional areas were surveyed to locate more populations of *C. cartwrightianus* in Gavrilos on Thera, Anafi, the uninhabited island of Christiana and finally the area of Akrotiri in Chania, Crete. The experimental plan involved the location, designation and study of the distribution of this species in particular regions and ecosystems, and the estimation of the demographic parameters of the studied populations. Moreover, the actions included the collection of field observations (associated flora, topographic and climatic data), the morphological and phenotypic characterization and the management and monitoring of the target species.

Population size and density were estimated with simple random sampling techniques (Elzinga *et al.*, 1998) using square quadrats of 1 m^2. Population structure was assessed through the direct counting of flowers in fragments of populations. Flowering scapes were used as a proxy for individuals. Random samples including corms and other vegetative parts were collected from 10–60 plants according to the size of each population. They were then transferred to pots and stored in a field gene bank at the Greek Gene Bank (National Agricultural Research Foundation) in Thermi-Thessaloniki to complement *in situ* conservation with *ex situ* measures. A hierarchical clustering analysis procedure of morphological traits for the discrimination of populations was carried out using Euclidean distances and Ward algorithm. To determine the optimal number of clusters, the upper tail approach was used for $\alpha = 0.1$ (Wishart, 1987). Finally, to obtain information for the ecogeographic survey and create a suitable environment for *ex situ* conservation, soil samples were taken from each site where populations were found. The measured components included texture, pH, electrical conductivity, $CaCO_3$, organic matter and nutrients. All statistical data were processed using the JMP program (ver. 9.0.0), and location data were determined using a Geographic Information System (GIS) (ArcGIS 9.3).

14.3 Results and Discussion

The area of distribution of *C. cartwrightianus* is relatively small, centred in the Cyclades islands and reaching south Euboea, Attica, and western Crete in the Chania area (Mathew, 1999). During the exploratory expeditions, five populations of *C. cartwrightianus* were found and studied in Taksiarchis and Gavrilos on Thera, Anafi, the uninhabited island of Christiana and finally the area of Akrotiri in Chania, Crete. Most of the sites recommended by the local people were selected because they have *C. cartwrightianus* populations that were deemed to merit protection before substantial losses occurred, as proposed by Meilleur and Hodgkin (2004) for the conservation of crop wild relatives (CWR). The studied populations on Thera and Anafi are particularly threatened from over-exploitation as inhabitants collect whole plants instead of only their stigmas. At the same time, the population on Crete suffers from habitat destruction due to urban development and extensive building construction.

Mean morphological traits values showed differences between the five studied populations. Mean corm diameter varied from 1.70 to 2.43 cm, length and width of tepals varied from 2.40 to 2.90 cm and 0.50–0.97 cm, respectively, number and width of leaves varied from 6 to 11 and 1.0–1.5 mm, respectively, while the length of red stigmas varied from 1.60 to 2.17 cm. The cluster analysis obtained two different clusters of populations (Fig. 14.3). According to the dendrogram, the Christiana population is classified in a separate group from the other four populations. This arises from the fact that this population had a different perianth morphology (larger length/width ratio) as well as larger corm size than the other populations.

The *C. cartwrightianus* populations found in Taksiarchis, Gavrilos and Akrotiri have experienced a rapid decrease in size in recent

Fig. 14.3. Hierarchical clustering analysis of morphological traits of *Crocus cartwrightianus* populations (Ward's method, upper tail for $\alpha = 0.1$).

years according to inhabitants, whereas the population found in Anafi has suffered a slower reduction. Population density of *C. cartwrightianus* varied from 1 to 10 plants/m^2 in the populations found on Thera, while population density on Anafi varied from 5 to 16 plants/m^2. In the other areas, population density was lower than 4 plants/m^2 (Fig. 14.4).

The populations were found in open rocky hillsides and sometimes in open areas in scrub from 60 to 380 m. The soils in the above areas had a sandy loam (SL) texture and only the soil in Akrotiri had a sandy clay loam (SCL) texture. They were characterized by neutral or low alkaline reaction (pH 7.07–7.95) and high quantity of organic matter. Electrical conductivity was in the range of 0.428–1.027 mS/cm, which is suitable for many plant species. In addition, $CaCO_3$ content was 0.9–1.3% (also an acceptable range for many plant species), while organic matter content was much higher in Gavrilos than on Anafi Island. P and K concentration varied greatly among the different sites and especially in the case of K (100–1670 ppm) (Table 14.1). Thus, the soil analysis indicates that *C. cartwrightianus* can adapt to different soil conditions. This information is useful for finding a suitable ecological niche to create a suitable environment for *ex situ* conservation.

Christiana is the only site in the studied areas located within the Natura 2000 Network of protected areas. Because all the other sites where the target species was found are not protected by any special nature protection system, significant new local initiatives should be undertaken to protect a sufficient amount of the area in which they occur, so as to allow the representation of viable populations that cover a sufficient sample of the genetic variation. Otherwise, alternative means of protection, including community participation, easements or habitat conservation planning (IPGRI, 2005) or a representative number of legally defined, smaller areas for microreserves should be considered. As almost all areas where *C. cartwrightianus* populations were found are private property, it is important to find ways to collaborate with landowners to promote *in situ* conservation to ensure the viability of these populations and protect them from the dangers of competitive development. In these cases, it may also be important to find financial support from governmental or regional public services or funds from European projects to promote genetic reserve conservation through payments to landowners (a measure that until now is only taken for the conservation of landraces that suffer from erosion), training, environmental education, and protection of the areas through fencing. In this sense, a new project, funded by the Prefecture of Cyclades, began at the end of 2009. The objective of this project is the evaluation, protection and rational utilization of *C. cartwrightianus*, as well as other local products and landraces of Thera.

In situ conservation is a complex and demanding form of conservation. It is an appropriate method for *Crocus* species, which are reproduced by corms and cannot properly be conserved in seed gene banks. According to Iriondo *et al.* (2008), the essence of any *in situ* plant genetic conservation project is found in its objectives. In some particular cases where the original status of the populations being conserved

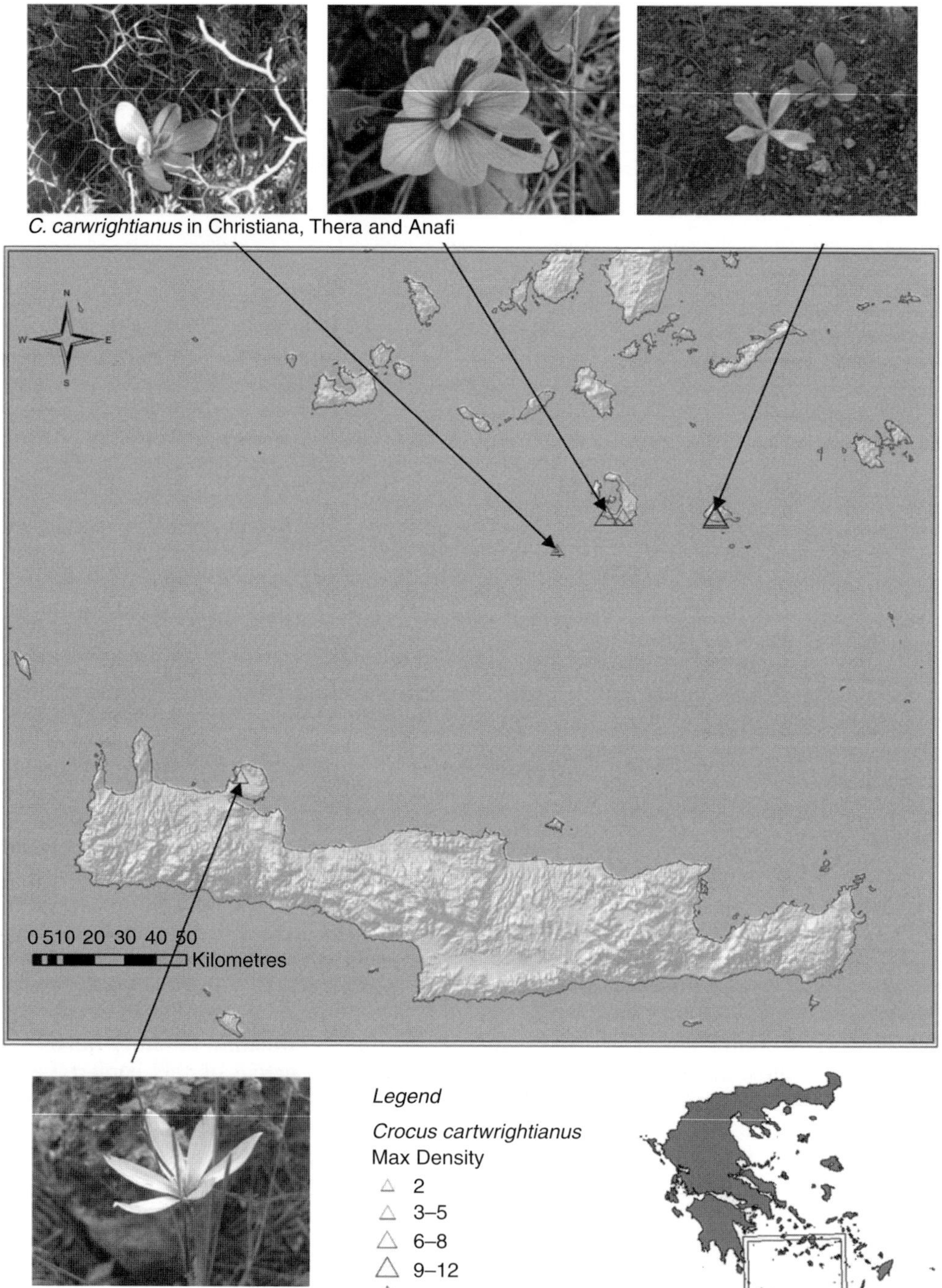

Fig. 14.4. Distribution and density of *Crocus cartwrightianus* populations found in the Cyclades islands and Crete.

Table 14.1. Analyses of soil samples from the different collection sites (from 0–30 cm depth).

	Mechanical texture	pH	EC (mS/cm)	Organic matter (%)	$CaCO_3$ (%)	P (ppm)	K (ppm)
Taksiarhis	SL	7.95	1.027	2.67	1.3	50.92	1670
Gavrilos	SL	7.85	0.589	4.54	1.3	39.96	170
Anafi	SL	7.07	0.469	1.00	1.3	23.26	100
Christiana	SL	7.55	1.002	1.47	0.9	20.08	420
Akrotiri	SCL	7.66	0.428	2.26	0.9	23.98	220

is not at its optimum, as in the case of *C. cartwrightianus*, or when the population has experienced a catastrophic event, the objectives may concentrate on achieving specific targets regarding population size, structure and genetic diversity. In this specific study the main objects were to determine the ecogeographic characteristics of *C. cartwrightianus* habitats, its morphological and phenotypic traits and demographic parameters to estimate its threat status and maintain initial levels of genetic diversity in the target populations. However, further studies are needed to understand its genetic structure and reproductive system. Characterization requires the study of genetic and chemical traits apart from morphological traits. With regard to the reproductive system, a study of sexually propagated offspring from individual plants and segregation patterns for basic morphological characteristics in relation to maternal plants should be carried out.

14.4 Conclusions

Crocus cartwrightianus shares many common characteristics with cultivated species and has similar uses. Thus, the materials that were collected are of great scientific interest and are a valuable genetic resource. However, there is a need further to study the commercial use of its products and the possibility of using its genes in the breeding of cultivated *C. sativus*.

Crocus cartwrightianus is currently under threat from habitat destruction, agricultural intensification, over-exploitation, urban development, climate change and lack of conservation attention. More emphasis should be given to promoting the genetic reserve conservation of populations of this target species in its natural habitats. In this sense, collaboration with local authorities and organizations will help. Apart from local interest, it is important to ensure the development of new laws for CWR and the development of agencies to manage and fund the conservation of CWR at the national level. Direct measures should be taken to promote *in situ* conservation, aiming to protect, manage and maintain rare or endangered wild species in their natural habitats.

The proposed plan for the monitoring and management of *C. cartwrightianus* populations that were found in the Greek islands involves:

- The detailed mapping of *C. cartwrightianus* populations in the studied areas, as well as in other areas of Greece.
- The regular monitoring and recording of demographic parameters of populations (number of populations, population size, spread of populations, etc.). Periodical observations and measurements of the ecological environment (soil, flora, climate and threats, etc.).
- Conservation actions in populations that are at risk in specific habitats (fencing or closing the site, funding landowners to protect the populations, creation of an information centre to promote public awareness of the importance of *C. cartwrightianus* and other CWR, environmental education, etc.).
- Complementing *in situ* conservation with *ex situ* measures by the sampling and safe maintenance of corms in field gene banks.
- Phenotypic characterization, agronomic evaluation and genetic study of identified populations.

- Development of a database with collection and passport data, field observations, population monitoring data, etc.

Until now, preliminary steps have been made in these directions, but there is still much progress to be made in the implementation of *C. cartwrightianus* conservation actions, the application of *in situ* techniques and the promotion of genetic reserve conservation of CWR populations in their natural habitats in general.

Acknowledgements

The authors would like to thank N. Stavropoulos (former director of the Greek Gene Bank) for training and advice. We are also grateful to S. Katsipis, K. Simoglou, K. Economakis and the personnel of the Union of Thera Products Corporations for facilitating our field survey, the personnel of the Soil Science Institute (NAGREF) for their help with soil analysis and T. Bouchounas for his help with GIS.

References

Elzinga, C.L., Salzer, D.W. and Willoughby, J.W. (1998) *Measuring and Monitoring Plant Populations*. Bureau of Land Management, Denver, Colorado.

Grilli Caiola, M., Caputo, P. and Zanier, R. (2004) RAPD analysis in *Crocus sativus* L. accessions and related *Crocus* species. *Biologia Plantarum* 48, 375–380.

Heywood, V.H. and Zohary, D. (1995) A catalogue of the wild relatives of cultivated plants native to Europe. *Flora Mediterranea* 5, 375–415.

IPGRI (2005) In situ *Conservation of Wild Plant Species, a Critical Global Review of Good Practices*. International Plant Genetic Resources Institute, Rome.

Iriondo, J.M., Ford-Lloyd, B., De Hond, L., Kell, S.P., Lefèvre, F., Korpelainen, H. and Lane, A. (2008) Plant population monitoring methodologies for the *in situ* genetic conservation of CWR. In: Iriondo, J.M., Maxted, N. and Dulloo, M.E. (eds) *Conserving Plant Genetic Diversity in Protected Areas*. CAB International, Wallingford, UK, pp. 88–123.

Mathew, B. (1999) Botany, taxonomy and cytology of *Crocus sativus* L. and its allies. In: Negbi, M. (ed.) *Saffron*. Harwood Academic Publishers, Amsterdam.

Maxted, N., Ford-Lloyd, B.V. and Hawkes, J.G. (1997) Complementary conservation strategies. In: Maxted, N., Ford-Lloyd, B.V. and Hawkes, J.G. (eds) *Plant Genetic Conservation: The* In Situ *Approach*. Chapman and Hall, London, pp. 20–55.

Meilleur, B. and Hodgkin, T. (2004) *In situ* conservation of CWRs: status and trends. *Biodiversity and Conservation* 13, 663–684.

Negbi, M. (1999) Saffron cultivation: past, present and future prospects. In: Negbi, M. (ed.) *Saffron:* Crocus sativus *L.* Harwood Academic Publishers, Australia, pp. 1–18.

Sarpaki, A. (2000) Plants chosen to be depicted on Theran wall paintings. Tentative interpretations. In: Sherratt, S. (ed.) *The Wall Paintings of Thera, Proceedings of the First International Symposium*. Pergamos Editions, Athens, pp. 657–680.

Tammaro, F. (1990) *Crocus sativus* L. cv Piano di Navelli (L' Aquila saffron): environment, cultivation, morphometric characteristics, active principles, uses. In: Tammaro, F. and Marra, L. (eds) *Proceedings of the International Conference on Saffron (*Crocus sativus *L.)*. L' Aquila, pp. 47–57.

Wishart, D. (1987) *CLUSTAN user manual*, 3rd edn. Program Library Unit, University of Edinburgh, Edinburgh.

15 Landraces: Importance and Use in Breeding and Environmentally Friendly Agronomic Systems

S. Ceccarelli

15.1 Introduction

Three of the global issues most frequently debated today are biodiversity in general and agrobiodiversity in particular, climate changes and hunger: the three problems are connected with each other and as such they should be dealt with.

It is now unequivocal that the climate is warming, as it is evident from observations of increases in global average air and ocean temperatures, widespread melting of snow and ice, and rising global average sea level. It is also very likely that in several areas the frequency and the intensity of drought as well as the variability of the climate would continue to increase to alarming levels. Some of the most profound and direct impacts of climate change over the next few decades will be on agricultural and food systems (Brown and Funk, 2008).

The industrialization of agriculture has caused an erosion of the diversity of crop varieties. Farms specialize in livestock or crops, reducing the number of species; fields are enlarged, reducing the extent of field margins and hedgerows; soil amendments enhance the uniformity of soils; and monocultures of genetically uniform individuals tend to dominate (Frison *et al.*, 2011).

Plant breeding has contributed greatly to the decrease of agricultural biodiversity, which can be quantified by the fact that barely more than 150 species are now cultivated; most of mankind now lives on no more than 12 plant species, with the four biggest staple crops (wheat, rice, maize and potato) taking the lion's share (Esquinas-Alcázar, 2010). Other examples from the World Conservation Monitoring Centre (1992) include:

- 74% of rice varieties in Indonesia descend from a common stock;
- 50% of the wheat crop in the USA represented by nine varieties;
- 75% of potato in the USA represented by four varieties;
- 50% of soybeans in US crops represented by six varieties;
- The number of rice varieties in Sri Lanka decreased from 2000 in 1959 to less than 100 today of which 75% descend from a common stock; and
- 62% and 74% of the rice varieties in Bangladesh and Indonesia, respectively, descend from a common stock.

Furthermore, the differences between collecting missions in Albania (1941 and 1993) and in south Italy (1950 and the late 1980s) showed high losses in genetic variability with levels of genetic erosion of 72.4% and 72.8%, respectively (Hammer *et al.*, 1996). In India, rice varieties have declined from an estimated 400,000 before colonialism to 30,000 in the mid-19th century with several thousand more lost after the Green Revolution in the 1960s; also Greece is estimated to have lost 95% of its broad genetic stock of traditional wheat varieties after being encouraged to replace local seeds with modern varieties developed by CIMMYT (Lopez, 1994). Lopez, quoted by Heal *et al.* (2004), also quotes a boast by Stalin to Churchill: 'We have improved beyond measure the quality of our wheat. We used to sow all varieties, but now we only cultivate the Soviet prototype. Any other cultivation than that is prohibited nation-wide.'

The evolution of plant breeding helps explain the progression of genetic erosion. For millennia plant breeding was done (not necessarily in the way we define it today) by farmers: implicit in that type of plant breeding was an emphasis on specific adaptation not only to the environment (climate and soil) but also to the uses so that it was obvious that the same farmer could select more than one variety of the same crop and that different farmers could select different varieties. In addition, when farmers moved from one place to another they took with them the seed of their crops contributing to the spreading of agriculture and to the adaptation of the crops to new climates. Over thousands of years this process (farmers' breeding) led to the formation of landraces. With the re-discovery of Mendel's work, two major changes took place. First, plant breeding was moved from farmers' fields to research stations and from farmers to scientists. What was done by very many farmers in very many different places started to be done by relatively few scientists in a relatively few places (the research stations), which with time became more and more similar to each other. Second, breeding for specific adaptation that was implicit in farmers' breeding, was gradually replaced by breeding for wide adaptation.

The best example of this change has been the development of the same high-yield varieties of common food crops in many countries, as a part of the Green Revolution (Porceddu *et al.*, 1988). The term Green Revolution was coined in March 1968 by William S. Gaud, the director of the US Agency for International Development (USAID), to indicate the outcome of a development strategy based on: (i) new crop cultivars; (ii) irrigation; (iii) fertilizers; (iv) pesticides; and (v) mechanization. Within that strategy, the new varieties were obtained by selecting for wide adaptation. Not only was this exactly the opposite of what farmers had done for millennia, but the term wide adaptation was somewhat misleading because it indicates wide 'geographical' adaptation rather than wide 'environmental' adaptation (Ceccarelli, 1989). In fact the agricultural environments in which these 'widely adapted' varieties were successful were actually very similar (high rainfall, good soil fertility, and chemical control of pests and diseases) or were made similar by adding irrigation water and fertilizers when farmers can afford them. This caused three major problems. First, the heavy use of chemicals soon began impacting the environment. Second, the poorest farmers and particularly those living in marginal environments were bypassed because they could not afford to purchase the chemicals needed to create the right environments for the new varieties – not all scientists agree on this, but most of the poor farmers do. The father of the Green Revolution, Norman Borlaug, pointed out recently that 'despite the successes of the Green Revolution, about two billion people still lack reliable access to safe, nutritious food, and 800 million of them are chronically malnourished' (Reynolds and Borlaug, 2006). Third, there was a dramatic decline in agricultural biodiversity because on one hand hundreds of genetically diverse local varieties selected by farmers over millennia for specific adaptation to their own environment and uses were displaced, and on the other hand the new varieties (despite having different names) were all very similar in their genetic constitution.

Eventually, and towards the end of the 19th century, plant breeding gradually went from predominantly public to predominantly private: one of the consequences was that not all crops were treated equally, and some became 'orphan crops', neglected by science. These include some important food crops such as banana, cassava and yam. In these changes there is no evidence that any use was made of, or any attention was paid to, the local knowledge accumulated by the farmers' communities over thousands of years. Following the privatization of breeding, another factor contributing to the loss of agrobiodiversity was the consolidation of the seed grain industry globally, leading to a more limited choice of seed varieties (Heal *et al.*, 2004): as of 2008, 49% of the global seed market was controlled by four companies, which also control 53% of the global pesticide market (Agrow News, 2008).

15.2 Diversity and Suicide

The major consequence of the dependence of modern agriculture on a small number of varieties for the major crops (Altieri, 1995) is that the main sources of food are more genetically vulnerable than ever before, i.e. food security is potentially in danger. A number of plant breeders have warned that conventional plant breeding by continuously crossing between elite germplasm lines would lead to the extinction of diverse cultivars and non-domesticated plants (Vavilov, 1992; Flora, 2001; Gepts, 2006; Mendum and Glenna, 2010) and climate change may exacerbate the crisis. Gepts (2006) claims that the current industrial agriculture system is 'the single most important threat to biodiversity'.

The key point is that a loss of genetic diversity may lead to significant risks for food supplies. A pathogen that attacks the predominant commercial variety of a food crop can inflict immense costs on society (Heal *et al.*, 2004). The three best known classic examples of this are the Irish potato famine of the 19th century, the southern corn leaf blight epidemic of 1970 in the USA, and Ug99, a virulent strain of stem rust (*Puccinia graminis triticii*), which attacked wheat for the first time in Uganda in 1999 and that is virulent to most wheat varieties and causes losses up to complete loss of the crop (Pretorius *et al.*, 2000; Singh *et al.*, 2006).

More recently, the loss of a significant fraction of the Asian rice crop to the grassy stunt virus illustrates the same point – the extreme vulnerability of a geographically extensive and genetically homogeneous crop to damage by a well-adapted predator (Heal *et al.*, 2004). Crop losses associated with genetic uniformity range from: US$1 billion (maize in the USA in 1970); 3 million t of rice destroyed in Indonesia in 1974; and 18 million citrus trees destroyed in Florida (USA) in 1984 (Thrupp, 2000). Such wide scale genetic erosion increases our vulnerability not only to new pests and diseases (Heal *et al.*, 2004) but applies equally well to climate changes as the current predominant uniformity does not allow the crops to evolve and adapt to the new environmental conditions. The expected increase of biofuel monoculture production may lead to increased rates of biodiversity loss and genetic erosion. Another serious consequence of the loss of biodiversity has been the displacement of locally adapted varieties that may hold the secret of adaptation to the future climate (Ceccarelli and Grando, 2000; Sarker and Erskine, 2006; Rodriguez *et al.*, 2008; Abay and Bjørnstad, 2009).

15.3 Benefits of Biodiversity

There is a wide consensus that agricultural biodiversity as such is an important asset that delivers substantial benefits in many different realms and that there is increasing evidence that diversity per se needs to be a central element of sustainable agricultural development (Frison *et al.*, 2011). Biodiversity provides many of the public goods on which mankind rely and therefore we do not have the right to deprive future generations of its economic and cultural benefits (Godfray *et al.*, 2010). A high degree of biodiversity is one of the salient features of traditional farming systems in the form of polycultures and/or agroforestry patterns. This strategy of minimizing risk by planting several species and varieties of crops makes the system more resilient to weather events, climate variability and change, and is more resistant to the adverse effects of pests and diseases, while at the same time stabilizing yields over the long term, promoting diet diversity and maximizing returns even with low levels of technology and limited resources (Altieri and Koohafkan, 2003). Women are central players as guardians of this biodiversity, fulfilling a role of utmost importance while exercising their unique criteria in determining desired traits of various food crops (Foresight. The Future of Food and Farming, 2011). The importance of biodiversity has been recognized by a recent report of the United Nations (De Schutter, 2009) that has defined crop genetic diversity as a global public good and is directly related to food security and in doing so has recognized the value of landraces particularly for the 1.5 billion individuals depending on small-scale farming.

15.4 The Value of Landraces

The value of landraces and their contribution to plant breeding has been recently summarized in the case of cereals by Newton *et al.* (2010). Examples of the contribution of landraces to plant breeding range from:

- Contributing useful traits for a more efficient nutrient uptake and utilization due to the fact that they developed mostly in environments with low nutrient availability, and may therefore represent a source of variation for selection of varieties adapted to low fertilizer input cropping systems;
- Contributing superior nutrition and quality characteristics such as protein content and structure, content of minerals (copper, iron, magnesium, manganese, phosphorous, selenium and zinc) in bread wheat; carotenoids in durum landraces and emmer; antioxidant compounds in maize landraces; tocols in bread wheat landraces, emmer and einkorn; antioxidants in landraces of black oats and blue, purple and red-pigmented maize kernels; recently it has been reported that vanillin (known for its strong flavouring power) was found exclusively in the free phenolics of Senatore Cappelli grain extract and it was hypothesized that its occurrence in an old wheat genotype may concur at conferring the peculiar sensory properties to ancient wheat-derived products (Dinelli *et al.*, 2009);
- Contributing resistance or tolerance to biotic stresses. This has been probably the most common and widespread contribution of landraces. In the case of all crops, the landraces have been used as donors of resistance genes to virtually all the most important diseases;
- Contributing useful genes for resistance to abiotic stresses: landraces have been assumed to be potential donors to abiotic stress resistance, particularly to drought and high temperatures. However, most of these studies, particularly in the last 15–20 years, have privileged genotyping over phenotyping; in most crops the number of QTLs identified is too large to be used in any form of molecular breeding. In barley, the genetic and molecular dissection of stress tolerance has led to the identification of either genomic regions involved in stress tolerance (major loci or QTLs) or DNA sequences known to play a role in molecular stress responses (stress-related genes, cis-acting elements and transcription factors). Although QTL analysis and gene cloning have been used to investigate the same stress responses, the relationship between QTLs and stress-related sequences is still far from understood and will remain a challenge for the near future (Cattivelli *et al.*, 2002). This conclusion, which leaves the solutions to the near future, is common to most genotypic studies. To date, hundreds of studies and reviews have reported QTLs that may play a role in mitigating the negative effects of abiotic stresses. However, MAS has contributed very little to the release of improved cultivars with greater tolerance to abiotic stresses, and this is perhaps partly because interactions between multiple

stresses are seldom considered (Collins *et al.*, 2008).

In the case of legumes there is also quite a considerable body of evidence suggesting the presence of useful genes both in landraces and wild relatives (Teshome *et al.*, 2001).

15.5 Landraces in Breeding Programmes

While there has been more emphasis on genotyping, phenotyping has also suffered from being often conducted as a one-time exercise often in environments that do not always allow a correct assessment of the value of landraces. An example of the importance of the evaluation environment is given in Fig. 15.1: when the comparison between barley landraces collected in Syria and Jordan and modern breeding material was conducted in a high rainfall research station (TH88), modern breeding material yielded significantly more than either landraces, whereas when the comparison was conducted in a low rainfall location the modern breeding material yielded significantly less than landraces (Ceccarelli and Grando, 2000). Even the comparison between landraces from different countries was affected by the evaluation site. One of the most important messages of the data shown in Fig. 15.1 concerns the choice of the selection environment. It is clear that, had the selection been done only under the high yielding conditions of a typically high-input research station, the landraces would have had a short life as breeding material as a consequence of genotype × environment (GE) interactions.

The material in Fig. 15.1 is part of a large collection made in 1981 (Weltzien, 1988), which was gradually evaluated revealing a large diversity between and within collection sites for many agronomically important characters (Ceccarelli *et al.*, 1987), disease reactions (van Leur *et al.*, 1989) and chloroplast polymorphism (Russell *et al.*, 2003). The evaluation of pure lines was not a separate activity, but it was conducted within the context of the ICARDA's barley breeding programme and the result of this evaluation was the identification of a number of successful pure lines: the first success was 'Arta', a pure line extracted from one of the two barley landraces grown in Syria, and officially released in 1994: the superiority of 'Arta' over the original landrace was particularly evident in the locations with the lowest yield levels (Fig. 15.2).

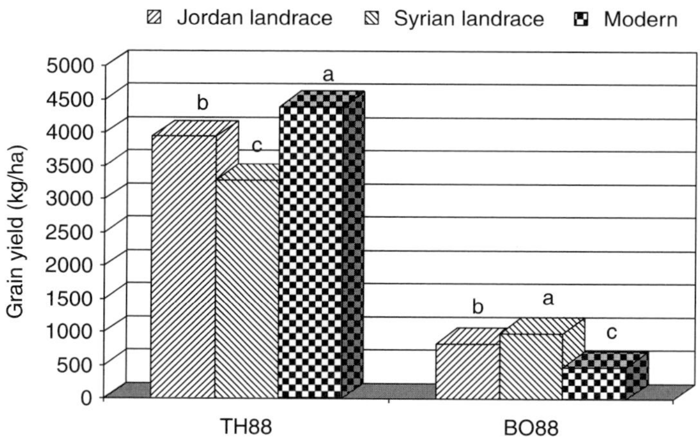

Fig. 15.1. Grain yield (kg/ha) of 86 Jordanian landraces, 92 Syrian landraces and 521 modern breeding lines measured in 1987/88 in two locations in Syria: Tel Hadya (TH88: 504.2 mm rainfall) and Bouider (BO88; 385.7 mm rainfall). Means followed by the same letter are not significantly ($P<0.05$) different based on *t*-test for samples of unequal size. (Redrawn from Ceccarelli and Grando, 2000.)

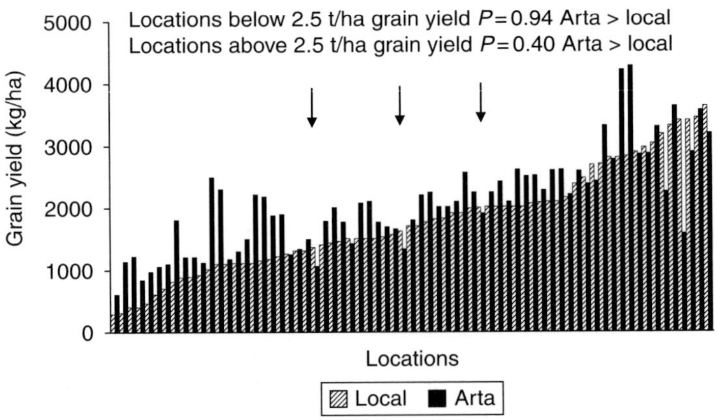

Fig. 15.2. Grain yield of 'Arta' (a pure line of barley derived by pure line selection from a Syrian landrace) compared with local barley in 69 farmers fields in five provinces of Syria in 1996. Each cultivar was grown on plots of 1 ha. The arrows show that only in three of the locations with an average yield of less than 2.5 t/ha (shown by an arrow), 'Arta' did not out yield the local barley.

Other examples of the usefulness of this approach were two other lines, 'Tadmor' and 'Zanbaka', which were not released by the official variety release system, but as we will see later they were eventually adopted by farmers thus highlighting the weaknesses of the existing variety release system. This approach has been used also in Ethiopia (Lakew *et al.*, 1997) where it led to a number of released varieties such as 'Shege', 'Demay', 'Misrach' and many others (Lakew and Assefa, 2011), and in Eritrea where it led to the release of three varieties, 'Tekonda', 'Shishai' and 'Rawa'.

The step which followed the identification of agronomically superior pure lines and sources of disease resistance within landraces was to start utilizing the superior pure lines as parental material in the breeding programme – as typically done in breeding programmes. An example of the value of this approach is given in Table 15.1, where 514 breeding lines unrelated to landraces (improved) and 525 pure lines extracted from landraces are compared to 214 lines derived from crosses between improved and landraces. The data were collected in a very dry site and year (Breda received 244 mm rainfall in 1995) where we measured grain yield, total biological yield, plant height and harvest index, and in a relatively wet site (Tel Hadya with 313 mm rainfall) where we measured yield potential. The landraces yielded, on average, more than the improved lines under stress and had a lower average yield potential confirming what was found earlier. Under stress, landraces and improved lines had a similar biological yield, but the landraces were, surprisingly, much shorter and had a higher harvest index – two characteristics usually associated with high yielding varieties. Crosses between landraces and improved germplasm generated breeding material equal to the landraces in terms of grain yield and total biological yield under stress and superior for plant height while maintaining a relatively high harvest index (Ceccarelli and Grando, 2000).

As in the case of pure line selection within landraces, the use of superior landraces/lines as parental material has been used in a number of countries. In Algeria, where the dominant barley variety ('Saida') is a population selected from a, probably lost, landrace, lines derived from crosses with 'Saida' are consistently superior to both the landrace and to the breeding material which does not include 'Saida' as parent (Fig. 15.3).

The superiority of the lines derived from crosses with landraces suggests that

Table 15.1. Grain yield (kg/ha), biological yield (kg/ha), plant height (cm) and harvest index in Breda (1995) and grain yield in Tel Hadya 1995 (kg/ha) of different types of breeding material (using data taken from Ceccarelli and Grando, 2000).

Breeding material	Grain yield (BR95)	Biological yield (BR95)	Grain yield (TH95)	Plant height (BR95)	Harvest index (BR95)
Improved (*n* = 514)					
Mean	591 ± 8	2559 ± 17	4125 ± 27	23.2 ± 0.2	22.8 ± 0.3
Max	1201	4504	5812	40.3	41.3
Min	69	1559	1375	14.8	3.24
Improved × Landraces (*n* = 214)					
Mean	775 ± 10	2678 ± 24	3883 ± 33	25.1 ± 0.3	29.1 ± 0.3
Max	1252	3658	5206	38	37.9
Min	259	1930	2630	16.9	11
Landraces (*n* = 525)					
Mean	752 ± 7	2549 ± 16	3657 ± 23	21.4 ± 0.1	29.8 ± 0.2
Max	1232	4027	5455	30.5	39.9
Min	320	1529	2250	13.1	16.5

BR95 = Breda 1995; TH95 = Tel Hadya 1995.

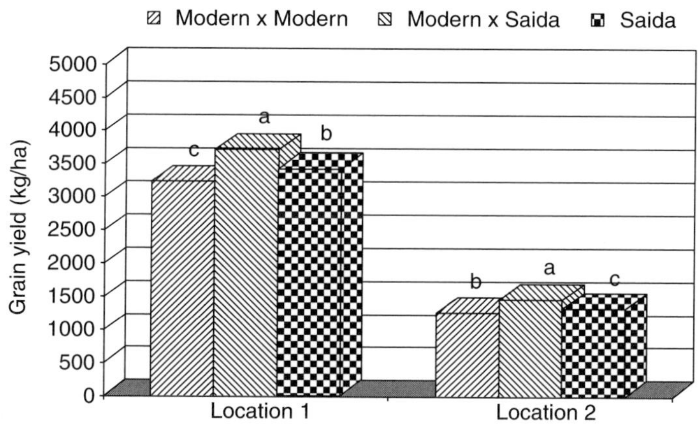

Fig. 15.3. Grain yield (kg/ha) of 16 breeding lines derived from crosses that did not include landraces, eight breeding lines derived from crosses with 'Saida' (a leading Algerian landrace of barley) and 'Saida' in two locations in Algeria in 2009 (means followed by different letters are significantly different based on *t*-test for groups of unequal size assuming unequal variances).

the strategy of using adapted germplasm in a breeding programme is to capitalize on their specific adaptation to drought and low-input conditions rather than to consider them as sources of new useful genes as is the case in most plant breeding programmes. Therefore, in breeding for stress environments, landraces should be regarded as recipients of few useful genes to be added to their adapted genetic background, rather than as donors of traits not available in 'elite germplasm' (Ceccarelli and Grando, 2000). This can be extrapolated to environmentally friendly agronomic systems such as organic agriculture by assuming that once external inputs are reduced landraces may prove useful in breeding programmes specifically designed for this type of agriculture.

15.6 Linking Knowledge, Conservation and Improvement

There is a global concern that the continuous decrease of biodiversity, combined with the urgent need to mitigate and to adapt to climate changes in a scenario of population increase that will probably reach over 9 billion by 2050, will cause an unprecedented pressure on the global food system over the next 40 years.

Appropriate new technologies have the potential to be very valuable for the poorest people in low-income countries. Farmer participation in technology development and participatory extension approaches have enabled novel technologies and practices to be learned directly and then adapted to particular agroecological, social and economic circumstances (Foresight. The Future of Food and Farming, 2011). Farmer participatory research, on-farm testing and farmer selection of plant materials will need increasingly to be embedded in research, extension and development institutions (Pretty *et al.*, 2010). There are grounds for optimism that agriculture can become a more powerful force for the reduction of hunger and poverty in the decades ahead – but agriculture needs to be repositioned within governments as a profession dedicated to multiple ends, of which hunger and poverty reduction are central. Such a repositioning would mean changing the formal and informal training of professionals in agricultural development (Foresight. The Future of Food and Farming, 2011). The same document indicates three important points.

1. Innovation in how to involve producers in improving yields sustainably is as important as innovation in research – there is still a need for far greater participation of producers in defining and monitoring success.
2. With much technology development taking place at greater distances from the farmer's plot, stronger mechanisms are needed to ensure that representatives of poor farmers and groups experiencing chronic hunger are included in local and national fora.
3. Smallholder farming has been long neglected. It is not a single solution, but an important component of both hunger and poverty reduction.

In addition to these recent papers, the International Treaty on Plant Genetic Resources for Food and Agriculture (FAO, 2009) in its Article 6 indicates specifically that:

> The sustainable use of plant genetic resources for food and agriculture may include such measures as: promoting, as appropriate, plant breeding efforts which, with the participation of farmers, particularly in developing countries, strengthen the capacity to develop varieties particularly adapted to social, economic and ecological conditions, including in marginal areas.

Also, the report of the United Nations (De Schutter, 2009) quoted earlier, recommends that donors and international institutions, including the Consultative Group on International Agricultural Research and FAO, should put farmers at the centre of research through participatory research schemes such as participatory plant breeding.

This widespread interest in participation has been recognized since the early 1980s by scientists (social scientists first and later biological scientists) and in the case of plant breeding has been implemented as participatory plant breeding (PPB), a process by which farmers are routinely involved in a plant breeding programme with opportunities to make decisions throughout (Halewood *et al.*, 2007).

The model of PPB we have implemented (Ceccarelli *et al.*, 2000; Ceccarelli and Grando, 2007), initially in Syria and then gradually in Tunisia, Morocco, Jordan, Egypt, Eritrea, Algeria, Yemen, Iran and Ethiopia in crops such as wheat, barley, lentil, chickpea and faba bean, combines modern science with the 'local knowledge', brings plant breeding back into farmers' hands – and not farmers back into breeding as a recent publication suggests (Almekinders and Hardon, 2006) – and also encourages a return to diversity.

The main feature of PPB is that farmers (or in general, users) are involved in designing and developing technologies – not just in testing the final products of scientific research as done in conventional (non-participatory) research. Specifically, there are several differences between conventional and participatory plant breeding: in conventional plant breeding – and only with few exceptions – new varieties are selected on research stations by breeders and the final products are tested on farm. Adoption occurs at the end of the breeding process. In PPB new varieties are selected in farmers' fields jointly by breeders and farmers and adoption occurs during the breeding process. In order to be fully participatory the programme needs to be inclusive with specific regard to women because particularly in low-income countries they play a critical role in agriculture, and agriculture plays a critical role in their livelihoods. Purposively empowering women and focusing on their unique challenges will bring much wider gains in terms of poverty and productivity (Foresight. The Future of Food and Farming, 2011).

Scientifically, conventional plant breeding and PPB are the same process but PPB differs in three key organizational aspects.

1. Trials are conducted in farmers' fields and managed by farmers.
2. Farmers participate as equal partners in the selection process.
3. The process can be duplicated independently in a large number of locations and of countries, with different methodologies and germplasm depending on the crop and the country.

PPB can impact positively on biodiversity because, being a highly decentralized process, it produces varieties which are different from country to country, from village to village within a country, and even within the same village depending, among other factors, on the age, wealth and gender of the farmers. In addition to increasing biodiversity in space PPB increases biodiversity in time because the process is cyclic and there is a rapid turnover of varieties thus creating a system which makes it difficult for pathogens to spread. Another dimension of the biodiversity generated by PPB is that the varieties selected by farmers are often not homogeneous, i.e. they are still genetically variable – like the landraces – in contrast to the majority of varieties produced by conventional breeding in which all the plants are genetically identical (pure lines, hybrids, clones).

Even though PPB has been practised for only 20 years, there are already indications of impacts at various levels:

- Adoption: many new varieties have already been adopted by farmers even though the programme is relatively new; in Syria more than 80 lines and/or populations have been named and adopted by farmers from the PPB trials since 2000, compared with seven varieties released by the conventional breeding programme in nearly 25 years. In some areas of Syria the adoption of the PPB varieties has reached 80% of the barley area. In Jordan and Algeria, the first PPB varieties (one in each country) are under multiplication to be submitted to the variety release committee; in Eritrea three food barley, ten bread wheat and two durum wheat varieties have been selected by farmers, in Yemen two varieties of barley and two of lentil have been adopted, in Egypt three barley varieties have been selected by farmers in the project area (the north-west coast). In Iran, at the end of the first PPB cycle, farmers selected four varieties and are currently testing various types of mixtures between them. Two aspects of the participatory selection process are: (i) the yield advantages, as high as 50–70% that are possible to achieve in low rainfall, drought-stressed areas *only* by changing the variety – in these areas conventional plant breeding was never able to introduce a new variety; and (ii) in most cases these yield advantages have been

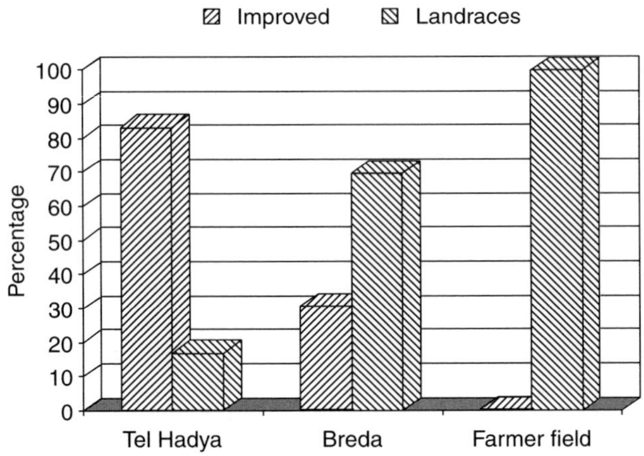

Fig. 15.4. Percentage of landraces and modern cultivars selected by farmers in a research station located in a high rainfall site (Tel Hadya), in a research station located in a low rainfall site (Breda) and in their own field in a low rainfall area.

obtained using landraces for which farmers have consistently expressed a strong preference, particularly in dry areas (Fig. 15.4).

- Institutional: in several countries, policy makers and scientists are showing much more interest in PPB as it is expected to generate more relevant results more quickly and at a lower cost.
- Farmers' skills and empowerment: the interactive nature of the PPB programmes has considerably improved farmers' knowledge, their ability to negotiate, and their dignity (Galié *et al.*, 2009). It is because of their skills and their increased self-confidence that farmers in a number of countries started exploiting the additional advantages of evolutionary plant breeding as described in the next section.
- Biodiversity: different varieties have been selected in different areas in each country, in response to different environmental constraints and users' needs. Interest in landraces has increased as indicated by the request of farmers in Syria, Jordan, Algeria and Iran to have access and to evaluate their landraces kept in the gene banks.

From the point of view of the global issues mentioned earlier, one of the advantages of PPB is that by matching one of the key recommendations of the *interim* report of the Special Rapporteur to the United Nations on the right to food ('Put farmers at the centre of research through participatory research schemes such as participatory plant breeding', p. 22) provides an increase of agricultural production directly in the farmers' fields making therefore those increases available and accessible.

Participatory plant breeding also has the ability of addressing the specific needs of family farms and to make them more productive thus alleviating poverty and meeting local and global food demand. This will shift the focus from large-scale industrial farming addressing the research themes for smallholdings which are very different from those of large-scale farming because they involve, for example, concepts such as crop rotation, complements of animals and plants, and the use of animal waste as fertilizer (Godfray *et al.*, 2010).

15.7 Combining Participation and Evolution

Many international organizations, recognizing the value of agrobiodiversity for the future of humankind, are promoting the conservation of local varieties and wild relatives of crops. The most frequent type of conservation is the *ex situ* conservation in gene banks; currently there are about 1500 gene banks, which hold more than 7 million seed samples (Fowler and Hodgkin, 2004). While gene banks are essential as the last resort where to rescue seed in case of natural disasters, they freeze not only seeds but also evolution at the time of the collection. This has suggested that landraces and wild relatives should also be conserved *in situ*, i.e. in their own native environment. Based on the evidence that evolutionary adaptation has occurred in a number of species in response to climate change both in the long term and in the short term, and on the recent demonstration (using experimental evolution) that while out-crossing populations are able to adapt rapidly to environmental changes, also a small amount of natural crossing (such as in self-fertilizing crops) allows adaptation to stress environments to develop (Morran *et al.*, 2009), we have attempted to make the process of *in situ* conservation more dynamic by combining participation and evolution in participatory–evolutionary breeding programmes (Murphy *et al.*, 2005; Phillips and Wolfe, 2005; Ceccarelli *et al.*, 2010). These programmes could represent a dynamic and inexpensive strategy, which will quickly enhance the adaptation of crops to climate change and that will combine better adapted varieties with the mitigation effects of eco-efficient management systems.

This idea was first proposed by Suneson (1956) as follows: 'the core features (of the evolutionary breeding method) are a broadly diversified germplasm and a prolonged subjection of the mass of the progeny to competitive natural selection in the area of contemplated use'.

We implemented the first participatory–evolutionary breeding programmes in 2008 by constituting a mixture of nearly 1600 barley F_2 representing the entire ICARDA's barley crossing programme of that year and hence including a wide range of germplasm from the wild progenitor, *Hordeum spontaneum*, to landraces from several countries and to modern breeding materials. The barley population was planted in 19 locations in five countries. This was followed in 2009 by a population of durum wheat consisting of a mixture of slightly more than 700 crosses, which was planted in four locations, and in 2010 by a population of nearly 2000 segregating populations of bread wheat, which was planted in two locations (one of which for seed multiplication). These populations will be left evolving in a multitude of environments, chosen by the farmers and characterized by single abiotic or biotic stresses or combinations of stresses and under different types of agronomic management (Fig. 15.5) with the expectation that the frequency of genotypes with adaptation to the conditions (climate, soil, agronomic practices and biotic stresses) of the locations where each year the population is grown will gradually increase. The simplest and cheapest way of implementing evolutionary breeding is for the farmers to plant and harvest in the same location. It is also possible and actually desirable, to plant samples in other locations affected by different stresses or different combinations of stresses by sharing the population with other farmers. For example, in Iran the barley population which was planted by five farmers in two provinces in 2008, spread to 50 farmers in four provinces in the cropping season 2010/2011.

However, the best way of exploiting the progressive better adaptation of the evolutionary populations is to consider it as an evolving source of new cultivars progressively better adapted to the evolving agronomic and climatic conditions: to do this farmers, by themselves or jointly with scientists, can use these evolving populations to select the most desirable plants, spikes, panicles, roots, tubers etc. – depending on the crops – and use them in participatory breeding programmes as described earlier.

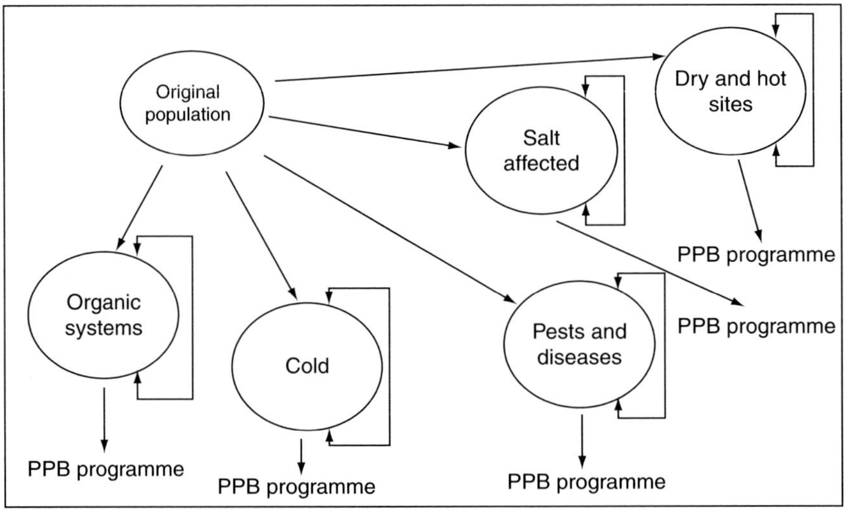

Fig. 15.5. Scheme of an evolutionary–participatory breeding programme.

While the population is evolving, the lines or sub-populations can be tested as pure lines (in the case of self-pollinated), clones (in the case of vegetatively propagated) or populations (in the case of cross-pollinated) in the participatory breeding programmes, or can be used as multi-lines, or a subsample of the population can be directly used for cultivation exploiting the advantages of genetic diversity described earlier. The key aspect of the method is that, while the lines are continuously extracted, evaluated and exploited, the population is left evolving for an indefinite amount of time, thus becoming a unique source of continuously better adapted genetic material directly in the hands of the farmers – a sort of evolving gene bank.

In Iran, the interest generated by the barley population has suggested the Iranian breeders to make their own bread wheat and durum wheat populations. The evolutionary bread wheat population, created by mixing Iranian breeding material, was distributed and planted in different regions of Kermanshah province and showed resistance to lodging and rust and out-yielded the most widely grown cultivar 'Sardari' (Haghparast, personal communicatio).

15.8 Conclusion

Global food security is threatened by the continuous decline of agrobiodiversity, by the unpredictability of the climate which is expected to increase and by the fact that increasing the accessibility and the availability of food for the most marginalized people is more urgent than increasing the food production per se. This is possible by reorganizing agricultural research through the incorporation of objectives such as the increase of agrobiodiversity, the participation of users and the adaptation of crops to the environment, as in environmentally friendly agronomic systems. This implies reconsidering the value of germplasm such as the landraces, which have represented the bulk of agricultural production until recently but that in high and unsustainable agricultural systems have been neglected. Yet, it is very likely that landraces, together with wild relatives, possess useful genes for adaptation to future climates: we argue that farmer-based programmes that facilitate the evolution of populations in an adaptive sense to both climatic conditions and agronomically friendly systems may reconcile an increase in available and accessible food production with an increase in agrobiodiversity.

References

Abay, F. and Bjørnstad, A. (2009) Specific adaptation of barley varieties in different locations in Ethiopia. *Euphytica* 167, 181–195.

Agrow News (2008) Available at: www.agrow.com/home/.

Almekinders, C. and Hardon, J. (eds) (2006) Bringing farmers back into breeding. Experiences with participatory plant breeding and challenges for institutionalization. *Agromisa Special* 5. Wageningen, the Netherlands, 140 pp.

Altieri, M.A. (1995) *Agroecology: the science of sustainable agriculture*. Westview Press, Boulder, Colorado.

Altieri, M.A. and Koohafkan, P. (2003). Enduring farms: climate change, smallholders and traditional farming communities. *Third World Network Environmental & Development Series* 6. Penang, Malaysia.

Brown, M.E. and Funk, C.C. (2008) Food security under climate change. *Science* 319(5863), 580–581.

Cattivelli, L., Baldi, P., Crosatti, C., Di Fonzo, N., Faccioli, P., Grossi, M., Mastrangelo, A.M., Pecchioni, N. and Stanca A.M. (2002) Chromosome regions and stress related sequences involved in resistance to abiotic stress in Triticeae. *Plant Molecular Biology* 48, 649–665.

Ceccarelli, S. (1989) Wide adaptation. How wide? *Euphytica* 40, 197–205.

Ceccarelli, S. and Grando, S. (2000) Barley landraces from the Fertile Crescent: a lesson for plant breeders. In: Brush S.B. (ed.) *Genes in the Field: On-farm Conservation of Crop Diversity*. IDRC, Boca Raton, Florida, pp. 51–76.

Ceccarelli, S. and Grando, S. (2007) Decentralized-participatory plant breeding: an example of demand driven research. *Euphytica* 155, 349–360.

Ceccarelli, S., Grando, S. and Van Leur, J.A.G. (1987) Genetic diversity in barley landraces from Syria and Jordan. *Euphytica* 36, 389–405.

Ceccarelli, S., Grando, S., Tutwiler, R., Baha, J., Martini, A.M., Salahieh, H., Goodchild, A. and Michael, M. (2000) A methodological study on participatory barley breeding. i. selection phase. *Euphytica* 111, 91–104.

Ceccarelli, S., Grando, S., Maatougui, M., Michael, M., Slash, M., Haghparast, R., Rahmanian, M., Taheri, A., Al-Yassin, A., Benbelkacem, A., Labdi, M., Mimoun, H. and Nachit, M. (2010) Plant breeding and climate changes. *Journal of Agricultural Science* 148, 1–11.

Collins, N.C., Tardieu, F. and Tuberosa, R. (2008) Quantitative trait loci and crop performance under abiotic stress: where do we stand? *Plant Physiology* 147, 469–486.

De Schutter, O. (2009) The right to food: seed policies and the right to food: enhancing agro biodiversity and encouraging innovation. *The interim report of the Special Rapporteur on the right to food*. Available at: www.srfood.org/.

Dinelli G., Carretero, A.S., Di Silvestro, R., Marottia, I., Fu, S., Benedettelli, S., Ghiselli, L. and Gutiérrez, A.F. (2009) Determination of phenolic compounds in modern and old varieties of durum wheat using liquid chromatography coupled with time-of-flight mass spectrometry. *Journal of Chromatography A* 1216, 7229–7240.

Esquinas-Alcázar, J. (2010) Protecting crop genetic diversity for food security: political, ethical and technical challenges. *Nature* 6, 946–953.

Flora, C. (2001) *Interactions between Agroecosystems and Rural Communities*. CRC Press, Boca Raton, Florida.

Food and Agriculture Organization of the United Nations (2009) *The International Treaty on Plant Genetic Resources for Food and Agriculture*. FAO, Rome, pp. 1–68.

Foresight. The Future of Food and Farming (2011) *Executive Summary*. The Government Office for Science, London.

Fowler, C. and Hodgkin, T. (2004) Plant genetic resources for food and agriculture: assessing global availability. *Annual Review of Environmental Resources* 29, 10.1–10.37.

Frison, E.A., Cherfas, J. and Hodgkin, T. (2011) Agricultural biodiversity is essential for a sustainable improvement in food and nutrition security. *Sustainability* 3, 238–253.

Galié, A., Hack, B., Manning-Thomas, N., Pape-Christiansen, A., Grando, S. and Ceccarelli, S. (2009) Evaluating knowledge sharing in research: the International Farmers' Conference organized at ICARDA. *Knowledge Management for Development Journal* 5, 107–125.

Gepts, P. (2006) Plant genetic resources conservation and utilization: the accomplishment and future of a societal insurance policy. *Crop Science* 46, 2278–2292.

Godfray, H.C.J., Beddington, J.R., Crute, I.R., Haddad, L., Lawrence, D., Muir, J.F., Pretty, J., Robinson, S., Thomas, S.M. and Toulmin, C. (2010) Food security: the challenge of feeding 9 billion people. *Science* 327, 812–818.

Halewood, M., Deupmann, P., Sthapit, B., Vernooy, R. and Ceccarelli, S. (2007) Participatory plant breeding to promote Farmers' Rights. *Bioversity International*, Rome, 7 pp.

Hammer, K., Knupffer, H., Xhuveli, L. and Perrino, P. (1996) Estimating genetic erosion in landraces – two case studies. *Genetic Resources and Crop Production* 43, 329–336.

Heal, G., Walker, B., Levin S., Arrow, K., Dasgupta, P., Daily, G., Ehrlich, P., Maler, K.-G., Kautsky, N., Lubchenco, J., Schneider, S. and Starrett, D. (2004) Genetic diversity and interdependent crop choices in agriculture. *Resource and Energy Economics* 26, 175–184.

Lakew, B. and Assefa, A. (2011) Advances and experiences in barley landrace improvement in Ethiopia. In: Mulatu, B. and Grando, S. (eds) *Barley Research and Development in Ethiopia, Proceedings of the 2nd National Barley Research and Development Review Workshop*. In press.

Lakew, B., Semeane, Y., Alemayehu, F., Gebre, H., Grando, S., van Leur, J.A.G. and Ceccarelli, S. (1997) Exploiting the diversity of barley landraces in Ethiopia. *Genetic Resources and Crop Evolution* 44, 109–116.

Lopez, P.B. (1994) A new plant disease: uniformity. *CERES* 26, 41–47.

Mendum, R. and Glenna, L.L. (2010) Socioeconomic obstacles to establishing a participatory plant breeding program for organic growers in the United States. *Sustainability* 2, 73–91.

Morran, L.T., Parmenter, M.D. and Phillips, P.C. (2009) Mutation load and rapid adaptation favour outcrossing over self-fertilization. *Nature* 462, 350–352.

Murphy, K., Lammer, D., Lyon, S., Carter, B. and Jones, S.S. (2005) Breeding for organic and low-input farming systems: an evolutionary–participatory breeding method for inbred cereal grains. *Renewable Agriculture and Food Systems* 20, 48–55.

Newton, A.C., Akar, T., Baresel, J.P., Bebeli, P.J., Bettencourt, E., Bladenopoulos, K.V., Czembor, J.H., Fasoula, D.A., Katsiotis, A., Koutis, K., Koutsika-Sotiriou, M., Kovacs, G., Larsson, H., Pinheiro de Carvalho, M.A.A., Rubiales, D., Russell, J., Dos Santos, T.M.M. and Vaz Patto, M.C. (2010) Cereal landraces for sustainable agriculture. A review. *Agronomy for Sustainable Development* 30, 237–269.

Phillips, S.L. and Wolfe, M.S. (2005) Evolutionary plant breeding for low input systems. *Journal of Agricultural Science* 143, 245–254.

Porceddu, E., Ceoloni, C., Lafiandra, D., Tanzarella, O.A. and Scarascia Mugnozza, G.T. (1988) Genetic resources and plant breeding: problems and prospects. In: Miller, T.E. and Koebner, R.M.D. (eds) *Proceedings of the Seventh International Wheat Genetics Symposium*. Institute of Plant Science Research, Cambridge, pp. 7–22.

Pretorius, Z.A., Singh, R.P., Wagoire, W.W. and Payne, T.S. (2000) Detection of virulence to wheat stem rust resistance gene Sr31 in *Puccinia graminis*. f. sp. *tritici* in Uganda. *Plant Disease* 84, 203.

Pretty, J., Sutherland, W.J., Ashby, J., Auburn, J., Baulcombe, D., Bell, M., Bentley, J., Bickersteth, S., Brown, K., Burke, J., Campbell, H., Chen, K., Crowley, E., Crute, I., Dobbelaere, D., Edwards-Jones, G., Funes-Monzote, F., Charles, H., Godfray, J., Griffon, M., Gypmantisiri, P., Haddad, L., Halavatau, S., Herren, H., Holderness, M., Izac, A.M., Jones, M., Koohafkan, P., Lal, R., Lang, T., McNeely, J., Mueller, A., Nisbett, N., Noble, A., Pingali, P., Pinto, Y., Rabbinge, R., Ravindranath, N.H., Rola, A., Roling, N., Sage, C., Settle, W., Sha, J.M., Shiming, L., Simons, T., Smith, P., Strzepeck, K., Swaine, H., Terry, E., Tomich, T.P., Toulmin, C., Trigo, E., Twomlow, S., Kees Vis, J., Wilson, J. and Pilgrim, S. (2010) The top 100 questions of importance to the future of global agriculture. *International Journal of Agricultural Sustainability* 8, 219–236.

Reynolds, M.P. and Borlaug, N.E. (2006) Applying innovations and new technologies for international collaborative wheat improvement. *Journal of Agricultural Science* 144, 95–110.

Rodriguez, M., Rau, D., Papa, R. and Attene, G. (2008) Genotype by environment interactions in barley (*Hordeum vulgare* L): different responses of landraces, recombinant inbred lines and varieties to Mediterranean environment. *Euphytica* 163, 231–247.

Russell, J.R., Booth, A., Fuller, J.D., Baum, M., Ceccarelli, S., Grando, S. and Powell, W. (2003) Patterns of polymorphism detected in the chloroplast and nuclear genomes of barley landraces sampled from Syria and Jordan. *Theoretical and Applied Genetics* 107, 413–421.

Sarker, A. and Erskine, W. (2006) Recent progress in the ancient lentil. *Journal of Agricultural Science* 144, 19–29.

Singh, R.P., Hodson, D.P., Jin, Y., Huerta-Espino, J., Kinyua, M.G., Wanyera, R., Njau, P. and Ward, R.W. (2006) Current status, likely migration and strategies to mitigate the threat to wheat production from race Ug99 (TTKS) of stem rust pathogen. *CAB Reviews: Perspectives in Agriculture, Veterinary Science, Nutrition and Natural Resources* 04, 1–13.

Suneson, C.A. (1956) An evolutionary plant breeding method. *Agronomy Journal* 48, 188–191.

Teshome, A., Brown, A.H.D. and Hodgkin, T. (2001) Diversity in landraces of cereal and legume crops. *Plant Breeding Reviews* 21, 221–261.

Thrupp, L.A. (2000) Linking agricultural biodiversity and food security: the valuable role of agro biodiversity for sustainable agriculture. *International Affairs* 76, 265–281.

Van Leur, J.A.G., Ceccarelli, S. and Grando, S. (1989) Diversity for disease resistance in barley landraces from Syria and Jordan. *Plant Breeding* 103, 324–335.

Vavilov, N.I. (1992) *Origin and Geography of Cultivated Plants*. Cambridge University Press, Cambridge.

Weltzien, E. (1988) Evaluation of barley (*Hordeum vulgare* L.) landraces populations originating from different growing regions in the Near East. *Plant Breeding* 101, 95–106.

World Conservation Monitoring Centre (1992) In: Groombridge, B. (ed.) *Global Biodiversity: Status of the Earth's Living Resources*. Chapman & Hall, London.

16 Landraces in Europe: An Approach towards Identifying Landrace-Rich Areas as a Priority for Protection

V. Negri, R. Barocco, L. Pacicco, F. Veronesi and R. Venanzoni

16.1 Introduction

With the aim to elaborate methods for planning landrace on-farm conservation activities, which at present do not exist, a panel of experts was established at the Applied Biology Department of Perugia University. The expert panel included botanists, agronomists and geneticists with specific knowledge of agrobiodiversity conservation problems. The panel discussed and defined the following items:

- The conservation aim;
- The criteria to be taken into account for giving priority among areas;
- The methods to be used in applying criteria;
- The definition of a strategy;
- The need to test the strategy on a case study; and
- The baseline data needed to carry out the testing of the strategy.

This chapter presents the most relevant issues of this discussion as well as the results obtained applying the developed strategy to a Central Italy agrobiodiversity case study.

16.2 Conservation Aim, Criteria and Methods Used to Establish Conservation Action

Considering that *in situ* conservation is defined by CBD and the International Treaty as 'the conservation of ecosystems and natural habitats and the maintenance and recovery of viable populations of species in their natural surroundings and, in the case of domesticated or cultivated species, in the surroundings where they have developed their distinctive properties' and that this definition was accepted by all the signatory countries, the aim of a conservation action focused on landraces was set as the 'Conservation and valorization of areas with high level of agrobiodiversity'.

The panel discussed and agreed that 'areas with high level of agrobiodiversity' were intended to be those areas that are richest in diversity for landraces, agroecosystems types and crop wild relatives (Negri *et al.*, 2010).

16.3 Definition of Criteria to be Taken into Account for Identifying Areas

Landrace diversity was the first criterion which the panel considered to be of fundamental importance: both the number of different species present in the considered area (i.e. the species richness) and the number of landraces belonging to each species in the same area (i.e. the landrace evenness) should be considered. All different species, irrespective of their economic or nutritional importance should be considered worthwhile to be conserved. For landraces, it is not possible to attribute any priority by taxa, as suggested for wild species (Maxted *et al.*, 2009), since they are cultivated populations. *De facto* this indicates they are all important and useful. Moreover, very often, landraces of species that at national or international level are not very important, could be the base for local economies and the source of profit for the farmers (see various authors in Veteläinen *et al.*, 2009).

The agroecosystem ecological diversity is another essential criterion to be considered, since it is usually associated with the level of inter- and intra-population diversity of each species and it determines the complexity level of relationships among living beings.

The presence of already existing protected areas nearby was also considered relevant. The presence of a protected area is per se an indicator of biodiversity (as protected areas are set where richness of species, rare or endemic species or rare ecosystems exist), in addition it can positively influence the setting of areas to be protected for landrace diversity and facilitate the valorization of products from landraces through the already existing circuits of naturalistic tourism and marks related to the protection of nature or sustainable agriculture (Barocco *et al.*, 2010a).

Finally, the presence of crop wild relatives (CWR) and the threat of extinction of landraces could be also considered useful criteria. CWR are essential for the introgression of new genetic variation into the cultivated gene pools and their presence in an area is a biodiversity value per se. However, it was noted that to use this criterion a detailed review of bibliographic records was required and possibly field surveys of the area, since updated CWR presence data are most generally lacking across Europe (Maxted *et al.*, 2007).

The threat of extinction can vary for each landrace depending on the characteristic of the landrace (i.e. the product destination) as well as socio-economic traits (i.e. farmer age, political context etc.). It is then difficult to be estimated. A model scheme to evaluate the risk of landrace loss has been proposed by a Technical Committee called to implement the Lazio Regional Law for the safeguard of agrobiodiversity (Porfiri *et al.*, 2009). This model could be usefully applied, but it needs a huge database to be collected before being used. The latter two criteria could then be applied only when comprehensive data are available.

16.4 How to Use Criteria

There may be problems in using the proposed criteria related to available sources of information and their possible use. Although landraces are very important for breeding, development of low input agricultural systems and the development of typical products, complete landrace inventories do not exist in any country of the world (Maxted and Kell, 2009), with the exception of Switzerland (B. Schierscher-Viret, pers. comm.). Where landrace data are present, they are often fragmentary and incomplete. The first step would then be to collect data on landrace presence and location. Then, diversity indices that give information on species richness and evenness (Shannon and Weaver, 1949; Simpson, 1949; Berger and Parker, 1970; Pielou, 1975; Lande, 1996) could be applied. However, in order to decide if to use only one or more than one index, it is necessary to evaluate their possible use for landraces and if they give similar or different information. The panel also agreed that the ecological diversity of the agricultural habitats could be only evaluated using proxies. To this purpose, information coming from different types and sources, such as those related to land use, climate, botanical formation, elevation, pluviometry and phytoclimatology of the sites where landraces are present, could be collected and used. However, also in this case, it is necessary to consider if the ecological information coming from the above mentioned sources are complementary or redundant. Data relative to the presence of protected areas, on the contrary, can be easily collected. Finally, it was noted all the collected information could be grouped in a global index, but that much research work is needed in this area.

16.5 Definition of a Strategy: a Mixed Model

The panel then discussed and agreed on a possible strategy to identify the areas richest in biodiversity and suitable to valorization and on possible methods to estimate parameters on which the strategy can be based (Barocco *et al.*, 2010b). In this stage, which was related to the practical definition of areas devoted to on-farm conservation, the panel of experts concluded that an operative scheme was needed and suggested a strategy (Fig. 16.1) that, taking into account the above mentioned criteria, considers the use of indexes and theoretical considerations in applying criteria as well as practical issues, namely:

1. The need to have a reliable database where landraces are listed and mapped on a defined area.
2. The need to apply the defined criteria in order of importance.
3. The need to discriminate among areas whether resources for conservation are limited. This can be done though a pragmatic, but rationally based, process. The areas which are most valuable for conservation can be identified by rating them on the base of estimate values of each proposed criteria in subsequent steps. For each criteria estimate a threshold is (arbitrary) defined, below which areas are not admitted to the following criteria estimates.
4. The opportunity to evaluate which areas contain the greatest diversity by using indices.

The need to test the proposed strategy and data sources was also stressed.

16.6 Applying the Strategy to Landrace Diversity in Central Italy: a Case Study

16.6.1 Landrace inventory creation, mapping and counting

In order to assess the applicability of the strategy and to carry out a wider testing of its efficiency, a detailed landrace inventory was initially created. Initially 1304 landraces (LRs) belonging to 52 genera and 76 species were listed from available official inventories (e.g Tuscany and Latium Regional Agencies for Agricultural Development websites and the inventory for Umbria), but for only 943 LRs data relative to the exact latitude and longitude of the site of cultivation were available (the other landraces were spread over a large territory). A database containing all data relative to these landraces was created. Each landrace cultivation site was highlighted by a point carrying information on: latitude, longitude, altitude of the landrace cultivation site, landrace accession number, landrace scientific and local name and the name of the farmer still maintaining it on farm. Finally, the database was imported into a GIS system. The GIS shape file was overlaid with a grid of 20 × 20km, and a procedure by GIS was applied that allowed the association with each cell with one or more data points that fell within it. This 'spatial join' found landraces in 105 quadrants. The points were then exported in KML (Keyhole Markup Language) format, which is able to maintain information related to each point to allow ease of use in Google Earth software. A system was developed that makes it possible to get landrace information by clicking on the corresponding points.

Finally, by using the counter function, the density, i.e. the number of points corresponding to each landrace cultivation site, was also calculated for each quadrant.

16.6.2 Working out how to apply criteria

The 105 quadrants where landraces were present have been initially selected on the base landrace density. A threshold of ten

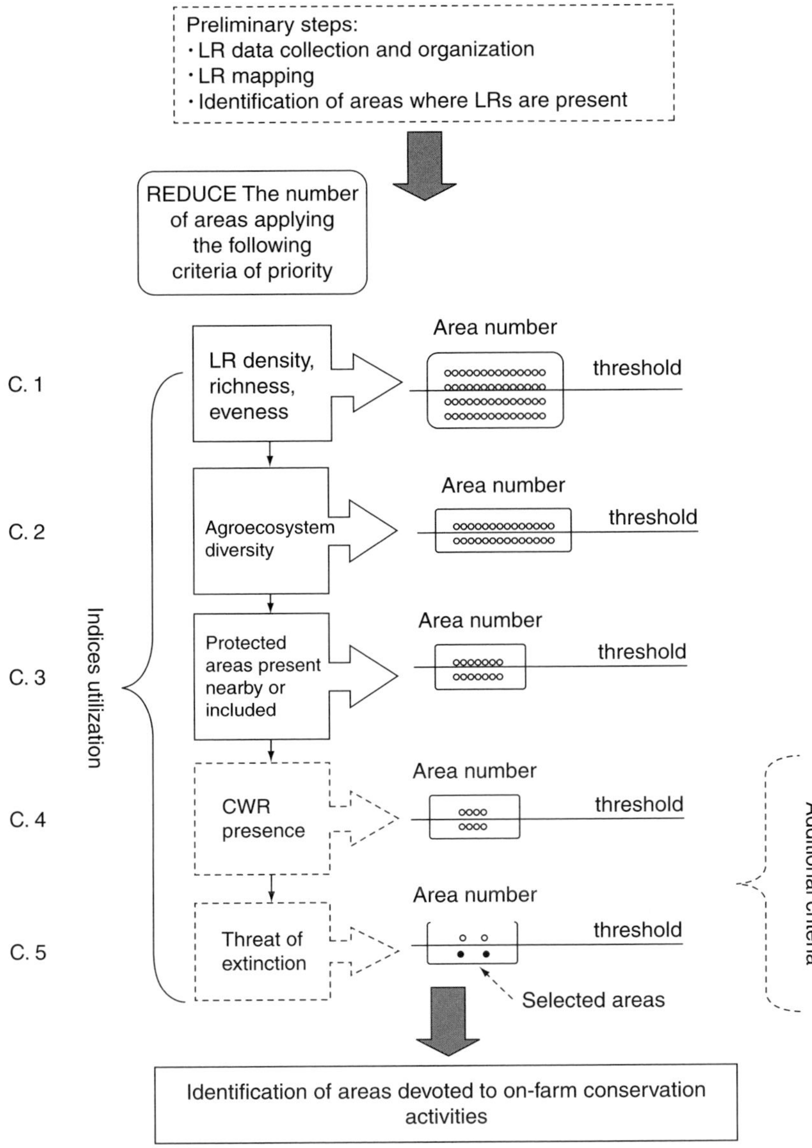

Fig. 16.1. Consolidated procedure for the choice of the most appropriated areas for LR conservation activities. Criteria are applied in sequence. For each level a threshold has to be defined below which areas are not admitted to the following level (area discrimination).

landraces (corresponding to about 10% of the maximum density recorded, i.e. 114 landraces per quadrant) was chosen for quadrant selection. The 18 quadrants containing ten or more landraces were then subjected to a diversity estimation by using a variety of diversity indices: Shannon index (H′) (Shannon and Weaver, 1949), Simpson diversity index (1-D) (Simpson, 1949), Pielou Evennes Index (J′) (Pielou 1975, 1996) and Berger-Parker dominance index (d) (Berger and Parker, 1970). However, since the index values were highly correlated (not shown), the Shannon index opportunely adapted to the case study was used (Siciliano *et al.*, 2009). The Shannon diversity index (H′) (Shannon and Weaver, 1949) has been calculated within each selected quadrant as described below:

$$H' = -\Sigma^{s} i = 1\ (Pi^{*} \ln Pi)$$

where *S* is the number of species and *Pi* is the LR frequency, i.e. number of LR belonging to each species / LR total number.

A threshold of Shannon index value of 1.2 was then arbitrary defined, below which areas were not admitted to the following discrimination step. This selection produced 14 quadrants for further diversity assessment.

These show the highest landrace density and diversity. In order to carry out a detailed agroecosystem diversity assessment to be able to apply the second criteria, several (pluviometric, phytoclimatic, geographical, land use) types of maps were initially overlapped to the 14 quadrants in order to find out the most valuable proxy. However, considered that only some of them gave complete information for the entire area of the quadrants, that available information was often obsolete, and the European scale of the project, it was decided that the CORINE land use map, which is available online for the entire European territory through the European Environment Agency website, was the most suitable source of information. The CORINE land use map was then overlapped to the previously created LR density map. CORINE land use classes are defined as follows: 1. Artificial surfaces; 2. Agricultural areas; 3. Forest and semi-natural areas; 4. Wetlands; 5. Water. For the protection and valorization of those areas that are richest in agrobiodiversity, the land use classes 2, 3 and 4 were considered the most relevant to be taken into account. Within each previously identified quadrant the number of different CORINE land use classes was counted and the percentage of each class, referred to the total number of classes present inside the quadrant, worked out. In order to find out the most agrobiodiversity-rich areas, a threshold value of 70% was chosen. On this base, nine quadrants were selected for the following discrimination step.

The presence of natural areas already protected (national or regional parks) inside or close to the areas to be subjected to on-farm conservation activities was then assessed by relying on the Ministry of Environment website. The application of this criterion finally reduced the number of areas to six.

The additional fourth criterion (i.e. CWR presence in the area) was also considered and applied by relying on available bibliographic data and field surveys, but did not further discriminate among areas since CWR were found to be present in all these areas. Finally, it was not possible to apply the fifth criterion to the case study due to the lack of reliable data. The selected areas are shown in Fig. 16.2 and consist of two main geographical units located in the hydrographic basin of Trasimeno Lake and the Apennine chains from Cucco to Sibillini mountains. In the first area the main category of land use is agricultural; the occurring landraces represent 26% of those inventoried in this study and the most important in terms of conservation and biological value are cowpea (*Vigna unguiculata* (L.) Walp. subsp. *unguiculata*) and turnip (*Brassica rapa* L.). In the second area the main category of land use is natural/semi-natural, and among the occurring landraces, representing 7% of those inventoried, the most important are black celery (*Apium graveolens L.*) and lentil (*Lens culinaris Medik*), from Trevi and Norcia villages, respectively.

16.7 Conclusions

The areas identified should be considered the most appropriate areas to be proposed to the National or Regional authorities as areas where to set or enhance political and economic actions in favour of landrace and agrobiodiversity conservation. Although the strategy can obviously be improved, it proved to be useful in identifying areas that include the maximum number of LRs and ecological diversity. If a suitable database were to be made available in the future, this

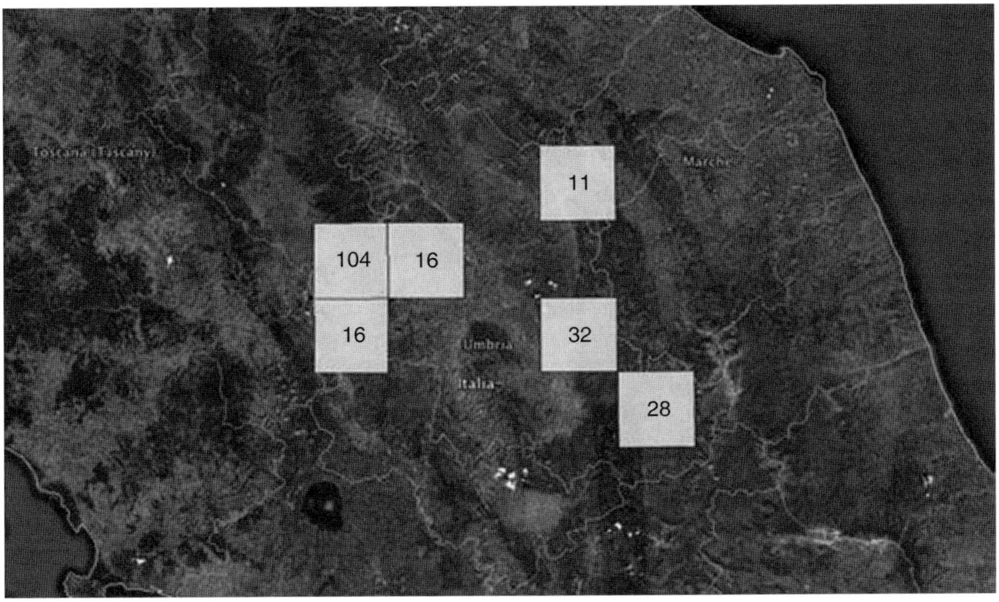

Fig. 16.2. The most appropriate areas for agrobiodiversity conservation in Central Italy as identified by the strategy. Numbers within the areas refer to the number of landraces inventoried.

model strategy could also be applied to other biogeographic regions in order to recommend other conservation areas.

Acknowledgements

Thanks are due to: Prof. L. Russi, Dr D. Donnini, Dr R. Torricelli, Dr V. Siciliano and Dr A. Battistini (Applied Biology Department, Perugia University) who took part in the expert panel, for having dedicated part of their time to this work; to Ms L. Stigter (Columbia University, New York) and Mr M. Cenci, who helped in inventorying LRs within their stage and MS thesis preparation, respectively.

References

Barocco, R., Venanzoni, R., Donnini, D., Pacicco, L. and Negri, V. (2010a) Towards planning the in situ conservation of local varieties (Landraces) and their wild relatives (CWR): The results of the project AEGRO (057_AGRIGENRES EC 870/2004). In: *Proceedings of 105th Congress of Italian Botanical Society Onlus*. Milano.

Barocco, R., Pacicco, L., Venanzoni, R., Veronesi, F. and Negri, V. (2010b) A comparison of strategies to sort out the most appropriate areas for on farm conservation of landraces. In: Symposium CWRs and Landraces in Europe. *'Towards the establishment of genetic reserves for crop wild relatives and landraces in Europe' and the joint meeting of the ECPGR In situ and On-farm Conservation Network and the final meeting of the EU project AGRI GENRES 057 – AEGRO*. University of Madeira, Funchal, 61 pp.

Berger, W.H. and Parker, F.L. (1970) Diversity of planktonic foraminifera in deep sea sediments. *Science* 168, 1345–1347.

Lande, R. (1996) Statistics and partitioning of species diversity, and similarity among multiple communities. *Oikos* 76, 5–13.

Maxted, N. and Kell, S.P. (2009) *Establishment of a Global Network for the In Situ Conservation of Crop Wild Relatives: Status and Needs*. FAO Commission on Genetic Resources for Food and Agriculture, Rome, 266 pp.

Maxted, N., Scholten, M.A., Codd, R. and Ford-Lloyd, B. (2007) Creation and use of a national inventory of crop wild relatives. *Biological Conservation* 140, 142–159.

Maxted N., Negri, V. and Veteläinen, M. (2009) Landrace inventories: needs and methodologies. In: Veteläinen, M., Negri, V. and Maxted, N. (eds) *European Landraces: On-farm Conservation, Management and Use*. Bioversity International, Rome, pp. 45–52.

Negri, V., Barocco, R., Pacicco, L., Veronesi, F. and Venanzoni, R. (2010) Landraces in Europe: an approach towards identifying landrace rich areas to be protected with priority. In: *CWRs and Landraces in Europe. 'Towards the establishment of genetic reserves for crop wild relatives and landraces in Europe' and the joint meeting of the ECPGR In situ and On-farm Conservation Network and the final meeting of the EU project AGRI GENRES 057 – AEGRO.* University of Madeira, Funchal 40 pp.

Pielou, E.C. (1975) *Ecological Diversity*. Wiley, New York.

Pielou, E.C. (1996) *An Introduction to Mathematical Ecology*. Wiley, New York.

Porfiri, O., Costanza, M.T. and Negri, V. (2009) *Landrace Inventories in Italy and the Lazio Region Case Study*. In: Veteläinen, M., Negri, V. and Maxted, N. (eds) *European Landraces: On-farm Conservation, Management and Use*. Bioversity International, Rome, pp. 117–123.

Shannon, C.E. and Weaver, W. (1949) *The Mathematical Theory of Communication*. University of Illinois Press, Urbana, Illinois.

Siciliano, V., Battistini, A., Barocco, R., Cenci, M., Venanzoni, R. and Negri, V. (2009) Definition of methodologies to develop genetic reserves for landraces in central Italy: model strategy for setting conservation areas in Europe. In: *Proceedings of 45th International Congress of the Italian Society of Vegetation Science (SISV) and Italian Federation of Phytosociology (FIP)*, Cagliari.

Simpson, E.H. (1949) Measurement of diversity. *Nature* 163, 688.

Veteläinen, M., Negri, V. and Maxted, M. (2009) *European Landraces: On-farm Conservation, Management and Use*. Bioversity International, Rome.

17 On-Farm Conservation of the Forage Species Timothy, Meadow Fescue and Red Clover: Generation of New Landraces in Norway

K. Daugstad

17.1 Introduction

A project funded by the Norwegian Genetic Resource Center started in 2003, the purpose being to make new landraces of the most common meadow species in Norway. These are the two grasses timothy (*Phleum pratense*) and meadow fescue (*Festuca pratensis*), and the legume red clover (*Trifolium pratense*). For each species the goal is to develop several local populations adapted to different climatic conditions and different farming systems. To achieve this we must restart the processes that developed our first landraces.

17.2 Short History of Forage Production in Norway

Before humans started cultivating the land there was limited area of grassland in Norway. Except for the grassy heathlands in the mountains, natural grassland only existed as small patches along the coast and rivers and around bogs. Most of the natural grassland in Norway today is a result of grazing ruminants and also an extended exploitation of wood for both building material and for firewood. In addition, winter feed harvested both on the farm and out-field was necessary to feed the animals during the long winter. Haymaking, grazing and traffic have favoured grass species in their competition with other plant communities.

The practise of cultivated leys and pastures is quite new in Norway. For instance the cultivation of timothy (*Phleum pratense*) started when the first seeds from England were introduced about 200 years ago. Vestad (1952) writes that the cultivation of timothy probably started in North America, with seeds brought from Scandinavia with the immigrants. Seeds of timothy were later introduced to England from North America in 1760, and further to other European countries. Since timothy is wild-growing in Europe it is not unlikely that the farmers had tried growing timothy before they obtained access to imported seed. This change from collection of forage from both outfields and infields, to cultivation of forage with ploughing and sowing, started in the early 1800s, but it took over a hundred years before it became common practise throughout the country.

The growing of the forage legume red clover (*Trifolium pratense*) has a somewhat similar story as timothy. Red clover is also wild-growing in Norway, but the wild populations are distinct from the cultivated populations, both in morphology and in time of flowering. The origin of the persistent cultivated red clover is not known, but has probably developed through a combination of natural selection of the non-persistent 'oceanic' red clover from southern and western parts of Europe crossing with wild adapted material (Wexelsen, 1937; Marum, 2009).

Meadow fescue (*Festuca pratensis*) has a shorter history of cultivation in Norway than timothy and red clover. Meadow fescue is earlier flowering and leafier than timothy, and well adapted to moist climate and frequent cuts. These qualities make meadow fescue more persistent and beneficial in modern agriculture with silage as the conservation method. Fjellheim *et al.* (2009) studied cultivars and populations of meadow fescue from Nordic and Baltic regions and concluded that the between-population diversity is limited. The wild-growing meadow fescue we find in Norway is probably naturalized due to spreading from cultivated meadows.

17.3 Norwegian Landraces of Forage Crops

The first seed of forage crops introduced to Norway in the early 1800s were not adapted to Norwegian conditions. However, it was common to produces one's own seed, either by setting aside part of a field for seed production or by collecting those seeds left lying on the hay barn floor. After several generations of seed production on the same farm or within a certain area, the population gradually became dominated by adapted plants. Most forage crops grown in Norway are out-breeding species. In out-breeding species a population consists of a large number of different genotypes, each with slightly different characteristics and growing value. Under a given climate and growing condition some genotypes will be better adapted than other genotypes. These adapted genotypes will produce more seed than the less adapted genotypes, resulting in an improvement in adaptation in the next generation.

Until 1950 several landraces of forage crops existed in practical use. This was especially the case for timothy (Vestad, 1952) and red clover (Wexelsen, 1951). Large scale seed production of timothy is known as far north as approximately 67°N (Nordland county) while seed of red clover probably was produced no further north than 64°N (Trøndelag county). The landraces developed after several generations of forage and seed production on the same location. These landraces were adapted to the harvest management and to the local climatic conditions. Since about 1950 the farmers have had access to better performing commercial varieties. This more or less put an end to the use of home-grown seed. At the same time the use of mineral fertilizers increased, thus diminishing the role of red clover in grassland farming.

The Nordic Genebank (now: Nordic Genetic Resource Center – NordGen) was founded in 1979. This was too late for the landraces, except for some landraces of red clover kept by the Norwegian Agricultural University. During the 1950s the university collected and evaluated landraces of timothy and red clover. Unfortunately only the most valuable, from that time's point of view, was conserved.

'Molstad' is the most famous red clover landrace. Seed production has been documented on the same farm for over 100 years, and the landrace was the main Norwegian red clover cultivar until 1990. 'Toten' and 'Leinum' are two other landraces of red clover. 'Grindstad' timothy is both a landrace and the most important timothy cultivar in Norway today. 'Grindstad' probably originated from seeds imported from Scotland in the 1860s (Marum and Daugstad, 2009). The reason it became such an important cultivar must be attributed to the method of seed propagation. This method is very similar to on-farm conservation, with 2 years of forage production, before seed harvest on the surviving plants in the third year. In meadow fescue there are no known landraces due to the short growing history. The old cultivar 'Løken' is the closest we can get a landrace. This cultivar originates from collections made on 'the Research Station for Mountain areas' (now part of Norwegian Institute for Agricultural and Environmental Research) shortly after the station was founded in 1918.

In the 1970s and the early 1980s extensive collection expeditions were conducted in the Nordic countries. Populations of forage species were collected in old meadows with long continuity to save some of the variability still existing. These populations were a major source for the national forage breeding programme and are also stored and maintained *ex situ* of NordGen.

17.4 Developing New Landraces with On-farm Conservation

Norwegian food production still depends on ruminants transforming roughage to human food. Today about 65% of the cultivated area is grassland. Non-cultivated areas are still used for grazing especially in the mountain areas, though in lesser extent than previous. Agriculture has undergone great changes, with intensification of management practice on one hand and with cultivated land abandoned and undergoing reforestation on the other. The number of farms declines and the remaining farms become larger. Today's agriculture most probably requires different landraces to those existing in 1950.

Marum (1999) launched the idea of creating new 'synthetic' landraces in different climatic zones and under different management systems and use them in an on-farm conservation system to create new adapted 'gene blocks'. The national forage breeding programme also tried the 'Grindstad-method' on a limited number of locations and with different breeding material. In 2003 the project 'On-farm conservation of the forage species timothy, meadow fescue and red clover – generation of new landraces in Norway' was initiated.

17.4.1 Gene pools

Since very few of the Norwegian landraces are conserved, the first step of the project was to select the material. The forage collection of NordGen was the main source, but also commercial cultivars were included. Even some cultivars with southern adaptation were included to meet the requirement of more climate-adapted material in the future warmer climate. From 2003–2006 the three wide 'starting populations' were made by crossing the material for two generations. The first generation of seeds were harvested on single plants, the second as bulk.

Timothy (Phleum pratense)

The main source for the timothy gene pool was a cross of 376 accessions from 1997. This cross was the result of a Nordic phenotypic evaluation project of populations in the forage collection of NordGen. Half of the evaluated accessions came from Norway, 27% from Finland, 16% from Sweden, 6% from Denmark and 1% from other countries. The project was conducted in four Nordic countries from 1994 to 1997. In addition, 20 cultivars were included: 'Vega' and 'LØTI8701' from Norway; 'Ragnar' and 'Alexander' from Sweden; 'Tuuka' from Finland; 'Dolina', 'RvP893' and 'RvP1121' from Belgium; 'Climax' and 'Richmond' from Canada; 'Comtal' and 'Liphlea' from the USA; 'Sobol' from the Czech Republic; 'CD18' from former USSR; 'Gintaras' and 'No 1532' from Lithuania; 'Tika' and 'Jogeva 54' from Estonia; and 'NOR1' and 'NOR2' from a Nordic project. The important cultivars 'Bilbo' from Denmark and 'Grindstad' and 'Engmo' from Norway were not included due to their high representation in the evaluation project 1994–1997. The field consisted of a total of 2000 genotypes, 400 from the new cultivars and 1600 from the Nordic evaluation project.

Red clover (Trifolium pratense)

The main source for the red clover gene pool was a cross made in 1998 of 283 accessions of red clover. This cross was part of a Nordic pre-breeding project (Helgadottir *et al.*, 2000). Of these accessions, 46 were from former USSR, 12 from outside Nordic countries, and the rest (225) from Nordic countries: 84 from Sweden, 73 from Norway,

45 from Finland and 23 from Denmark. For our project a 55:45 mixture was made of seed harvested in Norway and Denmark, the latter with two different harvest dates. In addition, nine varieties were included: 'Nordi', 'Lea' and 'Liv' from Norway; 'Rajah' from Denmark; and 'Pallas', 'Ares', 'Bjørn', 'Bjursele' and 'Jesper' from Sweden. The field consisted of a total of 2000 genotypes, 200 from the new cultivars and 1800 from the Nordic pre-breeding programme.

Meadow fescue (Festuca pratense)

The main source for the meadow fescue gene pool was 132 accessions from NordGen, selected in the way that all accessions available at the time the project started in 2003 were included. There were 87 accessions from Norway, 17 from Sweden, 18 from Denmark, nine from Finland and one from Iceland. In addition, 15 varieties were included: 'Norild', 'Vigdis', 'Salten', 'Fure' and 'Løken' from Norway; 'Laura' from Denmark; 'Arni/Kauni' from Estonia; 'Stella' and 'Darimo' from the Netherlands; 'Lifara' and 'Leopard' from Germany; 'Skrzeszowicka' and 'Skava' from Poland; 'Merifest' from Belgium; and 'Severodinskij' from former USSR. The field consisted of a total of 1800 genotypes.

In Fig. 17.1 each accession or cultivar counts as one, independent of the number of genotypes of each. In the meadow fescue gene pool each cultivar and accession are represented by the same number of genotypes (12). For timothy and red clover, the new cultivars are represented with more genotypes than each of the original accession. More detailed calculations are not possible due to limited control of how much each genotype contributes in the final seed lot produced in 2006.

17.4.2 Locations of experimental meadows

The seed harvested in 2006 was used to make seed mixtures containing 65% timothy, 25% meadow fescue and 10% red clover. During 2007 and 2008 seven locations spread all over Norway were sown with this mixture.

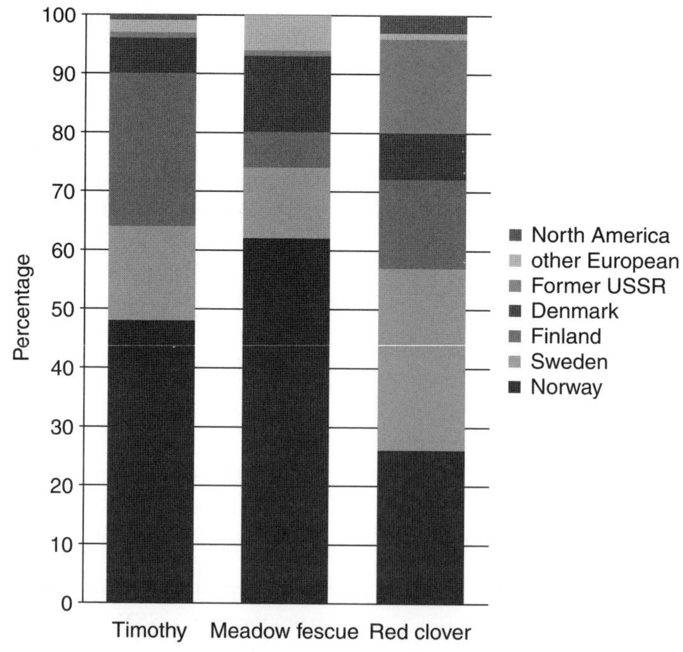

Fig. 17.1. Country of origin of the populations contributing to the three starting gene pools.

Table 17.1. Locations of experimental meadows.

Location	Height above sea level (m)	°N	°E	Annual precipitation (mm)	Annual temperature (°C)
Øsaker	40	59.32	11.04	853	6.1
Fureneset	20	61.29	5.04	2010	7.0
Løken	530	61.12	9.06	590	1.6
Brønstad	100	64.23	12.29	1000	4.3
Teigen	40	65.91	12.42	1020	5.3
Vågønes	40	67.28	14.45	1055	4.3
Flaten	20	69.93	23.25	400	1.3

These locations are the first generation of the actual on-farm conservation method, but shall also produce enough seed for several purposes. The locations are on both research stations and on farms (see Table 17.1).

The meadows were harvested for forage for 2 years. Management regime in the two forage years varied on the different locations. Type of and amount of fertilizer and the number of cuts depended on the local conditions and the main practice in the region. The third year (2010) seed was harvested on five of the locations. Commercial seed production is situated in south-eastern parts of Norway, due to dry summers and a long growing season. Proper seed maturation is a challenge in the regions with high rainfall and/or a short growing season. These areas are mainly in western (Fureneset) and northern (Teigen, Vågønes, Flaten) parts of Norway, but also in the mountain areas (Løken). At the same time these regions hopefully will give a stronger selection pressure. The experience so far is that regardless of the small seed yields the seed quality is satisfactory. The seed will be used to establish new meadows on the same locations, and also a sample of the seed will be stored. The stored seed will be available for both phenotypic and genotypic studies.

17.5 Follow-up of the Project

After 8 years with the project wide gene pools have been made and the first seed produced on five different locations. The first generations of seed production did not turn out like the meadows sown with poorly adapted seed from England in the 1800s. So far, an acceptable amount of red clover has survived, even in the northern-most location (Flaten), and for the grass species the winter survival has been very good. The project has so far depended on financial compensation to the research stations and the farmers. Due to the great costs of re-sowing meadows, farmers are genuinely interested in new persistent cultivars, but probably the interest will not survive loss of funding. The financial carrot for the farmers could be the prospects for future seed production of landraces. Seed production of species for restoration and reestablishment of vegetation after road construction and industry could be a niche. Maybe there will be more difficulties to develop seed production of common forage crops for a limited area as an economic sustainable industry. The temporary conclusion for the follow-up of the project is financial compensation to collaborating interested farmers. Agricultural schools and museums, and institutions with some governmental funding could be additional locations. Since the gene pools are made of Nordic material they should be used by institutions in other Nordic countries as well.

It is important to focus on the main goal, which is to develop useful landraces for future agriculture. But there is no contradiction at the same time to perform scientific studies of population genetics, conservation methods, phenotypic and genotypic characterization and plant breeding. For the last case a model with participating plant breeding could be evolved between the breeding companies and the on-farm conservationists.

References

Fjellheim, S., Pasakinskiene, I., Grønnerød, S., Paplauskiene, V. and Rognli, O.A. (2009) Genetic structure of local populations and cultivars of meadow fescue from the Nordic and Baltic regions. *Crop Science* 49, 200–210.

Helgadottir, A., Larsen, A., Marum, P., Fritsen, H., Lindvall, E. and Miettinen, E. (2000) Prebreeding of red clover (*Trifolium pratense* L.) for northern areas. *Acta Agriculturae Scandinavica Section B-Soil and Plant Science* 50, 187–190.

Marum, P. (1999) Should *in situ* conservation replace *ex situ* conservation of forage crops? *Botanica Lithuanica* (Suppl.)2, 99–104.

Marum, P. (2009) Molstad rødkløver – medalje for frødyrking. Available at: www.skogoglandskap.no/Artsbeskrivelser/moldstad (accessed 11 January 2011).

Marum, P. and Daugstad, K. (2009) Grindstad timothy: the landrace that became a major commercial variety. In: Veteläinen, M., Negri, V. and Maxted, N. (eds) *European Landraces: on-farm conservation management and use*. Bioversity technical Bulletin No. 15. Bioversity International, Rome, pp. 187–190.

Vestad, R. (1952) Norske timoteistammer og stammeforsøk i de forskjellige landsdeler. *Forskning og forsøk* 4, 55–78.

Wexelsen, H. (1937) Undersøkelser av norsk rødkløver. Stammeundersøkelser. *Tidsskrift for det norske landbruk. 2.hefte, februar* 1937, 41–67.

Wexelsen, H. (1951) Lokalstammer av norsk rødkløver. *Forskning og forsøk* 2, 185–191.

18 On-Farm Conservation and Participatory Maize Breeding in Portugal: An Overview

P. Mendes-Moreira and S.E. Pego

18.1 Introduction

Maize has evolved for more than five centuries in Portugal. During this period, maize contributed to altering the landscape (e.g. terraces, irrigating systems were built) and the improvement of livelihood (e.g. maize was available directly for human consumption as maize bread – 'Broa' – and indirectly by animal consumption). This period of time was especially important to generate diversity. The created diversity was driven by two main forces: (i) the environment (Portugal has a very diverse climate, mainly due to orography and by the Atlantic and Mediterranean influence); and (ii) the farmers' selections (e.g. plant and ear traits).

This genesis of diversity started to decrease as the American hybrids were introduced in Portugal after the Second World War. The FAO programmes for hybrid production in Europe had excellent adaptation to Portugal and breeding stations were established along the country from north to south, but only NUMI at Braga (NUMI – maize breeding station) survived for a long period. NUMI success was specially oriented for grain quality for human use as bread and early-maturing varieties adapted to highly intensified cropping systems. Silas Pego defined the breeding programme at NUMI as considering: on-station breeding, genetic resources, pre-breeding and participatory plant breeding.

The awareness of genetic erosion led Silas Pego to initiate the collection missions for maize in 1975. In the following years, a more in-depth collection supported by FAO/IBPGR and led by Rena Farias covered all the country in successive missions. These materials, together with the previous seed stock of the Maize Breeding Station NUMI, gave rise to the first long-term cold storage facilities that were the precursors of the Portuguese Plant Germplasm Bank (BPGV).

To bridge the gap between curators and breeders, pre-breeding techniques, such as *HUNTERS*, *Overlapping Index*, as well as test-crosses, were created or adapted to provide vital information from a breeder's perspective. In this way potential sources of germplasm could be used in a breeder's programme (Moreira *et al.*, 2008).

The awareness by the scientific community of genetic resources coevolution started in 1980 and 1990, i.e. the need for *in situ*/on-farm conservation. This topic was stressed by FAO throughout the 'Global Plan of Action' (FAO, 1996). Under this international frame, Silas Pego started in 1984 a participatory plant breeding (PPB) programme with CIMMYT support at Sousa Valley (VASO). The VASO programme was based on: (i) an integrant philosophy, i.e. intended to answer the problems of small farmers and increasing yield without losing the parameters defined by the farmer for bread-making quality, potential for polycropping systems and the use in sustainable agriculture; and (ii) the concepts of quantitative genetics in population improvement. Mass selection was applied both to landraces (e.g. 'Pigarro') and to 'Fandango', a synthetic population. Recurrent selection by S2 lines was also used for 'Pigarro'. On both populations, the yield component and pest and disease evaluations were performed and implicitly the quality for Broa (maize bread) (Vaz Patto *et al.*, 2009) was analysed. Seed samples were kept in cold storage from each selection cycle. To initiate the VASO programme, three main decisions had to be taken: (i) the location that better represented the region – a traditional maize area, where previous agro/sociologic/economics data existed. In addition, the support of a local elite farmers' association (CGAVS) that was committed to test the efficiency of an alternative project supposed to improve the local germplasm versus hybrids production, at least in certain specific circumstances; (ii) the farmers to work with – side by side, to whom the decision power will be allowed, and whose initial acceptance and enthusiasm were crucial; and (iii) the germplasm source to start from: 'Pigarro' (Moreira *et al.*, 2008) and 'Fandango' (Pego and Antunes, 1997; Moreira, 2006). These tacit choices imply a careful respect for the local traditional agriculture. While the breeder would apply his breeding methodologies, the farmers would continue a parallel programme with their own mass selection criteria. With this agreement, the breeder had to accept low input and intercropping characteristics, as well as accept and respect the local farmer as the decision maker. On the other hand, the farmer was able to compare the effectiveness of the two breeding systems (phenotypic recurrent selection and S2 lines recurrent selection). This allowed the farmer to base his decisions on solid grounds. Finally, due to the choice of local adapted germplasm, diversity and quality were considered as the priority traits.

The aim of this work is to present the main conclusions obtained from the evaluation of the yield gain and plant performance after 20 years of a PPB approach for 'Pigarro' (two breeding methodologies) and 'Fandango' (different cycles of mass selection initiated by the breeder and continued by the farmer). A determination of the best explanatory variables that explain yield was performed based on the collected data and applying MARS. Future perspectives are discussed.

18.2 Material and Methods

Mass selection on 'Pigarro' was done by the farmer for whom the aims of selection were to obtain bigger ears (larger and wider ears) while maintaining the kernel type (flint and white). Hence breeder selection intended to compare an alternative method (recurrent selection by S2 lines) and proceed with a systematic genetic improvement of maize populations, i.e. increase favourable alleles for both yield and two main physiologic traits: ear placement and stalk quality. This goal was intended to be achieved under the maintenance of its kernel quality standards and under a sustainable polycropping system with low inputs.

The 'Fandango' mass selection was divided in two phases: (i) the breeder's phase from cycle 1 to cycle 5; and (ii) the farmer's phase, after cycle 5. Both had different goals: breeder's for yield maximization, and farmer's for big ear size maximization.

Depending on selection method and population, 6–7 cycles were chosen for mass selection on 'Fandango' (two phases) and 'Pigarro' and three cycles of S2 recurrent selection cycles on 'Pigarro'. Trials started in 2005 and they were performed at four locations in Iowa, USA during 1 year and in Portugal, in two to three locations, in 3–4 years. Yield evaluation was done for Iowa and Portugal; for Portugal, morphological data (e.g. ear length and fasciation level) were also measured.

Data analyses included ANOVA and regression analyses. Multivariate Adaptative Regression Splines (MARS) was used instead of regression analyses when assumption of normality was not fulfilled. Both regression analyses and MARS allowed the evaluating of response to selection of the traits studied. To better understand what influences the yield, three methods for analysis have been used: MARS, Classification and Regression Trees (CART) and Random Forests (RF). More detailed information for 'Pigarro' and 'Fandango' can be found in Mendes-Moreira *et al.* (2008, 2009).

18.3 Results and Discussion

18.3.1 Improved populations under PPB

During 20 years of on-farm PPB some results and considerations can be recognized. The statistical analyses on 'Pigarro' response to mass selection indicate that after 20 cycles, ear, cob, medulla and raquis diameters, kernel and cob weight and fasciation significantly increased. On the contrary, thousand-kernel weight decreased significantly. Molecular analyses revealed that no effective loss of genetic diversity had occurred during the selective adaptation to the farmer's needs and to the regional growing conditions, but the genetic diversity maintained is not exactly the same (Vaz Patto *et al.*, 2008). The response to recurrent selection by S2 lines on Pigarro indicate that after three cycles of recurrent selection convulsion decreased significantly, while row number decreased as did the fasciation.

The 'Fandango' mass selection was divided in two phases: (i) the breeder phase from cycle 1 to cycle 5; and (ii) the farmer phase, after cycle 5. Both had different goals, the breeder for yield maximization and farmer for big ear size maximization. The results from response to mass selection in Portugal revealed that ear length significantly decreased and simultaneously plant and ear height, ear diameter, kernel row number and fasciation significantly increased as did convulsion. Nevertheless, for breeder cycles these traits were maintained. This selection also led to significant increases of days-to-silk and anthesis. Identical outcomes were observed in long-term divergent selection for ear length in maize (Hallauer, 1992). However, according to MARS (R^2=8.9% for all locations and R^2=80.5% for Lousada), the farmer selection contributed to grain moisture increase contrary to breeder selection. This tendency occurred also with days-to-silk and days-to-anthesis.

The lack of significant progress in yield for both 'Pigarro' and 'Fandango' (after C5) can be explained by low selection intensity due to the exclusion of stalk lodged plants in the basic units of selection. Specifically for 'Fandango' selection, plant and ear height significantly increased, which could mean less area available, i.e. competition in trials was more severe in advanced cycles and some plants did not produce ears. Probably for this reason a significant decrease in yield was observed at Iowa locations. In the case of recurrent selection by S2 lines, yield decrease could be related with fasciation expression decrease or even to the selection procedures for stalk and root lodging improvement. In order to better clarify this situation, more cycles of recurrent selection would be needed.

The fasciation evaluation indicates that the farmer explored fasciation for increased diameter and row number. This fact is significantly interesting at Lousada for: (i) 'Pigarro' (R^2=89.0) with 3.32% of gain per cycle/year; and (ii) 'Fandango' (R^2=78.8) with 4.66% of gain per cycle/year.

The RF, CART and MARS analyses were used to determine the best variables that explain yield. The response to both methods of selection for 'Pigarro' and 'Fandango' mass selection indicate kernel weight and ear weight as the most distinct traits for all methods. The secondary distinct traits were: (i) ear length and plant height for 'Pigarro' mass selection; (ii) ear height and thousand kernel weight for recurrent selection on 'Pigarro'; and (iii) rows numbers, number of kernels per row and ear length for 'Fandango' mass selection.

Both selection methods used in 'Pigarro' or different phases in selection for 'Fandango', suggested that mass selection is better than S2 recurrent selection due to the following reasons: (i) mass selection is a cheaper methodology, technically more accessible to farmers, which is a great advantage in the establishment of on-farm conservation programmes; (ii) one cycle of selection can be completed each summer, and *in situ*/on-farm conservation of the genetic diversity is effective (Vaz Patto *et al.*, 2008). This highlights its role as a backup system (complementary with *ex situ*) and a monitoring process for an effective on-farm conservation of diversity.

18.3.2 Socio-economics and anthropology

The VASO project throughout its praxis allowed the farmer and the breeder to compare breeding methodologies in loco, i.e. decisions based on knowledge (Pego and Antunes, 1997). It also encouraged local initiatives, such as the 'Sousa Valley Best Ear Annual Contest' (selection for big ears) by the local Farmers' Cooperative Association (Cooperativa Agrícola de Paredes). This has contributed to the recognition of the farmer by the community, but also has attracted new farmers and new germplasm to this programme that in this way could be identified and preserved on-farm by the same approach (Moreira *et al.*, 2008).

The anthropological and sociological objective of PPB suggests that more attention should be given to: (i) learning more about how plant breeding itself has been influencing farm changes and agricultural systems; for example, is on-farm plant breeding simply conventional plant breeding on farms, or is it a whole different kind of plant breeding approach for the future? (Powell, 2000); (ii) how on-farm conservation is managed to ensure genetic diversity and breeding success; (iii) the definition of 'yield' needs to be reconsidered and broadened to include the total yield of the polycropping system and not just the yield of a single crop per se (Pego and Antunes, 1997; Powell, 2000); and (iv) it is important for breeders to work with other people involved in the food production 'chain', such as traditional grain millers and also bakers (Powell, 2002).

At present, 48% of the farmers in Portugal are older than 65 (INE, 2010). This situation highlights the problem of successive agricultural policies pursued by the European Community, i.e. the disappearing of smallholder farming as a viable way of life in Portugal and the socio-economic 'pull' factors that remove younger generations from the farm (Powell, 2000; Vaz Patto *et al.*, 2007).

18.4 Future perspectives

Landraces, through their resilience to pest and diseases and abiotic stresses, nutrient uptake efficiency, their phytonutrients and micronutrient concentrations demonstrate adaptation to marginal conditions (e.g. protected areas and to climate changes, due to its diversity and long-term adaptation representing a valuable potential in organic and sustainable agriculture; Maxted *et al.*, 2002; Newton *et al.*, 2010). Furthermore, landraces are also 'the living masterpieces' of the interaction among human, genotypes and environment representing traditions (e.g. traditions, tastes, flavours) (Negri, 2005). Nevertheless, some of these landraces have a huge gap with modern cultivars' yield (e.g. maize). This fact led farmers to abandon their germplasm.

The participatory plant breeding approach (PPB) can be associated with *in situ* conservation of landraces contributing to their economically sustained presence in the farmers' fields. It can also contribute to define *in situ*/on-farm strategies that could help to design better synthetic hybrid populations for a new generation of low input and organic farming adapted to environmental changes and marginal areas.

The development of hybrid populations could also contribute to yield progress and to avoid the loss of some interesting germplasm. This approach can be applicable in a rural development strategy if economic benefits between associations for specialities (e.g. maize bread) and farmers could be achieved. This scenario is supported by hybrid populations' results from 2009 where yield was in average higher than 11 Mg/ha for two locations (data not published). This could be also of great importance to define heterotic groups that could enhance the breeding efforts (Tracy and Chandler, 2008). Current intellectual property rights do not protect farmers who are developing their own varieties, and this could also be a way to recognize the farmer conservation efforts.

The success of on-farm conservation programmes needs a cluster. A cluster where farmers' activities, such as food production, genetic resources conservation, environmental sustainability (e.g. soil and water management), forest protection (e.g. cleaning the forest to prevent fire) and landscape management could be rewarded by stakeholders. Some of these stakeholders can come from industry (e.g. milling industry), the plant-breeding sector (e.g. supporting conservation reserves – that can support co-evolutionary process and organic or low input niche markets) and tourism industry (e.g. rural tourism where specialities and traditional food are the major output). The multifunctional agriculture, as green care, social agriculture and urban agriculture, could be also an interesting promoter of on-farm conservation, because in many of these farms yield is not the main goal.

Finally, lessons from the VASO project could help us design new on-farm conservation projects not only for the Portuguese reality but also for developing countries where adaptation to small farmer's needs (e.g. maize quality for food, traditions) is outside the scope of multinational seed companies.

References

FAO (1996) *Global Plan of Action for the Conservation and Sustainable Utilization of Plant Genetic Resources for Food and Agriculture*. Food and Agriculture Organisation of the UN, Rome.

Hallauer, A.R. (1992) Recurrent selection in maize. In: Janick, J. (ed.) *Plant Breeding Reviews*. Wiley, London.

INE (2010) 2009 Agricultural census. Preliminary data.

Maxted, N., Guarino, L., Myer, L. and Chiwona, E.A. (2002) Towards a methodology for on-farm conservation of plant genetic resources. *Genetic Resources and Crop Evolution* 49, 31–46.

Mendes-Moreira, P., Pêgo, S., Vaz Patto, C. and Hallauer, A. (2008) Comparison of selection methods on 'Pigarro', a Portuguese improved maize population with fasciation expression. *Euphytica* 163, 481–499.

Mendes-Moreira, P., Vaz Patto, M.C., Mota, M., Mendes-Moreira, J., Santos, J.P.N., Santos, J.P.P., Andrade, E., Hallauer, A.R. and Pego, S.E. (2009) 'Fandango': long term adaptation of exotic germplasm to a Portuguese on-farm-conservation and breeding project. *Maydica* 54, 269–285.

Moreira, P.M. (2006) Participatory maize breeding in Portugal. A case study. *Acta Agronomica Hungarica* 54, 431–439.

Moreira, P., Pêgo, S., Vaz Patto, C. and Hallauer, A. (2008) Comparison of selection methods on 'Pigarro', a Portuguese improved maize population with fasciation expression. *Euphytica* 163, 481–499.

Negri, V. (2005) Agro-biodiversity conservation in Europe: ethical issues. *Journal of Agricultural and Environmental Ethics* 18, 3–25.

Newton, A.C., Akar, T., Baresel, J.P., Bebeli, P.J., Bettencourt, E., Bladenopoulos, K.V., Czembor, J.H., Fasoula, D.A., Katsiotis, A., Koutis, K., Koutsika-Sotiriou, M., Kovacs, G., Larsson, H., Pinheiro De Carvalho, M.A.A., Rubiales, D., Russell, J., Dos Santos, T.M.M. and Vaz Patto, M.C. (2010) Cereal landraces for sustainable agriculture. A review. *Agronomy and Sustainable Development* 30(2), 237–269.

Pego, S. and Antunes, M.P. (1997) Resistance or tolerance? Philosophy, may be the answer. *XIX Conference of the International Working Group on Ostrinia*, Guimarães, Portugal, pp. 303–341.

Powell, D. (2002) The Milling Landscape of Northwest Portugal. Unpublished Master of Science Thesis, California State University Davis, California.

Powell, J. (2000) *The Relationship Between Ideotypes, Knowledges and Practices in Plant Breeding Among Farmers and Scientists*. Society for Social Studies of Science (4S) and European Association for the Study of Science and Technology (EASST), Vienna, Austria.

Tracy, W.F. and Chandler, M.A. (2008) *The Historical and Biological Basis of the Concept of Heterotic Patterns in Corn Belt Dent Maize*. Blackwell Publishing, Oxford.

Vaz Patto, M.C., Moreira, P.M., Carvalho, V. and Pego, S. (2007) Collecting maize (*Zea mays* L. convar. *mays*) with potential technological ability for bread making in Portugal. *Genetic Resources and Crop Evolution* 54, 1555–1563.

Vaz Patto, M.C., Moreira, P.M., Almeida, N., Satovic, Z. and Pego, S. (2008) Genetic diversity evolution through participatory maize breeding in Portugal. *Euphytica* 161, 283–291.

Vaz Patto, M.C., Alves, M.L., Almeida, N.F., Santos, C., Mendes-Moreira, P., Satovic, Z. and Brites, C. (2009) Is the bread making technological ability of Portuguese traditional maize landraces associated with their genetic diversity? *Maydica* 54, 297–311.

19 Farm Seed Opportunities, Recommendations for On-Farm Conservation in Europe[1]

R. Bocci, V. Chable, G. Kastler and N. Louwaars

19.1 Introduction

Alongside conventional agriculture, other farming systems strongly connected to the 'terroir'[2] have been preserved and are now re-emerging in Europe based on growing market opportunities. These agricultural systems need a wide range of varieties – from landraces to old commercial varieties, farmers' varieties and populations – able to be adapted to diverse agronomical practices, social conditions and environments, particularly with the aim of increasing the resilience of the agroecosystems. Recently, in Europe the recognition of the limitations of modern commercial varieties – bred for conventional, high input agriculture – and the needs of organic farming have stimulated several innovative initiatives with regard to crop varieties, aimed at conservation, creating new forms of varieties, and changing plant breeding and seed production organization. But the large diversity of experiences and initiatives is not reflected in European laws and policies. European seed laws and policies had been conceived in order to modernize agricultural systems. This framework is based on the assumptions that seed systems follow a natural development pathway from farmers' production through government involvement towards a perfectly competitive private seed market, e.g. from the informal to the formal one. However, in the last years the European scenario has been changing due to the approval of the directives on 'conservation varieties' and the process of reviewing seed laws within the framework of the Better Regulation Strategy (FCEC, 2008).

Farm Seed Opportunities (FSO), a research project in the FP6 European Research Framework (2007–2009), was targeted to support the implementation of seed regulations on conservation varieties (directives 98/95/EC, 2008/62/EC, 2009/145/EC and 2010/60/EC) and to suggest complementary seed regulation scenarios taking into account the diversity of the European seed systems. The FSO project was a collaborative effort of farmers and scientists from France, Italy, the Netherlands, Spain, Switzerland and the UK. This paper, based on the Policy Recommendations of the project, highlights the role conservation varieties could have in Europe for agricultural diversity conservation and stresses the importance of appropriate laws and policies for the maintenance of vital informal seed systems even in Europe (Louwaars, 2007).

19.2 A New Concept of Conservation Varieties

The term 'conservation varieties' was first introduced in the EU Directive 98/95/CE, which included the policy objective of 'conservation' in the core of seed legislation. As stated in 17th preamble of the Directive, opening the official seed catalogue to conservation varieties – and thus marketing them as seed – was considered a means of reducing genetic erosion. In this regard the directive 98/95 implicitly acknowledges that seed regulations since the 1960s have contributed to the genetic erosion of agricultural diversity and must be amended. Since 1998, however, the road to acknowledge conservation varieties has been long and tortuous and the directive still lacks full application by Member States. In the 10 years spent developing new directives – giving the guidelines to Member States for the implementation of 98/95 – no less than 14 text revisions were discussed before it was passed by the Permanent Seed Committee, which shows that parties that have widely divergent interests do not easily reach an agreement. On the one hand, some saw a danger that regulating conservation varieties could undermine the main commercial system of introducing new varieties into the market; while others sought to open marketing possibilities to seed of varieties that were until then 'illegal' but of interest to other models of agriculture such as organic farming or bio-dynamics. Finally, in 2008, the Directive 2008/62/CE on agricultural species was approved; in November 2009 the Directive 2009/145/CE on vegetables saw the light and finally a comprehensive text on vegetatively propagated species and fodder plant mixtures was approved in 2010. Key features of a conservation variety within these directives are the concepts of 'landraces', 'local adaptation' and 'genetic erosion' (Bocci, 2009; Lorenzetti and Negri, 2009).

One of the intentions of any of the EU and national seed regulations is to guarantee seed quality. Good seed is important for every farmer. Purchased seed has to match the expectations of the buyer even though most quality factors cannot be identified by simply looking at the seed. The varietal identity has to be guaranteed and therefore the varietal uniformity should be within the expected range. The implications of these current practices, also included in conservation varieties directives, were particularly analysed during the FSO programme. The following are the main concerns that were identified.

1. *Region of origin.* Seed maintenance, production (except for vegetables) and marketing should be conducted in the identified region of origin of a conservation variety. However, throughout history, cultivated plants have travelled, such as potatoes and tomatoes from Latin America, carrots and onions from Asia and cabbages from Western Europe. Similarly, many old varieties, which may be thought to be local, may originate elsewhere. Hence, from a historical and ecological point of view there is no reason to restrict a plant genetic resource to a certain region. On the other hand, the name of a variety could be strictly correlated to a particular area; hence, it may be useful for some local communities to have a means to protect their patrimony from the global market (e.g. using tools such as Geographical Indications).

2. *Registration costs.* Seed producers of landraces and old varieties are mostly small enterprises that usually maintain and sell a wide range of crops and hundreds of varieties. Dutch authorities, for example, estimate that costs of registration and seed certification would amount to more than €1000 per variety. If the producers have to bear these costs, it would make it impossible for small enterprises to register all the varieties they maintain. Registration costs will then reduce instead of enhancing biodiversity on the seed market and the variety choice.

3. *Restriction of seed quantities.* Multiplication of each conservation variety is limited to 0.3–0.5% (depending on crop) of the total seed market of the crop concerned or the amount needed to sow 100 ha, whichever the greater quantity. These amounts may limit the commercial viability of multiplying and marketing conservation varieties, while there is no conceptual reason for restricting the hectarage. It appears that restrictions seem to be put in order to limit the market of conservation varieties and to prevent unfair competitiveness among seed industries by the use of the less restricted catalogue of these varieties.

4. *Registration requirements*: Distinctiveness, Uniformity and Stability. For description purposes varieties should be Distinct, Uniform and Stable. The uniformity requirement especially was considered a bottleneck. FSO therefore analysed current methods and standards for uniformity and conclude that a minimal description of the salient features of the conservation variety may be sufficient. In case a registrar may want to apply detailed descriptions for the registration of conservation varieties, then the same methods may be used as for conventional varieties with the exception that methods developed for cross-fertilizing crops may need to be applied for the description of genetically diverse self-fertilizing crops. Alternatively, a minimal description could be applied, describing the distinguishing features based on users' experiences. Strict uniformity standards should not be applied since the key objective of registering conservation varieties is to promote the sustainable use of diversity and that identification is a primary aim and not uniformity. Two issues need careful consideration: the inherent lack of stability of landraces may require a wide interpretation of the identity (description) of the landraces being considered as conservation variety, including an option to re-register a variety when it changes over time (e.g. as a result of climate change). Furthermore, the fact that current seed certification standards for uniformity are much stricter than the registration standards has to be dealt with in the implementing rules at the national level.

Finally, an analysis of the positions of countries (country representatives in the negotiations in Brussels) that led to the formulation of the directive revealed that the countries with a strong (conventional) seed industry have had a predominant position in the debate and not those that harbour the largest number (or hectarage) of potential conservation varieties.

19.3 Importance of Informal Seed Systems in Europe

FSO has painted a broad picture of the seed situations in Europe. The first result is that Europe is still full of diversity, at cultural, environmental, climatic and farming levels. Even if the formal seed system tends to impose its norms and modernization through regulations, it cannot address all the diversity of the European farming systems and the resulting farmers' needs.

FSO found that only two types of varieties will fall under the concept of conservation variety: traditional farmers' varieties/landraces and commercial varieties once registered in the catalogue, but for which the commercial interest declined. There are however other important categories for which seeds cannot be marketed at the moment, but for which it will be necessary to explore legislative openings in order to support the conservation and sustainable use of genetic resources in diverse farming systems. FSO studies identified the following categories:

- The varieties produced by farmers' and/or participatory plant breeding (Ceccarelli *et al.*, 2009) not in conformity with DUS requirements.
- The old varieties no longer registered in the Catalogue (there are factors that can make registering these varieties problematic: excessive registration costs, difficulty in proving the Value for Cultivation and Use (VCU), only limited marginal areas interested in growing them) and which do not have a precise geographical area of origin.
- Local varieties used as genetic resources in reintroduction programmes in different areas from gene banks (Louwaars, 2011) or other regions (Arndorfer *et al.*, 2008).
- Varieties – populations that have no historical link with a given territory and which cannot be registered in the official catalogue, having no correspondence with the DUS criteria.

These varieties may be important for increasing genetic diversity in the field – specifically in organic and low-input agriculture – and could play a key role also in facing climate change. All these categories are part of the European informal seed system (Lopez Noriega, 2009; Negri *et al.*, 2009; Osman and Chable, 2009; Veteläinen *et al.*, 2009).

Finding the right balance between formal and informal seed systems within the European context should be one of the objectives of the regional strategy for on-farm conservation of plant genetic resources for food and agriculture (PGRFA) (Visser, 2002). Such a strategy will also concretely address the implementation of the article 6 on sustainable use of PGRFA of the International Treaty on Plant Genetic Resources for Food and Agriculture (www.planttreaty.org), signed by the European Union and its members in 2004. This article is mandatory for Contracting Parties and applies to all crop species, not only to those listed in Annex I of the Treaty. Moreover, a good balance may facilitate the debate on Farmers' rights (Article 9) at regional and international level (Andersen, 2005) due to the fact that many actions included in Article 6 relate to Article 9:

1. The promotion of the use of local varieties and underutilized species can contribute to protecting traditional knowledge (Article 9.2 (a)).
2. Increasing farmers' options through participatory plant breeding could be considered a non-monetary benefit sharing measure (Article 9.2 (b)). Therefore an integrated on-farm strategy that includes informal seed systems and their varieties should consider the promotion of Participatory Plant Breeding (PPB) to help farmers to fulfil their needs, facilitating them in accessing the genetic resources and broadening the range of available species. All these are actions aiming to bring compensation in farmers' favour.
3. Promoting diversified agricultural systems would contribute to recognizing all farmers in the policy arena in agreement with Article 9.2 (c).
4. Finally, the relaxed regulations for conservation varieties enhances farmers' role on seeds exchange, reuse and sale in agreement with Article 9.3.

In this framework the on-farm strategy should allow the presence on the market of proximity (local market or direct sell) of the seeds of the varieties identified by FSO, and at the same time needs to avoid creating opportunities for the diffusion of poor quality varieties on commercial markets. To this goal the role of networks or associations could be a key element in order to set up a bridge between formal and informal seed systems. Since these are based on social norms, trust, reputation and governed by reciprocity, such networks could provide an alternative to the quality guarantees in the formal system (Lipper *et al.*, 2010). Therefore enhancing the role of social networks could improve the quality of the informal seed system. In this regard, the directives on conservation varieties open a new interesting possibility, for the first time allowing organizations to have a role within seed legislation (Article 34 of the directive 2009/145/CE and 21 of the directive 2008/62/CE).

Finally, we would like to stress the importance of such a strategy, also because 'it is impossible to replace farmers' seed systems completely and it would be unwise to try. Farmers' seed systems provide an important component of food security, a vital haven for diversity and space for further evolution of PGR' (FAO, 2009).

19.4 Conclusion

The continued use of landraces and other categories of local and genetically diverse varieties contributes to the objectives of the International Treaty on Plant Genetic Resources for Food and Agriculture (FAO, 2001). Such varieties also contribute to the resilience of agricultural production systems and to the development of markets for local and ecologically produced products. Conventional seed legislation outlaws the exchange or marketing of seed of such varieties. The directives on 'conservation varieties' of the European Union broaden opportunities to recognize different seed systems. However, several limitations are identified to the optimal use of genetic diversity in the field and much depends on the implementation of the European Union Directives by the Member States at the national level.

Notes

[1] This paper presents the main outcomes of the EU project 'Farm Seed Opportunities' – www.farmseed.net.

[2] A French word that refers simultaneously to the soil, climate and cultural values of an area, similar to the English notion of 'place'.

References

Andersen, R. (2005) The History of Farmers' Rights. The Farmers' Rights Project Background Study. Available at: www.fni.no/doc&pdf/FNI-R0805.pdf (accessed December 2010).

Arndorfer, M., Kajtna, B. and Vorderwuelbecke, B. (2008) Integrating *ex situ* and on-farm conservation approaches in the management of local vegetable diversity in Austria. *Acta Horticulturae* 817, 333–340.

Bocci, R. (2009) Seed legislation and agrobiodiversity: conservation varieties, Istituto Agronomico per l'Oltremare. *Journal of Agriculture and Environment for International Development* 103(1/2), 31–49.

Ceccarelli, S., Guimaraes, E.P. and Weltzien, E. (eds) (2009) *Plant Breeding and Farmer Participation*. Food and Agriculture Organization of the United Nations, Rome, 687 pp.

FAO (2001) International Treaty on Plant Genetic Resources for Food and Agriculture. Food and Agriculture Organization of the United Nations. Available at: www.fao.org/ag/cgrfa/itpgr.htm (accessed 21 October 2008).

FAO (2009) Strengthening seed systems: a contribution to the preparation of the Second Report on the State of the World's Plant Genetic Resources for Food and Agriculture. CGRFA, FAO, Rome.

FCEC (2008) Evaluation of the Community acquis on the marketing of seed and plant propagating material (S&PM). Food Chain Evaluation Consortium Final Report. European Commission Directorate General for Health and Consumers. Available at: http://ec.europa.eu/food/plant/propagation/evaluation/s_pm_evaluation_finalreport_en.pdf (accessed December 2010).

Lipper, L., Anderson, L. and Dalton, T.J. (eds) (2010) *Seed Trade in Rural Markets: implications for crop diversity and agricultural development*. Agricultural Development Economics Division, FAO, Rome.

Lopez Noriega, I. (2009) European legislation in support of home gardens conservation. In: Bailey, A., Eyzaguirre, P. and Maggioni, L. (eds) *Crop Genetic Resources in European Home Gardens*. Bioversity International, Rome.

Lorenzetti, F. and Negri, V. (2009) The European seed legislation on conservation varieties. In: Negri, V., Veteläinen, M., and Maxted, N. (eds) *European Landraces: on-farm conservation, management and use*. Bioversity International, Rome.

Louwaars, N.P. (2007) Seeds of Confusion; The impact of policies on seed systems. PhD Thesis, University of Wageningen, the Netherlands – with references – with summaries in English and Dutch.

Louwaars, N.P. (2011) Reconciling UPOV and integrated seed systems. *Seed Information* 40, 6–7. Available at: www.icarda.cgiar.org/News/Seed%20Info/SeedInfo_40/SeedInfo_40.pdf (accessed January 2010).

Negri, V., Maxted, N. and Veteläinen, M. (2009) European landraces conservation: an introduction. In: Veteläinen, M., Negri, V. and Maxted, N. (eds) *European Landraces: on-farm conservation, management and use*. Bioversity International, Rome.

Osman, A. and Chable, V. (2009) Inventory of initiatives on seeds of landraces in Europe, Istituto Agronomico per l'Oltremare. *Journal of Agriculture and Environment for International Development* 103(1/2), 95–130.

Veteläinen, M., Negri, V. and Maxted, N. (eds) (2009) *European Landraces: on-farm conservation, management and use*. Bioversity International, Rome.

Visser, B. (2002) An agrobiodiversity perspective on seed policies. In: Louwaars, N. (ed.) *Seed Policy, Legislation and Law: Widening a Narrow Focus*. The Haworth Press, London.

20 Portuguese Landraces: On-Farm Conservation, Management and Use

A.M. Barata, A. Reis, F. Rocha, V.R. Lopes, E. Bettencourt, J. Miranda, J.C. Dantas, O. Pinto-Carnide, M. Matos and V. Carnide

20.1 Introduction

Since the 1970s efforts have been made to mitigate the loss of variability of plant genetic resources (PGR) by establishing *ex situ* collections. In Europe, gene banks maintain approximately one-third of the world's *ex situ* crop germplasm collections (BISE, 2011). For more than three decades the Portuguese Genebank (BPGV) has been collecting, documenting, characterizing and conserving the diversity of crops and crop wild relatives (CWR) traditionally used and that occur in Portugal. BPGV maintains a total of 18,000 accessions from more than 100 species, including 1781 accessions of *Phaseolus vulgaris* L. (common bean), 258 of *Secale cereale* L. (rye) and 1913 of *Zea mays* L. (maize), as a result of systematic collecting missions, which have taken place since 1977 throughout the Portuguese mainland territory and the archipelagos of Azores and Madeira. (Bettencourt and Gusmão, 1981, 1982; Mota *et al.*, 1982; Farias, 1989, 1999; Farias *et al.*, 1992; Farias and Marcelino, 1993; Marcelino and Farias, 1994). Since the 1990s conservationists have believed that the conservation of plant genetic resources should be based on two complementary strategies – *ex situ* and *in situ* conservation – including on-farm approaches (Hawkes *et al.*, 2000). The Convention on Biological Diversity (CBD) (1992, 2002) specifically recognizes domesticated and cultivated species as an important component of global biological diversity.

Portuguese discoveries have greatly contributed to the introduction, exchange and dispersion of plant species. The most important plants involved in this exchange were maize, beans, potatoes, groundnuts, peppers, tomato and tobacco (Ferrão, 1992, 2005; Alves *et al.*, 1996). In Portugal, as in other European countries in the second half of the last century, important changes occurred in the productive systems and socio-economic context (Negri *et al.*, 2009). Agriculture became a market-oriented activity with replacement of landraces (Reis, 1997) and increased planting of monocultures based on few cultivars. Variety and seed certification systems associated with the application of plant breeders' rights, the creation of modern cultivars and governmental and European Community policies, as, for example, the European directives for agricultural crops, have also contributed to the decrease of the utilization of crop landraces. However, recently, the European Commission has adopted two Directives to promote a wider utilization of genetic diversity in agricultural systems and to improve the management of plant genetic resources, through the acceptance of agricultural and vegetable landraces and varieties to be grown and marketed (EC, 2008, 2009). In Portugal, important threats are the ageing of farmers and the depopulation of rural areas with consequent simplification of productive processes due to high cost and labour requirements and the loss of traditional knowledge associated with crops' production and utilization (Strecht, 2004; Frazão-Moreira and Fernandes, 2006).

BPGV, with their extensive experience in local varieties conservation and the promotion of traditional activities by educational approach (Barata *et al.*, 2008), defined the geographic area where conservation on-farm would have greater impact and potential due to the socio-economic characteristics, such as the practice, still common, of traditional agriculture and existing stocks of local varieties. The targeted area was a region that largely coincides with the National Park Peneda-Gerês (PNPG), situated in the north-west of the Iberian peninsula with an extensive area distributed over five Portuguese counties: Arcos de Valdevez, Melgaço, Ponte da Barca, Terras de Bouro and Montalegre (Fig. 20.1).

The agricultural landscape of this area is characterized by a pattern of small and fragmented farms which produce mainly for self-consumption, and is based on traditional agricultural practices. These traditional agroecosystems are not circumscribed to agricultural areas alone but do integrate vast mountain areas, important natural pasture lands and sources of bedding for animals and firewood (Fig. 20.2). Over the centuries, these traditional agroecosystems have been the cradle of a vast natural and cultural heritage and invaluable repositories of crop genetic diversity (Caldas, 1998; Altieri, 2004).

The general objectives were to develop strategies and methodologies for the conservation of wild species and of landraces and to encourage local farmers to become involved, in a participatory manner, in the conservation of this worthful genetic treasure associated with their traditional agroecosystems in the Peneda-Gerês area and to evaluate the potential genetic erosion,

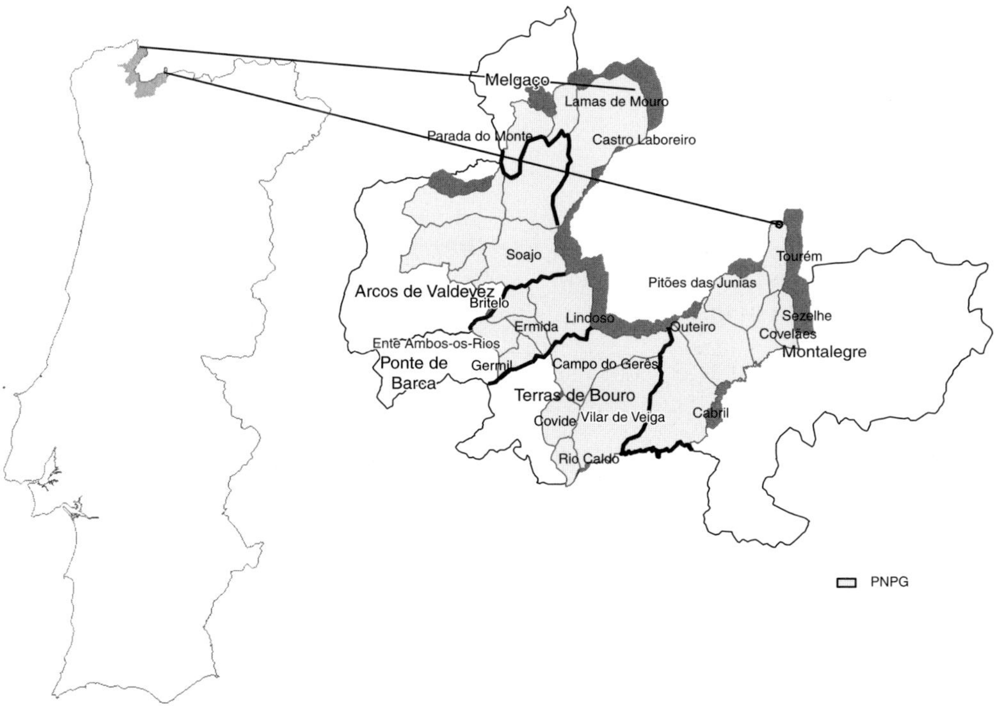

Fig. 20.1. Project territory – Peneda-Gerês National Park (PNPG) in Portugal.

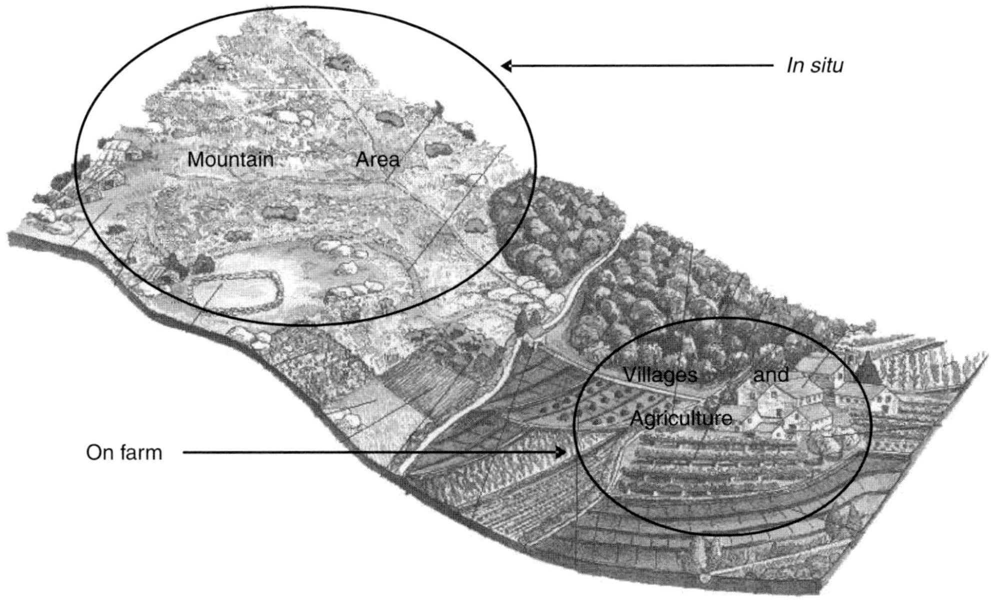

Fig. 20.2. Project approach.

through the re-collecting, as well as through the morphological and molecular characterization of this genetic material. The target species were: maize (*Zea mays* L.), rye (*Secale cereale* L.) and common bean (*Phaseolus vulgaris* L.). For the purpose of applying methodologies associated with the conservation strategy, the following actions were developed.

20.2 Survey and Characterization of the Region of Intervention

The workplan included:

- Survey of agricultural systems:
    - Farms' survey and characterization
    - Traditional varieties' current status and extent of production;
- Survey and characterization of households;
- Identification of accessions of rye, common bean and maize maintained in *ex situ* collections in BPGV and in ex-EAN (Estação Agronómica Nacional), collected in 1977–1998, in the five counties of PNPG.

20.2.1 Survey and geographic description

The target region is characterized by small and fragmented farms, where the available soil is scarce, the agro-system permits the full utilization of the available soil, where agricultural areas closer to villages are cultivated with maize, common bean (alone or intercropped with maize), and with cucurbits and grapevines around the fields' edges. During winter these agricultural plots are occupied by rye and forages, especially ryegrass. Currently, many of these agricultural areas, closer to the villages, are no longer cultivated with maize and are utilized to hay harvest, supported by the plant natural vegetation (Graça and Rita dos Santos, 2000).

From a socio-demographic analysis (PNPG, 2000), it is clear that all villages face a negative demographic evolution, which has been the tendency since the 1960s. The population decrease ranged between 10 and 30% depending on the county. The same study emphasizes that, in 1991, more than 50% of the villages in the Park's area had less than 100 residents, and 39.5% less than 50. This scenario reflects the trend that started long ago with the exodus from rural to urban areas. In terms of gender, and reflecting the effects of migration, which traditionally involved mostly males, 54% of the population is female. The vitality index (relation between the population over 64 and the population between 0 and 14 years of age) was, in 1981, 57%, rising to 112.7% in 1991. The changing in the population structure, which includes reduction in number of individuals and ageing, implies change in the traditional agricultural systems (Barata *et al.*, 2004; ICNB, 2010).

20.2.2 Traditional varieties' current status

The whole plant of maize was utilized: the grain to make the traditional bread called 'broa' and the tassel, leaves and straw, to animal feed, mostly autochthonous breeds. However, these practices, which require a high level of labour, have been progressively abandoned due to the reduction and ageing of the rural population, which, more than the introduction of hybrid varieties, mainly yellow maize hybrids, have negatively impacted and determined the abandonment of traditional agricultural practices, the associated crop landraces and dictated the cultural erosion exemplified by the loss of traditional knowledge imparted by the rural population.

Maize is still cultivated consociated with beans and ryegrass or as monocropping. Manual weeding and manuring is still practised, although without the removal of the leaves and tassel after pollination has occurred. Mounding, traditionally done to prevent lodging, today is not a frequent practice and irrigation is difficult because of lack of maintenance of the traditional irrigation systems. Hybrid maize, only used for animal feed (the bread is only from white maize landraces), suffers more from lack of water because heavy manuring leads to greater demands on water. Local landraces are well adapted to the traditional polycultural agrosystem, as they have a shorter cycle and have more palatable straw.

For rye, many mentioned that rye is not as much cultivated as it used to be in the old times, even when referring to places that, in the recent past, were considered as 'rye land'. In the 1990s rye was still a major crop, while today the majority of the farmers in this region cultivate only small plots, just enough for making the traditional 'broa' bread (a mixture of maize and rye, or rye only) for home consumption, and for the straw which is used to burn off pig hairs in the traditional slaughter way, as bedding for animals, to tie the maize straw and hay bundles, to cover the haystacks to protect them from the rain and to cover the soil, between the maize lines, to prevent water loss due to evaporation, thus maintaining the soil humid and fresh for longer. In Ponte da Barca and Montalegre counties, rye still carries importance for human consumption (bread) and folkloristic activities known as 'malhadas'.

Common bean is still widely grown, mostly using local varieties, for human consumption, entering in the preparation of many recipes in traditional and local gastronomy (Brás, 2000; Miranda and Reis, 2000; Reis and Miranda, 2000; Carvalho, 2004). The major problem of maintaining local varieties of this species comes from the abandonment of agricultural practices due to changes in population structure. It is still common practice to sow mixtures of local varieties, which, however, are separated at the time of utilization as they have different uses.

The results found are, in a way, similar to what other authors have encountered in different regions of the world, where the loss of landraces is due more to the abandonment of subsistence agriculture practices and alterations in population structure than to the introduction and use of modern varieties (Guarino *et al.*, 1991; Negri, 2003).

20.3 Development of Conservation Plans by Crop/Farming System

These plans involved the establishment of methods for conservation in the farmer's field, given the specificity of each area of intervention: agroecological conditions, crops, crop science, and cultural aspects and the socio-economic conditions of the populations involved. All species were observed and studied for their evolutionary process, derived from selection and adaptation by means of farmer interaction and intervention. Methods of intervention and monitoring of the evolution of species' populations in the targeted areas were developed and implemented in order to evaluate the benefits of the complementarity between the two conservation methods, on-farm and *ex situ*. The workplan consisted of:

- Survey and collecting missions in the same sites sampled earlier (same place and same farmer, if possible).
- Morphological characterization and data comparison analyses of the genetic material maintained *ex situ*, collected in 1977–1998 and the populations of these crops maintained in on-farm conditions.
- Assess the genetic variability of local populations of rye, maize and common bean, using molecular techniques.

The objectives were:

- To estimate genetic erosion by temporal comparison;
- To study the level of genetic variation;
- To evaluate the genetic diversity at different marker loci;
- To analyse the inter-population genetic structure.

20.3.1 Estimating genetic erosion by temporal comparison

In order to assess whether genetic erosion has occurred in landraces of rye, common bean and maize, populations that had been collected in the counties in the area of PNPG were identified. Only those populations that had been collected in the period 1977–1998 were selected. The re-collecting took place in 2005–2006 in all five counties. The studies of genetic erosion by temporal comparison were presented by Rocha *et al.* (2008), see Table 20.1.

20.3.2 Quantitative genetic variation and genetic diversity at marker loci

Characterization and morphological evaluation of common bean

In 2005–2006, 35 populations of common bean and a standard (var. 'Bencanta'), were characterized. The material had the following origin: Melgaço (8), Ponte da Barca (16), Arcos de Valdevez (8), Montalegre (2) and Terras de Bouro (1). The characterization was done in two locations using the traditional intercropping system of maize × common bean. The descriptors were the international standards for the species, in a total of 35 descriptors (2 flower, 14 pod, 10 seed and 9 plant descriptors). Observations were made in 20 plants per population.

The data analysis showed no significant difference between the material conserved on-farm and the material conserved *ex situ*. However, it was observed in the material maintained on-farm a slight

Table 20.1. Number of samples collected by species and by collecting mission, in the area of the PNPG, covering the PNPG counties and respective Simpson's Diversity Index (SDI) (Source: Rocha *et al.*, 2008).

	P. vulgaris		*S. cereale*		*Z. mays*	
Counties	1977–1998	2005–2006	1977–1998	2005–2006	1977–1998	2005–2006
Arcos de Valdevez	7	5	1	2	3	2
Melgaço	2	10	8	6	3	5
Montalegre	48	43	16	27	27	20
Ponte da Barca	16	13		5	18	9
Terras de Bouro	3	5	2		12	6
Total	76	76	27	40	63	42
SDI	0.546053	0.624654	0.554184	0.50375	0.693878	0.690476

decrease in the diversity of the seed and flower colour as well as a reduction of 2 weeks in the vegetative cycle. While the bean is a self-pollinating species and therefore with less risk of genetic contamination by the introduction of new varieties, there is clearly a danger of replacing the traditional varieties.

Characterization and morphological evaluation of maize

In 2005–2007, 92 maize populations, Ponte da Barca (25), Arcos de Valdevez (19), Montalegre (30), Terras de Bouro (13) and Melgaço (5), were characterized. The descriptors used were the international standards for the species in a total of 61 parameters (8 plant, 11 leaf, 11 flag, 16 ear and 15 seed descriptors). Observations were made in ten plants per population. The characterization and multiplication was done in two locations, using the traditional intercropping system of maize × common bean.

The more significant results were: populations showed different traits among populations from *ex situ* and on-farm conservation in morphological aspects. Length of grain and leaf increased in the on-farm populations; *ex situ* populations showed earliness, and variability among populations was smaller. Considering the material maintained on-farm it was observed that: Montalegre and Melgaço populations are the more early flowering, mainly from Melgaço, plants are smaller, have lower ear weight, smaller ears and less grains per row; Ponte da Barca and Terras de Bouro populations showed different characteristics: they are later, greater plant height, larger leaf area, heavier ears, longest grain, largest and longest ear and larger number of grains per row.

Molecular analysis of local populations of maize and common bean

Molecular analysis was performed using microsatellite markers in 72 local maize populations (40 conserved *ex situ* and the remaining on-farm) and 69 common bean local populations of the five counties of PNPG (18 conserved *ex situ* and the remaining on-farm). The common bean is of Andean gene pool although six populations from on-farm have both gene pools represented, Andean and Mesoamerican, however the Andean gene pool is the more important. The common bean populations were studied for 23 microsatellite and maize for 24 microsatellite markers. To assess the genetic erosion of populations we evaluated parameters of genetic diversity.

The genetic erosion concept defined by Maxted and Guarino (2006) was not observed in the area of intervention. Other published studies showed the same sort of results (Gómez *et al.*, 2005; Le Clerc *et al.*, 2005; Negri and Tiranti, 2005; Bitocchi *et al.*, 2009).

In maize, the average number of alleles per population was 1.07 and the average number of alleles per locus for all populations was 1.04. We can consider an increase of genetic variability, because on-farm was separate from *ex situ*: average number of alleles per population in *ex situ* was 0.62 and on-farm was 1.61.

In common bean, the average number of alleles per locus was 1.22 ranging between 1 and 5. The average number of alleles per population was also 1.22. The degree of diversity in the populations of common bean is different from county to county, with higher rates in Montalegre while Melgaço populations are closer to each other showing that gene flow with the outside will not be as intense as in other counties. The variability observed among and within counties for this species may be due to natural crosses as a result of the traditional agrosystem. The populations preserved in on-farm conditions have higher heterozygosity and the range in the number of alleles per locus and population was between 1.0 and 1.9, while in *ex situ* gene pool it ranged between 0.7 and 0.9.

Morphological and molecular analysis of regional rye populations

Morphological characterization was focused on 13 populations collected earlier (1977–1998) and re-collected in 2005–2006. These samples were from the same collecting sites of the accessions collected in the 1970s and 1980s. The descriptors were the international standards for the species in a total of 11 parameters. A great variation was observed both intra- and inter-populations/accessions independently of the collecting period.

Molecular studies were done in order to characterize the genetic variability and to evaluate the genetic erosion, using ISSR and SSR markers in six populations of 1977–1978 and the same populations collected on the same site in 2004–2005.

Eight SSR primer pairs, selected from mapped rye SSRs and nine ISSR primers from UBC primer Set #100/9 were used. The nine ISSRs produced 87 bands of which 54% were polymorphic. The data obtained with SSRs markers revealed the presence of a great number of alleles/loci in all populations studied and a high level of heterozygozity. The heterozygosity detected was similar in both groups of material, ranging from 0.33 and 0.59 in the accessions of the 1970s and 1980s and 0.34 and 0.53 in the material from 2005 to 2006. The number of alleles was equal (between 10 and 12) in the material from both collecting periods.

20.3.3 Inter-population genetic structure

Cluster analyses of the genetic pool from maize, common bean and rye, using Nei's coefficient (1972) and UPGMA method was performed with the NTSYSpc-2.0g software. Genetic variability among populations of maize and common bean maintained in *ex situ* and on-farm conditions was assessed. The actual genetic pool is different from the *ex situ* genetic pool, but it was observed that populations from the same origin were related. The maize and common bean populations in on-farm and *ex situ* conservation were different breeder's cultivars, which were analysed too. In rye it was not possible to do clusters according to the period or site of collecting (Fig. 20.3). Although the genetic structure of the material from the two collecting periods was different, a considerable genetic variation continues to exist, showing that during this period of more than 25 years genetic erosion did not occur in the

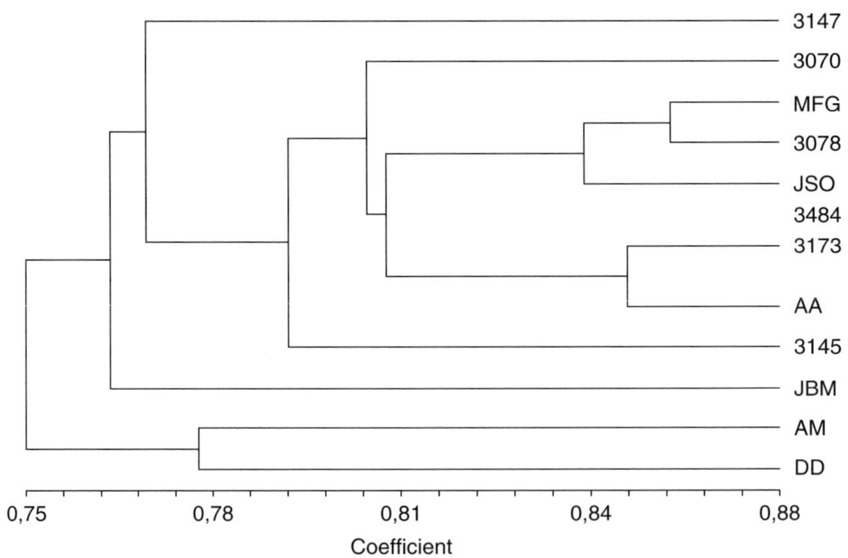

Fig. 20.3. Rye dendrogram obtained using ISSRs markers (rye populations codes from dendrogram – in *ex situ* and on-farm conditions: from 1977 and re-collected (3070 and MFG, 3078 and DD), from 1978 and 2005–2006 (3147 and JBM, 3173 and JSO, 3145 and AM) and of 1982 and re-collected (3484 and AA).

genetic material, while indicating the occurrence of evolution in the field.

20.4 Promoting the Use of the Products Subject to Conservation Recovery Plans

The main actions were to raise awareness and involvement of local communities, as key elements in a strategy for recovery and sustainable conservation of this unique and great value genetic resources. There was a continuous and persistent monitoring and advisory role to local farmers for the production of traditional varieties, and the development of tests for processing and manufacturing of maize bread using traditional techniques. Within the activities of the project, an old traditional watermill, completely abandoned and in ruins, was restored for processing the grain. In order to ascertain the region's traditional recipes, a regional survey and collation of traditional recipes as well as the survey and collation of the traditional technologies was carried out (e.g. oven temperature, length of baking), after which production has started. Identifying niche markets for these products was the next approach. Two of the products were included in the Slow Food Foundation catalogue of quality foods: 'Broa dos Arcos de Valdevez' and 'Feijão Tarrestre'.

20.5 Conclusions

The diversity of the target crops, the associated knowledge and resulting traditional products are at risk of disappearing for many reasons: farmer's age; depopulation as the result of the exodus from rural to urban areas and/or to other professional activities with consequent abandonment of the primary sector; and aggressive marketing that imposes hybrid varieties. As such, on-farm conservation is an important method for the conservation of these crops' genetic resources, complementary to *ex situ* conservation, in the area of PNPG.

The cultivation of maize, common bean and rye populations' landraces was proven to be an important income for the farmers of this region and continues to be important for maintaining natural and unique landscapes, imparting a profile to the region and the protected area, allowing their use as well as promoting touristic activities.

> The dynamic conservation in the original ecosystem and in the presence of a controlled selection pressure made by farmers, allows further evolution of landraces. The goal of this approach is not to preserve a given number of genes (*per se* diversity) but to maintain an agricultural system, which can generate new germplasm
>
> (Piergiovanni and Laghetti, 1999).

References

Altieri, M.A. (2004) Globally Important Ingenious Agricultural Heritage Systems (GIAHS): extent, significance, and implications for development. *Proceedings of the Second International Workshop and Steering Committee Meeting for the Globally Important Agricultural Heritage Systems (GIAHS) Project.* FAO, Rome, 7–9 June 2004.

Alves, J., Godinho, C., Varela, C., Bettencourt, E. and Gusmão, L. (1996) Early Portuguese Multidirectional Promotion of New Geographical Diversification Areas. Poster presented at the *International Technical Conference on Plant Genetic Resources*, Leipzig, Germany, 17–23 June 1996.

Barata, A., Carvalho, A. and Miranda, J. (2004) *Cadernos da Montanha Peneda-Soajo III.* ARDAL & CMAV, Arcos Valdevez, Portugal.

Barata, A., Silva, A. and Lopes, V. (2008) *Referencial de Formação na Área das actividades com tradição.* DRAPN, Mirandela, Portugal.

Bettencourt, E. and Gusmão, L. (1981) Maize collecting in the Azores. FAO/IBPGR *Plant Genetic Resources Newsletter* 46, 15–17.

Bettencourt, E. and Gusmão, L. (1982) Colheita de germoplasma de milho, nos Açores. I. S. Miguel e Terceira. *Agronomia Lusitana* 41(3–4), 241–258.

BISE (2011) http://biodiversity.europa.eu/topics/genetic-resources (accessed 22 December 2010).

Bitocchi, E., Nanni, L., Rossi, M., Rau, D., Bellucci, E., Giardini, A., Buonamici, A., Vendramin, G. and Papa, R. (2009) Introgression from modern hybrid varieties into landrace populations maize (*Zea mays* ssp. *mays* L.) in central Italy. *Molecular Ecology* 18, 603–621.

Brás, A. (2000) O itenerário técnico da consociação milhoxfeijão na Serra da Peneda. In: Carvalho, A. and Miranda, J. (eds) *Cadernos da montanha Peneda-Soajo II*. ARDAL & CMAV, Arcos Valdevez, Portugal, pp. 9–23.

Caldas, E.C. (1998) *A Agricultura na História de Portugal*. EPN, Lisboa.

Carvalho, A. (2004) *Património culinário de Arcos de Valdevez*. ARDAL & CMAV, Arcos de Valdevez, Portugal.

CBD (1992) *Convention on Biological Diversity*: text and Annexes. Secretariat of the Convention on Biological Diversity, Montreal, Canada.

CBD (2002) Global Strategy for Plant Conservation. Secretariat of the Convention on Biological Diversity, Montreal. Available at: www.cbd.int/decision/ (accessed 22 December 2010).

EC (2008) Commission Directive 2008/62/EC of 20 June 2008. *Official Journal of the European Union*. L 162/13. EN. 21.6.2008

EC (2009) Commission Directive 2009/145/EC of 26 November 2009. *Official Journal of the European Union*. L 312/44. EN. 27.11.2009

Farias, R.M. (1989) Recursos genéticos. Banco de Germoplasma de milho. Sua Majestade... o milho rei. *Revista da Feira do Milho*, Outubro, Braga, pp. 17–18.

Farias, R.M. (1999) Banco Português de Germoplasma Vegetal. In: Rena Farias and Botelho, A.C. (eds) *I Conferência Técnica sobre Recursos Genéticos Vegetais*. DRAEDM & BPGV, Braga, Portugal, pp. 23–29.

Farias, R.M. and Marcelino, F.R. (1993) O Banco Português de Germoplasma Vegetal guardião do tesouro nacional. *Vida Rural* 17/93. 1ª Quinzena de Setembro, 7–10.

Farias, R.M., Varandas, E. and Marcelino, F.R. (1992) O tesouro de Basto e a agricultura biológica. *Vida Rural* 8/92. 2ª Quinzena de Abril, 46–47.

Ferrão, J.E.M. (1992) *A aventura das plantas e os descobrimentos portugueses*. Edição: Instituto de Investigação Científica Tropical (IICT), Comissão Nacional para as Comemorações dos Descobrimentos Portugueses, Fundação José Berardo.

Ferrão, J.E.M. (2005) *A aventura das plantas e os descobrimentos portugueses*, 3rd Edn. IICT, Fundação Berardo, Chaves Ferreira-Publicações S.A., Lisboa, 288 pp.

Frazão-Moreira, A. and Fernandes, M.M. (eds) (2006) *Plantas e Saberes. No limiar da Etnobotânica em Portugal*. Colibri, Lisboa.

Gómez, O.J., Blair, M.W., Frankow-Lindberg, B.E. and Gullberg, U. (2005) Comparative study of common bean (*Phaseolus vulgaris* L.) landraces conserved *ex situ* in genebanks and *in situ* by farmers. *Genetic Resources and Crop Evolution* 52, 371–380.

Graça, L.L. and Rita Santos, H. (2000) *Cadernos da Montanha Peneda-Soajo I, ARDAL & DRAEDM*. Arcos de Valdevez, Portugal, 180 pp.

Guarino, L., Chadja, H. and Mokkadem, A. (1991) Wheat Collecting in Southern Algeria. Short Communications. *Rachis* 1991, 23–25.

Hawkes, J.G., Maxted, N. and Ford-Lloyd, B.V. (2000) *The* ex situ *Conservation of Plant Genetic Resources*. Kluwer, Dordrecht, the Netherlands, pp. 1–250.

ICNB (2010) Plano zonal do PNPG 2006. Available at: http://portal.icnb.pt/ICNPortal/ (accessed 22 December 2010).

Le Clerc, V., Bazante, F., Baril, C., Guiard, J. and Zhang, D. (2005) Assessing temporal changes in genetic diversity of maize varieties using microsatellite markers. *Theorical and Applied Genetics* 110, 294–302.

Marcelino, F.R. and Farias, R.M. (1994) Colheita de germoplasma vegetal na Madeira. *Vida Rural* 1589, Ano 41°, Abril, pp. 29–33.

Maxted, N. and Guarino, L. (2006) Genetic erosion and genetic pollution of crop wild relatives. In: Ford-Lloyd, B.V., Dias, S.R. and Bettencourt, E. (eds) *Genetic Erosion and Pollution Assessment Methodologies*. IPGRI, Rome, pp. 35–46.

Miranda, E.R. and Reis, A. (2000) Análise crítica da tecnologia de produção do feijão na consociação milho-feijão em zonas de montanha. In: Graça and Rita dos Santos (eds) *Cadernos da montnha Peneda I*. ARDAL & DRAEDM, Arcos de Valdevez, Portugal, pp. 146–159.

Mota, M., Farias, R.M., Teixeira, M. and Bettencourt, E. (1982) *Zea* and *Phaseolus* collecting in Portugal. I. FAO/IBPGR *Plant Genetic Resources Newsletter* 50, 21–22.

Negri, V. (2003) Landraces in central Italy: where and why they are conserved and perspectives for their on-farm conservation. *Genetic Resources and Crop Evolution* 50, 871–885.

Negri, V. and Tiranti, B. (2005) Molecular analysis for *ex situ* and on-farm conservation of common bean (*Phaseolus vulgaris* L.) Italian germplasm. Paper presented at *The role of Biotechnology*, Turin, 5–7 March 2005.

Negri, V., Maxted, N. and Veteläinen, M. (2009) European landrace conservation: an introduction. In: Veteläinen, M., Negri, V. and Maxted, N. (eds) *European Landraces: on-farm conservation, management and use*. Bioversity Technical Bulletin 15. Bioversity International, Rome, pp. 1–22.

Nei, M. (1972) Genetic distance between populations. *American Naturalist* 106, 283–292.

Piergiovanni, A.E. and Laghetti, G. (1999) The common bean landraces from Basilicata (Southern Italy): an example of integrated approach applied to genetic resources management. *Genetic Resources and Crop Evolution* 46, 47–52.

PNPG (2000) *Turismo de Natureza. Enquadramento Estratégico*. Instituto da Conservação da Natureza (ICN), Parque Nacional Peneda-Gerês.

Reis, A.V. (1997) *O mercado de sementes de forragens e de cereais no País e no Entre Douro e Minho*. DRAEDM, Braga, Portugal.

Reis, A. and Miranda, E.R. (2000) Caracterização e avaliação de ecótipos de feijão produzidos em regiões de montanha. In: Graça and Rita dos Santos (eds) *Cadernos da montanha Peneda I*. ARDAL & DRAEDM, Arcos de Valdevez, Portugal, pp. 161–167.

Rocha, F., Bettencourt, E. and Gaspar, C. (2008) Genetic erosion assessment through the re-collecting of crop germplasm. Counties of Arcos de Valdevez, Melgaço, Montalegre, Ponte da Barca and Terras de Bouro (Portugal). *Plant Genetic Resources Newsletter* 154, 6–13.

Strecht, A. (2004) Saberes e sementes tradicionais de Portugal. *O segredo da Terra*, no 8. EDIBIO, EDIÇÕES, LDA., Aveiro, Portugal.

21 What's in a Name: A Closer Look at Heritage Variety Definition

J.M. Preston, N. Maxted, R. Sherman, N. Munro and B.V. Ford-Lloyd

21.1 Introduction

The term heritage variety is part of the array of terminology used to refer to traditional crop varieties, which includes: landraces, primitive, folk, obsolete, farmer and heirloom varieties (Camacho Villa *et al.*, 2006). Although some terms are eponymous (such as farmer variety) or functional (such as obsolete variety) others are used without clear definition, and many are used interchangeably, both in the formal literature (for example, Rodriguez-Burruezo *et al.*, 2005, p. 453 refer to 'heirloom (traditional)' tomato varieties) and the less formal literature (for example, Thorness, 2009, The Royal Horticultural Society use the terms heritage and heirloom variety interchangeably).

Communication, conservation prioritization and the search for 'useful' genetic information/diversity for breeding requires a clarification of the terminology applied to specific sets of plant genetic resources (PGR) with characteristics held in common, so time and money may be directed effectively (Hawkes *et al.*, 2000). An artificial distinction where none exists in reality is not useful; however, if terms are not synonymous, characters identified under each term may affect potential use, for example, if the genetic profiles of the groups differ. These terms arguably refer to different sections of the suite of crop types that are each cultivated by humans, they have a distinct set of characteristics that define them, but some potential overlap is evident between certain terms.

In discussing the definition of a landrace, Zeven (1998), Camacho Villa *et al.* (2006), Tiranti and Negri (2007) and Berg (2009) highlighted the usefulness of term-clarification. The purpose of this chapter is to discuss and propose a definition of the term 'heritage variety' and its relationship to the term 'heirloom' with which it is sometimes considered as a direct synonym. These two terms are used widely by charities and seed-saving organizations, such as Garden Organic (UK), Seed Savers Exchange (USA), Irish Seed Savers (Ireland), gardeners as recorded by Watson (1997) and Stickland (1998), and seed companies like Thompson and Morgan (2011) and Thomas Etty Esq. (no date).

It is proposed that when we refer to heritage varieties we are referring to a specific subset of traditional crop varieties that are identified by users via consistently applied characteristics, namely historical origin, open pollination, and cultural/heritage value. The heritage variety will be discussed with reference to: historical origin, mode of breeding, genetic diversity, local genetic adaptation, and association with traditional farming systems.

21.2 Definitions and Terminology

Of the terminology used in association with traditional crop varieties, the definition of a landrace is the most explored. Recent papers (including Zeven, 1998; Camacho Villa *et al.*, 2006; Tiranti and Negri, 2007; Berg, 2009) have proposed definitions of the term landrace, with the view to aiding conservation of landrace diversity. Camacho Villa *et al.* (2006, p. 381) proposed the following definition:

> A landrace is a dynamic population(s) of a cultivated plant that has historical origin, distinct identity and lacks formal crop improvement, as well as often being genetically diverse, locally adapted and associated with traditional farming systems

This definition encompasses all of the traits included in alternative definitions, with the exception of the emphasis on cultural importance stressed by Tiranti and Negri (2007). They highlight the close association of landraces with the people who develop and grow them, and their role in traditions and culture. Camacho Villa *et al.* (2006) emphasize that the presence of all six (seven if local cultural importance is added) characteristics is not necessary to define a landrace, as the exact mix of characteristics will differ between crops and contexts.

Some of the terms are functional definitions, such as obsolete variety, which refers to those varieties that are no longer commercially available and have been superseded by 'elite' varieties (Hawkes *et al.*, 2000; Skovmand *et al.*, 2001). Identity of breeder is often used in the nomenclature; for example, farmer's variety (where the farmer may be breeding for his/her own personal use or for commercial purposes) (Zeven, 2000) or garden race where the gardener is the putative breeder (Zeven, 1998, 2002). Perhaps the broadest term is that of traditional variety itself, this being anything that is not a 'modern variety' and is associated with traditional cultivation practice, seed management and breeding techniques (Rhoades and Nazarea, 1999; Camacho Villa *et al.*, 2006). A modern variety is then one that is genetically definable and results from commercial breeding strategies.

21.3 Elements that Define Heritage Varieties

21.3.1 Mode of breeding

Heritage varieties are likely to be of non-homogeneous breeding origin and, as their custodianship has changed over time, the precise origin of many of these varieties has been lost. However, many heritage varieties are ex-commercial, for example those UK varieties not commercially traded following the implementation of the Seed (National List of Varieties) Act of 1973 (Stickland, 2008). They have been subject to definite human selection through directed seed-saving (from plants with desired characters) or crossing to select for specific phenotypic characters such as colour, size and shape (Science and Advice for Scottish Agriculture, no date). This selection for particular crop types distinguishes heritage varieties from other traditional crop varieties, where human selection is at a very low level (Zeven, 2000), using mass selection, or, more stringently, where selection is absent (Berg, 2009), with landraces being simply seed-saved each year, and new adapted genotypes mixed in. Berg (2009) uses degree of selection to distinguish between landraces and 'folk varieties'; the latter is subject to human selection, including for particular traits, resulting in a narrower definition of a landrace that would exclude many entities and varieties included in both the Camacho Villa *et al.* (2006) and Zeven (1998) definitions. Heritage varieties, as described here, would not be landraces according to the Berg (2009) definition; they would, however, be included in the Camacho Villa *et al.* (2006) definition, as the latter states that not all characteristics in the definition have to be present in order to be recognized as a landrace.

21.3.2 Historical origin

Historical origin encompasses both temporal and spatial aspects of landrace development (Camacho Villa *et al.*, 2006). Stickland's (1998) research and variety summaries suggest that many heritage varieties were developed and popularized in the 1800s. The exact length of cultivation history is not standardized, for example Thorness (2009) states that these varieties have been grown since before World War I, however, the most commonly used length of cultivation period is a minimum of 40–50 years (N. Munro, personal communication, 2010; Thompson and Morgan, 2011). This is in contrast to the length of cultivation period of other traditional crop varieties, which is relatively long; they have been grown ‘since time immemorial’ or ‘for many centuries’ (von Runker, 1908, and Cholton, personal communication, both in Camacho Villa *et al.*, 2006, p. 375).

The spatial aspect of historical origin relates to the cultivation of that landrace in a specific geographic location. Heritage varieties are often developed in one particular location and then distributed elsewhere: if developed by a breeder through an associated seed company (Stickland, 1998), by a farmer through family or other local farmers (Zeven, 1998, 1999), or by a home gardener or allotment holder to family and friends (Stickland, 1998). In the case of seed from companies, the seed origin would be the area the company is located in, rather than the location at which the varieties are actually grown by customers (Kell *et al.*, 2009).

21.3.3 Open pollination

Open pollination is proposed to be one of the three main characters of heritage varieties, as identified by users (Stickland, 1998; Thorness, 2009; Dyfi Valley Seed Savers, 2010; Garden Organic, 2010; Irish Seed Savers, 2011). In common with most traditional crop varieties, heritage varieties are open-pollinated, meaning that they are not hybrids and breed true, except where gene flow has unintentionally occurred from another variety, and thus can be seed-saved. Although this is the same as for other traditional crop varieties, it is a key feature identified by users to distinguish heritage varieties from modern varieties, and it is important because it provides further distinction between modern and more traditional breeding techniques. Some heritage varieties may originally have been early hybrids but have since been stabilized and continue as open-pollinated varieties (Watson, 1996).

21.3.4 Level of genetic diversity

There are concerns regarding the loss of plant genetic resource (PGR) diversity due to: replacement of traditional crop varieties with modern cultivars (Hawkes *et al.*, 2000; Negri *et al.*, 2009); a reduction in the number of varieties relied upon for food; along with legislation prohibiting the sale of unlisted varieties, which has resulted in a reduction in the availability of some varieties, particularly heritage varieties (Stickland, 2008). However, meta-analyses suggest that genetic diversity rates in crop cultivars have recovered since a decrease in the 1960s, and overall no reduction in regional genetic diversity has been found (van de Wouw *et al.*, 2010). Yet the importance of traditional crop varieties, including heritage varieties, as potential sources of genetic diversity and rare alleles for future breeding must be recognized. Previous studies have found traditional crop varieties to contain high levels of genetic diversity, such as in *Phaseolus vulgaris* (Tiranti and Negri, 2007), *Phaseolus coccineus* (Sicard *et al.*, 2005), *Solanum lycopersicon* (Terzopoulos and Bebeli, 2008) and *Daucus carota* (Shim and Jorgensen, 2000).

Genetic diversity is proposed as one of the characters that can be used to distinguish heritage varieties from other traditional crop varieties, finding heritage varieties on the spectrum in between landraces and modern varieties. Landraces can have the appearance of highly variable populations, such that they may not be strictly referred to as ‘cultivars’ (Zeven, 1998; Camacho Villa *et al.*, 2006); the application of some breeding, particularly for selection of desired characters (Astley and Munro, personal communication, in Camacho Villa *et al.*, 2006, p. 376) in heritage varieties, means that heritage varieties may not demonstrate this attribute. The genetic diversity of heritage varieties can be problematical to unearth, due to the past uses of the term or lack thereof. Varieties, fitting the heritage variety definition proposed here, have been investigated, often as ex-commercial varieties or by date of cultivation. For example, Shim and Jorgensen (2000, p. 228) compared 'old' varieties of *D. carota* (carrot) to wild and modern varieties. These were open-pollinated varieties released between 1976 and 1978 and were found to have relatively high within-population genetic diversity compared with recent cultivars, which can be attributed to breeding history. Archak *et al.* (2002, p. 1140) referred to 'old local cultivars' of *S. lycopersicon* (tomato) from India, which were found to be more genetically diverse than varieties released since the 1990s, due to breeding for uniformity of specific plant and fruit types. Although only few studies are found that specifically investigated heritage variety genetic diversity, there are accounts of heritage varieties being used as the basis of improved varieties, such as *Phaseolus coccineus* (runner bean) variety 'Prizewinner', introduced by Suttons of Reading in 1892; it has since been improved for disease resistance and released as the modern variety 'Enorma' (Stocks, 2008).

21.3.5 Local genetic adaptation

Although local adaptation is not proposed as a defining character of heritage varieties, some users do highlight as an important feature that seed be adapted to local climatic or edaphic conditions (Dyfi Valley Seed Savers, 2010; Irish Seed Savers, 2011). Since the adoption of National Lists, many heritage varieties, previously supplied by 'local' seed companies with their own selection criteria, are now seed-saved by individuals and seed-saving organizations.

Local genetic adaptation arises as a result of repeated cycles of planting, harvesting and selection over extended periods of time, particularly in marginal environments (Camacho Villa *et al.*, 2006). Local adaptation is cited as a character of some heritage varieties (Stickland, 1998; Dyfi Valley Seed Savers, 2010; Irish Seed Savers, 2011); due to the necessity of extensive field trials to determine the evaluative characters of crops, much of the evidence for adaptation is anecdotal.

Franks *et al.* (2007) found that genetic diversity allows crop adaptation to environmental change to occur in very few generations. This suggests that the length of cultivation proposed here for heritage varieties (40–50 years) is sufficient time for varieties to be under selection pressure and adapt if grown in a particular location. However, quantifying these changes is problematical: details of seed sources can be lost and conserved seed samples small (and thus vulnerable to genetic bottlenecks and founder effects (Prada, 2009), so masking adaptation); long-term seed storage in *ex situ* collections can lead to genetic drift (Hawkes *et al.*, 2000; Prada, 2009); and evaluation trials in different locations over time would be necessary to explore this further, but it would be a valid avenue to explore (Prada, 2009).

21.3.6 Association with traditional grower/gardener systems

Heritage varieties are identified by users of the term as being of heritage or cultural value (Stickland, 1998; Irish Seed Savers, 2011). Similarly, the association between people and landraces can be related to the use of the variety in specific personal traditions and habits, or preference for characters not found in modern varieties and hence for the landrace itself, rather than with the farming system (Camacho Villa *et al.*, 2006). The importance of this tight intertwining of biological and cultural heritage is strongly argued by Negri (2005).

Heritage varieties provide important links with the past (Stickland, 1998) such as local customs and festivals, family recipes, and can be connected specifically with places or names. For example 'Brighstone bean', is a variety of *P. vulgaris* (French bean) grown by gardeners on the Isle of Wight, which has its own local story of origin as it is said to have washed ashore from a shipwreck in the late 1800s (Stickland, 2008).

Heritage value can be associated with personal or common good value. For example, for varieties gardeners have grown in the past or for particular traits they value a variety has personal value. The most prominent of these traits is taste preference of heritage over modern varieties (Russo, 2008; Kell *et al.*, 2009), but a wide range of other characters such as unusual colours/shapes and diversity of maturation time, to avoid gluts, are also valued (Kell *et al.*, 2009). While at the common good value, many growers find the concept of conserving heritage for historical/cultural value to be of importance (Negri, 2003; Kell *et al.*, 2009) and so grow with the aim of being directly involved in the conservation of these varieties. People often start growing heritage varieties for personal reasons, then become interested in the biological diversity conservation aspects (Jordan, 2007).

21.4 Proposed Definition of a Heritage Variety

The discussion of characteristics associated with the term heritage variety, in the context of traditional crop varieties, has confirmed the importance of three key traits most often identified by users with heritage varieties: open pollination, cultivation history of 40–50 years or more and the heritage and cultural value of the varieties to growers. The discussion also highlighted that some characteristics identified in landrace definitions may be absent or not yet adequately assessed in heritage varieties (degree of formal improvement, level of genetic diversity and local adaptation). It can be argued therefore that heritage varieties are a subset of traditional crop varieties that can be consistently identified with the proposed definition:

> A traditional crop variety that has historical origin of over 40 years, is open-pollinated and is of cultural/heritage value to its users.

21.5 The Case of Heirloom Varieties

This chapter has so far focused on the term heritage variety; however, the terms heritage variety and heirloom variety are often used interchangeably. The term heirloom is particularly used in the USA, and it is for this reason it has been omitted from discussion thus far in this chapter, as definitions in the Europe and USA appear to differ. Many sources use the term heirloom to describe varieties that would fit the above definition of heritage variety; for example, Taylor's Guide to Heirloom Vegetables (Watson, 1996) defines heirloom using the characteristics identified above for a heritage variety (open-pollinated, cultivated for over 50 years, with a history of its own). Some sources offer no description, such as Gonçlaves *et al.* (2008, p. 1289) who refer to 'traditional (heirloom) seeds'. There is certainly considerable overlap between the two terms: both refer to open-pollinated varieties, derived from moderate levels of classical breeding (not modern-bred or genetically engineered), and are of significant cultural importance, these characters being highly valued by users. However, it could be argued that heirlooms have the additional character of never having been available in seed catalogues, as they are closely tied to family members or close family associates, being bred by gardeners, and are exchanged along these lines, outside of the commercial seed trade (Watson, 1996; DeMuth, 1998). These heirlooms have a strong identity often linked with the breeder (or selector) by a name or the locality of development.

Both Watson (1996) and DeMuth (1998) recognize the dilemma of inclusion of commercial (or ex-commercial) varieties within the definition of an heirloom and recommend using the wider definition for general use (tallying with the one proposed for heritage varieties, above) as it is more inclusive, with 'true' heirlooms being those that have not been sold and are handed down in families or communities. Watson (1996) opts for the broader definition (tallying with that of heritage variety) as he argues to do otherwise ignores the valuable contributions of professional breeders and explorers; DeMuth (1998) argues that since many varieties are poorly documented and changes arise in the plants over time, the origin of variety can be impossible to determine.

This suggests that heritage variety and heirloom are used widely as direct synonyms; however, it can be useful to distinguish between the two as their genetic profiles may differ. The genetic character of 'true' heirloom varieties is unknown and may be different to that of heritage varieties. The original source of seed for heirlooms is usually unknown; many will originally have been commercial varieties seed-saved and possibly selected from by gardeners. This could potentially represent a significant bottleneck. Others may have been developed from landraces and undergone selection for specific characters. With time and genetic diversity (and restrictions in reproductive biology (Zeven, 1998)), both heirlooms and heritage varieties that are not maintained or selected can lose their improvement (reflected in changing allele and genotype frequencies), through forces such as outcrossing, mutation and natural selection (Parlevliet, 2007), potentially becoming secondary landraces (also known as creole varieties) (Mayr, 1937 in Zeven, 1998).

A proposed definition of an heirloom variety therefore, is simply an extension of the heritage variety definition:

> A traditional crop variety that has historical origin of over 40 years, is open pollinated, is of cultural/heritage value to its users, that has been developed, maintained and transferred through families and communities rather than commercial seed trade.

21.6 Discussion

We have proposed that heritage varieties are part of a suite of important PGR, critical for sustaining food security, they constitute a subset of traditional crop varieties that at least partially overlap with the broad definition of heirlooms. Kell *et al.* (2009) state when reporting their UK landrace survey that it is prudent to use the widest definition of a landrace to encompass as much diversity for conservation as possible; therefore even though heritage varieties may be less heterogeneous than other traditional crop varieties and have some formal improvement, they are still an important constituent of traditional crop diversity. Heritage varieties are at risk for multiple reasons, including restrictive seed legislation and replacement with modern, improved, varieties (Kell *et al.*, 2009). Conservation of both heritage varieties and heirlooms is important for cultural reasons (such as growers' choice and conservation as cultural artefacts) and, in the face of potential genetic erosion, as a source of novel genetic material for breeders to use.

The application of the traditional crop variety terminology matters to users of the seed (conservationists, growers, breeders) and may have legislative implications in the future (such as with reference to European seed legislation). Therefore, we suggest an artificial classification of crop variety terminology based on terminology usage, that attempts to distinguish between and indicates the relationship between traditional and modern crop varieties (including obsolete crop varieties and current crop varieties), and within traditional crop varieties between commercial/farm varieties (including landrace, heritage and farmer's varieties) and non-commercial/garden (including heirloom and garden varieties) (see Fig. 21.1). The classification is proposed as an aid to clearer terminology use and it is suggested that clearer usage of agreed terminology might help promote conservation of traditional crop varieties themselves. However, as implied in the title of this chapter, if the Shakespearian quotation is continued, 'What's in a name? that which we call a rose by any other name would smell as sweet', definitions of heritage varieties and heirlooms are merely a tool to assist distinction, counter examples of usage are likely to exist. But it is hoped that by agreeing a more concrete definition of terminology it will be practically easier to plan strategically and

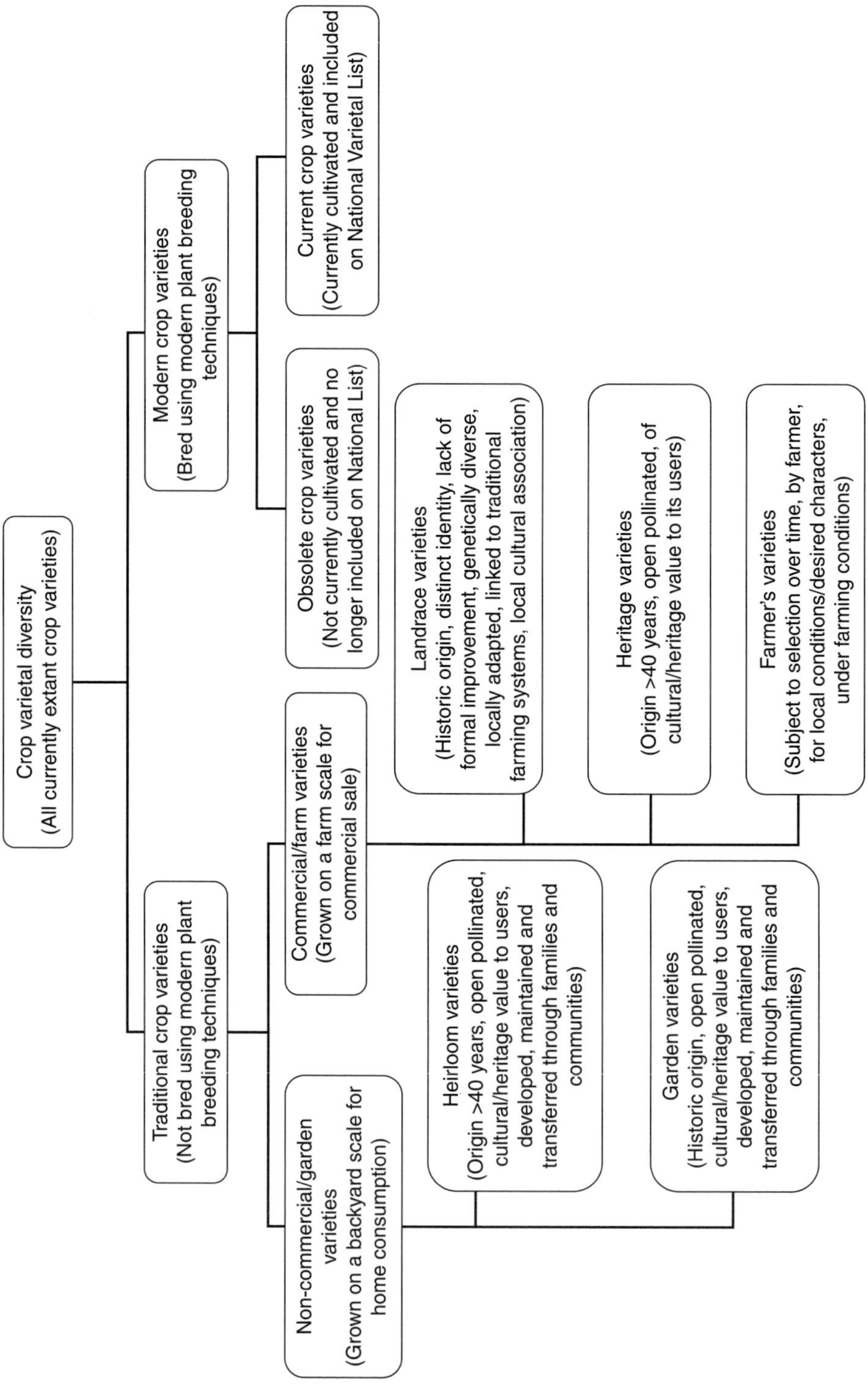

Fig. 21.1. Proposed artificial classification of traditional crop variety terminology.

implement necessary traditional crop variety conservation actions before diversity is lost and definitions themselves become superfluous.

Acknowledgements

The authors wish to thank The Lillian Goldman Charitable Trust for its financial support.

References

Archak, S., Karihaloo, J.L. and Jain, A. (2002) RAPD markers reveal narrowing genetic base of Indian tomato cultivars. *Current Science* 82(9), 1139–1143.

Berg, T. (2009) Landraces and folk varieties: a conceptual reappraisal of terminology. *Euphytica* 166, 423–430.

Camacho Villa, T.C., Maxted, N., Scholten, M. and Ford-Lloyd, B. (2006) Defining and identifying crop landraces. *Plant Genetic Resources: Characterization and Utilization* 3(3), 373–384.

DeMuth, S. (compiler) (1998) Vegetables and fruits: a guide to heirloom varieties and community-based stewardship, Vol. 1. Annotated bibliography. Special reference brief series 98-05. Available at: www.nal.usda.gov/afsic/AFSIC_pubs/heirloom/heirloom.htm#heirbk10 (accessed 12 June 2011).

Dyfi Valley Seed Savers (2010) Welsh vegetable project. Available at: www.dyfivalleyseedsavers.org.uk/images/stories/seed_search/Welsh_Veg_project_final_report_finalV3.pdf (accessed 30 March 2011).

Franks, S.J., Sim, S. and Weis, A.E. (2007) Rapid evolution of flowering time by an annual plant in response to a climate fluctuation. *Proceedings of the National Academy of Sciences* 104(4), 1278–1282.

Garden Organic (2010) Garden Organic's Heritage Seed Library. Available at: www.gardenorganic.org.uk/hsl/hsl.php (accessed 10 August 2010).

Gonçalves, L.S.A., Rodrigues, R., Amaral Júnior, A.T., Karasawa, M. and Sudré, C.P. (2008) Comparison of multivariate statistical algorithms to cluster tomato heirloom accessions. *Genetics and Molecular Research* 7(4), 1289–1297.

Hawkes, J.G., Maxted, N. and Ford-Lloyd, B.V. (2000) *The ex-situ Conservation of Plant Genetic Resources*. Kluwer Academic Publishers, Dordrecht, the Netheralnds.

Irish Seed Savers (2011) About Irish Seed Savers. Available at: www.irishseedsavers.ie/about-biodiversity-conservation.php (accessed 28 February 2011).

Jordan, J. (2007) The heirloom tomato as cultural object: investigating taste and space. *Sociologia Ruralis* 47(1), 20–41.

Kell, S.P., Maxted, N., Allender, C., Astley, D., Ford-Lloyd, B.V. and contributors (2009) Vegetable Landrace Inventory of England and Wales. Unpublished report. The University of Birmingham, UK.

Negri, V. (2003) Landraces in central Italy: where and why are they conserved and perspectives for their on-farm conservation. *Genetic Resources and Crop Evolution* 50, 871–885.

Negri, V. (2005) Agro-biodiversity conservation in Europe: ethical issues. *Journal of Agricultural and Environmental Ethics* 18, 3–25.

Negri, V., Maxted, N. and Veteläinen, M. (2009) European landrace conservation: an introduction. In: Veteläinen, M., Negri, V. and Maxted, N. (eds) *European Landraces: On-farm Conservation, Management and Use. Bioversity Technical Bulletin No. 15*. Bioversity International, Rome, pp. 1–22.

Parlevliet, J.E. (2007) How to maintain improved cultivars. *Euphytica* 153, 353–362.

Prada, D. (2009) Molecular population genetics and agronomic alleles in seed banks: searching for a needle in a haystack? *Journal of Experimental Botany* 60(9), 2541–2552.

Rhoades, R.E. and Nazarea, V.D. (1999) Local management of biodiversity in traditional agroecosystems. In: Collins, W.W. and Qualset, C.O. (eds) *Biodiversity in Agroecosystems*. CRC Press, Boca Raton, Florida, pp. 215–236.

Rodriguez-Burruezo, A., Prohens, J., Rosello, S. and Nuez, F. (2005) 'Heirloom' varieties as sources of variation for the improvement of fruit quality in greenhouse-grown tomatoes. *Journal of Horticultural Science and Biotechnology* 80(4), 453–460.

Royal Horticultural Society (2010) Heirloom veg set to return. Available at: www.rhs.org.uk/Plants/News/Heirloom-veg (accessed 13 March 2011).

Russo, V.M. (2008) Is older better? *International Journal of Vegetable Science* 14(2), 95–97.

Science and Advice for Scottish Agriculture (no date) Scottish landraces and traditional varieties. Available at: www.scottishlandraces.org.uk/what_are_landraces.htm (accessed 28 February 2011).

Shim, S.I. and Jorgensen, R.B. (2000) Genetic structure in cultivated and wild carrots (*Daucus carota* L.) revealed by AFLP analysis. *Theoretical Applied Genetics* 101, 227–233.
Sicard, D., Nanni, L., Porfiri, O., Bulfon, D. and Papa, R. (2005) Genetic diversity of *Phaseolus vulgaris* L. and *P. coccineus* L. landraces in central Italy. *Plant Breeding* 124, 464–472.
Skovmand, B., Reynolds, M.P. and DeLacy, I.H. (2001) Mining wheat germplasm collections for yield enhancing traits. *Euphytica* 199, 25–32.
Stickland, S. (1998) *Heritage Vegetables*. Gaia Books Limited, London.
Stickland, S. (2008) *Back Garden Seed Saving* (updated edn). Eco-logic Books, UK.
Stocks, C. (2008) *Forgotten Fruits: A guide to Britain's traditional fruit and vegetables*. Random House, London.
Terzopoulos, P.J. and Bebeli, P.J. (2008) DNA and morphological diversity of selected Greek tomato (*Solanum lycopersicon* L.) landraces. *Scientia Horticulturae* 116, 354–361.
Thomas Etty Esq (no date) *Thomas Etty Esq*. Available at: www.thomasetty.co.uk (accessed 30 March 2011).
Thompson and Morgan (2011) Heritage seeds, what is a heritage variety? Available at: www.thompson-morgan.com/news (accessed 28 February 2011).
Thorness, B. (2009) *Edible Heirlooms: heritage vegetables for the maritime garden*. Skipstone Press, Washington.
Tiranti, B. and Negri, V. (2007) Selective microenvironmental effects play a role in shaping genetic diversity and structure in a *Phaseolus vulgaris* L. landrace: implications for on-farm conservation. *Molecular Ecology* 16, 4942–4955.
Van de Wouw, M., van Hintum, T., Kik, C., van Treuren, R. and Visser, B. (2010) Genetic diversity trends in twentieth century crop cultivars: a meta analysis. *Theoretical Applied Genetics* 20 (6), 1241–1252.
Watson, B. (1996) *Taylor's Guide to Heirloom Vegetables*. Houghton Mifflin, Boston.
Zeven, A.C. (1998) Landraces: a review of definitions and classifications. *Euphytica* 104, 127–139.
Zeven, A.C. (1999) The traditional inexplicable replacement of seed and seed ware of landraces and cultivars: a review. *Euphytica* 110, 181–191.
Zeven, A.C. (2000) Traditional maintenance breeding of landraces: 1. Data by crop. *Euphytica* 116, 65–85.
Zeven, A.C. (2002) Traditional maintenance breeding of landraces: 2. Practical and theoretical considerations on maintenance of variation of landraces by farmers and gardeners. *Euphytica* 123, 147–158.

22 On-Farm Conservation of Plant Genetic Resources in Lazio Region, Italy. Implementation of the Regional Act 1st March 2000 N°15"

M.T. Costanza, I. Barbagiovanni Miracolo, P. Taviani, S. Paoletti, R. Rea, L. Lelli, J.H. Garzia, O. Porfiri, P. Nardi and M. Tanca

22.1 Introduction

The Region of Lazio, according to the Convention on Biological Diversity, implemented by Italy with the Act 124 (14/02/1994), has issued the Regional Act 15 (1 March 2000), 'Protection of autochthonous genetic resources of agricultural interest'. The Act entrusts the Regional Agency for the Development and Innovation of Agriculture in Lazio (ARSIAL) for its implementation in Italy.

ARSIAL is the official public body in charge for the application and implementation of Lazio Region's agricultural regional policies, in particular ARSIAL is involved in: agriculture and biodiversity activities (Regional Act 15, 2000); regional monitoring of Rural Development Plan; monitoring organic farm control organizations; GMO control for food security (Lazio Region is 'GMO free'); management of Regional Fauna Observatory in the Lazio Region; and protection of agricultural production quality (PDO and PGI).

Since 2010 ARSIAL has been involved in 'REVERSE', a European interregional cooperation project. REVERSE, 'Regional Exchanges and policy making for the protection and valorisation of biodiversity in Europe', responds to priority 2 of INTERREG IV C: environment and risk prevention, sub-theme 'biodiversity and preservation of natural heritage' and it consists of an exchange of experiences between the 14 European Partners (from seven European countries) that are involved in the protection of biodiversity. Its aim is to promote biodiversity on a European scale, by favouring positive action in territories (Reverse, 2010).

The Lazio Region is located centrally in the Italian peninsula, in the Mediterranean bio-geographical region. It is characterized by a complex morphology and a great bioclimate variability: Lazio's area is 17.236 km^2 (mountain 26%; hill 54%; plain 20%) and its population of about 5,500,000 inhabitants has a non-homogeneous distribution. The Natura 2000 Network recognized 183 Sites of Community Interest (SCI) spread on an area of 143,163 ha and 42 Special Areas of Conservation (SAC) spread on an area of 375,154 ha (Rete Natura 2000, 2008).

In Lazio the area used for agriculture is 40% of the total area (about 1,700,000 ha) managed by 102,572 farmers, usually elderly people (61.5% ≥65 years) and in Italy the average farm area is about 7.4 ha. Organic methodology is applied on 10% (70,000 ha) of regional farms. Major crops are represented by cereals, olive and grapevine, wood trees, vegetables (tomato, potato, carrot, watermelon) and fruit trees (kiwi, peach, sweet cherry, apple) (ISTAT, 2000; Centauri and Pasquarelli, 2003; Sabbatini, 2009). Lazio is very rich in agricultural environments and the impact of the new agricultural techniques in the last century has produced a progressive loss of agricultural diversity.

22.2 Regional Legislative Protection for On-farm Conservation

In the 1990s several Italian regions decided to promote regional acts to protect agrobiodiversity. First was the region of Toscana (in 1997 Regional Act 50 and in 2004 Regional Act 64) followed by the regions of Lazio (2000), Umbria (2001), Friuli-Venezia Giulia (2002), Marche (2003), Emilia Romagna (2008) and Basilicata (2008) (Lorenzetti *et al.*, 2009). Some regions do not have only the Act, but also the operative plan to support *in situ* conservation (Rete Semi Rurali, 2009).

The first article of the Lazio Regional Act 1st of March 2000, n°. 15, 'Protection of autochthonous genetic resources of agricultural interest' (Regione Lazio, 2000), gives the definition of the subject: 'autochthonous genetic resources of agricultural interest, including wild plants that mixed with cultivated species, such as species, races, varieties, populations, cultivars, ecotypes, and clones for which there is an economic, scientific, environmental, or cultural interest, threatened by genetic erosion' when their origin is in Lazio Region or introduced and integrated into the agroecosystem of Lazio for at least the past 50 years or disappeared from the region and collected in botanical gardens, breeding farms, experimental institutions, public or private genetic banks, and research centres of other regions or countries. The fifth article defines that 'the heritage of the genetic resources of these plants and animals belongs to the indigenous local communities, within which the benefits must be distributed equally'.

The Act entrusts ARSIAL to manage the operating tools for its application: the Regional Voluntary Register (RVR) and the Conservation and Safety Network (Network). The RVR is the official repertoire (register) of the Lazio Region, where the protected genetic resources, plants and animals, are registered, subject to the opinion of two scientific commissions, one for the plant sector and one for the animal sector. Anyone that owns, grows, or breeds plants or animals registered in the RVR may become a member of the Network. The members of the Network may be public or private institutions, associations of interest and single or associated farmers (fourth article). The goals of this action are: to reduce the 'genetic erosion threat' of autochthonous genetic resources of agricultural interest; to promote on-farm and *in situ* conservation; to improve 'plant and animal genetic resources' through a development of an economic interest of their food products; information and dissemination.

22.3 Plant Genetic Resources Protection

Plant genetic resources protection is described below and animal genetic resources activities are excluded. The preliminary step is a census, on the entire regional territory, of plant genetic resources of agricultural interest threatened by genetic erosion. The census is an ongoing activity and it consists of: material collection (seeds, fruits, etc.) of accessions, agronomical data and biographic research, including the examination of archives to verify the historical autochthony (Costanza *et al.*, in press). The local population is involved in the research with on-farm interviews, through ethnographic methods, on local knowledge linked to biodiversity, in collaboration with the Psychology Department of University of Roma 'La Sapienza'. All the accessions found are accurately described (ARSIAL database) in the agroecosystem in which they have been conserved, as: area of cultivation, agronomic practices used and geographical location (GPS).

The second step is represented by morpho-physiologic characterization of the accessions grouped in species. For this action ARSIAL has partnership agreements with different scientific institutions. Characterization of fruit trees and olive germplasm is carried out in collaboration with the Plant Production Department (University of Viterbo) and the Departments of Agricultural Research Council (CRA-Centre for Fruit Tree Research in Roma and CRA-OLI in Spoleto) using UPOV-IPGRI descriptors. Characterization of grapevine germplasm is carried out in Lazio Region's 'Agricultural Quality' project for the valorization of regional quality products and in collaboration with the Departments of Agricultural Research Council (CRA-VIT in Conegliano Veneto) using OIV descriptors. For herbaceous germplasm characterization the activity is carried out by 'Seeds Plants Operative National Programme – Lazio' (2007–2010 financed by MiPAAF of Italy) and in collaboration with the Italian Public Organization for Seeds Control (ENSE, nowadays INRAN) using UPOV-CPVO descriptors.

Every landrace is recognized as an autochthonous genetic resource by a Plant Scientific Commission. The pronouncement of the Commission is based on the resource's morpho-physiologic characterization results together with its agronomical remarks and its historical data, reported on ARSIAL's application form for the registration of the resource in RVR. At the same time the '*in situ* area' is defined and the evaluation of the 'genetic erosion threat' is assessed for every resource registered. The setting of criteria to establish the level of genetic erosion has been a crucial step in the application of the law (Porfiri *et al.*, 2009). Table 22.1 shows the five indicators chosen: existence of the product on the market; presence of the landrace on the catalogues of a seed company or nurseries; numbers of farmers still cultivating the landrace; cultivated areas of the landrace in comparison to the total regional area for that crop; and trend of new cultivation areas dedicated to that specific landrace. Each indicator is associated to other conditions to attribute a risk score (1 = low; 2 = medium; 3 = high) and the sum of different values gives a total level of erosion, with the following classification of the erosion risk: 'low risk' as total value ≤9; 'medium risk' as total value 10–13; and 'high risk' as total value ≥14. The presence of only one indicator with a score equal to 3 was sufficient to consider the landrace as under threat.

Each owner of a resource registered in RVR can join the Network to maintain on farm its landrace and cultivate it for production and/or for seed production. ARSIAL coordinates this operative system giving all technical support free of charge. All the plant genetic resources registered are listed in the Rural Development Programme of the Region of Lazio, EU fond, Action 214.9 – Plant Agricultural Biodiversity Protection (Regione Lazio, 2009) and every year this list is updated with new resources registered on RVR. To grant the economic support to those who want to cultivate the landraces protected in their '*in situ* area', owners of the resource have to be members of the Network. The genetic resource can be withdrawn from the RVR, whenever it does not fulfil the law specifications any longer.

22.4 Current Status

Currently registered in the plant section of the RVR are 172 landraces: 138 of 13 trees species and 34 of 14 herbaceous species (Table 22.2). The most frequent tree species: *Malus domestica* Bork, *Vitis vinifera* L. and *Pyrus communis* L., with 36, 32 and 23 local varieties, respectively; for the herbaceous germplasm there are 14 landraces of *Phaseolus vulgaris* L. The landraces list of RVR (Tables 22.3, 22.4 and 22.5) reports local name, conservation *in situ* distribution area and genetic erosion threat for every resource registered in Lazio.

In particular, Table 22.3 shows 93 landraces of fruit tree germplasm identified in Lazio home gardens and small farms, their names expressing the strong link between the fruit and their territory (Pavia *et al.*, 2009a). The majority of fruit tree autochthonous germplasm of Lazio was identified by a 2-year project called 'Individuation, recovery and characterization of local autochthonous fruit tree germplasm at risk of genetic erosion' (Regione Lazio *et al.*, 2003; Pavia *et al.*, 2009b). The collection field for *ex situ* conservation of these resources is located at the Centre for Fruit Tree Research in Roma; a new

Table 22.1. Evaluation of genetic erosion threat: risk level and score are estimated for each indicator.

Indicator	Description	Risk level	Score
Presence of the product on the market	Markets and/or producer's cooperatives sector: main variety in a certain DOC, IGT, PDO and PGI certified production	Low	1
	Niche market: locally limited cultivated areas Market section: secondary varieties in a DOC, IGT, PDO and PGI certified production	Medium	2
	Only some fruits/few seeds available for consumption or research No product on the market	High	3
Presence in the catalogues of the seed companies/ nurseries	Fruits: presence in variety list A, B and C Vegetables and plants: listed in the National register of varieties Grapevine: listed in the National and Regional register of wine varieties	Low	1
	Grapevine: under registration in the National and Regional register of wine varieties Propagation materials available at a few nurseries	Medium	2
	Fruits: not registered in the variety list Vegetables and plants: not registered in the National register of varieties Grapevine: not registered in the National and Regional register of wine varieties No propagating material available out of the maintaining farm	High	3
Number of cultivating farmers	>100	Low	1
	30–100	Medium	2
	<30	High	3
Areas under cultivation (as percentage of the total regional area for the species)	>5%	Low	1
	1–5%	Medium	2
	<1% Isolated plants or home garden cultivations	High	3
New dedicated area trend	New areas cultivated to landrace present	Low	1
	No new areas dedicated to landrace present	High	3

ARSIAL field in Montopoli in Sabina (Roma district) is a work in progress; some accessions of fruit tree landraces, collected in Colle San Magno (Frosinone district), are planted in ARSIAL's field of Alvito (FR). Phytosanitary control activity of tree material is carried out in collaboration with the Departments of Agricultural Research Council (CRA-PAV, in Roma) for *ex situ* germplasm conservation (collection fields).

In Table 22.4, autochthonous grapevine germplasm (32) and olive landraces (13) are reported; several accessions of olive and grapevine genetic resources protected and registered in RVR were characterized by molecular markers analysis (SSR markers analysis reported in Pandolfi *et al.*, 2010a). In particular, olive germplasm collected in Lazio (Catta *et al.*, in press) has been characterized in a overview work (Pandolfi *et al.*, 2010b): every landrace, of large and small distribution, is identified by morphologic markers and agronomic data together with chemical and sensorial analysis of its oil. Olive autochthonous germplasm identified in Lazio is represented by ancient trees integrated in its agroenvironment and currently deteriorated. Specific landscapes of the Mediterranean region have been shaped by ancient cultivation of olive – at the present time restoration of these landscapes is possible by on-farm and *in situ* conservation of autochthonous olives. ARSIAL collection field for *ex situ* conservation of olive germplasm is in Montopoli in Sabina.

ARSIAL project 'Agricultural Quality' for the valorization of regional quality

Table 22.2. List of 172 autochthonous plant genetic resources of agricultural interest of Lazio registered in the RVR, arranged by number of landraces for each species.

Tree species	Landraces	Herbaceous species	Landraces
Malus domestica Bork	36	*Phaseolus vulgaris* L.	14
Vitis vinifera L.	32	*Lens culinaris* Med	3
Pyrus communis L.	23	*Lycopersicum esculentum* Mill.	3
Prunus avium L.	14	*Triticum turgidum* L. Subsp.	2
Olea europea L.	13	*Cynara dicoccum* Schubler *scolymus* L.	2
Punica granatum L.	4	*Allium sativum* L.	2
Prunus persica (L.) Batsch.	4	*Apium graveolens* L.	1
Prunus insititia L.	4	*Cucurbita pepo* L.	1
Corylus avellana L.	3	*Foenicum vulgare* Mill.	1
Prunus armeniaca L.	2	*Zea mays* L.	1
Crategus azarolus L.	1	*Capsicum annum* L.	1
Prunus cerasus L.	1	*Cicer arietinum* L.	1
Castanea sativa L.	1	*Phaseolus coccineus* L.	1
		Fragaria vesca L.	1
13 species	138	14 species	34

Table 22.3. Lazio autochthonous fruit-tree germplasm (93 local varieties) registered in Lazio RVR.

Crop	Species	Landrace name	Genetic erosion threat	Conservation *in situ* distribution area: municipality (district)
Apple	*Malus domestica* Bork	Agre di Sezze	High	Lazio
		Agre di Viterbo	High	Lazio
		Appia	High	Lazio
		Bebè	High	Poggio Mirteto (RI)
		Calvilla	High	Lazio
		Capo d'Asino	High	Lazio
		Cerina (Zitella, Gelata)	Medium	Lazio
		Cipolla	High	Lazio
		Cocoine	High	Lazio
		Dolce di Sezze	High	Lazio
		Francesca	High	Lazio
		Francesca di Castelliri	High	Castelliri and Sora (FR)
		Mbriachella	High	Roma and Rieti district
		Nana	High	Lazio
		Paoluccia	High	Lazio
		Paradisa	High	Lazio
		Pianella (Rosa)	High	Lazio
		Pontella	High	Lazio
		Rosa	High	Lazio
		Rosa gentile	High	Roma district
		Rosa piatta ciociara	High	Frosinone district
		S. Giovanni	High	Viterbo district
		Spugnaccia	High	Viterbo district
		Velletrana	High	Subiaco and Velletri (RM)
		Verdona	High	Provincia di Rieti
		Verdonica	High	Provincia di Rieti
		Zuccherina o Gelata	High	Lazio

Continued

Table 22.3. Continued.

Crop	Species	Landrace name	Genetic erosion threat	Conservation *in situ* distribution area: municipality (district)
		Fragola	High	Lazio
		Gaetana	High	Lazio
		Maiolina	High	Lazio
		Prata	High	Lazio
		Rosetta o Rosone	High	Lazio
		S. Agostino	High	Lazio
		Sublacense	High	Roma district
		Tonnorella	High	Lazio
		Limoncella	Medium	Lazio
Pear	*Pyrus communis* L.	Abitiritir	High	Alatri (FR) and neighbouring municipality
		De lu Prete	High	Grisciano (RI) and neighbouring municipality
		Del Principe	High	Soriano al Cimino (VT)
		Monteleone	Medium	Castiglione in Teverina, Bolsena, Acquapendente (VT)
		Angina o Ancina	Medium	Roma and Latina district
		Baccelli	High	Genazzano (RM) and neighbouring municipality
		Barocca – Invernale di S. Vito	High	Genazzano (RM) and neighbouring municipality
		Biancona	High	Roma and Latina district
		Bottiglia	Medium	Alatri (FR)
		Campana	High	Borbona (RI)
		Cannella	High	Grisciano (RI)
		Castrese	High	Roma and Latina district
		Cocozzola (Cucuzzara, Zucchina)	High	Roma district
		Di Posta	High	Frosinone district
		Di S. Cristina (Peruzza)	High	Bolsena (VT)
		Fegatella	High	Roma and Latina district
		Pero-melo	High	Roma and Latina district
		Rossa di Maenza	High	Maenza (LT) and neighbouring municipality
		Sellecca	High	Alatri, Ferentino (FR) and neighbouring municipality
		Spadona di Castel Madama	High	Castel Madama (RM) and neighbouring municipality
		Spina (Spinacarpi, Coccia d'Asino, Casentina)	Medium	Lazio
		Trentonce	High	Borbona (RI)
		Tunnella	High	Roma and Latina district
Apricot	*Prunus armeniaca* L.	Di Monteporzio	Medium	Frascati, Monteporzio Catone, Colonna, Montecompatri (RM)
		S. Maria in Gradi – AL1	High	Provincia di Viterbo
Azzeruolo	*Crataegus azarolus* L.	Azerole Rosso	Medium	Lazio

Continued

Table 22.3. Continued.

Crop	Species	Landrace name	Genetic erosion threat	Conservation *in situ* distribution area: municipality (district)
Chestnut	*Castanea sativa* Mill.	Marrone Premutico (Primatico, Primaticcio)	Medium	Viterbo district and Manziana Municipality (RM)
Cherry-sweet	*Prunus avium* L.	Bella di Pistoia	High	Rieti, Viterbo and Roma district
		Biancona	High	Rieti, Viterbo and Roma district
		Buonora	High	Rieti, Viterbo and Roma district
		Core (Durona)	High	Rieti, Viterbo and Roma district
		Crognolo	High	Rieti, Viterbo and Roma district
		Graffione	Medium	Rieti, Viterbo and Roma district
		Maggiolina	High	Rieti, Viterbo and Roma district
		Morona	High	Rieti, Viterbo and Roma district
		Ravenna a gambo corto	High	Rieti, Viterbo and Roma district
		Ravenna a gambo lungo	High	Rieti, Viterbo and Roma district
		Ravenna precoce	High	Rieti and Roma district
		Ravenna tardiva	High	Rieti and Roma district
		Petrocca	High	Montelibretti (RM) and neighbouring municipality
		Lingua de Fori	High	Montelibretti (RM) and neighbouring municipality
Pomegranate	*Punica granatum* L.	Di Gaeta MG1	High	Latina district
		Di Gaeta MG2	High	Latina district
		Di Formia MG3	High	Latina district
		Di Formia MG4	High	Latina district
Hazelnut	*Corylus avellana* L.	Barrettona	High	Viterbo district
		Casamale o nostrale (Comune di Sicilia)	High	Viterbo district
		Rosa (Nocchia R.)	High	Viterbo district
Peach	*Prunus persica* L.	Ala	High	Velletri (RM) and neighbouring municipality
		Reginella Pesca Uovo (Early Crawford)	High	Roma and Viterbo district
		Reginella II	High	Roma district
		Tardiva di San Vittorino	High	Tivoli (RM) and neighbouring municipality
Plum	*Prunus insititia* L	Coscia di Monaca di Ponzano Romano	Medium	Ponzano Romano (RM)
		Di Gallinaro	Medium	Sora (FR)
		Recinella	High	Giuliano di Roma (FR) and neighbouring municipality
		S. Giovanni	Medium	Arce (FR)
Sour cherry	*Prunus cerasus* L.	Nana dei Castelli	High	Castelli Romani municipalities (RM)

Table 22.4. Autochthonous grapevine and olive germplasm registered in Lazio RVR.

Crop	Species	Landrace name	Genetic erosion threat	Conservation *in situ* distribution area: municipality (district)
Grapevine	*Vitis vinifera* L.	Abbuoto n.[1]	Medium	Viterbo, Roma and Latina district
		Aleatico n.[1]	Low	Viterbo, Roma, Latina and Rieti district
		Bombino bianco b.[1]	Low	Lazio
		Bombino nero n.[1]	Medium	Roma and Frosinone district
		Cannaiola di Martan[1]	Medium	Marta, Bolsena, Tuscania (VT)
		Capolongo b.[1]	Medium	Frosinone district
		Greco b.[1]	Medium	Lazio
		Greco bianco b.[1]	Medium	Viterbo, Roma and Latina district
		Greco nero n.[1]	Medium	Viterbo, Roma and Latina district
		Lecinaro n.[1]	Medium	Frosinone district
		Maturano b. (Motulano)[1]	Medium	Frosinone district
		Moscato di Terracina[1]	Low	Roma, Latina and Frosinone district
		Nero Buono n.[1]	Low	Latina and Roma district
		Olivella nera n.[1]	Medium	Frosinone district
		Pampanaro b.[1]	Medium	Frosinone district
		Passerina b.[1]	Low	Roma and Frosinone district
		Pecorino b.[1]	Medium	Rieti district
		Rosciola r.[1]	Medium	Roma district
		Verdello b.[1]	Low	Viterbo and Rieti district
		Albarosa[2]	High	Grottaferrata (RM)
		Angelica[2]	High	Frosinone district
		Nerone[2]	High	Agosta, Canterano, Cervara di Roma, Gerano, Marano Equo, Rocca Canterano, Subiaco (RM)
		Nostrano[2]	High	Piglio (FR)
		Uva dei vecchi[2]	High	Montefiascone (VT)
		Cesanese di Castelfranco[2]	High	Rieti district
		Maturano nero (Motulano)[2]	High	Frosinone district
		Pedino[2]	High	Montefiascone (VT)
		Romanesco[2]	High	Montefiascone (VT)
		Uva Mecella[2]	High	Pecosolido (FR)
		Pizzutello bianco b. (den. locali: Pizzutello di Tivoli, Dito di Donna)[3]	Low	Roma and Latina districts
		Pellegrina[4]	Medium	Lazio
		Pizzutello nero[4]	High	Roma and Latina districts
Olive	*Olea europea* L.	Oliva dei Monti	High	Monti Lucretili
		Palmuta	High	Tivoli, S Polo dei Cavalieri, Marcellina and neighbouring municipality (RM)
		Rappaiana	High	Tivoli, Marcellina and neighbouring municipality (RM)
		Romana	Medium	Tivoli, Marcellina and neighbouring municipality (RM)
		Roscetta Gagliarda	High	Tivoli, Marcellina and neighbouring municipality (RM)

Continued

Table 22.4. Continued.

Crop	Species	Landrace name	Genetic erosion threat	Conservation *in situ* distribution area: municipality (district)
		Rosciola Nostrana	High	Tivoli, Marcellina and neighbouring municipality (RM)
		Rotonda di Tivoli	High	Tivoli, Marcellina and neighbouring municipality (RM)
		Salvia Montelibretti	Medium	Montelibretti, Palombara Sabina, Neroli (RM)
		Sbuciasacchi	High	Tivoli, Marcellina and neighbouring municipality (RM)
		Sirole cl. Soratte	Low	Civitella S.Paolo, Fiano Romano, Filacciano, Nazzano, Ponzano Romano, Rignano, S. Oreste, Torrita Tiberina (RM)
		Marina	Medium	S. Donato Val Valcomino, Alvito, Gallinaro, Settefrati (FR)
		Minutella Casarè	Medium	Priverno, Sonnino, Itri (LT)
		Vallanella	Medium	Priverno, Sonnino, Itri (LT)

[1] Grapevine listed in the National and Regional register of wine varieties.
[2] Grapevine under registration in the National and Regional register of wine varieties.
[3] Grapevine listed in the National and Regional register of table grapevine varieties.
[4] Grapevine under registration in the National and Regional register of table grapevine varieties.

Table 22.5. Lazio autochthonous herbaceous germplasm registered in Lazio RVR.

Crop	Species	Landrace name	Genetic erosion threat	Conservation *in situ* distribution area: municipality (district)
Common bean	*Phaseolus vulgaris* L.	Fagiolo a Pisello	High	Colle di Tora (RI)
		Fagiolina Arsolana	High	Arsoli, Marano Equo, Vivaro Romano, Riofreddo, Vallinfreda, Vallepietra (RM)
		Fagiolo Cioncone	High	
		Fagiolo Regina di Marano Equo	Medium	
		Fagiolo Cappellette di Vallepietra	High	
		Fagiolo Romanesco di Vallepietra	High	
		Pallino di Vallepietra	High	
		Fagiolo Ciavattone piccolo	High	Viterbo district
		Fagiolo di Gradoli o del Purgatorio	Medium	
		Fagiolo Giallo	High	
		Fagiolo Solfarino	High	
		Fagiolo Verdolino	High	
		Fagiolo Cannellino di Atina	Low	Atina, Casalattico, Casalvieri, Gallinaro, Picinisco, Villa Latina (FR)
		Fagiolo Borbontino	Medium	Borbona (RI)
Spain bean	*Phaseolus coccineus* L.	Fagiolone di Vallepietra	High	Arsoli, Marano Equo, Vivaro Romano, Riofreddo, Vallinfreda, Vallepietra (RM)

Continued

Table 22.5. Continued.

Crop	Species	Landrace name	Genetic erosion threat	Conservation *in situ* distribution area: municipality (district)
Lentil	*Lens culinaris* L.	Lenticchia di Onano	Medium	Onano (VT)
		Lenticchia di Rascino	Medium	Fiamignano e Petrella Salto (RI)
		Lenticchia di Ventotene	Medium	Ventotene (LT)
Chick pea	*Cicer arietinum* L.	Cece di Canepina	High	Canepina (VT)
Strawberry	*Fragaria vesca* L.	Fragolina di Nemi	High	Nemi and neighbouring municipality (RM)
Tomatoes	*Lycopersicum esculentum* L.	Pomodoro Scatolone di Bolsena	High	Bolsena (VT)
		Pomodoro Spagnoletta di Formia e Gaeta	Medium	Itri, Gaeta, Formia, Minturno, Castelforte, Spigno Saturnia, SS Cosma e Damiano (LT)
		Pomodoro da secca di Minturno	High	Minturno, Formia e Castelforte (LT)
Pepper	*Capsicum annum* L.	Peperone Cornetto di Pontecorvo	Low	Pontecorvo, Esperia, S. Giorgio a Liri, Pignataro Interamna, Villa S. Lucia, Piedimonte S. Germano, Aquino, Castrocielo, Roccasecca, San Giovanni Incarico (FR)
Zucchini	*Cucurbita pepo* L.	Zucchino di Cerveteri tipo Romanesco	High	Cerveteri (RM)
Celery	*Apium graveolens* L.	Sedano Bianco di Sperlonga	Low	Fondi e Sperlonga (LT)
Fennel	*Foenicum vulgare* L.	Finocchio di Tarquinia	High	Tarquinia, Monte Romano, MontHigh di Castro e Tuscania (VT), Allumiere e Civitavecchia (RM)
Artichoke	*Cynara cardunculus* var. *scolymus* L.	Carciofo di Campagnano	Medium	Roma, Viterrbo and Latina district
		Carciofo di Castellamare	Medium	Roma, Viterrbo and Latina district
Garlic	*Allium sativum* L.	Aglio Rosso di Castelliri	Medium	Castelliria and Isola Liri (FR)
		Aglio Rosso di Proceno	Medium	Proceno (VT)
Emmer	*Triticum dicoccum* Schrank.	Farro dell'alta Valle del Turano e della Valle dell'Aniene		Cinto Romano, Riofreddo, Vallinfreda and Vivaro (RM)
		Farro dell'Alta Valle del Tronto	Medium	Leonessa and Amatrice (RI)
Maize	*Zea mays* L.	Mais Agostinella	High	Vallepietra (RI)

products was carried out on Lazio grapevine germplasm; through their genetic characterization and wine tests, many grapevine autochthonous genetic resources (registered in RVR) were listed in the National and Regional register of wine varieties or in the National and Regional register of table grapevine varieties (Table 22.4). It represents an attempt to promote their cultivation (Costacurta *et al.*, in press). Collection field for *ex situ* conservation of grapevine germplasm is in ARSIAL's field of Velletri (Roma district).

Table 22.5 shows 34 herbaceous landraces registered in RVR and these landraces are represented in *ex situ* collection of ARSIAL germplasm seed bank with 130 accessions collected by Lazio farmers. The distribution of herbaceous landraces depends on crop: cereals and legumes are concentrated in small areas (hills and mountains) with several groups of farmers, vegetables such as artichoke and fennel have large cultivation areas (intensive agriculture areas) but very few farmers for on-farm conservation of these landraces. Some examples of their valorization are represented by: 'Sedano bianco di Sperlonga' celery (PGI quality standard certification), 'Cannellino di Atina' bean (PDO quality standard certification) and 'Cornetto di Pontecorvo' pepper (looking forward for EU recognition as a PDO certified product).

Currently there are enrolled in the Network 255 farmers (plant resources owners) for *in situ* active conservation of plant genetic resources (1 September 2010) among the districts of: Frosinone 97, Viterbo 72, Rieti 39, Latina 39 and Roma 8. The entity of grants to the resource keepers for on-farm conservation are: €250–300/ha for cereals, €500–600/ha for vegetables, €800–900/ha for trees and €70–90/plant for a single tree up to a maximum of five tree-plants per local variety.

22.5 Next Steps

Beyond the activities related to the application of the Regional Act 15/2000, in 2005 a programme on soil biological fertility and microbial diversity monitoring of Lazio Region was started, in collaboration with Departments of Agricultural Research Council (CRA-RPS, in Rome).

ARSIAL started a partnership agreement with Professor of Cultural Anthropology (V. Padiglione, Psychology Department of University of Roma 'La Sapienza') in anthropology investigation on social capital knowledge and practices linked to plant genetic resource owners; it represents a tool to understand the complex system of on-farm landrace conservation planned by every farmer and the relationship among farmers. ARSIAL's future actions will focus on:

- Re-integration and valorization of landraces;
- Promotion of traditional food products;
- Enhancement of farmers' network;
- Participatory plant breeding methodology promotion;
- Research on food products' nutritional properties;
- Phytosanitary and genetic tests on resources;
- Identification of soil microbial communities characteristic of landraces field; and
- Strategic and political recommendations, intended for decision makers, in order to create new perspective for local policies to protect biodiversity (Reverse project).

References

Catta, M., Pica, G., Petruzzelli, A., Di Ferdinando, S. (in press) Valorizzazione degli oli tipici e monovarietali del Lazio. In: *Atti del VIII Convegno Nazionale 'La Biodiversità – una risorsa per sistemi multifunzionali' Lecce 21–23 April 2008.*

Centauri, L. and Pasquarelli, P. (2003) Principali colture nella Regione Lazio – anno 2003. Available at: www.agricoltura.regione.lazio.it/binary/agriweb/agriweb_allegati_schede_informative/Statistiche_2003.1190975174.pdf (accessed 8 September 2010).

Costacurta, A., Moretti, S., Casadei, G., Pica, G. (in press) Recupero e valorizzazione di vitigni autoctoni del Lazio. In: *Atti del VIII Convegno Nazionale 'La Biodiversità – una risorsa per sistemi multifunzionali' Lecce 21–23 April 2008.*

Costanza, M., Barbagiovanni, I., Nardi, P., Paoletti, S. (in press) Tutela delle risorse genetiche autoctone di interesse agrario del Lazio: quadro delle attività svolte da Arsial nell'ambito delle L.R. 1 marzo 2000, n. 15. In: *Atti del VIII Convegno Nazionale 'La Biodiversità – una risorsa per sistemi multifunzionali' Lecce 21–23 April 2008.*

ISTAT (2000) V censimento generale dell'agricoltura Presentazione dei dati definitivi Lazio. Available at: www.agricoltura.regione.lazio.it/binary/agriweb/agriweb_allegati_schede_informative/5_censimento_generale_dell_agricoltura_ottobre_2000_dati_del_Lazio.1190975320.pdf (accessed 8 September 2010).

Lorenzetti, F., Lorenzetti, S. and Negri, V. (2009) The Italian Laws on Conservation Varieties and the National Implementation of Commission Directive 2008/62 EC. In: Veteläinen, M., Negri, V. and Maxted, N. (eds) *European Landraces: on-farm conservation management and use*. Bioversity Technical Bulletin No. 15. Bioversity International, Rome, pp. 300–304.

Pandolfi, S., De Angelis, S., Paoletti, A., Baldoni, L., Cultrera, G.M.N., Mariotti, R., Perri, E. and Pannelli, G. (2010a) *Varietà ed ecotipi di olivo del Lazio*. Tipolitografia CSR, Roma, pp. 227–230.

Pandolfi, S., De Angelis, S., Paoletti, A., Baldoni, L., Cultrera, G.M.N., Mariotti, R., Perri, E. and Pannelli, G. (2010b) *Varietà ed ecotipi di olivo del Lazio*. Tipolitografia CSR, Roma, pp. 51–55, 93–226.

Pavia R., Barbagiovanni, I.M., Della Strada, G., Piazza, M.G., Engel, P. and Fideghelli, C. (2009a) Autochthonous Tree Fruit Germplasm at risk of genetic erosion found in Home Gardens in Latium (Italy). In: Bailey, A., Eyzaguirre, P. and Maggioni, L. (eds) *Proceeding of a Workshop Lubiana, Slovenia, 3–4 October 2007. Crop genetic resources in European home gardens*. Bioversity International, Rome, pp. 21–25.

Pavia, R., Barbagiovanni, I.M., Bignami, C., Cristofori, V. and Piazza, M.G. (2009b) La biodiversità frutticola del Lazio. *Notiziario Risorse Genetiche Vegetali* 3–4/2009, 7–10.

Porfiri, O., Costanza, M. and Negri, V. (2009) Landrace Inventories in Italy and the Lazio Region Case Study. In: Veteläinen, M., Negri, V. and Maxted, N. (eds) *European Landraces: on-farm conservation management and use*. Bioversity Technical Bulletin No. 15. Bioversity International, Rome, pp. 117–123.

Regione Lazio (2000) Legge Regionale n.15/2000. Available at: www.arsialweb.it/cms/index.php?option=com_docman&task=cat_view&gid=66&Itemid=100 (accessed 8 September 2010).

Regione Lazio (2009) Programma di Sviluppo Rurale del Lazio per il periodo 2007 – 2013. 717. 517–537. Available at: www.agricoltura.regione.lazio.it/binary/agriweb/agriweb_allegati_schede_informative/PSR_2007_2013_LAZIO_HC_DEF_consolidato_19_11_09.pdf (accessed 8 September 2010).

Regione Lazio, Istituto Sperimentale per la Frutticoltura – Roma, Università degli Studi della Tuscia - Viterbo (2003) *Individuazione, recupero e caratterizzazione del germoplsma frutticolo autoctono laziale a rischio di erosione genetica,* Tipolitografia New Interstampa, Rome.

Rete Natura 2000 (2008) Cartografia. Available at: www.regione.lazio.it/web2/contents/ambiente/cartografia_tab.php (accessed 8 September 2010).

Rete Semi Rurali (2009) Leggi Regionali e Valorizzazione delle varietà vegetali e delle razze locali, stato dell'arte 2008. Available at: www.semirurali.net/modules/documentazione/index.php?content_id=3 (accessed 8 September 2010).

Reverse (2010) Project description. Available at: http://reverse.aquitaine.fr (accessed 8 September 2010).

Sabbatini, M. (2009) *Evoluzione e prospettive dell'agricoltura del Lazio Statistiche ufficiali e informazioni ausiliarie al 2007*. Franco Angeli s.r.l., Milano, Italy.

23 On-Farm Conservation in Industrialized Countries: A Way to Promote Dynamic Management of Biodiversity within Agroecosystems

M. Thomas, E. Demeulenaere, C. Bonneuil and I. Goldringer

23.1 Introduction

In the context of global change, it is important to maintain adaptability of plant genetic resources (Cooper *et al.*, 2001). Studying the evolutionary mechanisms occurring not only in the centres of diversity but also where diversified crop populations are grown in heterogeneous environments seems crucial because these populations may also correspond to adaptable varieties (Hawtin *et al.*, 1996). For this reason, this chapter considers on-farm conservation in France, an industrialized farmland outside the centres of diversity. Initially, in pre-industrial Europe, several genetically diverse and locally adapted landraces were grown on farm and they were adapted to the various needs of the farmers and users (Berg, 2009) as it is the case in countries where traditional farming is still present (Elias *et al.*, 2001; Alvarez *et al.*, 2005; Barnaud *et al.*, 2007; Allinne *et al.*, 2008). Yet, since industrial transition replaced original landraces by modern varieties (Bonneuil *et al.*, 2006; Arndorfer *et al.*, 2009; Berg 2009), only a few landraces and old varieties are still grown in rare traditional (Negri *et al.*, 2000; Negri, 2003) or organic farming systems (Bocci and Chable, 2008). These populations are named folk varieties following Berg (2009), who defined a folk variety as:

> a farmers' variety that is selected and maintained for one or more distinctive properties. It may be fairly uniform for the selected traits, but otherwise diverse and therefore responsive to new selection. It differs from modern varieties by having an unknown (or unclear) origin, more inherent diversity, and less varietal stability.

While agriculture is highly industrialized in France, some folk varieties (also called population-variety for their genetic heterogeneity) are still grown and seeds are exchanged through informal social networks (Bonneuil and Demeulenaere, 2006; Thomas *et al.*, 2011).

In this chapter, we analyse the role of this social network as an actor in the sustainable management of crop genetic diversity, and we examine how this system is taken into account in the framework of the French policy for genetic resources.

To this aim: (i) we first provide a brief description of folk varieties maintained in this social network; (ii) we summarize results obtained on the folk variety named 'Rouge de Bordeaux' showing the link between genetic diversity and population structure and the social knowledge on farmers practices and seed exchanges; and (iii) we briefly describe the stand taken by France on on-farm conservation (acknowledgement and policy) and we discuss the consequences for the sustainability of this conservation strategy.

23.2 Folk Varieties Maintained in the French Social Network

In the French case study, ethno-botanical surveys showed that seed exchanges mainly occur between farmers but also between the national gene bank and farmers (Bonneuil and Demeulenaere, 2006; Demeulenaere *et al.*, 2008). The 30 surveys achieved between 2005 and 2007 detected at least 700 varieties of bread wheat maintained in this social network, which were further classified into four types of varieties: (i) landraces; (ii) old varieties (before 1980); (iii) modern varieties (after 1980); and (iv) mixtures. In general, farmers were growing pure varieties in small plots (1–10 m^2) to characterize them at the phenotypic level and mixtures in larger fields to improve the stability of their production (Enjalbert *et al.*, 2011). Using a snow-bowl approach, around 200 actors were identified in 2005, among which 92.5% were farmers or amateurs. The rest correspond to people working for conservation or research institutes (5%) or breeding companies (2.5%). It is important to note that this distribution is not representative of seed exchange activities among and within actor categories (Bonneuil and Demeulenaere, 2006).

In 2008, to better characterize the population-varieties maintained in this social network and to assess the role of environmental conditions and agronomical practices in their evolution, we completed seed-exchange characterization by conducting 33 phone calls with farmers already recognized as the most active farmers in the social network. Then, we focused on the population-varieties that were the most often shared within the social network. Farmers who agreed to participate in the study and the national gene bank sent us seed samples of 15 population-varieties without any selection of the seeds. These 15 population-varieties fell into three classes among the four already mentioned (five landraces, seven old varieties and three mixtures, see Table 23.1 for more details). A 'common garden' trial was carried out for their phenotypic characterization. The objective was to assess under the same environmental conditions the phenotypic divergence of the different versions of the same initial population-variety. These versions differed in their history (different farmers' practices and different environments). Each population was grown in a 1 m^2 plot with around 35 plants per plot. Quantitative data scored at the plot level included heading date, disease sensitivity and grain production. Plant and spike size were measured on the main tiller. Grain production components such as the number of grains per spike and thousand kernel weight were measured on the same spike. Each measured spike was classified based on qualitative colour scores, the spikelet density (compact, semi-compact or loose) and the presence or absence of awns. A picture was taken for one spike per plant measured under controlled lighting conditions to assess differences in terms of spike colour. Analyses are still in progress.

Table 23.1. List of population-varieties of bread wheat studied in the context on-farm conservation in France.

Variety	Type	Number of samples
Automne rouge barbu	Landrace	6
Blanc de la Réole	Landrace	9
Blé du Lot	Landrace	10
Touzelle anone	Landrace	8
Touzelle blanche barbue	Landrace	7
Noe	Old Variety	7
Rouge de Bordeaux	Old Variety	18
Chiddam d'automne	Old Variety	8
Prince Albert	Old Variety	9
Concorde	Old Variety	17
Dattel	Old Variety	11
Talisman	Old Variety	8
Mélange de James	Mixture	14
Mélange de Jean-François	Mixture	5
Mélange de Touselle	Mixture	26
Total 15	3	163

23.3 The 'Rouge de Bordeaux' Case Study

Based on the 30 interviews, it was possible to identify ten farmers who were growing a common bread wheat (*Triticum aestivum* L.) folk variety named 'Rouge de Bordeaux' (RDB), which appeared in France during the middle of the 19th century (Vilmorin-Andrieux Cie, 1880). Thus, these ten farmers were growing the same variety under contrasting cultural practices, in locations spread over the French territory (see Fig. 23.1), and submitted to stochastic environmental conditions. Because some farmers conserved their seed samples during several years or because they were growing the same variety according to different practices, the ten farmers provided us with a total of 18 populations of RDB (Table 23.2, Fig. 23.1). The French gene bank provided us with an additional sample, which will be further considered as a 'reference' sample.

Based on the social information collected during the survey, we drew a network of the seed exchanges among all actors identified. In this network, a node represents an RDB population characterized by a location (farmer name), a year and if necessary by a practice. A vertical line on Fig. 23.2 represents one or several multiplication events and a slanting or a horizontal line represents a diffusion event (upper part of Fig. 23.2). In parallel, we studied the genetic diversity of the 19 (18 + the ref. sample) RDB populations (~30 individuals per population) using 19 SSR neutral markers located on 19 different chromosomes. The within-population genetic

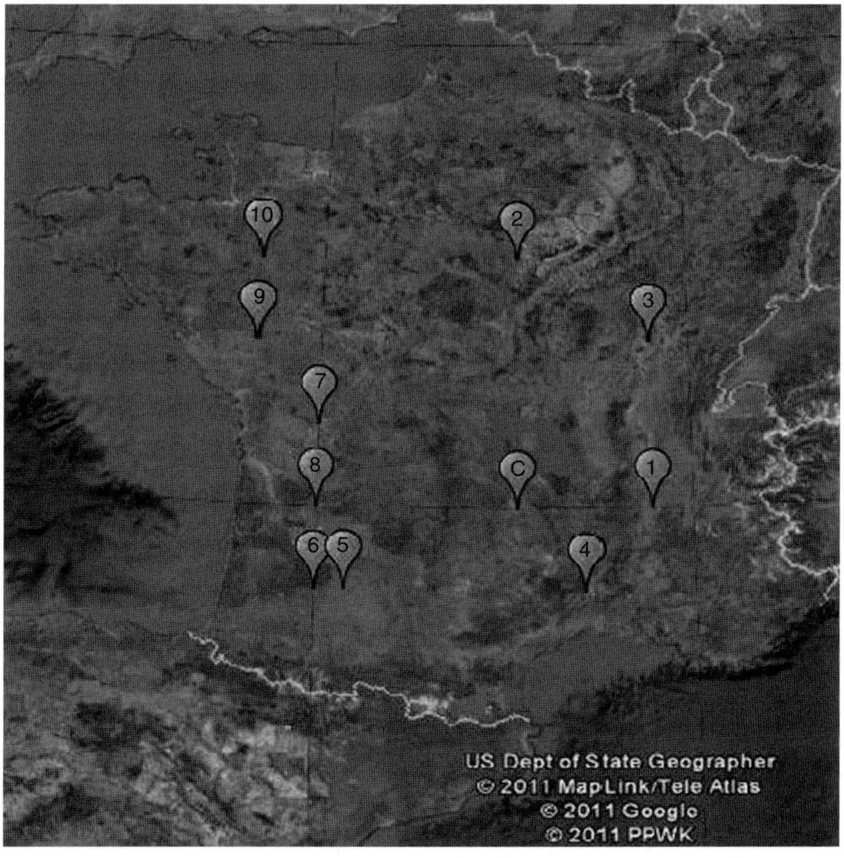

Fig. 23.1. Location of the ten farms (1 to 10) and the national gene bank (c) involved in the RDB study.

Table 23.2. Description of the 19 sampled populations of the 'Rouge de Bordeaux' folk variety.

Population name	Variety	People	Location	Practice	Origin	Receipt Year	Sampling Year	Number of Generations	Coordinates Longitude	Latitude	Altitude (m)
RBAim_06	RDB	Aimé	1	production		1991	2006	14	5.815359	45.153889	588
RBBC_06	RDB	Basile	2	conservation		1998	2006	8	3.813602	48.621333	78
RBBI_06	RDB	Basile	2	isolation		1998	2006	8	3.813602	48.621333	78
RBBE_03	RDB	Basile	2	selection		1998	2003	5	3.813602	48.621333	78
RBBM_03	RDB	Basile	2	conservation		1998	2003	5	3.813602	48.621333	78
RBDe_03	RDB	Denis	3	conservation		1999	2003	4	5.26975895	47.560589	296
RBDe_06	RDB	Denis	3	conservation		1999	2006	7	5.26975895	47.560589	296
RBEmi_06	RDB	Emile	4	production		1998	2006	8	4.506487	44.092741	225
RBGas_03	RDB	Gaston	5	production		1998	2003	5	0.42594	44.255174	64
RBGas_06	RDB	Gaston	5	production		1998	2006	8	0.42594	44.255174	64
RocGas_05	RDR	Gaston	5	selection		1998	2005	7	0.42594	44.255174	64
RBJul_06	RDB	Jules	6	multiplication		2000	2006	6	0.525513	44.353705	86
RBLuc_06	RDB	Lucien	7	multiplication		2005	2006	1	0.666243	46.154473	33
RBMar_06	RDB	Marcel	8	multiplication		2004	2006	2	0.221064	45.843305	97
RBPN2_06	RDB	Paul	9	multiplication	N2	2005	2006	1	1.133151	47.01239	90
RBPN3_06	RDB	Paul	9	multiplication	N3	2004	2006	2	1.133151	47.01239	90
RBRe_03	RDB	Remi	10	multiplication		2003	2004	1	4.506487	44.092741	225
RBGB_03	RDB	Gene bank	C	conservation		1984	2003	11	3.142604	45.775126	336
RocGB_04	RDR	Gene bank	C	conservation		2003	2004	1	3.142604	45.775126	336

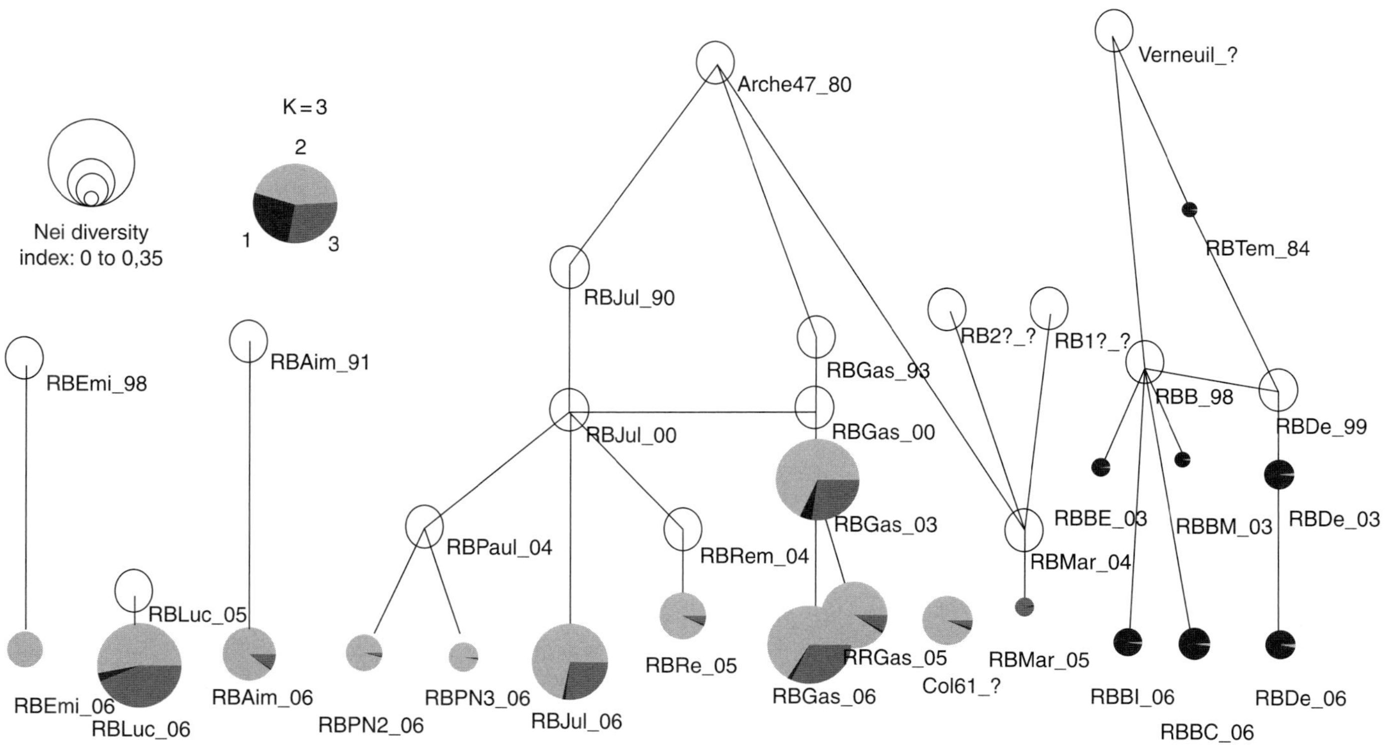

Fig. 23.2. Partial seed exchange and multiplication network of the 'Rouge de Bordeaux' (RDB) folk variety. A vertical line represents one or several multiplication events and a slanting or horizontal line represents a diffusion event. Studied RDB populations are coloured grey. Size of pies is proportional to within-population genetic diversity based on Nei's unbiased index. Population structure inferred by Instruct software was represented for the most consistent number of ancestral populations, which is K=3.

diversity ranged from 0% to 36% depending on populations (proportional to the size of pies on the bottom of Fig. 23.2). We inferred the underlying population genetic structure of our dataset by using the Baysian algorithm implemented in Instruct software (Gao *et al.*, 2007). The most consistent structure was obtained for K=3 ancestral populations (called groups in the following to avoid confusion with the 'real' populations). These three groups were clearly not evenly distributed among all sampled populations. Seven populations including the gene bank reference sample (right side of Fig. 23.2) were mostly composed of one ancestral group (group I: pies in dark grey in Fig. 23.2), while the other populations were more likely composite populations composed by one, two or three ancestral groups in variable frequency (the rest of the pies in Fig. 23.2). Consistently, the left branch of the network was often associated with larger within-population diversity while the right branch was associated with reduced within-population genetic diversity. These results confirm the trend observed in a previous study relying only on 13 SSR markers (Demeulenaere *et al.*, 2008). It can be noted that this branch includes all populations that derived from samples that have been conserved *ex situ* for a certain period either by breeders or by the gene bank. This can be related to the sampling effects that may arise during the collection or during regeneration (Parzies *et al.*, 2004). Thus, the knowledge of the social network (i.e. seed exchange and multiplication events) was highly consistent with the genetic diversity structure observed. Moreover, the genetic data allowed us to assign three populations whose origins were unknown to one branch of the network (left side of Fig. 23.2).

These results gave evidence that on-farm conservation is complementary to *ex situ* conservation because a higher genetic diversity was found to be maintained in the whole system compared to what would be conserved in the gene bank only for a particular variety (RDB). These findings also illustrate that the ancestral diversity of this population-variety is probably more reliably represented when considering the whole social network. For these reasons, it seems important to consider these informal social networks for a sustainable management of agrobiodiversity in industrialized countries.

23.4 Critical View on the French Position for On-farm Management

The previous sections highlighted the key role of the emergent social network in the dynamics of seed exchanges. These seed flows strongly influence the structure of crop genetic diversity maintained in this system. However, these initiatives have not found any real space in the French official framework for plant genetic resources. The French national charter for genetic resources management, approved in 1998 by the French administration, states that 'economic development has for long led to a division of labour which makes seed production a specialized activity.' It goes on to say that, in this context, 'on-farm conservation will probably not play a significant role' (BRG, 1998). In line with this view, no representative of the French on-farm conservation sector has been appointed in the international arena dealing with on-farm conservation. In particular, the absence of French representatives in the *in situ* and on-farm section of the European Cooperative Programme for Plant Genetic Resources (ECPGR) induces a gap between the different actors involved in this field. For instance, it becomes crucial now for France to be represented in this ECPGR section, so that it can take part in the European agenda of *in situ* on-farm conservation and also officially communicate about on-farm conservation initiatives occurring in France (see Maxted *et al.*, Chapter 43, this volume). Yet, it seems that things are changing. In February 2008, the French Ministries for Ecology and Sustainable Development and for Higher Education and Research launched the Foundation for Research on Biodiversity (FRB). This structure is in charge of developing and supporting research on biodiversity (wild and domesticated) as well as disseminating, teaching and developing the attractiveness of the results. In December 2009, the FRB officially recognized the role of farmers, gardeners and amateurs as actors of the genetic resources management.

23.5 Conclusion

This paper described an emergent farmer-led on-farm management of agrobiodiversity in France. We showed that a large set of varieties are grown and exchanged in this social network composed of farmers, curators and plant breeders, scientists and other amateurs. The genetic diversity maintained at the level of population-variety scale (metapopulation) was found to be larger than in the *ex situ* compartment (considering only the maintained genetic diversity in the national gene bank). This case study highlights the importance of considering the complementarity between *in situ* and *ex situ* conservation of genetic resources. We also demonstrated that the observed genetic structure was mainly explained by seed exchanges and farmer practices. These results strengthened the need to consider and value the role of social networks in on-farm conservation. But until now, the French regulatory body for genetic resources has not recognized this integrated management of the genetic resources.

References

Allinne, C. *et al.* (2008) Role of seed flow on the pattern and dynamics of pearl millet (*Pennisetum glaucum* [L.] R. Br.) genetic diversity assessed by AFLP markers: a study in south-western Niger. *Genetica* 133(2), 167–178.

Alvarez, N. *et al.* (2005) Farmers' practices, metapopulation dynamics, and conservation of agricultural biodiversity on-farm: a case study of sorghum among the Duupa in sub-sahelian Cameroon. *Biological Conservation* 121(4), 533–543.

Arndorfer, M., Kajtna, B. and Vorderwülbecke, B. (2009) Integrating *ex situ* and on-farm conservation approaches in the management of local vegetable diversity in Austria. In: *1st International Symposium on Horticulture in Europe*. G.R. Dixon, Wien, Austria, pp. 333–340.

Barnaud, A. *et al.* (2007) Local genetic diversity of sorghum in a village in northern Cameroon: structure and dynamics of landraces. *Theoretical and Applied Genetics* 114(2), 237–248.

Berg, T. (2009) Landraces and folk varieties: a conceptual reappraisal of terminology. *Euphytica* 166(3), 423–430.

Bocci, R. and Chable, V. (2008) Peasant seeds in Europe: stakes and prospects. *Cahiers Agricultures* 17(2), 216–221.

Bonneuil, C. and Demeulenaere, E. (2006) Une génétique de pair à pair? L'émergence de la sélection participative. In: *Les sciences citoyennes. Vigilance collective et rapport entre profane et scientifique dans les sciences naturalistes*. F. Charvolin, A. Micoud et L. K. Nyhart, dir, pp. 122–147.

Bonneuil, C. *et al.* (2006) Innover autrement? La recherche face à l'avènement d'un nouveau régime de production et de régulation des savoirs en génétique végétale. Quelles variétés et semences pour des agricultures paysannes durables? *Dossiers de l'environnement de l'INRA* 30, 29–51.

Bureau des Ressources Génétiques (BRG) (1998) *Charte Nationale pour la gestion des ressources génétiques*, 98 pp.

Cooper, H.D., Spillane, C. and Hodgkin, T. (2001) Broadening the genetic base of crops: an overview. In: *Broadening the Genetic Base of Crop Production*. CAB International, Wallingford, UK, pp. 1–23.

Demeulenaere, E. *et al.* (2008) Étude des complémentarités entre gestion dynamique à la ferme et gestion statique en collection : cas de la variété de blé Rouge de Bordeaux. *Les Actes du BRG* 7, 117–138.

Elias, M. *et al.* (2001) Traditional management of cassava morphological and genetic diversity by the Makushi Amerindians (Guyana, South America): perspectives for on-farm conservation of crop genetic resources. *Euphytica* 120(1), 143–157.

Enjalbert, J. *et al.* (2011) Dynamic management of crop diversity: from an experimental approach to on-farm conservation. *Comptes Rendus Biologie*, In press.

Gao, H., Williamson, S. and Bustamante, C.D. (2007) A Markov Chain Monte Carlo approach for joint inference of population structure and inbreeding rates from multilocus genotype data. *Genetics* 176(3), 1635–1651.

Hawtin, G., Iwanaga, M. and Hodgkin, T. (1996) Genetic resources in breeding for adaptation. *Euphytica* 92(1), 255–266.

Negri, V. (2003) Landraces in central Italy: where and why they are conserved and perspectives for their on-farm conservation. *Genetic Resources and Crop Evolution* 50(8), 871–885.

Negri, V. *et al.* (2000) Characterisation of thirteen cowpea landraces from Umbria (Italy). Strategy for their conservation and promotion. *Genetic Resources and Crop Evolution* 47(2), 141–146.

Parzies, H., Spoor, W. and Ennos, R. (2004) Inferring seed exchange between farmers from population genetic structure of barley landrace Arabi Aswad from Northern Syria. *Genetic Resources and Crop Evolution* 51(5), 471–478.

Thomas, M. *et al.* (2011) Seed exchanges, a key to analyze crop diversity dynamics in farmer-led on-farm conservation. *Genetic Resources and Crop Evolution*, Accepted.

Vilmorin-Andrieux Cie (1880) *Les meilleurs blés – Description et culture des principales variétés de froments d'hiver et de printemps*. Available at: http://museum.agropolis.fr/pages/documents/bles_vilmorin/index.htm (accessed 14 November 2010).

24 A Second Look at the European Strategic Approach to Conserving Crop Landraces

M. Veteläinen, V. Negri and N. Maxted

24.1 Introduction

The authors of this chapter recently published a proposal for a European Strategic approach to conserving crop landraces (Veteläinen *et al.*, 2009). This paper – an overview prepared as the joint effort by European on-farm conservation experts – was presented as a summary and further analysis at the joint meeting of the ECPGR *In situ* and On-farm Conservation Network and the EU project AGRI GENRES 057 – AEGRO in September 2010 in Funchal, Madeira.

During the formulation of the 'Strategic approach' the underlying information gap on the diversity of landraces (LR) managed on-farm was clear, as well as the inadequacy of current landraces (LR) *in situ* and *ex situ* conservation and the ongoing and extent of the threat faced by European LR. It is worth re-stressing that LR are highly threatened, if not the most threatened element of plant genetic resources (Negri *et al.*, 2009). Despite the urgent need to solve conservation issues, the framework for the 'Strategic approach' also includes political, economic and sociological (perhaps even anthropological) factors that enable maintainers to sustain LR and to allow them to continue to evolve as a critical genetic resource for future food security. It is recognized that a strategic approach for conserving and using LR in Europe is required to meet the changing needs and demands of future generations. Further, that the strategic approach needs to encompass an integrated multi-level methodology due to the complexity of the issue involved. Figure 24.1 summarizes those factors that are discussed more in detail in the following paragraphs.

24.2 Issues Considered when Formulating the 'Strategic Approach'

Conservation of cultivated, but threatened, LR is the most important issue to be solved in order to secure their further revitalization and sustained use. The strategic approach has to be considered both from the crop and the farmer's perspective. In terms of crop conservation the activities involved include: (i) inventory methodology development for LR; (ii) completion of national or regional inventories; (iii) threat, extinction and genetic erosion assessments; (iv) prioritization of crop gene pools and regions for on-farm conservation; (v) organization of an *ex situ* back-up system for LR; and (vi) local seed depositories in order to secure continuation of their on-farm management. Due to the nature of the on-farm conservation and

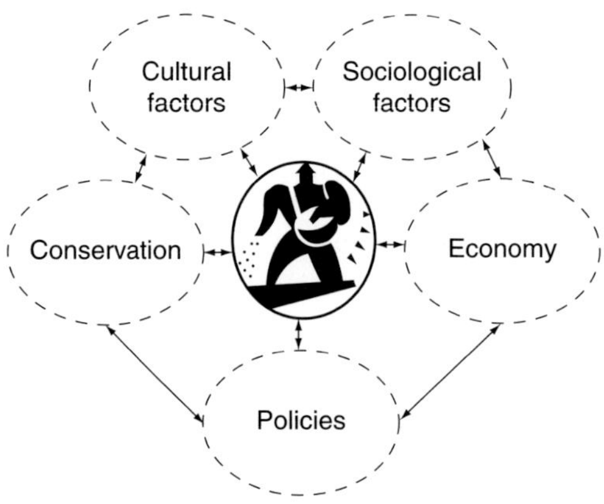

Fig. 24.1. Different factors considered in the strategy development.

management of LR, farmers play the key role, therefore the strategic approach must include: (i) incentives for maintainers to continue to maintain LR cultivation; (ii) knowledge collation and transfer between maintainers of LR cultivation and utilization practices; and (iii) action to ensure on-farm conservation/management of LR is included in local/national/regional agroenvironmental schemes. It is also self evident that from the strategic point of view the utility of LR for different stakeholders has to be made more evident at all levels to ensure maintenance is underpinned and this should be pursued through the strategic approach.

It should then foresee actions that strengthen LR-based product development and product labelling systems for farmer income improvement as well as actions that exploit links between LR and environment friendly production systems and ecotourism for the entire community benefit. Finally, there is more hope of success for on-farm conservation of LR when the combined interests of plant breeding and LR maintainers are addressed together. Novel approaches in plant breeding such as high-throughput sequencing, other genomic and GIS techniques are showing that entire LR diversity has not yet been fully exploited in plant breeding (see previous chapter in this volume). Thus, creating an active link between LR maintainers, conservationists and plant breeders would promote both provision of novel breeding material for plant breeders and in depth knowledge on the adaptive genes to be targeted and preserved within the on-farm system for conservationists.

In addition, possibilities of participatory plant breeding should be explored along with conservation when breeding for traits connected to new biotic and abiotic stresses associated with changing environmental conditions. This may meet the dual needs of maintaining LR populations that sustain local agro-ecotypic adaptation to local environmental conditions and meet farmers' needs in terms of development aspirations. To address these demands, the EC-funded FP7 research project Strategies for Organic and Low-input Integrated Breeding and Management (SOLIBAM, see www.solibam.eu) plans to develop new valuable populations from variable material like LR and to study LR change under different management and environmental conditions.

There are a number of areas of research that are needed to improve our knowledge when strengthening conservation efforts both from the biological/genetic and socio-economic point of view. It is clear that we need a better understanding of present LR diversity and population dynamics in order to manage on-farm conservation of LR in an effective manner. In particular, we need to understand the likely impact of climate change on LR diversity when targeting conservation areas and management practices. Again the farmer perspective and examination of socio-economic factors that drive on-farm maintenance of LR are the key questions to be included on the top of important strategic research issues.

On-farm conservation and management of LR is affected by national and EU-level decision making and legislation on seed production. For example, quite recently there has been much debate on the effectiveness of the EU Commission Directives 2008/62/EC ('agricultural landraces and varieties which are naturally adapted to the local and regional conditions and threatened by genetic erosion and for marketing of seed potatoes of those landraces and varieties'), 2009/145/EC ('for acceptance of vegetable landraces and varieties which have been traditionally grown in particular localities and regions and are threatened by genetic erosion and of vegetable varieties with no intrinsic value for commercial crop production but developed for growing under particular conditions and for marketing of seed of those landraces and varieties') and 2010/60/EU ('for marketing of fodder plant seed mixtures intended for use in the preservation of the natural environment'). These directives may only favour maintenance of those LR for which a seed market exists or can be developed, while for most of the LR (e.g. horticultural LR maintained in home gardens or small farms) this is not the case. In addition, it is a framework that does not fully take into account the nature of LR as variable and evolving populations, but merely 'squeezes' LR seed production in the system developed for varieties bred by scientific plant breeding. Much of the future success of LR production within on-farm systems is not only likely to depend on the national interpretation and implementation of these directives, but also on the European and national policies in favour of diversity conservation *per se*. It is, then, important that a common European strategic approach on LR should help prevent these kinds of legislative contradictions and perverse incentives to LR diversity maintenance. The strategic approach should also include goals that result in 'real' and 'effective' legislative protection of the on-farm conservation sites as well as the LR diversity they contain, in such including those LR that are maintained and used in a limited context. Here land use and planning may play a critical role, i.e. protection for narrowly adapted LR or on-farm sites that may also be attractive for other competing land uses.

Public awareness on the importance and nature of LR needs to be achieved at all levels of society. The strategic approach should include an action plan that utilizes different means to promote the multitude of values associated with LR diversity, its maintenance and sustainable exploitation. Only when wide public and professional understanding exists, will the long-term preservation of LR be secured.

There is very little information on the socio-economic and cultural value of LR in Europe, not to speak about their value in adapting conventional crop production for warming climate. However, it is clear that farmers who still manage LR in their fields should be encouraged to continue to do so in the future. A major step would be inclusion of management of genetic resources in regional development, agribusiness and environmental schemes. Furthermore, examination of commercial value of LR, enhancement of niche markets, as well as the development of new types of market chains and novel markets would undoubtedly strengthen the future on European LR. The bottom line being, we need to support local seed production, exchange networks and market chains.

Even cultural aspects should be taken into consideration in the strategic approach. LR that are still in cultivation are often connected to local traditions and as such have significant cultural heritage value. This is a yet another strengthening factor for conservation and continued on-farm management of LR. Therefore, the multidisciplinary approach to conserve and develop LR as the part of the maintenance of our common European culture is necessary.

Finally, cooperation needs to be strengthened between all actors that deal with LR and their on-farm management. The picture emerging today is quite fragmented when it comes to cooperation at the national level between LR maintainers, farmer organizations, gene banks, research institutes and NGOs. The strategic approach needs to link these different actors together and ensure information flow to all directions. Many national programmes for plant genetic resources are already enhancing networking, but due to limited resources building up the 'national on-farm conservation programmes' is far from complete. However, at the European level there is an existing network within the framework of the European Co-operative Programme for Plant Genetic Resources (ECPGR). The ECPGR *In situ* and On-farm Conservation Network is currently leading a new EC-funded FP7 research project

(Novel characterization of crop wild relative and landrace resources as a basis for improved crop breeding – PGR Secure), which plans to develop and promote a common European conservation methodology development and data banking for LR diversity. It is desirable that in the longer term the *In situ* and On-farm Conservation Network continues to disseminate good practices and experiences on on-farm LR conservation between all European countries. In the future, inter-regional cooperation should also strengthen the on-farm LR conservation globally.

24.3 Conclusion

Due to complexity and interrelatedness of the issues it needs to be concluded that the preparation and implementation of the 'Strategic Approach for conservation and use of European landraces' needs to be incorporated into national and European policy and conservation action. However, in order to create an effective strategic approach greater public awareness, conservation security and legislative protection for European LR diversity needs to be led at the European level. In our previous paper (Veteläinen *et al.*, 2009) we proposed targets as the basis for the strategic approach and these are summarized in Table 24.1, a clearer distinction with suggested national and European level responsibilities is presented. We hope that the time-bound targets will be helpful both at the national and European level in meeting our overall goal of LR diversity conserved with on-farm systems across Europe and used by European farmers and plant breeders.

Table 24.1. Targets and responsibilities for national and European strategy for conservation of landrace diversity.

Target/sub-target	Responsibility (X) National	Responsibility (X) European	Explanation for European tasks	Time-bound target
1. Prepare conservation and use strategic action plan including:				2012
– landrace inventory	x		PGR Secure[1]	
– landrace and farmer survey	x			
– conservation *ex situ*/*in situ*	x	x	Aegis principles[2]	
– sustainable use	x			
– integration of the plan in national/regional planning	x	x	EU level integration	
– development of the concept for on-farm conservation network		x	PGR Secure	
2. Create national and European priority landrace lists and identify priority sites for on-farm conservation[3]	x	x	ECPGR *In Situ* and On-farm Working Group: Establishment of the on-farm conservation network in the Europe	2015/2018
3. Develop effective means of systematically conserving landrace diversity *in situ*:				2012
– raise awareness among conservationists on the need of on-farm conservation and appropriate management strategies	x	x	ECPGR *In Situ* and On-farm Working Group	
– enhance farmer management of landraces	x			

Continued

Table 24.1. Continued.

Target/sub-target	Responsibility (X) National	Responsibility (X) European	Explanation for European tasks	Time-bound target
– integrate on-farm conservation and farmer management of landraces in a system that promotes landrace use	x			
4. Develop effective means of systematically conserving landrace diversity *ex situ*:				
– carry out gap analyses in gene bank collections and fill gaps when possible	x	x	PGR Secure	2014
– promote the establishment of community seed banks for local landrace diversity	x			
5. Establish a European mechanism/ clearing house		x	ECPGR common landrace descriptors	2012
6. Establish protocols for landrace information management and dissemination	x	x	Inclusion of the landrace suitable descriptors in EURISCO[4]	2015
7. Assess landrace threat and climate change:				2015
– establish protocols for assessing landrace threat status (according to the model of the IUCN Red List Criteria)		x	ECPGR *In Situ* and On-farm Working group	
– develop methods to assess impacts of climate change on landraces		x	European research community	
8. Ensure public awareness, effective security and legislative protection for European landraces	x	x	EU level regulations require European level coordination	Continuously
9. Promote sustainable utilization of European landrace diversity	x	x	European research community: developing novel approaches to characterization and evaluation of landrace diversity	Continuously

[1] Novel characterization of crop wild relative and landrace resources as a basis for improved crop breeding: website to be developed, see www.ecpgr.cgiar.org/Networks/Insitu_onfarm/insitu_onfarm.htm.
[2] A European Genebank Integrated System: http://aegis.cgiar.org/documents/constitutional_documents.html.
[3] Formal on-farm conservation apart from the on-farm management carried out by farmers.
[4] A web-based catalogue that provides information about *ex situ* plant collections maintained in Europe: http://eurisco.ecpgr.org/static/index.html.

References

Negri, V., Maxted, N. and Veteläinen, M. (2009) European landrace conservation: an introduction. In: Veteläinen, M., Negri, V. and Maxted, N. (eds) *European Landraces: On-farm conservation, Management and Use*. Bioversity Technical Bulletin 15. Bioversity International, Rome, pp. 1–22.

Veteläinen, M., Negri, V. and Maxted, N. (2009) A European strategic approach to conserving crop landraces. In: Veteläinen, M., Negri, V. and Maxted, N. (eds) *European Landraces: On-farm conservation, Management and Use*. Bioversity Technical Bulletin 15. Bioversity International, Rome, pp. 305–325.

25 Comparing Genetic Diversity within a Crop and Its Wild Progenitor: A Case Study for Barley

B. Kilian, H. Özkan, S. Shaaf, S. Hübner, R.K. Pasam, R. Sharma, K. Neumann, W. Weißgerber, F.A. Konovalov, J. Keilwagen, S. Friedel, H. Knüpffer, M. von Korff, G. Coupland and A. Graner

25.1 Introduction

Cereals constitute over 50% of the crop production worldwide. Modern varieties have been selected for adaptation to intensive cropping in non-limiting conditions, as demonstrated by the continuous yield improvement from 1920 to 1990. However, their reduced diversity poses limitations to their further genetic improvement, particularly under climate change. Crop wild relatives and local varieties (landraces) are most likely to contain novel, unique and high levels of genetic diversity needed to sustain innovations in breeding programmes (Reidsma *et al.*, 2009). In Europe, barley is the second most important cereal crop and grown on an area of over 105 million ha in 2008 (www.FAOSTAT.fao.org). Worldwide, barley is mainly used for animal feed and malting, but also has gained regional importance in human nutrition.

Barley (*Hordeum vulgare* L.) ($2n=2x=14$) is an annual, predominantly self-pollinated crop with seven chromosome pairs. It is one of the oldest crops and has been domesticated from its wild progenitor, *H. spontaneum* K. Koch, in the Fertile Crescent some 10,000 years ago. Barley varieties are either two-rowed or six-rowed, based on type of ear. Winter and spring forms are recognized.

25.2 Present-day Distributions and Habitats

Wild barley, *H. spontaneum*, occurs in the eastern Mediterranean, western Asia and reaches Turkmenistan and Afghanistan in the east (Harlan and Zohary, 1966). Single populations are found in secondary habitats, e.g. in Morocco. Wild barley grows in diverse environments and has the ability to colonize disturbed habitats. Our current knowledge on wild barley distribution and natural habitats is far from being complete. We have a good overview of wild barley stands from Israel and south-east Turkey, and fairly good information for Syria and Jordan. However, good passport data and detailed GPS based distribution maps are available only for a few wild barley collections (see below). Large gaps exist east of Turkey. Few *H. spontaneum* individuals have been sampled from Iraq, Iran and central Asian countries. The present status in Cyprus, Egypt, Libya and Morocco is unknown.

25.3 Brief Introduction to Barley Domestication

Wild barley grains have been found in several pre-agricultural Pre-Pottery Neolithic sites.

The earliest evidence is from Ohalo II, located at the shore of the Sea of Galilee, where 21,000 years old wild grain remains were found in large amounts (Kislev *et al.*, 1992). Evidently, wild barley has been collected from nature long before cultivation and domestication. The earliest carbonized remains of domesticated barley are of the two-row type, but six-row types appear already at Ain Ghazal around 9000–8500 years ago (Willcox, 1998). In prehistoric times barley was the main crop in Mesopotamia and a primary cereal in ancient Egypt and Greece (Harlan, 1995).

The most important traits modified during domestication are the brittle rachis and spike morphology. Subsequent modifications included seed size, seed shape, plant height, awn roughness, tillering, seed dormancy, photoperiod or vernalization requirement. The corresponding adaptive changes provided the basis for the spread of domesticated barley out of the Fertile Crescent (for more details see Kilian *et al.*, 2009).

Several opinions exist on the domestication history of barley. Recent archaeobotanical and molecular data imply that two-rowed and six-rowed genotypes may have different, independent origins. The particular matter concerning single versus multiple origins of barley is complicated by the fact that: (i) multiple independent historical and ongoing introgression of genes from wild relatives to cultivated varieties (and vice versa) can mimic multiple domestication events; and (ii) splitting of domesticated genotypes into two alternative groups based on two- versus six-rowed ears, hulled versus naked caryopsis, western versus eastern varieties might have followed the domestication process, rather than being coeval with it (see references in Zohary and Hopf, 2000; Bothmer *et al.*, 2003; Kilian *et al.*, 2009). Thus, barley domestication history still remains to be fully resolved

25.4 *Ex Situ* Germplasm Resources for Crop Improvement: New Strategies and Challenges

One option to sustain wild relatives and landraces is the development of *ex situ* collections. The conservation of *ex situ* plant genetic resources is rather fragmented, largely because it is still mainly based on individual national programmes. Many European national plant genetic resources (PGR) programmes cooperate in the frame of the European Cooperative Programme for Plant Genetic Resources (ECPGR). For barley, there are more than 200 collections worldwide with ca. 450,000 accessions, and of these 175,000 accessions are in European gene banks (Knüpffer, 2009). Duplication of germplasm between collections is an issue. *Ex situ* germplasm collections for wild barley are rather limited due to the relatively large workload for maintaining wild plants. Current re-sequencing studies showed that introgression from wild to cultivated germplasm and vice versa may occur in gene bank material. Examples for well-documented wild barley collections are given below. Another problem is that gene bank accessions often are genetically heterogeneous, but core collections, with their associated genotyping and phenotyping studies, are based on single seed purified material. Best gene bank practices have to be developed, and the botanical status of many barley accessions needs to be revised. Deep genetic and phenotypic characterization of genetic resources by high-throughput techniques, including full genome sequencing will increasingly become available. Concomitantly large amounts of data need to be integrated with the current documentation systems.

25.5 New Ways to Look at Diversity of *In Situ* Genetic Resources

An optimal strategy of *in situ* conservation should not only enable the conservation of existing natural variation, but also allow the continuation of natural diversification processes. The major parameters affecting natural variation are effective population size, breeding strategy (selfing/outcrossing), time (i.e. stability) and environmental heterogeneity (niche availability). It is therefore important to emphasize these parameters in management planning for *in situ* conservation.

25.6 Germplasm Collections of Wild and Landrace Barleys to Study Genetic Diversity

Methods are required to optimize access to genetic diversity contained in large germplasm collections. The core collection concept (Frankel, 1984) was developed to facilitate access to the diversity available in large collections. Glaszmann *et al.* (2010) suggested implementing the core collection concept through 'core reference sets'. A crop core reference set is to be understood as 'a set of genetic stocks that are representative of the genetic resources of the crop and are used by the scientific community as a reference for an integrated characterization of its biological diversity'. The value of a formalized reference set will emerge from its use by the largest number of scientists. To serve as a standardized resource, core collections must be transformed into genetic stocks that have been purified (homogeneous/stabilized) and taxonomically classified to facilitate practical choices for comparative phenotyping studies.

25.6.1 Towards core reference sets for barley: three large *ex situ* collections of wild barley

The Barley1K collection (B1K; 1020 wild barleys from 51 sites) captures barley diversity at multiple levels in Israel from microsite to country-wide (Hübner *et al.*, 2009). This collection allows the analysis of the relationship between ecology and genetic make-up of the plants because transects of environments varying in temperature, soil and precipitation along a wide spatial range are available. The whole collection has been genotyped by SSRs, and 230 lines have subsequently been genotyped by BOPA1 (see below).

A second wild barley collection named 'World Barley Diversity Collection' (WBDC) has been developed (Steffenson *et al.*, 2007). The collection comprises 313 *H. spontaneum* accessions from different countries. The germplasm has been mostly sampled from gene banks and along roadsides. The collection has been phenotyped and genotyped using BOPA1 and BOPA2 assays (3072 SNPs, see below).

The third collection has been established at IPK Gatersleben in cooperation with the MPIPZ in Cologne. This collection combines the advantages of the two abovementioned collections (Kilian *et al.*, unpublished). We first made an inventory of wild barley relatives and barley landraces stored *ex situ* in gene banks and private collections. From these, accessions were selected based on phenotypic data, molecular markers and passport data. Furthermore, we selected material from three well-documented resources: (i) the Badr *et al.* (2000) collection that includes the ICARDA wild barley collection; (ii) a Turkish collection (Özkan *et al.*, unpublished); and (iii) the B1K collection (see above). The resulting collection consists of 579 wild *H. spontaneum*, 109 barley landrace and 128 cultivar samples derived from single seed descent (SSD). Several traits were phenotyped under greenhouse and field conditions. Molecular marker data are available for the whole collection.

The *H. spontaneum* subset of the International Barley Core Collection (BCC) that includes 70 accessions from 16 countries (Knüpffer and van Hintum, 2003) was not taken into account due to its small size.

25.7 Molecular Diversity within Barley and its Wild Progenitor

A series of studies has been undertaken to study barley diversity by genotyping or re-sequencing approaches (e.g. Graner *et al.*, 2003). Until recently, AFLP or SSR molecular markers were mainly considered for diversity studies in barley. Meanwhile, whole genome genotyping using SNPs and large-scale re-sequencing of a range of candidate genes using next-generation sequencing approaches are state-of-the-art.

25.7.1 Genome-wide association mapping in barley

Early association mapping studies in barley were hampered by the availability of a limited amount of mapped markers and thus were mainly based on re-sequencing candidate genes (Stracke *et al.*, 2009; Haseneyer *et al.*, 2010). The development of a comprehensive set of SNP markers that can be interrogated in a highly multi-parallel fashion ushered the era of genome-wide association studies (Waugh *et al.*, 2009; Ramsay *et al.*, 2011; Russell *et al.*, 2011; Pasam *et al.*, submitted). Several germplasm collections including wild and landrace barley have been genotyped using custom-made OPAs (oligo-pool assays) by Illumina GoldenGate technology. Most commonly the BOPA1 marker information (1536 SNPs) has been used to study population structure and patterns of linkage disequilibrium (LD) (Close *et al.*, 2009). SNP markers significantly associated with traits are being used to identify genomic regions that harbour candidate genes for these traits (various collaborative barley projects). It is relatively easy to detect marker-trait associations in cultivar populations which have extensive linkage disequilibrium (5–10 cM). Conversely, populations with low LD are supposed to provide high-resolution associations (landraces, <5 cM; wild barley, <1 cM) but the numbers of markers needed to find significant associations are high. This rapid decay in LD in populations of wild and landrace germplasm is a key generic problem with allele mining in primitive germplasm. Furthermore, ascertainment bias of SNP discovery complicates the situation in landraces and wild germplasm (Moragues *et al.*, 2010).

Higher marker coverage is required in order to identify candidate genes more efficiently in diverse collections. Therefore, the Scottish Crop Research Institute in Dundee, TraitGenetics GmbH (Germany) and IPK Gatersleben developed a 9K Infinium iSELECT array for barley. This chip contains 7864 bi-allelic SNPs coming from next generation sequence information from a broad range of barley cultivars (Waugh *et al.*, unpublished).

25.7.2 Allele mining at candidate genes

Plant accessions from wild or locally adapted landrace gene pools are commonly considered to contain a rich repertoire of alleles that have been left behind by the selective processes of domestication, cultivation and breeding that generated today's elite cultivated gene pool. The most effective strategy for determining allelic richness at a given locus is currently to determine its DNA sequence in a representative collection of individuals. Re-sequencing candidate genes for agriculturally important traits is in the focus of interest. Several studies considering wild and cultivated germplasm have been dedicated to *PpdH1*, the major locus affecting photoperiodic response on the short arm of chromosome 2H in barley (Jones *et al.*, 2008; Kilian *et al.*, unpublished). Kilian *et al.* re-sequenced an 898 bp fragment of *PpdH1* for 704 homozygous lines and detected 50 haplotypes in wild barley compared to only 12 haplotypes in cultivated material (Fig. 25.1). This emphasizes the necessity of using available wild resources for discovery of new alleles for crop improvement purposes.

25.8 Outlook: Tools and Resources for Barley Research that Will Become Available

25.8.1 The barley reference genome sequence

Steps towards a complete reference genome sequence for barley are now being taken. This is challenging because of the large genome size of 5.1 Gbp and the large number

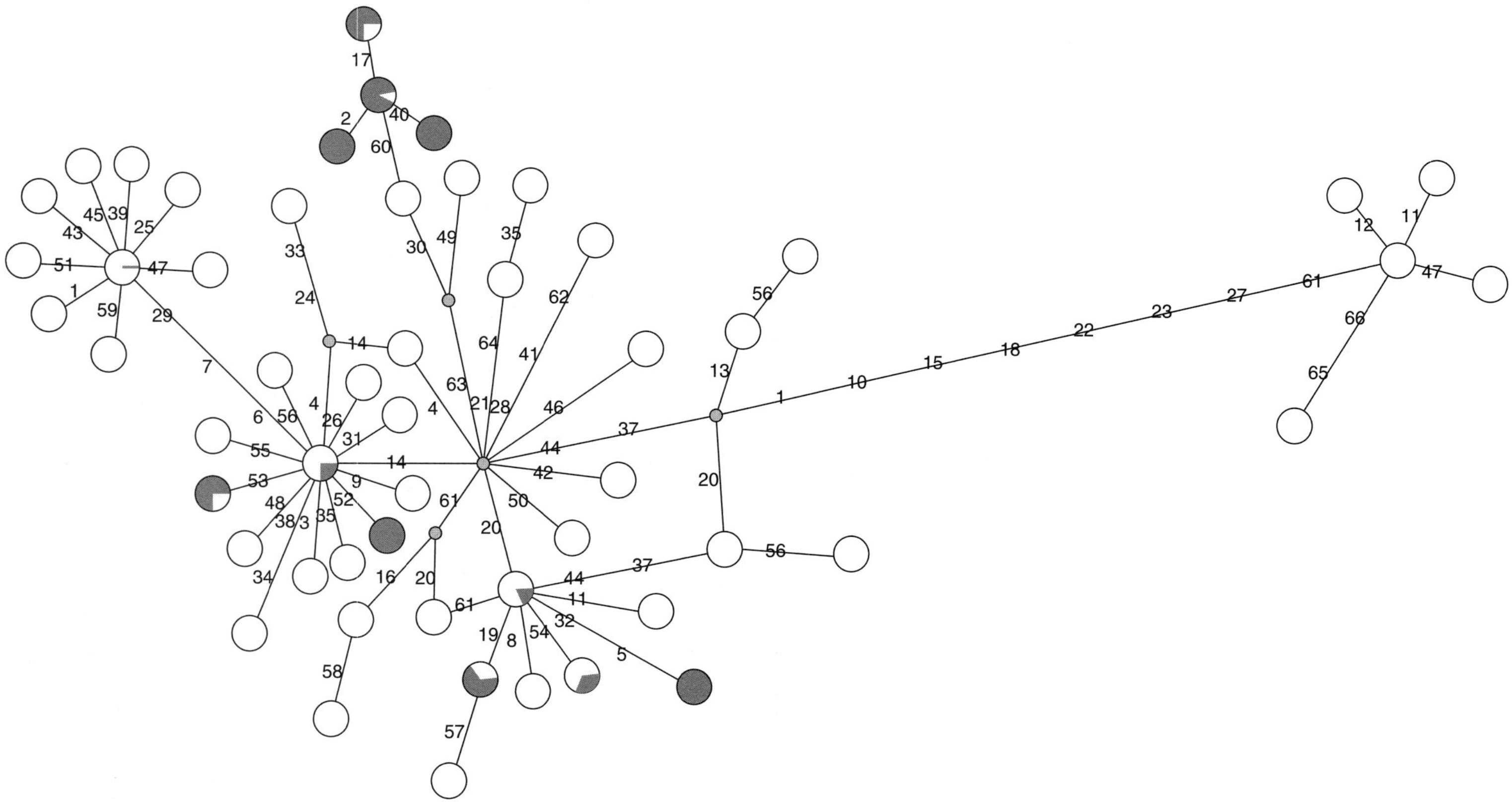

Fig. 25.1. Median-joining network of an 898 bp fragment of *PpdH1* for 704 barley accessions. Haplotypes found in cultivated barleys are indicated by filled circles. Gaps in the sequence alignment were not considered.

of repetitive regions (Doležel *et al.*, 1998). Recently, Mayer *et al.* (2011) reported a high-resolution sequence-based gene map of barley. About 32,000 genes are predicted for barley. *De novo* sequencing of the barley genome will lead to accelerated SNP discovery and will enable low-coverage whole genome re-sequencing studies for wild barley.

Acknowledgements

We thank Kerstin Schäfer, Ute Krajewski, Christiane Kehler, Birgit Dubsky, Heike Harms and the group V&G at the IPK in Gatersleben for excellent technical assistance. We are grateful to Peter Schreiber, Jürgen Marlow and Michael Grau for their support in characterizing our collections.

References

Badr, A., Müller, K., Schäfer-Pregl, R., El Rabey, H., Effgen, S., Ibrahim, H.H., Pozzi, C., Rohde, W. and Salamini, F. (2000) On the origin and domestication history of barley (*Hordeum vulgare*). *Molecular Biology and Evolution* 17, 499–510.

Bothmer, R. von, Hintum, Th. van, Knüpffer, H. and Sato, K. (eds) (2003) *Diversity in barley* (*Hordeum vulgare*). Elsevier, Amsterdam.

Close, T.J., Bhat, P.R., Lonardi, S., Lonardi, S., Wu, Y., Rostoks, N., Ramsay, L., Druka, A., Stein, N., Svensson, J.T., Wanamaker, S., Bozdag, S., Roose, M.L., Moscou, M.J., Chao, S., Varshney, R.K., Szücs, P., Sato, K., Hayes, P.M., Matthews, D.E., Kleinhofs, A., Muehlbauer, G.J., DeYoung, J., Marshall, D.F., Madishetty, K., Fenton, R.D., Condamine, P., Graner, A. and Waugh, R. (2009) Development and implementation of high-throughput SNP genotyping in barley. *BMC Genomics* 10, 582.

Doležel, J., Greilhuber, J., Lucretti, S., Meister, A., Lysák, M.A., Nardi, L. and Obermayer, R. (1998) Plant genome size estimation by flow cytometry: inter-laboratory comparison. *Annals of Botany* (London) 82, 17–26.

Frankel, O.H. (1984) Genetic perspective of germplasm conservation. In: Arber, W., Illmensee, K., Peacock, W.J. and Starlinger, P. (eds) *Genetic Manipulation: impact on man and society*. Cambridge University Press, Cambridge, pp. 161–170.

Glaszmann, J.C., Kilian, B., Upadhyaya, H.D. and Varshney, R.K. (2010) Accessing genetic diversity for crop improvement. *Current Opinion in Plant Biology* 13, 167–173.

Graner, A., Bjornstad, A., Konishi, T. and Ordon, F. (2003) Molecular diversity of the barley genome. In: von Bothmer, R., van Hintum, Th., Knüpffer, H. and Sato, K. (eds) *Diversity in Barley* (Hordeum vulgare). Elsevier Science, Amsterdam, the Netherlands, pp. 121–141.

Harlan, J.R. (1995) *The Living Fields: Our agricultural heritage*. Cambridge University Press, Cambridge.

Harlan, J.R. and Zohary, D. (1966) Distribution of wild wheats and barley. *Science* 153, 1074–1080.

Haseneyer, G., Stracke, S., Piepho, H.-P., Sauer, S., Geiger, H.H. and Graner, A. (2010) DNA polymorphisms and haplotype patterns of transcription factors involved in barley endosperm development are associated with key agronomic traits. *BMC Plant Biology* 10, 5.

Hübner, S., Hoffken, M., Oren, E., Haseneyer, G., Stein, N., Graner, A., Schmid, K. and Fridman, E. (2009) Strong correlation of the population structure of wild barley (*Hordeum spontaneum*) across Israel with temperature and precipitation variation. *Molecular Ecology* 18, 1523–1536.

Jones, H., Leigh, F.J., Mackay, I., Bower, M.A., Smith, L.M.J., Charles, M..P, Jones, G., Jones, M.K., Brown, T.A. and Powell, W. (2008) Population-based resequencing reveals that the flowering time adaptation of cultivated barley originated east of the Fertile Crescent. *Molecular Biology and Evolution* 25, 2211–2219.

Kilian, B., Özkan, H., Pozzi, C. and Salamini, F. (2009) Domestication of the Triticeae in the Fertile Crescent. In: Feuillet, C. and Muehlbauer, G.J. (eds) *Genetics and Genomics of the Triticeae*. Plant Genetics and Genomics: Crops and Models 7, Springer Science+Business Media, LLC, New York, pp. 81–119.

Kislev, M.E., Nadel, D. and Carmi, I. (1992) Epipalaeolithic (19,000 BP) cereal and fruit diet at Ohalo II, Sea of Galilee. *Review of Palaeobotany and Palynology* 73, 161–166.

Knüpffer, H. (2009) Triticeae genetic resources in *ex situ* genebank collections. In: Feuillet, C. and Muehlbauer, G.J. (eds) *Genetics and Genomics of the Triticeae*. Plant Genetics and Genomics: Crops and Models 7, Springer Science+Business Media, LLC, New York, pp. 31–79.

Knüpffer, H. and van Hintum, Th. (2003) Summarised diversity – the Barley Core Collection. In: von Bothmer, R., van Hintum, Th., Knüpffer, H. and Sato, K. (eds) *Diversity in barley (Hordeum vulgare)*. Elsevier, Amsterdam, pp. 259–267.

Mayer, K.F.X., Martis, M., Hedley, P.E., Šimková, H., Liu, H., Morris, J.A., Steuernagel, B., Taudien, S., Roessner, S., Gundlach, H., Kubaláková, M., Suchánková, P., Murat, F., Felder, M., Nussbaumer, T., Graner, A., Salse, J., Endo, T., Sakai, H., Tanaka, T., Itoh, T., Sato, K., Platzer, M., Matsumoto, T., Scholz, U., Doležel, J., Waugh, R. and Stein, N. (2011) Unlocking the barley genome by chromosomal and comparative genomics. *The Plant Cell* doi:10.1105/tpc.110.082537.

Moragues, M., Comadran, J., Waugh, R., Milne, I., Flavell, A.J. and Russell, J.R. (2010) Effects of ascertainment bias and marker number on estimations of barley diversity from high-throughput SNP genotype data. *Theoretical and Applied Genetics* 120, 1525–1534.

Pasam, R.K., Sharma, R., Haseneyer, G. *et al.* (submitted) Genome wide association studies for agricultural traits in a worldwide spring barley collection.

Ramsay, L., Comadran, J., Druka, A., Marshall, D.F., Thomas, W.T.B., Macaulay, M., MacKenzie, K., Simpson, C., Fuller, J., Bonar, N., Hayes, P.M., Lundqvist, U., Franckowiak, J.D., Close, T.J., Muehlbauer, G.J. and Waugh, R. (2011) *INTERMEDIUM-C*, a modifier of lateral spikelet fertility in barley, is an ortholog of the maize domestication gene *TEOSINTE BRANCHED 1*. *Nature Genetics* 43, 169–172.

Reidsma, P., Ewert, F., Lansink, A.O. and Leemans, R. (2009) Vulnerability and adaptation of European farmers: a multi-level analysis of yield and income responses to climate variability. *Regional Environmental Change* 9, 25–40.

Russell, J., Dawson, I.K., Flavell, A.J., Steffenson, B., Weltzien, E., Booth, A., Ceccarelli, S., Grando, S. and Waugh, R. (2011) Analysis of >1000 single nucleotide polymorphisms in geographically matched samples of landrace and wild barley indicates secondary contact and chromosome-level differences in diversity around domestication genes. *New Phytologist* doi: 10.1111/j.1469-8137.2011.03704.x.

Steffenson, B.J., Olivera, P., Roy, J.A., Jin, Y., Smith, K.P. and Muehlbauer, G.J. (2007) A walk on the wild side: mining wild wheat and barley collections for rust resistance genes. *Australian Journal of Agricultural Research* 58, 532–544.

Stracke, S., Haseneyer, G., Veyrieras, J.-B., Geiger, H., Sauer, S., Graner, A. and Piepho, H.-P. (2009) Association mapping reveals gene action and interactions in the determination of flowering time in barley. *Theoretical and Applied Genetics* 118, 259–273.

Waugh, R., Jannink, J.L., Muehlbauer, G.J. and Ramsay, L. (2009) The emergence of whole genome association scans in barley. *Current Opinion in Plant Biology* 12, 218–222.

Willcox, G. (1998) Archaeobotanical evidence for the beginnings of agriculture in southwest Asia. In: Damania, A.B., Valkoun, J., Willcox, G. and Qualset, C.O. (eds) *The Origins of Agriculture and Crop Domestication*. Proceedings of the Harlan Symposium. ICARDA, Aleppo, pp. 25–38.

Zohary, D. and Hopf, M. (2000) *Domestication of Plants in the Old World*. Oxford University Press, Oxford.

26 Using Neutral Genetic Diversity to Prioritize Crop Wild Relative Populations: A Portuguese Endemic Case Study for *Dianthus cintranus* Boiss. & Reut. Subsp. *barbatus* R. Fern. & Franco

J. Magos Brehm, B.V. Ford-Lloyd, N. Maxted and M.A. Martins-Loução

26.1 Introduction

A species-targeted conservation strategy should, whenever possible, include information on the genetic diversity of the target taxa to maximize the intra-specific diversity and potentially useful genes conserved. The development of a national conservation strategy starts with an inventory of taxa followed by the establishment of species priorities for conservation (Maxted *et al.*, 2007), then at species level, deciding which populations of each priority taxon should be targeted for *in situ* and *ex situ* conservation actions (Fig. 26.1). In the latter respect, molecular markers are commonly used in assessing genetic diversity *per se* and are widely used in assisting conservation-related studies (e.g. Rottenberg and Parker, 2003; Eckstein *et al.*, 2006; Watson-Jones *et al.*, 2006).

In Portugal, the development of a conservation strategy for crop wild relatives (CWR) was initiated in order to ensure their persistence for future use. An inventory of the mainland Portuguese CWR was created (available from www.jb.ul.pt; see also Magos Brehm *et al.*, 2008a) which comprises 2262 taxa, about 77% of the Portuguese flora, of which the great majority (ca. 93%) are native to the country. Of these, 32% of taxa are wild relatives of aromatic and medicinal species, 30% are food crops and 21% ornamentals. Around 6% of the Portuguese CWR are endemic to the country, and over 11% are endemic to the Iberian peninsula. From the species for which there were readily available national distribution data (slightly less than 50%), about 36% of taxa occur in one to three administrative regions. Less than 1% is actively conserved *in situ*, around 12% are conserved *ex situ*, about 14% are thought to be threatened using IUCN Red List categories and criteria and 6% are protected by national or international legislation (Magos Brehm *et al.*, 2008a).

Given the relatively large numbers of Portuguese CWR and the limited financial resources allocated to conservation, priorities were established based on different criteria and different prioritization schemes (see Magos Brehm *et al.*, 2010). Twenty CWR were identified as the highest priorities for conservation in Portugal and include, among others, wild relatives of the crop genus *Dianthus* (Magos Brehm *et al.*, 2010).

The aim of this study is: (i) to evaluate the inter- and intra-population genetic diversity of one Portuguese endemic priority CWR (*Dianthus cintranus* Boiss. & Reut. subsp.

 Agrobiodiversity Conservation: Securing the Diversity of Crop Wild Relatives and Landraces (eds N. Maxted *et al.*)

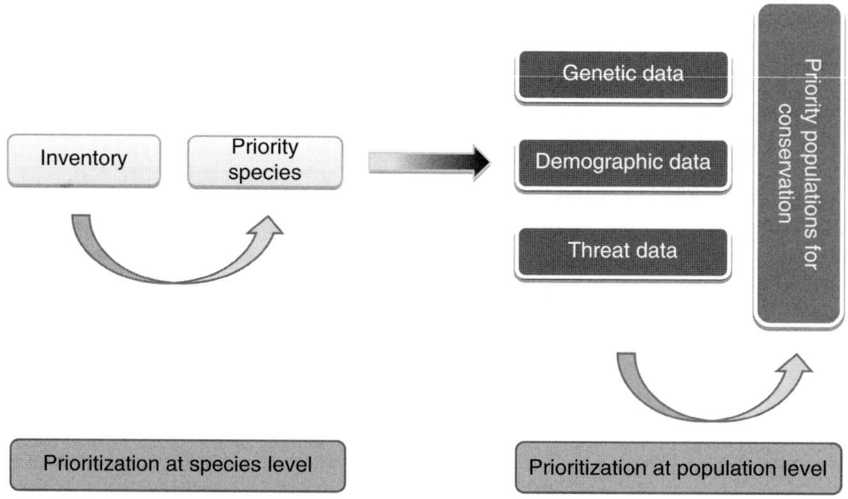

Fig. 26.1. General scheme for assigning population priorities for conservation.

barbatus R. Fern. & Franco) throughout its distribution area as a means of obtaining genetic baseline information for future conservation; and (ii) to use genetic, demographic and threat data in order to prioritize populations for conservation.

26.2 *Dianthus cintranus* Boiss. & Reut. subsp. *barbatus* R. Fern. & Franco

Dianthus cintranus subsp. *barbatus* is a Portuguese endemic wild relative of ornamental carnations. It presents a latitudinal distribution from Cabo da Roca up to the centre of Portugal (Fig. 26.2) and occurs in outcrops of mainly limestone (Espírito-Santo *et al.*, 1997). It is not actively protected *in situ* but part of its distribution is known to occur within formal protected areas. There are no *ex situ* accessions in gene banks and it is not legally protected. The taxon has been assessed as Endangered after the 2001 IUCN Red List Categories and Criteria (see Magos Brehm *et al.*, 2008b; Magos Brehm, 2009). Main threats to populations include fires, invasive species, construction, trampling, grazing and trash deposition (Magos Brehm, 2009).

Nothing is known about the breeding system of this *Dianthus* taxon, but there are several references to other species of the same genus. Most of the *Dianthus* spp. studied so far are predominantly pollinated by Coleoptera, Lepidoptera, Hymenoptera and Diptera (see Müller, 1873; Knuth, 1898; Meusel and Mühlberg, 1979; Erhardt, 1988, 1990, 1991; Jennersten, 1988; Ebert and Rennwald, 1993; Ebert *et al.*, 1994; Erhardt and Jaggi, 1995; Collin *et al.*, 2002; Bloch *et al.*, 2006). Nevertheless, there are some references to inbreeding in *Dianthus* spp. (e.g. Erhardt and Jäggi, 1995; Colin and Shykoff, 2003). There is little information on seed dispersal in *Dianthus* spp. However, it is known that their capsule shape and position facilitates the seed release by the wind (The seed site, 2008). No genetic studies have been previously carried out for this particular taxon, although some studies have been undertaken for other *Dianthus* spp. (e.g. Smulders *et al.*, 2000, 2003; de Benedetti *et al.*, 2003; Wen *et al.*, 2003).

26.3 Materials and Methods

Genetic diversity was examined by amplified fragment length polymorphism (AFLP) (Vos *et al.*, 1995), which is an appropriate and reliable technique when

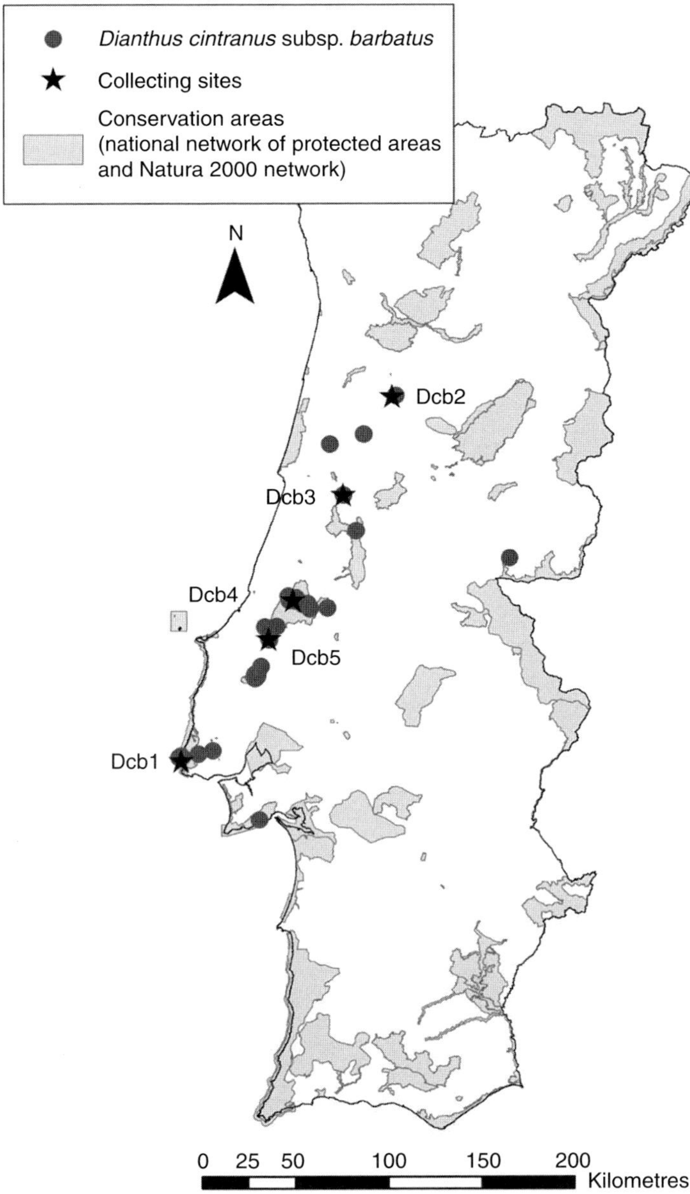

Fig. 26.2. *Dianthus cintranus* subsp. *barbatus* distribution map and collecting sites.

no previous genetic information is available and when dealing with small or rare populations, because it can detect small genetic differences and only small amounts of DNA are needed (Mueller and Wolfenbarger, 1999; Palacios and González-Candelas, 1999).

26.3.1 DNA samples and AFLP

Leaf samples were collected from five different populations throughout the taxon distribution range (see Fig. 26.2 and Table 26.1), between March and June 2006. Twenty randomly selected individuals were sampled

Table 26.1. Collecting sites of *D. cintranus* subsp. *barbatus*. Rainfall, temperature, soil type and pH were extracted using a geographic information system from Agência Portuguesa do Ambiente (2007) data.

Site code	Location	Lat/long	Collection date (2006)	Conservation area? (Y/N)	Rainfall (mm)	Temp. (°C)	Soil type	Soil pH
Dcb1	Sintra: entre Camarinheiras e a Ponta do Rebolo	38°45′53″N/009°29′23″W	3 June	Y	500–600	12.5–15.0	Cambisol	Mainly acid
Dcb2	Tondela: junto à estrada da Vila do Caramulo para Cadraço, antes do desvio para Bezerreira/Varzielas, do lado NW da estrada	40°33′49″N/008°11′24″W	24 June	N	2000–2400	10.0–12.5	Cambisol	Mainly acid
Dcb3	Condeixa-a-Nova: A NW do marco geodésico Cruto	40°04′43″N/008°29′45″W	24 June	N	1000–1200	12.5–15.0	Luvisol	Mainly neutral
Dcb4	Porto de Mós: Cabeço da Fórnea	39°33′36″N/008°48′33″W	25 June	Y	1400–1600	15.0–16.0	Luvisol	Mainly neutral
Dcb5	Serra dos Candeeiros: perto do marco geodésico Conde, do lado E da estrada	39°22′11″N/008°57′30″W	25 June	Y	800–1000	15.0–16.0	Luvisol	Mainly alkaline

per population. Fresh and young green leaves were chosen and stored with silica-gel until DNA extraction took place. The genomic DNA was extracted using a modified 4% CTAB protocol (Gawel and Jarret, 1991). The AFLP protocol followed that of Applied Biosystems (2005) with minor modifications, which was based on the protocol described by Vos *et al.* (1995). Preselective amplification was performed using *Eco*RI 5′-GACTGCGTACCAATTCA-3′ and *Mse*I 5′-GATGAGTCCTGAGTAAC-3′ adapter-directed primers, each with a single selective nucleotide (Vos *et al.*, 1995). Following a screening of ten primer combinations, two primer pairs were chosen for the final selective amplification: *Mse*I 5′-GATGAGTCCTGAGTAACAC-3′ and *Eco*RI 5′-GACTGCGTACCAATTCAAG-3′, *Mse*I 5′-GATGAGTCCTGAGTAACTT-3′ and *Eco*RI 5′-GACTGCGTACCAATTCAGG-3′. Both preselective and final selective amplifications were performed on a GeneAmp® PCR System 9700 thermal cycler. Samples were then processed on a MegaBACE™ 1000 DNA analysis system (Amersham Biosciences). The cleanup of the samples and the addition of the ROX size standard MEGABACE™ ET 550-R allowing precision sizing between 50 and 550bp was performed prior to genotyping. The resulting electropherograms were analysed using the MegaBACE™ Fragment Profiler Software Suite v2.2 (Amersham Biosciences) and automated peak calling was obtained. In addition, the electropherograms were visually reviewed in order to clarify ambiguities and prevent false identification of polymorphisms.

26.3.2 Genetic data analysis

A binary matrix of presences (1) and absences (0) of peaks on the electropherograms for each sample was built. Only strong (more than 100 RFU) reproducible bands ranging between 56 and 430bp (CTTXAGG) and 56 and 440bp (CACXAAG) were scored, resulting in a total of 650 polymorphic loci.

Descriptive statistics included estimates of allele frequencies following a Bayesian approach suggested by Zhivotovsky (1999), percentage of polymorphic loci/population at 5% level, expected heterozygosity (or genetic diversity) according to Lynch and Milligan (1994) for dominant markers, and obtained with AFLP-SURV v. 1.0 (Vekemans, 2002). The number of private fragments, defined here as the alleles unique to a single population (Maguire *et al.*, 2002), was determined as an additional diversity measurement.

Wright's F_{ST} (1951) (according to Lynch and Milligan, 1994) was estimated with AFLP-SURV 1.0 (Vekemans, 2002) in order to clarify the population genetic structure. Dendrograms were obtained by agglomerative hierarchical clustering using the Unweighted Pair Group Method with Arithmetic Mean (UPGMA) algorithm (Michener and Sokal, 1957; Sneath and Sokal, 1973). This analysis was conducted in NEIGHBOR, a program of PHYLIP v. 3.66 (Felsenstein, 2004). Support for the clusters was evaluated with 1000 bootstrap replicated pairwise genetic distance matrices based on Nei's (1978) genetic distance (after Lynch and Milligan, 1994) produced in AFLP-SURV v. 1.0 (Vekemans, 2002). CONSENSE program, from PHYLIP v. 3.66 software (Felsenstein, 2004), was then used to compute consensus trees from the multiple trees obtained with NEIGHBOR by the majority-rule consensus tree method (Tuimala, 2004).

A Principal Coordinate Analysis (PCoA) conducted in GENALEX v. 6.0 (Peakall and Smouse, 2006) on a genetic distance matrix, where distances between individuals were estimated as squared Euclidean distance (Excoffier *et al.*, 1992) defined for AFLP after Huff *et al.* (1993), was performed in order to assess the dimensionality of data and visualize the dispersion of individual plants in relation to the first three axes of variation (Gower, 1966). Hierarchical analysis of molecular variance (AMOVA) (Excoffier *et al.*, 1992) was carried out in GENALEX v. 6.0 (Peakall and Smouse, 2006) and based on Euclidean pairwise genetic distances calculated according to Huff *et al.* (1993) that follows the methods of Excoffier *et al.* (1992), Huff *et al.* (1993), Peakall *et al.* (1995) and

Michalakis and Excoffier (1996). Population genetic structure was also inferred using a Bayesian clustering method, suggested by Pritchard *et al.* (2000), in STRUCTURE v. 2.2 (http://pritch.bsd.uchicago.edu/). Exploratory analyses were undertaken with a 500,000 MCMC generations after a 500,000 iterations burnin period, and using both the non-admixture (each individual comes solely from one population or another) and the admixture model (each individual has some fraction of each population) (Pritchard *et al.*, 2000).

The final analyses were based on the admixture ancestry model with correlated allele frequencies. The probability for a range of values of *K* populations ranging from 1 to 5 (i.e. the total number of populations sampled) was examined and the number of genetic populations (*K*) deduced by posterior probabilities [LnP(D)] using 1,000,000 MCMC generations after a burnin period of 500,000 iterations. The Dirichlet parameter (α), a measure of the degree of admixture (when α is close to 0, most individuals are essentially from one population or another, while $\alpha>1$ implies that most individuals are admixed), was also obtained (Falush *et al.*, 2003). The best fit model was selected based on the posterior probability of data [LnP(D)] and the value of α. The number of genetic populations (*K*), proportion of admixture (proportion of membership of each individual in each cluster) was calculated and the significance of membership was evaluated with 90% probability intervals following Beaumont *et al.* (2001), Hansen (2002) and Albert *et al.* (2006) suggestions.

Isolation by distance, 'the process by which geographically restricted gene flow generates a genetic structure, because random genetic drift is occurring locally' (Hardy and Vekemans, 1999), was inferred from a Mantel test (Mantel, 1967) between pairwise F_{ST} values (obtained from AFLP-SURV v. 1.0 (Vekemans, 2002)(transformed to $F_{ST}/(1–F_{ST})$ and the log-transformed geographic distance separating the populations according to Rousset (1997) in GENALEX v. 6.0 (Peakall and Smouse, 2006) with a significance test using 9999 permutations.

26.3.3 Selection of populations for conservation

Genetic as well as demographic and threat data were combined in order to establish priorities among the populations of the target taxon following the methodology suggested by Delgado *et al.* (2008) with few modifications. Different criteria were used to characterize each population and data were subsequently standardized to 1. Genetic data included expected heterozygosity, total number of polymorphic loci, both previously obtained with AFLP-SURV 1.0 (Vekemans, 2002), the number of 'common and local' alleles according to the Marshall and Brown (1975) classification system adapted to the small number of populations and individuals sampled in this study, and inter-population genetic distance obtained previously with the PHYLIP v. 3.66 software (Felsenstein, 2004). Population size estimates collected during the field survey were categorized into different size classes as well as the number of threats affecting each population were used as additional criteria. Standardization to 1 was performed by giving the highest priority to smallest populations, highest number of threats, largest expected heterozygosity, number of polymorphic loci, 'common and local' loci, and those populations with higher inter-genetic distance (further way from the root of the dendrogram). These standardized indices were then integrated in a sum per population and transformed into percentages using the highest score as the reference value of 100%, as suggested by Delgado *et al.* (2008) based on the Vane-Wright *et al.* (1991) theoretical principles. Populations with higher percentage values were the priorities for conservation.

26.4 Results

26.4.1 Descriptive statistics and population genetic structure and differentiation

Two primer combinations generated 651 loci: *Mse*I-CAC/*Eco*RI-AAG (318 loci) and

*Mse*I-CTT/*Eco*RI-AGG (333). Out of the total loci, 17.20% are private to single populations, where Dcb1 presents the higher value and Dcb5 the lowest (Table 26.2). The number of polymorphic loci ranges from 302 in Dcb5 to 346 in Dcb3. Expected heterozygosity (*Hj*) varies from 0.149 (Dcb5) to 0.170 (Dcb3) and is relatively uniform among populations. Total expected heterozygosity (*Ht*) for all five populations of this taxon is moderate ($Ht=0.163$), level of inter-population diversity is very low ($Hb=0.006$), whereas the expected heterozygosity within all five populations is relatively high ($Hw=0.157$).

Very low but significant level of population differentiation was detected as indicated by very small F_{ST} (0.038). Both the AMOVA and the PCoA are in agreement with the F_{ST} results. The AMOVA shows an extremely high contribution of the within population component of the genetic variance (even for outbreeding species) for *D. cintranus* subsp. *barbatus* (92%) and a very low between population component (8%). The PCoA (where 63.99% of the total variance is explained by the first three coordinates) (Fig. 26.3) shows that Dcb3 separates from the rest of the populations, while Dcb1 and Dcb2 cluster together and Dcb4 and Dcb5 make another cluster in all three principal coordinate combinations.

The UPGMA dendrogram (Fig. 26.4) shows no obvious geographic pattern amongst the populations of *D. cintranus* subsp. *barbatus*. In fact, the Mantel test did not show a significant relationship between pairwise genetic distances, measured as $F_{ST}/(1-F_{ST})$, and log-transformed geographic distance ($P=0.267$), meaning that there is no geographic restriction to gene flow. These results are further supported by the Bayesian analysis conducted in STRUCTURE. The highest posterior probability was obtained for two clusters ($K=2$, Table 26.3). Using the admixture model, the degree of admixture for each individual in one of the two genetic clusters ($K=2$) was obtained. The results show a fairly high level of admixture with a Dirichlet parameter of 0.432 (Table 26.3). The average degree of admixture is approximately 0.30 of membership to the cluster 1 and 0.70 to the cluster 2, which means that most individuals have a higher proportion of cluster 2 rather than 1 (Table 26.4, Fig. 26.5). Only 10% of individuals have a membership to the cluster 1 of >0.8, whilst 53% have a proportion of cluster 2 of >0.8. Hence the remaining individuals (37%) are fairly admixed. Two main genetic clusters were then identified (Fig. 26.5), one enclosing mainly populations Dcb4 and Dcb5 (with 94% and 96% of membership to cluster 2), and the other one with a mixture of individuals from populations Dcb1, Dcb2 and Dcb3 with more or less the same proportion of membership of both clusters 1 and 2 (Table 26.4).

Table 26.2. Descriptive population genetic analysis based on allele frequencies for *D. cintranus* subsp. *barbatus*.

Pop.	Sample size	Class pop. size[a]	# PL[b]	% PL[b]	Exp. hetero. (*Hj*)[c]	SE (*Hj*)	% PA (absolute numbers)
Dcb1	20	51–100	344	52.8	0.161	0.006	5.069 (33)
Dcb2	20	11–50	339	52.1	0.154	0.006	3.840 (25)
Dcb3	20	11–50	346	53.1	0.170	0.007	3.533 (23)
Dcb4	20	11–50	308	47.3	0.151	0.006	2.458 (16)
Dcb5	20	11–50	302	46.4	0.149	0.006	2.304 (15)

POP: population, PL: polymorphic loci, Exp. Hetero: expected heterozygosity, SE: standard error of expected heterozygosity, PA: private alleles.

[a]population size estimates were categorized into size classes.

[b]at 5% level.

[c]expected heterozygosity or Nei's gene diversity under Hardy-Weinberg genotypic proportions.

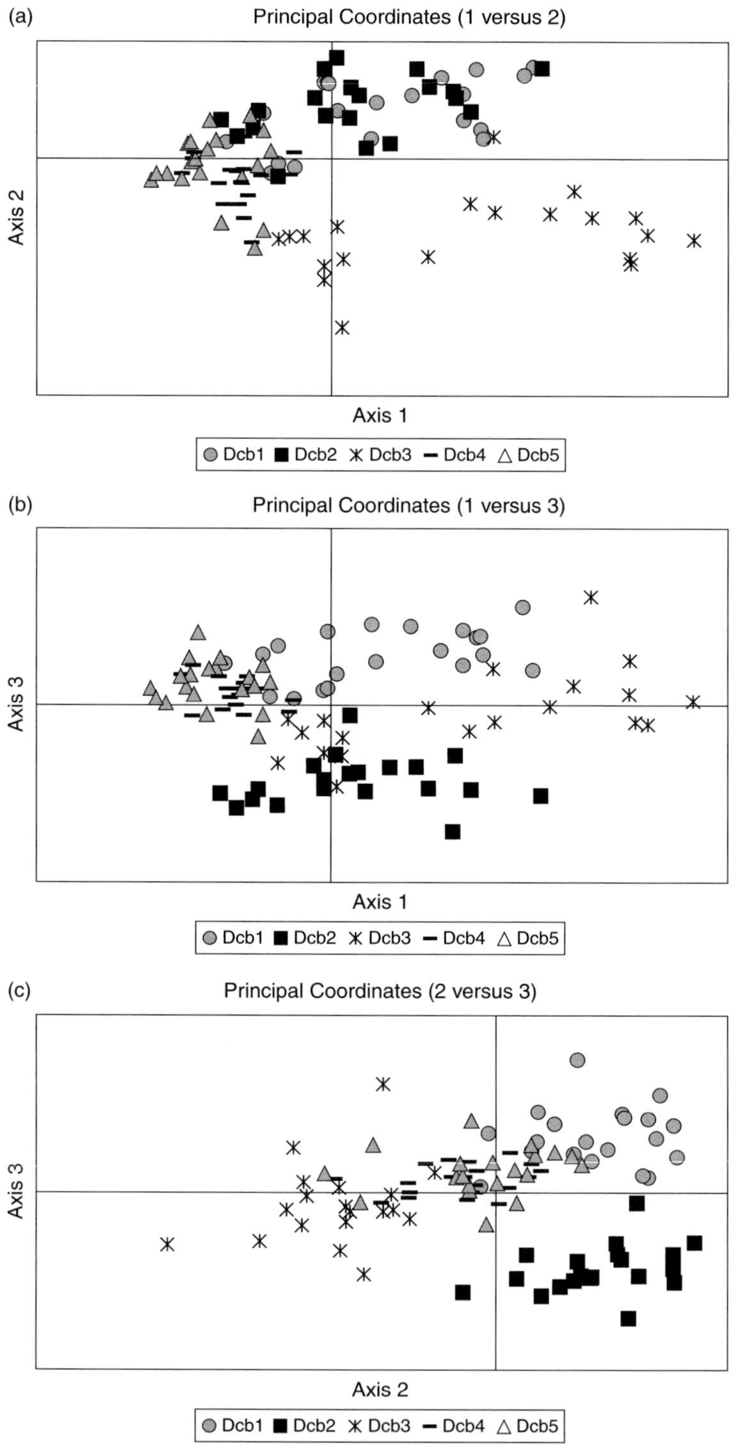

Fig. 26.3. PCoA plot of *D. cintranus* subsp. *barbatus* sampled populations based on genetic distances between individuals: (a) coordinates 1 and 2; (b) coordinates 1 and 3; (c) coordinates 2 and 3.

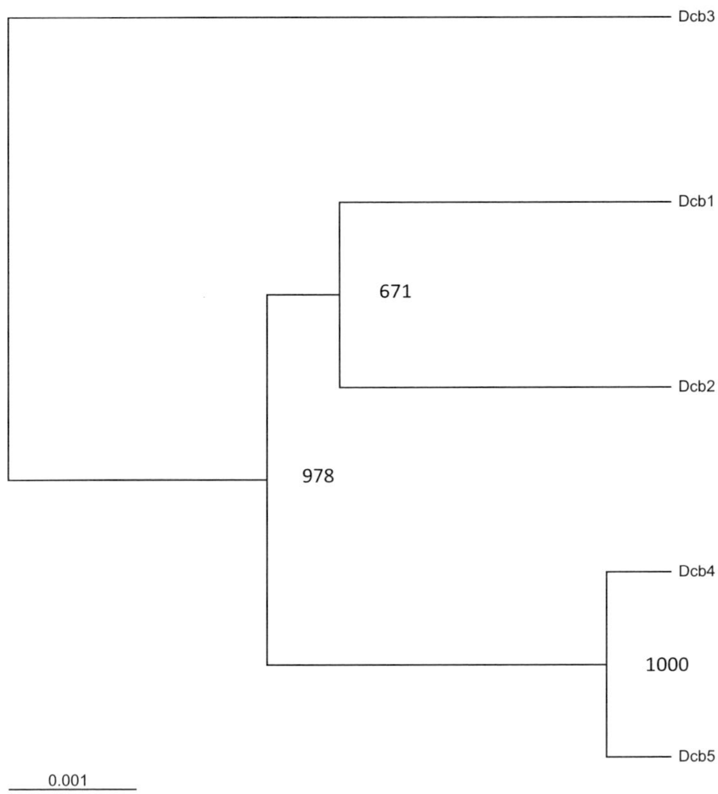

Fig. 26.4. UPGMA dendrogram illustrating the genetic relationships between the five populations of *D. cintranus* subsp. *barbatus*. Bootstrap values based on 1000 replications are shown. The scale indicates genetic distance.

Table 26.3. Estimated Ln Prob of data, posterior probabilities of *K* and Dirichlet parameter (α) assuming five genetic clusters for *D. cintranus* subsp. *barbatus*.

K	ln *P*(*X*|*K*)	*P*(*X*|*K*)	α
1	–21886.7	~0	na
2	–21196.5	~1	0.432
3	–21492.2	~0	0.547
4	–41154.1	~0	0.260
5	–21434.4	~0	0.156

na: not applicable

Table 26.4. The average proportion of membership for each cluster compared with sampled populations of *D. cintranus* subsp. *barbatus*. Underlined, the proportion of the cluster with more representation in each sampled population.

Populations	Cluster 1	Cluster 2
Dcb1	0.44	0.57
Dcb2	0.38	0.62
Dcb3	<u>0.57</u>	0.43
Dcb4	0.06	0.94
Dcb5	0.04	<u>0.96</u>
Average	0.30	0.70

26.4.2 Prioritizing populations for conservation

Genetic data were complemented with population size and threat data in order to prioritize populations for the *in situ* and *ex situ* conservation of the priority taxon.

Table 26.5 summarizes the genetic data for each of the populations of *D. cintranus* subsp. *barbatus*. Dcb3 is the population with more criteria having the highest values. Table 26.6 summarizes the population size and threats affecting populations of each species under

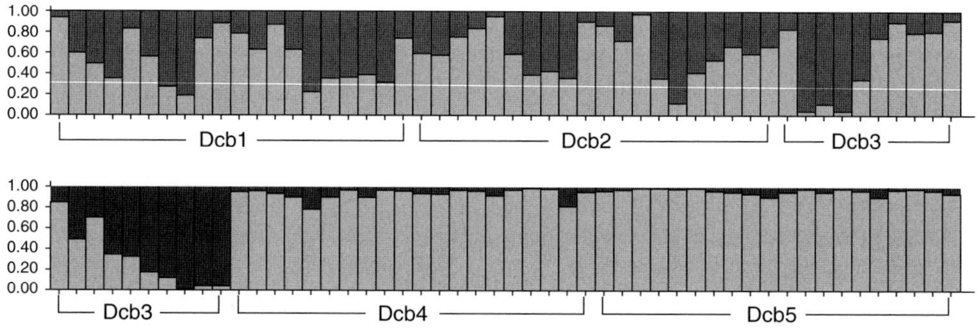

Fig. 26.5. Bayesian clustering of 100 individuals of *D. cintranus* subsp. *barbatus* using STRUCTURE software under admixture model with correlated frequencies between populations. Population genetic structure assuming $K=2$ (posterior probability ~1). For each individual (*x*-axis), the proportion of membership (*y*-axis) to the inferred cluster is indicated by bars of different colours: cluster 1 (dark grey) and 2 (light grey).

Table 26.5. Genetic data obtained for *D. cintranus* subsp. *barbatus* included in the analysis. Underlined, the highest values of each criterion.

Pop. ID	Hj^a	D	# PL[b]	# cl
Dcb1	0.161	0.003	344	0
Dcb2	0.154	0.003	339	3
Dcb3	<u>0.170</u>	<u>0.005</u>	<u>346</u>	<u>37</u>
Dcb4	0.151	0.001	308	16
Dcb5	0.149	0.001	302	17

Pop., population; *Hj*, expected heterozygosity; *D*, genetic distance; PL, polymorphic loci; cl, common and localized alleles.

[a]expected heterozygosity or Nei's gene diversity under Hardy-Weinberg genotypic proportions.

[b]at 5% level.

Table 26.6. Population size estimates and threat information for *D. cintranus* subsp. *barbatus*.

Pop.	Pop. size	Threats	Notes
Dcb1	51–100	Invasive species, trampling, fires, construction?	Invasive species (*Carpobrotus edulis*); trampling by fisherman, climbers; there have been fires in the area
Dcb2	11–50	Fires, climate change*	–
Dcb3	11–50	None	–
Dcb4	11–50	Grazing, fires	There have been fires in the area
Dcb5	11–50	Fires, trash deposition, stone quarry?	This area is consecutively being burned; there is a stone quarry nearby

*see Magos Brehm, 2009.

study. Populations of *D. cintranus* subsp. *barbatus* are generally small, ranging from 11 to 50 plants. Dcb1, which is located between Camarinheiras and Ponta do Rebolo (Sintra), is the largest population with 51–100 individuals (see Fig. 26.2 for the map of collecting sites). Dcb1 is also the population facing more threats: invasive species (such as *Carpobrotus edulis*), trampling, fires, and possibly construction. Dcb5, in Serra dos Candeeiros is the next most threatened population, its major threats include fires, trash deposition and perhaps activities related to a stone quarry located nearby. Dcb2 was the only population that was threatened by climate change (see Magos Brehm, 2009).

Table 26.7. Application of Vane-Wright *et al.* (1991) method using genetic, population size, threat standardized data for setting conservation priorities of *D. cintranus* subsp. *barbatus*. Populations are in decreasing order of priority.

Pop.	Pop. size St	Threats St	*Hj* St	*D* St	# PL St	# cl St	Sum	Priority (%)
Dcb3	1.00	0.00	1.00	1.00	1.00	1.00	5.00	100.00
Dcb5	1.00	0.75	0.88	0.10	0.87	0.46	4.06	81.20
Dcb2	1.00	0.50	0.91	0.50	0.98	0.08	3.97	79.40
Dcb1	0.50	1.00	0.95	0.50	0.99	0.00	3.94	78.80
Dcb4	1.00	0.50	0.89	0.10	0.89	0.43	3.81	76.20

The final relative priorities for population conservation were then obtained following the methodology described: Dcb3 (Condeixa-a-Nova) is the priority population for conservation. It is followed by Dcb5 (Serra dos Candeeiros), Dcb1 (between Camarinheiras and Ponta do Rebolo, Sintra), Dcb4 (Cabeço da Fórnea, Porto de Mós) and Dcb2 (Tondela) with very similar priority scores (Table 26.7).

26.5 Discussion

26.5.1 Conservation of *Dianthus cintranus* subsp. *barbatus*

Total expected heterozygosity for all five populations of this taxon is moderate and comparable to those obtained by Hamrick *et al.* (1979) for endemic, perennial and out-crossing plant species but slightly lower than those found in other out-breeding endemic species with AFLP (e.g. Gaudeul *et al.*, 2000). The very low level of inter-population diversity for *D. cintranus* subsp. *barbatus* and the higher value of expected heterozygosity within all five populations are in accordance with the predominantly out-crossing mating system found in this genus (Gottlieb, 1977; Brown, 1979; Loveless and Hamrick, 1984). Low values of total expected heterozygosity obtained with *D. cintranus* subsp. *barbatus* can have resulted from the low number of individuals of the natural populations, which varies from a minimum of 10 to a maximum of 100. Often, rare and endemic species show genetic impoverishment, which has been associated with the small size of their populations and their isolation (Hamrick and Godt, 1990; Barrett and Kohn, 1991; Ellstrand and Elam, 1993; Meffe and Carroll, 1994; Godt and Hamrick, 1995) and, consequently, to some degree of inbreeding (Dudash and Fenster, 2000), random genetic drift and population bottlenecks (Young *et al.*, 1996). Alternatively, this lack of genetic variability may also be due to strong, directional natural selection driven by uniformity of habitats (Babbel and Selander, 1974; Waller *et al.*, 1987). Also, a decline in the number of individuals of *D. cintranus* subsp. *barbatus* throughout its area of occurrence has been reported (Espírito-Santo *et al.*, 1997), which then supports these results. This same type of scenario has been previously reported in the literature (e.g. Fleishman *et al.*, 2001; Neel and Ellstrand, 2001; Mattner *et al.*, 2002; Wang *et al.*, 2004; Godt *et al.*, 2005; López-Pujol *et al.*, 2006). In fact, Oostermeijer *et al.* (2003) emphasizes that inbreeding redistributes alleles from heterozygous to homozygous combinations rather than leading to a loss of alleles and hence, loss of diversity.

There is no evidence to support any strong geographical structure from the genetic data (the Mantel tests were not significant). There is no correlation between genetic and geographic distances suggesting that there is no significant geographic restriction to gene flow among the populations. Additionally the UPGMA tree did not show any particular geographic pattern for the taxon, which reinforces the hypothesis that either long distance gene flow can occur throughout the distribution area, there was a high level of ancestral gene flow or the populations shared their ancestors. Bayesian analysis identified two genetic clusters among the populations: Dcb4 and Dcb5 (the geographically closest populations) are mainly represented by one genetic cluster, while the remaining populations have more or less equal proportions of the two genetic clusters.

All the results suggest that gene flow or some other homogenization factor is playing a role in determining the population genetic structure of this taxon. Nevertheless, the low between-population genetic differentiation supports the out-breeding nature of *D. cintranus* subsp. *barbatus* (Hamrick and Godt, 1996), which is in concordance with other studies that reported a predominantly insect-pollinated mating system in *Dianthus* spp. (e.g. Meusel and Mühlberg, 1979; Erhardt, 1988; Erhardt and Jaggi, 1995).

Given the out-breeding nature of this taxon as well as its type of seed dispersal mechanism, which enables the migration of small numbers of seeds further away from the parental plants, genetic divergence among populations is likely to be diminished. Such a high degree of genetic homogenization among populations can thus be attributed to high levels and long distance contemporary gene flow, high levels of ancestral gene flow or common ancestry (see Muir and Schlötterer, 2005), or longevity (see Hamrick *et al.*, 1979). However, a small but significant population differentiation was observed and several factors might have resulted in this differentiation: (i) low level of inbreeding – a few authors reported self-compatibility or some degree of inbreeding in *Dianthus* spp. (e.g. Erhardt and Jäggi, 1995; Collin and Shykoff, 2003); (ii) selection operating at a few loci (see Muir and Schlötterer, 2005); and (iii) habitat fragmentation in co-ancestral populations where populations are undergoing genetic differentiation.

Dianthus cintranus subsp. *barbatus* populations are genetically very homogeneous, with moderate values of genetic diversity, which can have resulted from the low number of individuals per population (maybe resultant by the many threats that threaten all populations) and consequently inbreeding. Low but significant levels of population differentiation were detected and the within population component of the genetic variance is extremely high. There is not a correlation between genetic and geographic distances suggesting that there is no significant geographic restriction to gene flow among the populations. Given the genetic results, only one population needs to be conserved ensuring the protection of the majority of genetic diversity. When combining both genetic, population size and threat data, Dcb3 (in Condeixa-a-Nova) is the priority population where a genetic reserve could be established. It presents the highest values of expected heterozygosity, genetic distance, number of polymorphic loci and commonly local alleles. It is a small population with a maximum of 50 plants, outside a conservation area but no threats were detected. In practice, the conservation of populations outside protected areas is more difficult to achieve. However, a genetic reserve should be implemented here for *D. cintranus* subps. *barbatus* and the long-term stability of this site and *in situ* conservation of the target population should be ensured by an appropriate legal framework (see Iriondo *et al.*, Chapter 10, this volume). From the *ex situ* conservation point of view, we also suggest that all studied populations of *D. cintranus* subsp. *barbatus* should be targeted for *ex situ* conservation given that it is a rare Portuguese endemic, with low numbers of plants per population.

26.5.2 Neutral versus adaptive diversity

Typically, conservation biology aims at conserving the maximum biological diversity possible. However, along with species diversity the conservation of diversity within a taxon (gene conservation) has also been identified as being equally important (Jump *et al.*, 2008). The genetic diversity available within a species represents the evolutionary potential allowing the species to evolve and to adapt to a changing environment (Lynch, 1996; Barrett and Schluter, 2008). However, even if not all the genetic diversity is potentially adaptive, some proportion is, and when subjected to environmental change, the value of the genetic diversity is likely to be proportional to its amount (Ledig *et al.*, 1986). Since it is difficult to predict the direction of the evolutionary pathway driven by environmental changes, we cannot identify which genetic variants are likely to be more valuable to the persistence of a taxon in their natural environment. Therefore, we have to assume that all genetic variation is equally important (Humphries *et al.*, 1995).

Nevertheless, two distinct but complementary components of genetic variation have been identified (McKay and Latta, 2002; Moritz, 2002). The first is related to the functional diversity, which has resulted from adaptive evolution due to natural selection (Bonin *et al.*, 2007). The second involves neutral alleles, which result from neutral evolutionary forces such as migration, mutation and genetic drift (Bonin *et al.*, 2007). The relative importance of adaptive versus neutral variation in conservation genetics has been vastly debated over the years (e.g. Bowen, 1999; Fraser and Bernatchez, 2001; Merilä and Crnokrak, 2001; Reed and Frankham, 2001; McKay and Latta, 2002; Holderegger *et al.*, 2006).

Adaptive variation refers to alleles (or quantitative traits) that affect fitness (Holderegger *et al.*, 2006). They are the primary targets of natural selection and reflect the species' potential ability to adapt to changing environments (e.g. Falconer and Mackay, 1996; McKay and Latta, 2002; van Tienderen *et al.*, 2002). Adaptive genetic variation is evaluated in quantitative genetic experiments under controlled and uniform environmental conditions (Holderegger *et al.*, 2006). The assessment of adaptive variation assessment is very time consuming (Bonin *et al.*, 2007) and quantitative traits involved in adaptation are sometimes difficult to find (Lynch, 1996). Moreover, adaptive variation is the result of environmental and genetic factors (Falconer and Mackay, 1996) and large sample sizes are required (which are usually unavailable in threatened populations) in order to understand the contribution of these components to the overall variation (Bonin *et al.*, 2007).

Neutral genetic diversity refers to those alleles that have no direct effect on species fitness and are not affected by natural selection (Holderegger *et al.*, 2006). They do not directly provide information on the adaptive or evolutionary potential of populations or taxa (Holderegger *et al.*, 2006). This type of genetic diversity can be assessed using a wide range of molecular markers. The assessment of neutral genetic variation has been frequently used as a shortcut to infer global genetic diversity and to support strategies for the conservation of threatened taxa (e.g. Palacios and González-Candelas, 1999; Rottenberg and Parker, 2003; Eckstein *et al.*, 2006; Watson-Jones *et al.*, 2006). Molecular markers are a fast and relatively cheap technique, which allows the study of gene flow, migration and dispersal.

The topic on whether a correlation between neutral and adaptive variation exists has been debated and conclusions do not always agree. Some authors have found that neutral and adaptive genetic diversity and differentiation are positively correlated (e.g. Merilä and Crnokrak, 2001; Pearman, 2001), whereas other studies indicate that measurements of neutral diversity have a very limited prediction ability of quantitative variation (e.g. Reed and Frankham, 2001, 2002) and thus cannot be used as a surrogate of adaptive genetic data, at least for some traits (Holderegger *et al.*, 2006).

Within the context of genetic conservation under a climate change threat, Lefèvre (2007) emphasized that gene conservation strategies should focus on the adaptive capacity of populations (and species) by considering their 'individual plasticity' (i.e. their ability to respond to different environmental conditions), their adaptive genetic diversity and the occurrence of natural selection that acts upon them, as well as their ability to disperse. Adaptive variation assessment is therefore particularly important since it allows the identification of the components of genetic diversity responsible for the adaptation of populations to different conditions. Nevertheless, the reason to use neutral molecular markers in this study was two-fold: (i) a positive correlation between neutral and adaptive genetic diversity was assumed, as supported by the findings of Merilä and Crnokrak (2001) and Pearman (2001); (ii) the use of neutral molecular markers is a quick technique of assessing genetic diversity, in contrast to that of the analysis of quantitative traits.

Acknowledgements

Joana Magos Brehm was financed by the Fundação para a Ciência e Tecnologia (FCT, Portugal) with grant number SFRH/BD/16508/2004. The authors are grateful to Serene Hargreaves for providing some guidance in the genetic data analysis.

Notes

[1] Marshall and Brown (1975) developed a two-way classification system of alleles based on their frequency in each population (common or rare) and distribution across populations (widespread over many populations or localized to just a few). Marshall and Brown (1975) and Brown and Hardner (2000) defined any allele occurring in ≥25% of populations as a widespread allele and those occurring in <25% of populations as a localized allele. Marshall and Brown (1975) also classified the alleles according to their average frequency in a population as common ($P \geq 0.05$) or rare ($P < 0.05$). Given that only 5 populations and 20 plants per population were sampled in this study, an adaptation of this classification system was used. A widespread locus was defined to occur in ≥50% of populations (≥3 populations) whereas any allele occurring in ≤2 populations was considered a localized allele. A common allele were those occurring at a frequency of >0.05 and a rare allele occurring at a frequency of ≤0.05. Four classes of alleles were then defined: (i) common and widespread (population frequency $P > 0.05$, and occurring in ≥3 populations); (ii) common and local (population frequency $P > 0.05$, and occurring in ≤2 populations); (iii) rare and widespread (population frequency $P \leq 0.05$, and occurring in ≥3 populations); (iv) rare and local (population frequency $P \leq 0.05$, and occurring in ≤2 populations). Marshall and Brown (1975) argued that the 'common and local' category is the most important in terms of conservation because it includes those alleles that confer adaptation to local conditions. 'Common and widespread' alleles are everywhere so they will inevitably be conserved regardless of the conservation strategy; 'rare and widespread' alleles will be conserved depending on the total number of sampled plants if *ex situ* accessions are to be sampled or if the conservation area includes most of the population in an *in situ* approach; 'rare and local' class includes very rare variants and recent or deleterious mutants, which are extremely difficult to collect but a fraction will always be included in any conservation strategy (Marshall and Brown, 1975).

References

Albert, V., Jonsson, B. and Bernatchez, L. (2006) Natural hybrids in Atlantic eels (*Anguilla anguilla, A. rostrata*): evidence for successful reproduction and fluctuating abundance in space and time. *Molecular Ecology* 15, 1903–1916.

Applied Biosystems (2005) *AFLP® Plant Mapping – Protocol.* Available at: www3.appliedbiosystems.com/cms/groups/mcb_support/documents/generaldocuments/cms_040959.pdf (accessed February 2011).

Babbel, G.R. and Selander, R.B. (1974) Genetic variability in edaphically restricted and widespread plant species. *Evolution* 28, 619–630.

Barrett, R.D.H. and Schluter, D. (2008) Adaptation from standing genetic variation. *Trends in Ecology and Evolution* 23, 38–44.

Barrett, S.C.H. and Kohn, J.R. (1991) Genetic and evolutionary consequences of small population size in plants: implications for conservation. In: Falk, D.A. and Holsinger, K.E. (eds) *Genetics and Conservation of Rare Plants*. Oxford University Press, New York, pp. 3–30.

Beaumont, M., Barratt, E.M., Gottelli, D., Kitchener, A.C., Daniels, M.J., Pritchard, J.K. and Bruford, M.W. (2001) Genetic diversity and introgression in the Scottish wildcat. *Molecular Ecology* 10, 319–336.

Bloch, D., Werdenberg, N. and Erhardt, A. (2006) Pollination crisis in the butterfly pollinated, wild carnation *Dianthus carthusianorum*? *New Phytologist* 169, 699–706.

Bonin, A., Ehrich, D. and Manel, S. (2007) Statistical analysis of amplified fragment length polymorphism data: a toolbox for molecular ecologists and evolutionists. *Molecular Ecology* 16, 3737–3758.

Bowen, B.W. (1999) Preserving genes, species, or ecosystems? Healing the fractured foundations of conservation policy. *Molecular Ecology* 8, S5–S10.

Brown, A.D.H. (1979) Enzyme polymorphisms in plant populations. *Theoretical Population Biology* 15, 1–42.

Brown, A.D.H. and Hardner, C.M. (2000) Sampling the gene pools of forest tree species for *ex situ* conservation. In: Young, A., Boshier, D. and Boyle, T. (eds) *Forest Conservation Genetics: Principles and Practice.* CSIRO Publishing Australia and CABI Publishing, Melbourne, pp. 185–196.

Collin, C.L. and Shykoff, J.A. (2003) Outcrossing rates in the gynomonoecious–gynodioecious species *Dianthus sylvestris* (Caryophyllaceae). *American Journal of Botany* 90(4), 579–585.

Collin, C.L., Pennings, P.S., Rueffler, C., Widmer, A. and Shykoff, J.A. (2002) Natural enemies and sex: how seed predators and pathogens contribute to sex-differential reproductive success in a gynodioecious plant. *Oecologia* 134, 94–102.

De Benedetti, L., Burchi, G., Bruna, S., Mercuri, A. and Schiva, T. (2003) Use of molecular markers to improve cut flowers longevity in carnation. *Acta Horticulturae (ISHS)* 624, 343–348.

Delgado, P., Eguiarte, L.E., Molina-Freaner, F., Alvarez-Buylla, E.R. and Piñero, D. (2008) Using phylogenetic, genetic and demographic evidence for setting conservation priorities for Mexican rare pines. *Biodiversity and Conservation* 17, 121–137.

Dudash, M.R. and Fenster, C.B. (2000) Inbreeding and outbreeding depression in fragmented populations. In: Young, A.G. and Clarke, G.M. (eds) *Genetics, Demography and Viability of Fragmented Populations.* Cambridge University Press, Cambridge, pp. 35–53.

Ebert, G. and Rennwald, E. (1993) *Die Schmetterlinge Baden-Württembergs (Band 3: Tagfalter 1).* Ulmer, Stuttgart.

Ebert, G., Esche, T., Herrmann, R., Hofmann, A., Lussi, H.G., Nikusch, I., Speidel, W., Steiner, A. and Thiele, J. (1994) *Die Schmetterlinge Baden-Württembergs (Band 3: Nachtfalter 1).* Ulmer, Stuttgart.

Eckstein, R.L., O'Neill, R.A., Danihelka, J., Otte, A. and Köhler, W. (2006) Genetic structure among and within peripheral and central populations of three endangered floodplain violets. *Molecular Ecology* 15, 2367–2379.

Ellstrand, N.C. and Elam, D.R. (1993) Population genetic consequences of small population size: implications for plant conservation. *Annual Review of Ecology and Systematics* 24, 217–242.

Erhardt, A. (1988) Pollination and reproduction in *Dianthus silvester* Wulf. In: Cresti, M., Gori, P. and Pacini, E. (eds) *Sexual Reproduction in Higher Plants.* Springer, Berlin, pp. 351–356.

Erhardt A. (1990) Pollination of *Dianthus gratianopolitanus* (Caryophyllaceae). *Plant Systematics and Evolution* 170, 125–132.

Erhardt, A. (1991) Pollination of *Dianthus superbus* L. *Flora* 185, 99–106.

Erhardt, A. and Jaggi, B. (1995) From pollination by Lepidoptera to selfing: the case of *Dianthus glacialis* (Caryophyllaceae). *Plant Systematics and Evolution* 195, 67–76.

Espírito-Santo, M.D. (coord.), Capelo, J., Costa, J.C., Lousã, M., Monteiro, A., Moreira, I., Neto, C., Sousa, E. and Vasconcelos, T. (1997) *Distribuição geográfica e estatuto de ameaça das espécies da Flora a proteger em Portugal Continental.* Relatório Final. Instituto Superior de Agronomia. Departamento de Protecção das Plantas e de Fitoecologia, Lisboa.

Excoffier, L., Smouse, P.E. and Quattro, J.M. (1992) Analysis of molecular variance inferred from metric distances among DNA haplotypes: application to human mitocondrial DNA restriction sites. *Genetics* 131, 479–491.

Falconer, D.S. and Mackay, T.F. (1996) *Introduction to Quantitative Genetics.* Longmans Green, Harlow, UK.

Falush, D., Stephens, M. and Pritchard, J.K. (2003) Inference of population structure: extensions to linked loci and correlated allele frequencies. *Genetics* 164, 1567–1587.

Felsenstein, J. (2004) *PHYLIP (Phylogeny Inference Package) version 3.6.* Distributed by the author. Department of Genome Sciences, University of Washington, Seattle.

Fleishman, E., Launer, A.E, Switky, K.R., Yandell, U., Heywood, J. and Murphy, D.D. (2001) Rules and exceptions in conservation genetics: genetic assessment of the endangered plant *Cordylanthus palmatus* and its implications for management planning. *Biological Conservation* 98, 45–53.

Fraser, D.J. and Bernatchez, L. (2001) Adaptive evolutionary conservation: towards a unified concept for defining conservation units. *Molecular Ecology* 10, 2741–2752.

Gaudeul, M., Taberlet, P. and Till-Bottraud, I. (2000) Genetic diversity in an endangered plant, *Eryngium alpinum* L. (Apiaceae), inferred from amplified fragment length polymorphism markers. *Molecular Ecology* 9, 1625–1637.

Gawel, N.J. and Jarret, R.L. (1991) A modified CTAB DNA extraction procedure for *Musa* and *Ipomoea*. *Plant Molecular Biology Reporter* 9, 262–266.

Godt, M.J.W. and Hamrick, J.L. (1995) Allozyme variation in two Great Smoky Mountain endemics: *Glyceria nubigena* and *Rugelia nudicaulis*. *Journal of Heredity* 86, 194–198.

Godt, M.J.W., Caplow, F. and Hamrick, J.L. (2005) Allozyme diversity in the federally threatened golden paintbrush, *Castilleja levisecta* (Scrophulariaceae). *Conservation Genetics* 6, 87–99.

Gottlieb, L.D. (1977) Electrophoretic evidence and plant systematics. *Annual Missouri Botanical Garden* 65, 164–180.

Gower, J.C. (1966) Some distance properties of latent root and vector methods used in multivariate analysis. *Biometrika* 53, 325–338.

Hamrick, J.L. and Godt, M.J.W. (1990) Allozyme diversity in plant species. In: Brown, A.H.D., Clegg, M.T., Kahler, A.L. and Weir, B.S. (eds) *Plant Population Genetics, Breeding and Genetic Resources*. Sinauer Associates, Sunderland, Massachusetts, pp. 43–63.

Hamrick, J.L. and Godt, M.J.W. (1996) Effects of life history traits on genetic diversity in plant species. *Philosophical Transactions of the Royal Society of London, Series B, Biological Sciences* 351, 1291–1298.

Hamrick, J.L., Linhart, Y.B. and Mitton, J.B. (1979) Relationships between life history characteristics and electrophoretically detectable genetic variation in plants. *Annual Review of Ecology and Systematics* 10, 173–200.

Hansen, M.M. (2002) Estimating the long-term effects of stocking domesticated trout into wild brown trout (*Salmo trutta*) populations: an approach using microsatellite DNA analysis of historical and contemporary samples. *Molecular Ecology* 11, 1003–1015.

Hardy, O.J. and Vekemans, X. (1999) Isolation by distance in a continuous population: reconciliation between spatial autocorrelation analysis and population genetics models. *Heredity* 83(2), 145–154.

Holderegger, R., Kamm, U. and Gugerli, F. (2006) Adaptive vs. neutral genetic diversity: implications for landscape genetics. *Landscape Genetics* 21, 797–807.

Huff, D.R., Peakall, R. and Smouse, P.E. (1993) RAPD variation within and among natural populations of outcrossing buffalograss *Buchloe dactyloides* (Nutt) Engelm. *Theoretical and Applied Genetics* 86, 927–934.

Humphries, C.J., Williams, P.H. and Vane-Wright, R.I. (1995) Measuring biodiversity value for conservation. *Annual Review of Ecology and Systematics* 26, 93–111.

Jennersten, O. (1988) Pollination in *Dianthus deltoides* (Caryophyllaceae): effects of habitat fragmentation on visitation and seed set. *Conservation Biology* 2, 359–366.

Jump, A.S., Marchant, R. and Peñuelas, J. (2008) Environmental change and the option value of genetic diversity. *Trends in Plant Science* 14(1), 51–58.

Knuth, P. (1898–1905) *Handbuch der Blütenbiologie, I Bd. 1898 (Einleitung und Literatur). II Bd. 1898 (Die bisher in Europa und im arktischen Gebiet gemachten blütenbiologischen Beobachtungen. 1. Ranunculaceae bis Compositae, 2. Lobeliaceae bis Gnetaceae, 1899); III Bd. (Die bisher in außereuropa ïschen Gebieten gemachten blütenbiologischen Beobachtungen. 1. Cycadaceae bis Cornaceae, 1904, 2. Clethraceae bis Compositae. Nachträge, Rückblick, 1905)*. Engelmann, Leipzig.

Ledig, F.T. (1986) Heterozygosity, heterosis, and fitness in outbreeding plants. In: Soulé, M.E. (ed.) *Conservation Biology: The Science of Scarcity and Diversity*. Sinauer Associates, Sunderland, Massachusetts, pp. 77–104.

Lefèvre, F. (2007) Conservation of forest genetic resources under climate change: the case of France. In: Koskela, J., Buck, A. and Teissier du Cros, E. (eds) *Climate Change of Forest Genetic Diversity. Implications for Sustainable Forest Management in Europe*. Bioversity International, Rome, pp. 95–101.

López-Pujol, J., Zhang, F.M. and Ge, S. (2006) Plant biodiversity in China: richly varied, endangered and in need of conservation. *Biodiversity and Conservation* 15(12), 3983–4026.

Loveless, M.D. and Hamrick, J.L. (1984) Ecological determinants of genetic structure in plant populations. *Annual Review of Ecology and Systematics* 15, 65–95.

Lynch, M. (1996) A quantitative–genetic perspective on conservation issues. In: Avise, J. and Hamrick, J. (eds) *Conservation Genetics: Case Histories From Nature*. Chapman and Hall, New York, pp. 471–501.

Lynch, M. and Milligan, B.G. (1994) Analysis of population genetic-structure with RAPD markers. *Molecular Ecology* 3, 91–99.

Magos Brehm, J. (2009) Conservation of Wild Plant Genetic Resources in Portugal. PhD thesis, University of Birmingham, UK.

Magos Brehm, J., Maxted, N., Ford-Lloyd, B.V. and Martins-Loução, M.A. (2008a) National inventories of crop wild relatives and wild harvested plants: case-study for Portugal. *Genetic Resources and Crop Evolution* 55, 779–796.

Magos Brehm, J., Mitchell, M., Maxted, N., Ford-Lloyd, B.V. and Martins-Loução, M.A. (2008b) IUCN Red Listing of crop wild relatives: is a national approach as difficult as some think? In: Maxted, N., Ford-Lloyd,

B.V., Kell, S.P., Iriondo, J.M., Dulloo, E. and Turok, J. (eds) *Crop Wild Relative Conservation and Use*. CAB International, Wallingford, UK, pp. 211–242.

Magos Brehm, J., Maxted, N., Martins-Loução, M.A. and Ford-Lloyd, B.V. (2010) New approaches for establishing conservation priorities for socio-economically important plant species. *Biodiversity and Conservation* 19, 2715–2740.

Maguire, T.L., Peakall, R. and Saenger, P. (2002) Comparative analysis of genetic diversity in the mangrove species *Avicennia marina* (Forsk.) Vierh. (Avicenniaceae) detected by AFLPs and SSRs. *Theoretical and Applied Genetics* 104, 388–398.

Mantel, N.A. (1967) The detection of disease clustering and a generalized regression approach. *Cancer Research* 27, 209–220.

Marshall, D.R. and Brown, A.H.D. (1975) Optimum sampling strategies in genetic conservation. In: Frankel, O.H. and Hawkes, J.G. (eds) *Crop Genetic Resource for Today and Tomorrow*. Cambridge University Press, Cambridge, pp. 53–80.

Mattner, J., Zawko, G., Rossetto, M., Krauss, S.L., Dixon, K.W. and Sivasithamparam, K. (2002) Conservation genetics and implications for restoration of *Hemigenia exilis* (Lamiaceae), a serpentine endemic from Western Australia. *Biological Conservation* 107, 37–45.

Maxted, N., Scholten, M.A., Codd, R. and Ford-Lloyd, B.V. (2007) Creation and use of a national inventory of crop wild relatives. *Biological Conservation* 140, 142–159.

McKay, J.K. and Latta, R.G. (2002) Adaptive population divergence: markers, QTL and traits. *Trends in Ecology and Evolution* 17, 285–291.

Meffe, G.K. and Carroll, C.R. (1994) *Principles of Conservation Biology*. Sinauer, Sunderland, Massachusetts.

Merilä, J. and Crnokrak, P. (2001) Comparison of genetic differentiation at marker loci and quantitative traits. *Journal of Evolutionary Biology* 14, 892–903.

Meusel, H. and Mühlberg, H. (1979) Unterfamilie Silenoideae (Lindl.) A. Br. In: Hegi, G. (ed.) *Illustrierte Flora von Mitteleuropa*, Band 3. Parey, Berlin.

Michalakis, Y. and Excoffier, L. (1996) A generic estimation of population subdivision using distances between alleles with special reference for microsatellite loci. *Genetics* 142, 1061–1064.

Michener, C.D. and Sokal, R.R. (1957) A quantitative approach to a problem in classification. *Evolution* 11, 130–162.

Moritz, C. (2002) Strategies to protect biological diversity and the evolutionary processes that sustain it. *Systematic Biology* 51, 238–254.

Mueller, U.G. and Wolfenbarger, L. (1999) AFLP genotyping and finger-printing. *Trends in Ecology and Evolution* 14, 389–394.

Muir, G. and Schlötterer, C. (2005) Evidence for shared ancestral polymorphism rather than recurrent gene flow at microsatellite loci differentiating two hybridizing oaks (*Quercus* spp.). *Molecular Ecology* 14, 549–561.

Müller, H. (1873) *Die Befruchtung der Blumen durch Insekten und die gegendeitigen Anpassungen beider*. Wilhelm Engelmann, Liepzig.

Neel, M.C. and Ellstrand, N.C. (2001) Patterns of allozyme diversity in the threatened plant *Erigeron parishii* (Asteraceae). *American Journal of Botany* 88, 810–818.

Nei, M. (1978) Estimation of average heterozygosity and genetic distance from a small number of individuals. *Genetics* 89, 583–590.

Oostermeijer, J.G.B., Luijten, S.H. and den Nijs, J.C.M. (2003) Integrating demographic and genetic approaches in plant conservation. *Biological Conservation* 113, 389–398.

Palacios, C. and González-Candelas, F. (1999) AFLP analysis of the critically endangered *Limonium cavanillesii* (Plumbagincaea). *Journal of Heredity* 90(4), 485–489.

Peakall, R. and Smouse, P.E. (2006) GENALEX 6: genetic analysis in Excel. Population genetic software for teaching and research. *Molecular Ecology Notes* 6, 288–295.

Peakall, R., Smouse, P.E. and Huff, D.R. (1995) Evolutionary implications of allozyme and RAPD variation in diploid populations of dioecious buffalograss *Buchloe dactyloides*. *Molecular Ecology* 4, 135–147.

Pearman, P.B. (2001) Conservation value of independently evolving units: sacred cow or testable hypothesis? *Conservation Biology* 15, 780–783.

Pritchard, J.K., Stephens, M. and Donnelly, P. (2000) Inference of population structure using multilocus genotype data. *Genetics* 155, 945–959.

Reed, D.H. and Frankham, R. (2001) How closely correlated are molecular and quantitative measures of genetic variation? A meta-analysis. *Evolution* 55, 1095–1103.

Reed, D.H. and Frankham, R. (2002) Correlation between fitness and genetic diversity. *Conservation Biology* 17, 230–237.
Rottenberg, A. and Parker, J.S. (2003) Conservation of the critically endangered *Rumex rothschildianus* as implied from AFLP diversity. *Biological Conservation* 114, 299–303.
Rousset, F. (1997) Genetic differentiation and estimation of gene flow from F-statistics under isolation by distance. *Genetics* 145, 1219–1228.
Smulders, M.J.M., Rus-Kortekaas, Y. and Vosman, B. (2000) Microsatellite markers useful throughout the genus *Dianthus*. *Genome* 43, 208–210.
Smulders, M.J.M., Noordijk, Y., Rus-Kortekaas, Y., Bredemeijer, G.M.M. and Vosman, B. (2003) Microsatellite genotyping of carnation varieties. *Theoretical and Applied Genetics* 106, 1191–1195.
Sneath, P.H.A. and Sokal, R.R. (1973) *Numerical Taxonomy*. Freeman, San Francisco.
The seed site (2008) Seed dispersal. Available at: http://theseedsite.co.uk/sdwind.html (accessed February 2011).
Tuimala, J. (2004) *A Primer to Phylogenetic Analysis Using Phylip Package*, 2nd edn. Center for Scientific Computing, Espoo, Finland.
Van Tienderen, P.H., de Haan, A.A., van der Linden, C.G. and Vosman, B. (2002) Biodiversity assessment using markers for ecologically important traits. *Trends in Ecology and Evolution* 17, 577–582.
Vane-Wright, R.I., Humphries, C.J. and Williams, P.H. (1991) What to protect? Systematics and the agony of choice. *Biological Conservation* 55, 235–254.
Vekemans, X. (2002) *AFLP-SURV version 1.0*. Distributed by the author. Laboratoire de Génétique et Ecologie Végétale. Université Libre de Bruxelles, Belgium.
Vos, P., Hogers, R., Bleeker, M., Reijans, M., van de Lee, T., Hornes, M., Frijters, A., Pot, J., Peleman, J., Kuiper, M. and Zabeau, M. (1995) AFLP: a new technique for DNA fingerprinting. *Nucleic Acids Research* 23, 4407–4414.
Waller, D.M., O'Malley, D.M. and Gawler, S.C. (1987) Genetic variation in the extreme endemic *Pedicularis furbishiae* (Scrophulariaceae). *Conservation Biology* 1, 335–340.
Wang, Z.F., Hamrick, J.L. and Godt, M.J.W. (2004) High genetic diversity in *Sarracenia leucophylla* (Sarraceniaceae), a carnivorous wetland herb. *Journal of Heredity* 95, 234–243.
Watson-Jones, S.J., Maxted, N. and Ford-Lloyd, B.F. (2006) Population baseline data for monitoring genetic diversity loss for 2010: a case study for *Brassica* species in the UK. *Biological Conservation* 132, 490–499.
Wen, W., YouMing, C., HuiYu, Z. and MinRen, H. (2003) Genetic diversity of *Dianthus chinensis* L. and *D. caryophyllus* L. with RAPD. *Journal of Nanjing Forestry University* 27(4), 2–74.
Wright, S. (1951) The genetical structure of populations. *Annual Eugenics* 15, 323–354.
Young, A.G., Boyle, T. and Brown, A.H.D. (1996) The population genetic consequences of habitat fragmentation for plants. *Trends in Ecology and Evolution* 11, 413–418.
Zhivotovsky, L.A. (1999) Estimating population structure in diploids with multilocus dominant DNA markers. *Molecular Ecology* 8, 907–913.

27 The Challenge of *In Situ* Conservation of Crop Wild Relatives in the Biotechnology Era – A Case Study of Wild Rice Species

Bao-Rong Lu

27.1 Introduction

With the continued increase in global population, the declining availability of farming land, the increasing shortage of water and the losses of rural labour to the urban centres, the world's food security faces a great challenge. Global climate change that may result in a lot of uncertainties for agriculture has worsened the situation. To meet the increasing demand for food worldwide, we need to significantly enhance crop productivity. Effective exploitation and utilization of plant genetic resources, particularly those in the same gene pools (GP) as the crop wild relatives (CWR) will provide more opportunities to meet the increasing demand. Usually, the CWR gene pool includes: (i) the direct ancestral species from which the crops were domesticated; (ii) genetically closely related wild species; (iii) the conspecific weedy types of crops; and (iv) the natural hybrids and progeny between wild and crop species.

During the millions of years of adaptive evolution in the changing environments, CWR have accumulated abundant genetic diversity in their gene pools. Many traits unique to CWR could be beneficial to crop species. Therefore, serving as a vast genetic reservoir, CWR can provide elite genetic resources for the genetic improvement of crop varieties by offering beneficial genes. In fact, CWR have already played a very important role in enhancing crop production, particularly when modern breeding practices and changes of crop cultivation management have caused great losses of genetic diversity in crop species (Lu, 1996). The production of hybrid rice that involves the introduction of the male sterility (MS) gene from the wild ancestor of rice, *Oryza rufipogon*, well exemplifies the successful use of CWR (for review, see Wu, 1990; Wang and Sun, 1996). The grassy stunt virus resistance gene in rice varieties was also introduced from the annual wild rice (*O. nivara*) (for review, see Wu, 1990; Wang and Sun, 1996). There are many more examples of introducing disease resistance genes in other cereal crop varieties from CWR. In addition to introducing useful genes resistant to biotic and abiotic stresses, CWR can also broaden the genetic background of crop varieties. Therefore, effective conservation of CWR is essential for the continued availability and potential utilization of these valuable genetic resources in the wild gene pool.

27.2 *In situ* Conservation of Wild Rice Species in China

Rice is one of the world's most important crops, providing staple food for nearly one half of the global population. The productivity of rice is closely associated with world food security. There are two cultivated rice species: Asian cultivated rice (*O. sativa*) that was domesticated in Asia and is cultivated and consumed worldwide, and African cultivated rice (*O. glaberrima*) that has its origin in West Africa and remains only locally important in some areas (Chang, 1976). More than 90% of Asian cultivated rice is grown and consumed in Asia where more than one half of the world population lives (Lu, 1996).

In addition to the two crop species, there are more than 20 wild relative species in the genus *Oryza*, and about 50 other wild species in the tribe Oryzeae of the grass family (Poaceae). Species in the genus *Oryza* include both diploids ($2n=2x=24$) and tetraploids ($2n=4x=48$), and ten different genome types, i.e. the AA, BB, CC, BBCC, CCDD, EE, FF, GG, JJHH and HHKK genomes (Vaughan, 1994; Ge *et al.*, 1999), widely distributed in pan-tropics and subtropics of the world (Table 27.1).

Wild *Oryza* species with different genome types have significant reproductive isolation, making the accessibility of them as genetic resources for rice breeding difficult. For example, wild *Oryza* species containing the AA-genome have relatively high sexual compatibility, complete chromosome

Table 27.1. Chromosome number, genome and distribution of species in *Oryza*.

Series Species	2*n*	Genome	Distribution
Ser. Meyerianae			
O. granulata	24	GG	South and South-east Asia
O. neocaledonica	24	GG	New Caledonia, Asia
Ser. Ridleyanae			
O. longiglumis	48	HHJJ	Indonesia and PNG, Asia
O. ridleyi	48	HHJJ	South Asia
Ser. Schlechterianae			
O. schlechteri	48	HHKK	PNG* and Indonesia, Asia
Ser. Brachyanathae			
O. brachyantha	24	FF	Africa
Ser. Latifoliae			
O. alta	48	CCDD	South and Central America
O. eichingeri	24, 48	CC [a]	South Asia and East Africa
O. grandiglumis	48	CCDD	South and Central America
O. malampuzhaensis	48	BBCC	India, Asia
O. latifolia	48	CCDD	South and Central America
O. minuta	48	BBCC	The Philippines and PNG, Asia
O. officinalis	24, 48	CC [a]	Asia, Australia
O. punctata	24	BB	Africa
O. rhizomatis	24	CC	Sri Lanka, Asia
O. schweinfurthiana	48	BBCC	Africa
Ser. Australianses			
O. australiensis	24	EE	Australia
Ser. Sativae			
O. barthii.	24	AA	Africa
O. glaberrima	24	AA	West Africa
O. glumaepatula	24	AA	South and Central America
O. longistaminata	24	AA	Africa
O. meridionalis	24	AA	Australia
O. nivara	24	AA	Asia
O. rufipogon	24	AA	Asia, Australia
O. sativa	24	AA	Worldwide (cultivated)

PNG, Papua New Guinea.

pairing in meiosis of F_1 interspecific hybrids, and relatively high pollen and panicle fertility of the F_1 hybrids (Naredo *et al.*, 1997, 1998). Therefore, the AA-genome wild relatives, including weedy rice (*O. sativa* f. *spontanea*), which is the same biological species as the crop and occurs in most major rice-producing regions of the world, are the most accessible genetic resources. All the wild species in tribe Oryzeae compose the primary, secondary and tertiary gene pools valuable for genetic improvement of rice (Lu and Snow, 2005). The conservation of genetic diversity in the rice gene pool is essential for sustainable rice production.

There are three wild rice species in China: *O. rufipogon*, *O. officinalis* and *O. granulata*, in addition to weedy rice (Lu and Snow, 2005), distributed in the southern parts of China (Fig. 27.1). The conservation and use of wild rice in China can be traced back to as early as the 1920s after Dr E.D. Merrill first found *O. rufipogon* in Lofu Mountain of Guangdong Province. Prof Ding Ying collected the same species at more sites in Guangdong in 1926 and later Chinese scientists found the other two wild species *O. officinalis* and *O. granulata* (Wu, 1990). Seed samples of the wild rice were conserved in gene banks and used for breeding and research. The well-known Chinese rice variety 'Zhongshan No. 1' tolerant to cold and other abiotic stresses was bred by Prof Ding Ying in 1931 through wide hybridization with *O. rufipogon*.

During 1978–1982, the Chinese Academy of Agricultural Sciences organized a nation-wide extensive survey and collection for wild rice species. More than 5000 wild rice samples, representing four taxa in *Oryza* and seven species in the related genera of the Oryzeae, were collected in different provinces. The collected wild rice samples have been deposited in the Chinese National Genebank (Beijing) as well as gene banks and seed storage facilities of provincial and local agricultural agencies, for the *ex situ* conservation of wild rice germplasm. In addition, more than 2000 accessions of wild rice samples are also conserved as living stocks in the National Wild Rice Nurseries in Guangdong and Guangxi provinces. A few *in situ* conservation sites for *O. rufipogon* have also been established in China, including in Dongxiang (Jiangxi Province), Caling (Hunan Province) and Zhengcheng (Guangdong Province) (Wang and Sun, 1996). Recently, the central and provincial governments paid more attention to the *in situ* conservation of wild rice species, and more conservation sites have been set up by national and provincial conservation programmes (Q.W. Yang, pers. comm.).

However, the survival and continued availability of wild rice populations in China have been under severe threat during the past five decades, owing to habitat deterioration caused by human disturbance, e.g. agricultural land expansion, water drainage, animal (buffalo) grazing, and road and house construction. According to the unpublished data collected by the Chinese Academy of Sciences (CAS) in 1994 (D.Y. Hong, pers. comm.), nearly 80% of the *O. rufipogon* populations recorded during the above-mentioned conservation activities

Fig. 27.1. Distribution of *Oryza rufipogon, O. officinalis* and *O. granulata* in China. The intensity of the colour indicates the relative richness of the wild species in different provinces.

in the 1970s were no longer in existence. The size of existing *O. rufipogon* populations was considerably reduced. A similar situation was also reported for many other CWR in China.

Very importantly, one factor which also significantly affects the continued availability and genetic integrity of wild rice populations has not caught the attention of conservationists for CWR. This is the extensive introgression of crop alleles to populations of CWR through recurrent gene flow. This will become one of the greatest challenges for *in situ* conservation of CWR, particularly in the biotechnology era.

27.3 Challenge of Crop–Wild Introgression to *In Situ* Conservation of Wild Rice Species

Following the definition in its broad sense, introgression indicates 'the transfer of genes between genetically distinguishable populations' (Rieseberg and Carney, 1998). Introgression can happen between populations of wild species that have a certain genetic affinity (wild–wild). It can also happen between a cultivated species and its close wild relatives (crop–wild). Introgression is considered to have great significance in affecting genetic diversity (Anderson and Stebbins, 1954; Ellstrand and Schierenbeck, 2000) and differentiation of plant populations or species (Ellstrand *et al.*, 1999). It is proven that natural introgression of a cultivated species with its wild relatives has played an important role in their evolutionary process (Ellstrand *et al.*, 1999). The gradual introgression of crop-specific alleles through recurrent gene flow into local populations of a wild relative species may considerably alter the intrinsic genetic composition of these populations and alter the diversity patterns of the wild gene pool (e.g. Song *et al.*, 2003a). Such crop-to-wild introgression may enhance adaptability of a wild population (e.g. Ellstrand and Schierenbeck, 2000), but also contribute to the extinction of a local wild population (Kiang *et al.*, 1979). Consequently, crop–wild introgression may considerably influence the strategies and outcomes of CWR conservation.

For example, the wild ancestor of cultivated rice *O. rufipogon* is cross-compatible with cultivated rice and extensively sympatric with the distribution of cultivated rice, in addition to the situation that the wild rice is usually surrounded by rice fields. As a consequence, gene flow from cultivated rice can occur consistently, and the crop–wild introgression has a potential to alter the genetic composition of the nearby wild populations. In a previous study, we estimated introgression of cultivated rice to *O. rufipogon* based on the analyses of 17 selected rice SSR primer pairs involving 139 rice varieties (86 *indica* and 53 *japonica* ecotypes) and 336 wild individuals from 11 *O. rufipogon* populations in China. Consanguinity of cultivated rice in *O. rufipogon* populations was detected according to the admixture model using the STRUCTURE program. It is found from these analyses that four wild populations, DX-P1, DX-P2, GZ-P2 and HL-P, contained some rare alleles that were commonly present in the rice varieties studied, and that the four wild populations scattered among the varieties showed a significant affinity, compared with the other wild populations. This finding supports the contention of substantial crop-to-wild gene flow and introgression when wild and cultivated taxa occur in proximity. Crop-to-wild introgression may have accumulative impacts on the genetic variation of wild populations, leading to significant differentiation in wild species. In addition, effective measures should be taken to avoid introgression from cultivated rice that would significantly influence *in situ* conservation of wild rice (Song *et al.*, 2006). Our further studies on genetic diversity and differentiation of *O. rufipogon* populations distributed in Gaozhou of Guangdong Province in China, have also indicated evidence of crop-to-wild introgression that has significantly influenced the genetic composition of the wild populations (X. Jin and B.R. Lu, unpublished data). All the observations and research results demonstrate the challenge that crop–wild introgression poses to the success of *in situ* conservation of wild rice species.

27.4 Transgene Flow and its Further Impacts on *In Situ* Conservation of CWR

The rapid development of biotechnology has greatly promoted the research and development of genetically modified (GM) crops worldwide. Consequently, a large number of transgenes conferring diverse traits have been successfully transferred into crop varieties through transgenic biotechnology. These transgenes include disease and insect resistance (Bock, 2007), virus resistance (Shepherd *et al.*, 2007), herbicide resistance (Toyama *et al.*, 2003), and salt and drought tolerance (Tang *et al.*, 2006). The great success in transgenic biotechnology has had a tremendous impact on the world crop production and cultivation patterns of agricultural species (James, 2009). However, the extensive environmental release and cultivation of GM crop varieties has also aroused considerable biosafety concerns and debates worldwide (Ellstrand, 2003; Stewart *et al.*, 2003).

Recently, widespread attention has been paid to the introgression of genes from GM crops into their wild relatives, mainly due to the great concerns on environmental biosafety caused by the extensive environmental release of GM crops. In the context of *in situ* conservation of CWR, concerns of the spreading of transgenes, that could confer strong selective advantages, into wild populations may also arouse unwanted evolutionary impact on populations of CWR (Lu and Snow, 2005; Lu and Yang, 2009). This will make the *in situ* conservation of CWR more challenging. We believe that the study of introgression between a (transgenic) crop and its wild relatives that are targeted for *in situ* conservation allows an insight into the demographic and evolutionary impacts of CWR. The knowledge is particularly useful for designing effective *in situ* conservation strategies for CWR in the biotechnology era.

In principle, the magnitude of impacts caused by transgene escape to CWR can be determined by: (i) the quantity of transgenes that have flowed into CWR; and (ii) the transgenic traits that have or do not have evolutionary advantages under natural selection. If the transgene conveys selectively advantageous or disadvantage traits, the flow of such a transgene into CWR may change their fitness. Transgenes that can enhance the fitness of CWR with favourable traits, such as pest resistance, drought tolerance and enhanced growth ability, would persist and quickly spread in the populations of CWR through introgression. This process will promote the fast increase in frequencies of transgenes, causing potential losses of genetic integrity of the CWR gene pool. On the contrary, transgenes that reduce the fitness of CWR may lead to the extinction of local CWR populations by the so-called swarm effect (Ellstrand and Elam, 1993). A large influx of fitness-reducing transgenes can contribute to population declines or even local extinction of small, isolated populations of CWR that occur near the crop (Haygood *et al.*, 2003). Such a swarm effect has already occurred in wild rice (*O. rufipogon*) through crop–wild introgression, even without the inclusion of transgenes, resulting in the extinction of local wild rice populations (Kiang *et al.*, 1979). (Trans)gene flow and introgression from cultivated rice to wild and weedy rice is widely observed in nature (Song *et al.*, 2003b; Chen *et al.*, 2004; Wang *et al.*, 2006). The long-term impacts of such introgression of crop genes (including transgenes) on the evolutionary and ecological potential of CWR populations need to be understood for the effective *in situ* conservation of CWR.

27.5 Conclusion

The gene pools of CWR include important genetic resources for the improvement of crop varieties, via the transfer of agronomically beneficial genes from CWR in plant breeding. This can significantly increase the productivity of crops and enhance their tolerance to biotic and abiotic stresses under the changing environments. However, due to the continued increase in human population, environmental deterioration, agricultural land uses and other human-mediated disturbances to the environments, the long-term availability and diversity of CWR are under severe threat. Among the threats, the introgression of crop alleles to populations of CWR through recurrent gene flow is an important but often neglected factor that can significantly affect the *in situ* conservation of CWR. The crop–wild introgression can considerably affect the evolutionary potential of CWR populations through genetic swamping or genetic sweeping if the introgressed alleles confer strong selective advantage. This will lead to the extensive losses of genetic diversity and integrity of CWR or result in the extinction of small, isolated local populations of CWR. The rapid development and extensive environmental release (cultivation) of GM crops worldwide may worsen such a situation because the introgression of transgenes from GM crop to populations of CWR on a large scale may accelerate the procedure in which the losses of genetic diversity/integrity or local extinction of CWR populations become evident. Well-designed research should be carried out to facilitate our profound understanding of the relevant problems and to help us to design effective strategies for the *in situ* conservation of CWR.

Acknowledgements

This study was funded by the '973' programme (no. 2011CB100401) of Ministry of Sciences and Technology, and NSFC (nos 30730066, 30871503).

References

Anderson, E. and Stebbins, G.L. (1954) Hybridization as an evolutionary stimulus. *Evolution* 8, 378–388.

Bock, R. (2007) Plastid biotechnology: prospects for herbicide and insect resistance, metabolic engineering and molecular farming. *Current Opinion in Biotechnology* 18, 100–106.

Chang, T.T. (1976) The origin, evolution, cultivation, dissemination and diversification of Asian and African rice. *Euphytica* 25, 425–441.

Chen, L.J., Lee, D.S., Song, Z.P., Suh, H.S. and Lu, B.-R. (2004) Gene flow from cultivated rice (*Oryza sativa*) to its weedy and wild relatives. *Annals of Botany* 93, 67–73.

Ellstrand, N.C. (2003) Current knowledge of gene flow in plants: Implications for transgene flow. *Philosophical Transactions of the Royal Society B: Biological Sciences* 358: 1163–1170.

Ellstrand, N.C. and Elam, D. (1993) Population genetic consequences of small population size: implications for plant conservation. *Annual Review of Ecology Evolution and Systematics* 24, 217–242.

Ellstrand, N.C. and Schierenbeck, K.A. (2000) Hybridization as a stimulus for the evolution of invasiveness in plants? *Proceedings of the National Academy of Sciences USA* 97, 7043–7050.

Ellstrand, N.C., Prentice, H.C. and Hancock, J.F. (1999) Gene flow and introgression from domesticated plants into their wild relatives. *Annual Review of Ecology and Systematics* 30, 539–563.

Ge, S., Sang, T., Lu, B.-R. and Hong, D.Y. (1999) Phylogeny of rice genomes with emphasis on origins of allotetraploid species. *Proceedings of the National Academy of Sciences USA* 96, 14400–14405.

Haygood, R., Ives, R.A. and Andow, D.A. (2003) Consequences of recurrent gene flow from crops to wild relatives. *Proceedings of the Royal Society Series B* 270, 1879–1886.

James, C. (2009) Global status of commercialized biotech/GM Crops. *ISAAA Brief No. 41*. ISAAA, Ithaca, New York.

Kiang, Y.T., Antonvics, J. and Wu, L. (1979) The extinction of wild rice (*Oryza perennis formosa*) in Taiwan. *Journal of Asian Ecology* 1, 1–9.

Lu, B.-R. (1996) Diversity of the rice genepool and its sustainable utilization. In: Zhang, A. and Wu, S. (eds) *Proceedings of the International Symposium on Floristic Characteristics and Diversity of East Asian Plants*. China Higher Education Press-Berlin: Springer-Verlag, Beijing, pp. 454–460.

Lu, B.-R. and Snow, A.A. (2005) Gene flow from genetically modified rice and its environmental consequences. *BioScience* 55, 669–678.

Lu, B.-R. and Yang, C. (2009) Gene flow from genetically modified rice to its wild relatives: assessing potential ecological consequences. *Biotechnology Advances* 27, 1083–1091.

Naredo, M.E.B., Juliano, A.B., Lu, B.-R. and Jackson, M.T. (1997) Hybridization of AA genome rice species from Asia and Australia I. Crosses and development of hybrids. *Genetic Resources and Crop Evolution* 44, 17–23.

Naredo, M.E.B., Juliano, A.B., Lu, B.-R. and Jackson, M.T. (1998) Taxonomic status of *Oryza glumaepatula* Steud. II. Hybridization between New World diploids and AA genome species from Asia and Australia. *Genetic Resources and Crop Evolution* 45, 205–214.

Rieseberg, L.H. and Carney, S.E. (1998) Plant hybridization. *New Phytologist* 140, 599–624.

Shepherd, D.N., Mangwende, T., Martin, D.P., Bezuidenhout, M., Kloppers, F.J., Carolissen, C.H., Monjane, A.L., Rybicki, E.P. and Thomson, J.A. (2007) Maize streak virus-resistant transgenic maize: a first for Africa. *Plant Biotechnology Journal* 5, 759–767.

Song, Z.P., Xu, X., Wang, B., Chen, J.K. and Lu, B.-R. (2003a) Genetic diversity in the northernmost *Oryza rufipogon* populations estimated by SSR markers. *Theoretical and Applied Genetics* 107, 1492–1499.

Song, Z.P., Lu, B.-R., Zhu, Y.G. and Chen, J.K. (2003b) Gene flow from cultivated rice to the wild species *Oryza rufipogon* under experimental field conditions. *New Phytologist* 157, 657–665.

Song, Z.P., Zhu, W.Y., Rong, J. and Lu, B.-R. (2006) Evidences of introgression from cultivated rice to *Oryza rufipogon* (Poaceae) populations based on SSR fingerprinting: implications for wild rice differentiation and conservation. *Evolutionary Ecology* 20, 501–522.

Stewart, C.N., Halfhill, M.D. and Warwick, S.I. (2003) Transgene introgression from genetically modified crops to their wild relatives. *Nature Reviews Genetics* 4, 806–817.

Tang, W., Newton, R.J., Lin, J.X. and Charles, T.M. (2006) Expression of a transcription factor from *Capsicum annuum* in pine calli counteracts the inhibitory effects of salt stress on adventitious shoot formation. *Molecular Genetics and Genomics* 276, 242–253.

Toyama, K., Bae, C.H., Kang, J.G., Lim, Y.P., Adachi, T., Riu, K.Z., Song, P.S. and Lee, H.Y. (2003) Production of herbicide-tolerant zoysiagrass by *Agrobacterium*-mediated transformation. *Molecules and Cells* 16, 19–27.

Vaughan, D.A. (1994) *The Wild Relatives of Rice: a genetic resources handbook*. International Rice Research Institute, IRRI, Los Baños, Philippines.

Wang, F., Yuan, Q.H., Shi, L., Qian, Q., Liu, W.G., Kuang, B.G., Zeng, D.L., Liao, Y.L., Cao, B. and Jia, S.R. (2006) A large-scale field study of transgene flow from cultivated rice (*Oryza sativa*) to common wild rice (*O. rufipogon*) and barnyard grass (*Echinochloa crusgalli*). *Plant Biotechnology Journal* 4, 667–676.

Wang, X.K. and Sun, C.Q. (eds) (1996) *The Origin and Differentiation of Chinese Cultivated Rice*. China Agricultural University Press, Beijing.

Wu, M.S. (ed.) (1990) *Collection of Research Articles on Wild Rice Germplasm*. Chinese Science Technical Press, Beijing.

28 European Crop Wild Relative Threat Assessment: Knowledge Gained and Lessons Learnt

S.P. Kell, N. Maxted and M. Bilz

28.1 Introduction

The IUCN Red List Categories and Criteria (IUCN, 2001) have been widely applied to assess the relative risk of extinction (or threatened status) of vascular plant species and the resulting Red List assessments have been published in the IUCN Red List of Threatened Species, as well as in national Red Lists. The 2010 IUCN Red List of Threatened Species includes 12,510 vascular plants, of which 8487 (68%) are classified as threatened, 1128 (9%) as Near Threatened, 1846 (15%) as Least Concern and the remainder as Data Deficient or Lower Risk/conservation dependent (IUCN, 2010). We do not know how many of these species are crop wild relatives (CWR), but an analysis of the 2004 IUCN Red List of Threatened Species revealed that of the CWR that occur in Europe and the Mediterranean, only 161 species were included and of these, only one (*Olea europaea* subsp. *cerasiformis*) is a wild relative of a major food crop and 16 are wild relatives of minor food crops – all of them being tree species (Kell *et al.*, 2008). Analysis of the 2006 IUCN Red List of Threatened Species showed that the overall number of wild relatives from the Euro-Mediterranean region had increased to 223 but that still only one wild relative of a major food crop was included and only 19 of minor food crops – all but one of these (*Allium rouyi*) being tree species.

One reason for the lack of CWR taxa included in the IUCN Red List of Threatened Species is that many of the plant taxa listed in the 1997 IUCN Red List of Threatened Plants have not yet been evaluated against the revised Red List Criteria (IUCN, 2001) – re-evaluation of the CWR included would be beneficial, as well as a thorough review of CWR included in national Red Lists (Kell *et al.*, 2008). The latter recommendation has been partially addressed by a recent initiative of Botanic Gardens Conservation International (BGCI), who developed a consolidated list of 1917 European threatened plant taxa based mainly on national Red Lists and species distribution data (Sharrock and Jones, 2009). Of the taxa included, 112 are CWR species found in 32 genera, including eight wild relatives of major food crops (*Brassica* and *Hordeum* spp.) and 50 wild relatives of minor food crops (*Allium*, *Avena*, *Beta*, *Brassica*, *Daucus*, *Fragaria*, *Lactuca*, *Pisum*, *Prunus*, *Pyrus* and *Vicia* spp.).

As noted by Heywood (2009), the lack of an up-to-date regional Red List not only means we do not know how many plants

 Agrobiodiversity Conservation: Securing the Diversity of Crop Wild Relatives and Landraces (eds N. Maxted *et al.*)

are threatened in Europe, but also that it has been a 'serious obstacle' to tackling some of the targets of the Global Strategy for Plant Conservation. A recent initiative of the International Union for the Conservation of Nature and Natural Resources (IUCN) and the European Commission set out to begin to redress this by undertaking regional Red List assessments of 2000 vascular plant species as a component of the first published European Red List (see http://ec.europa.eu/environment/nature/conservation/species/redlist/). Three plant groups were selected for inclusion in this initiative – CWR, aquatic plants and policy species (i.e. species listed in the annexes of the Habitats Directive, Bern Convention, CITES and the EU Wildlife Trade Regulation). As a contribution to the 2000 species to be assessed, regional Red List assessments of 591 CWR species were undertaken both for Europe and for the geographical area defined by the 27 EU member states.

This chapter summarizes the procedure used to select the CWR species for inclusion in the European Red List and the process and results of undertaking the regional assessments using the IUCN Red List Categories and Criteria (IUCN, 2001).

28.2 Selection of CWR Species for Assessment

Due of the large number of CWR species present in Europe, a clear process of target taxon selection was needed to maximize impact in terms of raising awareness about the importance of European CWR and their threatened status; therefore, wild relatives of a list of priority crops were selected based primarily on food and economic security in Europe. Species were selected from the CWR Catalogue for Europe and the Mediterranean (the CWR Catalogue) (Kell *et al.*, 2005), which contains taxon and distribution data from Euro+Med PlantBase (2006). At the time of production of the species list, the taxonomic and distribution data in Euro+Med PlantBase (www.emplantbase.org/home.html) had been revised for several families; including three of the largest families – Compositae, Poaceae and Rosaceae.[1] These revised data were combined with the 2006 dataset for the remaining families to form the basis for species selection, as well as the taxonomic standard for the CWR list. The taxon selection process (Kell *et al.*, in prep.) is outlined below in five steps.

28.2.1 Step 1: CWR native to Europe

The IUCN Red List Categories and Criteria should only be applied to wild populations inside their natural range, or to populations resulting from benign introductions[2] (IUCN, 2001); therefore, the first step in the target taxon selection procedure was to select CWR native to Europe. In the unrevised (2006) Euro+Med PlantBase dataset, each occurrence record is either recorded as 'Status Unknown' or if the status is known, a coding system is used in three fields – 'native', 'introduced' and 'cultivated'. For these records, taxon occurrences recorded in the 'native' field as 'native', 'assumed to be native' or 'doubtfully native' were selected as well as those recorded as 'formerly native (extinct)' (see Table 28.1).[3] In the revised (2009) dataset, a new field is used ('Summary Status') to record the status of each taxon occurrence. For these data, all occurrences recorded as 'native', 'native: doubtfully native', 'native: formerly native' and 'native: presence questionable' were selected. The list of CWR native to Europe contains 19,345 species; this includes CWR of agricultural and horticultural crops, forestry species, ornamentals, and medicinal and aromatic plants.

28.2.2 Step 2: CWR of human and animal food crops

Data from three primary sources were used to select a list of priority crop genera containing wild relatives native to Europe – the

Table 28.1 Codes for recording native status in Euro+Med PlantBase (Euro+Med PlantBase Secretariat, 2002).

Code	Value	Explanation
N	Native	The taxon is native (autochthonous) within the area concerned (as contrasted with 'introduced' and 'cultivated' defined below).
S	Assumed to be native	Assumed to be native to the area concerned.
D	Doubtfully native	There is doubt as to whether the status of the plant in the area concerned is native or not.
E	Formerly native (extinct)	The plant is native, doubtfully native or assumed to be native in the area concerned and has become extinct as such.
A	Not native	The plant is definitely not native.
F	Recorded as native in error	The plant has been recorded as native in the area concerned but all such records have been disproved or discounted.

CWR Catalogue for Europe and the Mediterranean (Kell *et al.*, 2005), GRIN Taxonomy for Plants (USDA, ARS, National Genetic Resources Program, 2009) and Mansfeld's World Database of Agricultural and Horticultural Crops (Hanelt and IPK Gatersleben, 2001; IPK Gatersleben, 2003). Genera containing cultivated taxa used for human and animal food were initially selected as priority groups. Within the human food crop group, cultivated taxa with the use types 'cereals', 'fruits', 'nuts', 'oil/fat', 'pseudocereals', 'pulses', 'seeds', 'starch', 'sugar' and 'vegetables' were selected.[4] This selection excludes beverage bases, gums/mucilages and any taxa identified as being of direct use potential (i.e. not as gene donors). The animal food crop group includes forage and fodder crops. This list contains 262 genera, within which there are 7324 CWR species native to Europe. Of these, 5955 are wild relatives of human food crops (found in 185 genera) and 2332 are wild relatives of forage crops (found in 146 genera); 955 species are wild relatives of both food and forage crops – these species are found in 62 genera. Although these species were selected on the basis of their potential as gene donors to human and animal food crops, some genera also include taxa cultivated for other purposes (e.g. medicinal, ornamental); therefore, the CWR assessed may have wider value as gene donors beyond food crops.

28.2.3 Step 3: CWR of high priority human food crops

The list of CWR of human and animal food crops encompasses a large number of species (7324) and it was therefore necessary to narrow down this list further by selecting the highest priority species. The first step was to select CWR of a number of human food crops that are particularly important to Europe in terms of production quantity and/or value. In terms of production quantity, there are 18 crops or crop groups of which Europe produced an average of >1 Mt in the 5 years from 2003 to 2007 that have CWR native to Europe that may be important for crop improvement: wheat, sugarbeet, barley, grapes, rapeseed, apples, oats, cabbages (and other brassicas), rye, olives, carrots and turnips, onions, peaches and nectarines, peas, lettuce and chicory, pears, plums and sloes, and strawberries (Fig. 28.1). Note that there are other economically important crops excluded from this list (e.g. potato) that have wild relatives in Europe, but they are very distant wild relatives – the centre of diversity of the potato gene pool being in South America – and are therefore not considered a priority in terms of their potential as gene donors for crop improvement. Figure 28.2 shows the average value of crops or crop groups produced in Europe over 5 years from 2004 to 2008 that have CWR native to Europe which may be important

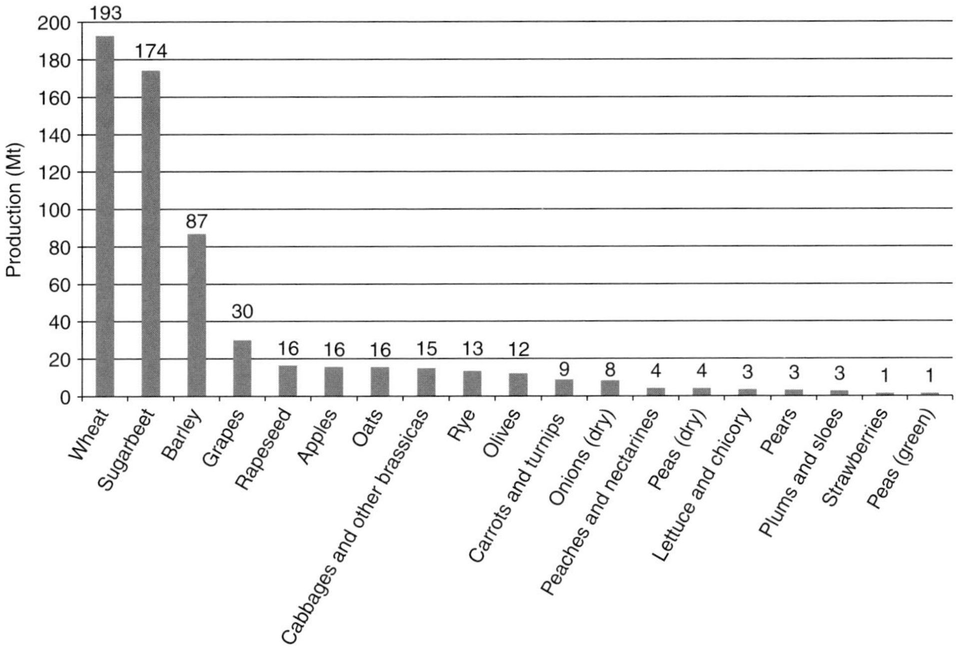

Fig. 28.1. Crops/crop groups of which Europe produced an average of >1 Mt in five years from 2003 to 2007 that have CWR native to Europe which may be important for crop improvement (Kell *et al.*, in prep). Data source: FAOSTAT (FAO, 2009).

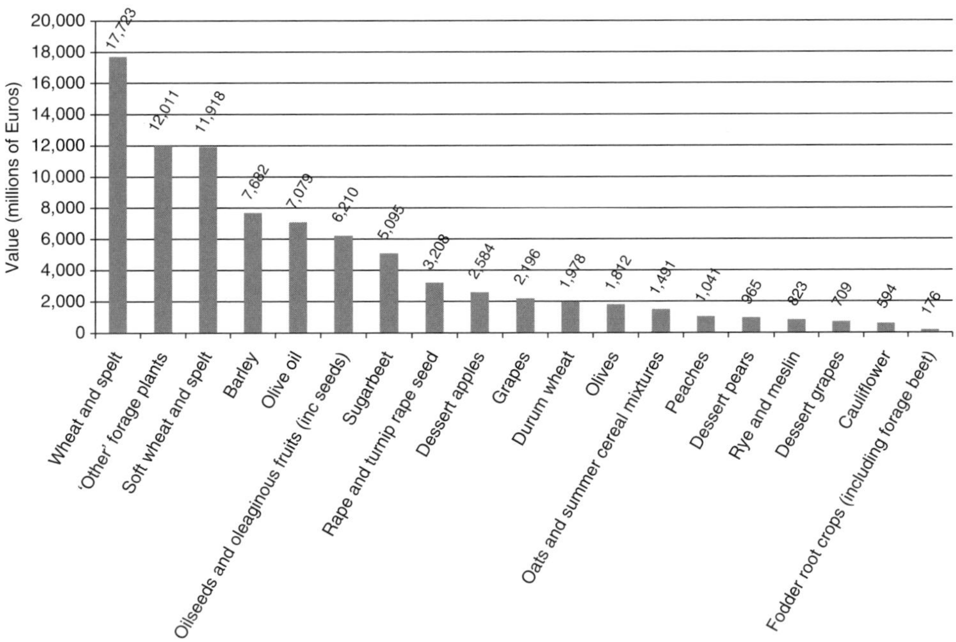

Fig. 28.2. The average value (millions of Euros) of crops/crop groups produced in Europe over 5 years from 2004 to 2008 that have CWR native to Europe which may be important for crop improvement (Kell *et al.*, in prep). Data source: Eurostat (European Communities, 1995–2009).

for crop improvement. All of the crops or crop groups included in this analysis are also included in the priority list of human food crops based on production quantity.

This preliminary selection of 18 crops or crop groups are found within 19 genera within which there are 279 recorded as native to Europe. All the species within this group were included in the European Red List due to their high potential economic importance as gene donors to human food crops.

28.2.4 Step 4: CWR of animal food crops

The production quantity and economic value data that are available for human food crops are not readily available for animal food crops on an individual crop basis; therefore, it is not possible to prioritize animal food crops according to these criteria. However, of the 279 CWR species identified in the high priority human food CWR group, 106 are wild relatives of forage and/or fodder crops, as well as human food crops; therefore, CWR of a number of animal food crops are included in this list.

28.2.5 Step 5: CWR of other human and animal food crops

To add to the high priority list of 279 species described above, Annex I of the International Treaty on Plant Genetic Resources for Food and Agriculture (ITPGRFA) was also used as the basis for species selection. This is a list of Plant Genetic Resources for Food and Agriculture (PGRFA) established according to criteria of food security and interdependence and includes 78 genera containing human or animal food crops. Fifty-nine of these genera contain taxa that are native to Europe, some of which are already included in the high priority CWR list defined above.

Annex I of the ITPGRFA is divided into two lists: (i) human food crops; and (ii) forages. The human food crop list mainly lists entire genera because the CWR of these crops are recognized as being important for food security. The forage list only includes specific species because (i) the crops are mainly selections from populations of wild species and the CWR are less likely to be used for crop improvement; and (ii) many of the forage genera contain a very large number of species; for example, *Festuca* contains 204 species native to Europe.

Additional human food crop genera listed in Annex I of the ITPGRFA that were included in the list of CWR to be assessed are: *Asparagus* – 21 species, brassica complex (*Armoracia*, *Barbarea*, *Camelina*, *Crambe*, *Diplotaxis*, *Eruca*, *Isatis*, *Lepidium*, *Raphanus*, *Rorippa* and *Sinapis*)[5] – 121 species, *Cicer* – four species, *Lathyrus* – 18 species (only those in Gene Pools (GP) 1b and 2 and Taxon Groups (TG) 1b and 2 – see Maxted *et al.*, 2006), *Lens* – five species, wheat complex (*Agropyron* and *Elymus*)[6] – 17 species, and *Vicia* – 20 species (GP1b, TG1b, GP2, TG2 and four species for which data were readily available). *Lathyrus* and *Vicia* species were limited to the close wild relatives only, due to the large number of species included in these genera.

Fifty-two of the forage species listed in Annex I of the ITPGRFA are native to Europe. These were all included for assessment as their continued existence in the wild is important for the future of these crops; thus, knowing their conservation status in the wild is important to inform conservation planning. In addition, all *Medicago* species native to Europe were included on the basis of data availability.

This selection resulted in a list of 596 species; however, some of these were removed as they are hybrids which are generally not included in the IUCN Red List. Later in the project, some additional species were added by experts at a European CWR Red List workshop; these included five species in the genus *Sinapidendron*, which is related to brassica crops and endemic to the Madeira archipelago, and some recently described species of *Crambe* endemic to the Canary Islands. The final list of CWR species for assessment comprised 591 species in 25 crop gene pools/ groups (Kell *et al.*, in prep.) (Table 28.2),

Table 28.2. Overview of the list of CWR species selected for inclusion in the European Red List.

Crop gene pool/group	Genus (or genera)	Total no. of species in gene pool/group[a]	No. of species assessed[b] (% of gene pool/group)
Brassica complex	*Armoracia, Barbarea, Brassica, Camelina, Crambe, Diplotaxis, Eruca, Isatis, Lepidium, Raphanus, Rorippa, Sinapidendron, Sinapis*	506	142 (28%)
Onion, leek, garlic etc.	*Allium*	750	118 (16%)
Legume forages	*Astragalus, Hedysarum, Lotus, Lupinus, Medicago, Melilotus, Onobrychis, Ornithopus, Securigera, Trifolium*	3469	93 (3%)
Wheat	*Aegilops, Agropyron, Elymus, Triticum*	213	36 (17%)
Lettuce	*Lactuca*	130	27 (21%)
Faba bean/vetch	*Vicia*	160	23 (14%)
Asparagus	*Asparagus*	120	19 (16%)
Grass pea	*Lathyrus*	160	19 (12%)
Stone fruits and almond	*Prunus*	200	16 (8%)
Grass forages	*Agrostis, Alopecurus, Arrhenatherum, Festuca, Lolium, Phalaris, Phleum, Poa*	1210	14 (1%)
Oat	*Avena*	25	13 (52%)
Carrot	*Daucus*	22	12 (55%)
Pear	*Pyrus*	15	11 (73%)
Cultivated beets	*Beta, Patellifolia*	13	10 (77%)
Barley	*Hordeum*	32	8 (25%)
Lentil	*Lens*	5	5 (100%)
Apple	*Malus*	40	5 (13%)
Chickpea	*Cicer*	44	4 (9%)
Chicory	*Cichorium*	6	3 (50%)
Strawberry	*Fragaria*	330	3 (1%)
Rye	*Secale*	3	3 (100%)
Other forages	*Atriplex, Salsola*	380	2 (1%)
Garden pea	*Pisum*	3	2 (67%)
Olive	*Olea*	33	2 (6%)
Grape	*Vitis*	65	1 (2%)
Total		7933	591 (7%)

[a]Data primarily sourced from Mabberley (2008).
[b]Including species assessed as Not Applicable.

188 of which are endemic to Europe. Although it is possible to apply the IUCN Red List Categories and Criteria at subspecific level, all assessments were undertaken at species level as stipulated by the contractual arrangements of the project.

28.3 The Red List Assessment Process

Assessment of the threatened status of species using the IUCN Red List Categories and Criteria (IUCN, 2001) is essentially a two-step process:

1. Data of seven types are collated and documented: (i) taxonomic; (ii) distribution; (iii) population; (iv) habitat and ecology; (v) use and trade; (vi) threats; and (vii) conservation actions (Box 28.1). These data are gathered from a number of sources, including taxon experts, published and grey literature, databases and websites.
2. The taxon is evaluated against the IUCN Red List Criteria and the Red List Category is selected.

There are five main Red List Criteria: (A) population reduction, (B) geographic range, (C) small population size and decline, (D) very small or restricted population and (E) quantitative analysis indicating the probability of extinction. Each main criterion includes a number of sub-criteria against which the species is evaluated. If the species meets the criteria in at least one of the main classes, it is assigned one of the threatened categories, Critically Endangered (CR), Endangered (EN) or Vulnerable (VU). If the species meets the criteria in more than one main class, it is assigned the highest category of threat but the less threatened category according to the other criterion or criteria is also documented. If the species does not meet any of the criteria A–E needed to evaluate it as threatened, another category is selected; these are Extinct (EX), Extinct in the Wild (EW), Regionally Extinct (RE), Near Threatened (NT), Least Concern (LC), Data Deficient (DD) or Not Applicable (NA) (for definitions of the categories, see IUCN, 2001).

The two criteria primarily used to assess the European CWR species as threatened were B (geographic range) and D (very small or restricted population) because detailed population data were rarely available to apply Criterion A or C, and quantitative data did not exist to apply Criterion E. The majority of threatened species were assessed under Criterion B which is based on the extent of occurrence (EOO) and/or area of occupancy (AOO) of the species (see IUCN, 2001 for definitions). To assess a species as threatened using Criterion B, two of three sub-criteria must be met:

1. The population is severely fragmented and/or it occurs in a small number of highly threatened locations.
2. There is a continuing decline in the EOO, AOO, area, extent and/or quality of habitat, number of locations or subpopulations, or number of mature individuals.
3. There are extreme fluctuations in the EOO, AOO, number of locations or subpopulations, or number of mature individuals.

Sub-criteria 1 and 2 were used most frequently and under sub-criterion 2, many assessments were based on a decline in the area, extent and/or quality of habitat (sub-criterion 2(iii)) because it is often possible to infer that a species' habitat is declining due to the reported threats, even if an assessor does not have first-hand experience from visiting the sites.

Following data collation, application of criteria and selection of the Red List Category, the rationale to justify the assessment is documented, highlighting the key issues to support the assessment and explaining why the species qualifies for the assigned category, and finally, the assessor and contributor names are added. Each assessment is edited to data consistency standards and then reviewed and evaluated by at least two experts – the names of the reviewers are also published with the assessment.

The data and the selected category are entered into IUCN's Species Information Service (SIS – an online information management system for documenting species accounts and Red List assessments). All data sources are fully referenced and when the assessments are published, they present an account that summarizes the taxonomy and nomenclature of the species, where it occurs, what the trends are in population size, what the species' habitat and ecological requirements are, what threats it faces, the Red List status, and current or needed conservation actions.

The assessment process involved the collaboration of more than 70 experts who have good knowledge of the national flora of their country and/or of a particular taxonomic group. A key part in the process was

Box 28.1 Summary of data types collated to undertake the European Red List assessments

Taxonomy

- Nomenclature (taxon name, authority, synonyms etc.).
- Recent taxonomic changes, any current taxonomic doubts or debates about the validity or identity of the species, or issues of synonymy.
- A note of any subspecific taxa.
- Crop(s) the species is related to (common and scientific names) and information on the degree of relationship of the wild relative to the crop (where known) using the Gene Pool concept (Harlan and de Wet, 1971) or Taxon Group concept (Maxted *et al.*, 2006).

Distribution and occurrence

- A summary of the current information available for the geographic range of the species within Europe. If the species has part of its natural distribution range outside Europe, a brief note of its global distribution is also included.
- Country occurrences (and sub-national unit(s) where applicable) recorded using built-in descriptors in SIS. Only occurrences within the species' native European range were recorded, or cases where the origin or presence is uncertain.
- A map showing the distribution of the species.

Population

- A summary of the information available for size and trend (i.e. increasing, decreasing or stable) of the European population of the species. If the population is severely fragmented, this was also recorded.
- Information about sizes and trends of subpopulations or populations of subspecific taxa, or trends in particular areas of the species' European range were also included when available.
- Where no quantitative information on population sizes or trends were available, if possible it was noted whether the species is common, abundant, or rare, etc. If there really was no information at all about the population, this was noted.

Habitats and ecology

- A summary account of the suitable habitats and ecological requirements of the species, highlighting any potential traits that may be of interest for crop improvement (e.g. drought resistance, salt tolerance).
- Comments on the area, extent and/or quality of habitat; in particular, whether the habitat is thought to be stable or declining.
- The habitat(s) in which the species occurs are also documented using IUCN's Habitats Classification Scheme.

Use and trade

- A summary account of the information available for any utilization and/or trade of the taxon (local, national and international trade).
- A note of any known or potential uses of the species as a gene donor for crop improvement.

Threats

- Major threats that have affected the species in the past, those that are affecting the species now, or those that are likely to affect the species in the future.
- The main reason for the threat, the scale of the threat, and the stress this places on the species are also recorded where the information is available.
- Threats are also documented using IUCN's Threats Classification Scheme.

Conservation

- Conservation actions currently in place (if any) and realistic actions needed to mitigate the threats causing declines (if any). This includes information on both *in situ* and *ex situ* conservation measures.
- Conservation actions are also documented using IUCN's Conservation Actions Classification Scheme.

a 5-day Red List workshop involving 26 experts and a team of facilitators, during which many of the assessments were drafted. The remaining work was undertaken through email correspondence and completion and editing of the assessments was undertaken mainly by three members of staff of the coordinating institutes.

28.4 The Threatened Status of European CWR

Out of the 591 CWR species for which regional assessments were carried out, 19 were assessed as Not Applicable (NA)[7] as they were either considered by experts not to be native to Europe (i.e. they were introduced after AD 1500) or only had a marginal distribution in the region. One species, *Allium jubatum*, which is native to Asiatic Turkey and Bulgaria, was assessed as RE – according to Mathew (1996), it has not been found in Bulgaria since its original collection in 1844. Of the remaining 571 species assessed, 313 (55%) were assessed as LC, 166 (29%) as DD, 26 (5%) as NT, 22 (4%) as VU, 25 (4%) as EN and 19 (3%) as CR (Fig. 28.3).

28.4.1 Threatened and Near Threatened species

Figure 28.4 shows that of the 25 crop gene pools/groups for which the European CWR were assessed, at least 14 contain regionally and/or globally threatened (CR, EN or VU) or Near Threatened (NT)[8] species (92 species in total, of which 65 are endemic to Europe), the highest number occurring in the brassica complex which in total contains 137 species native to and with a significant proportion of the global population in Europe. At least 8–50% of the species assessed in each of these crop gene pools/groups are threatened or NT (Fig. 28.5) and these percentages are likely to increase when the Data Deficient species are re-evaluated. Note that none of the crop gene pools/groups are endemic to Europe; therefore, this is not a comparison of the threatened status between entire crop gene pools/

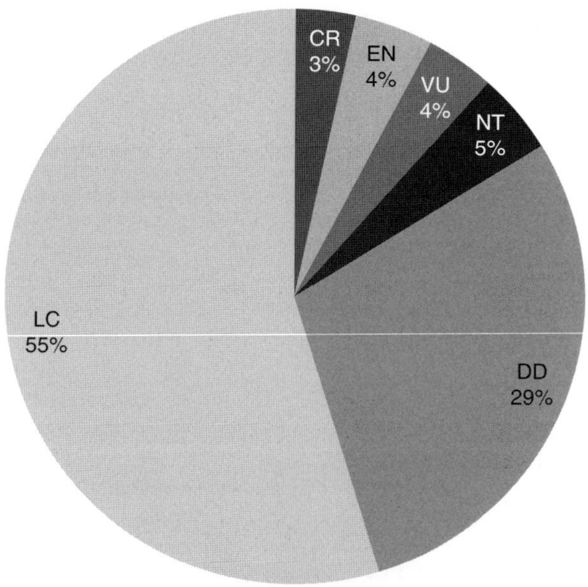

Fig. 28.3. IUCN Red List categories assigned to 571 European CWR (regional assessments). (LC=Least Concern; DD=Data Deficient; NT=Near Threatened; VU=Vulnerable; EN=Endangered; CR=Critically Endangered.)

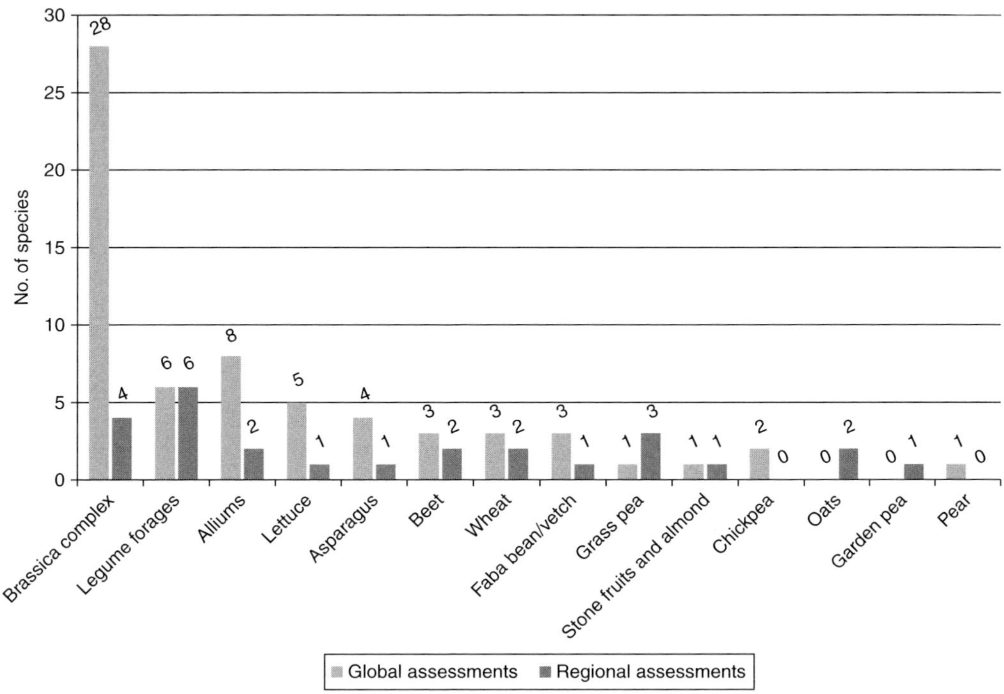

Fig. 28.4. The numbers of globally and regionally threatened (CR, EN or VU) or Near Threatened (NT) species, out of the sample assessed in 14 crop gene pools/groups.

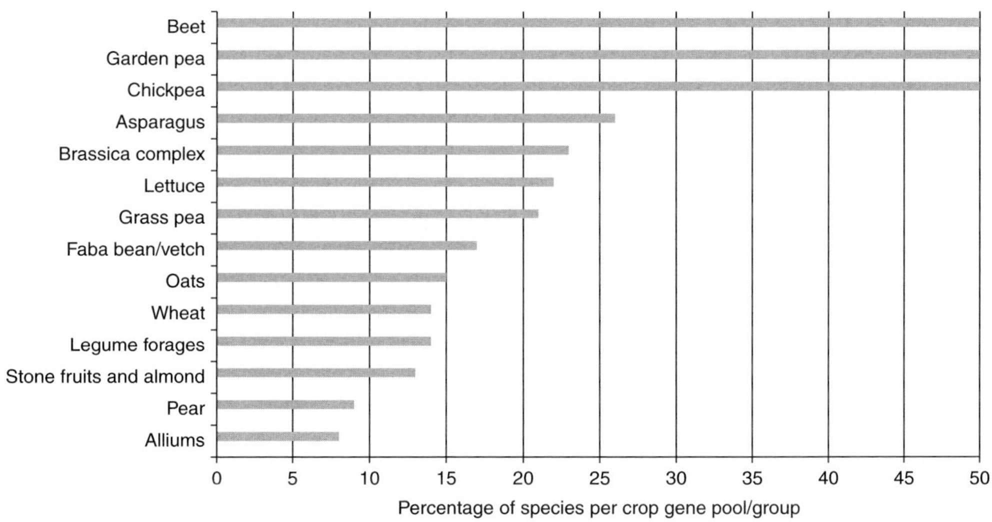

Fig. 28.5. The percentages of globally and regionally threatened (CR, EN or VU) or Near Threatened (NT) species, out of the sample assessed in 14 crop gene pools/groups, excluding species assessed as Not Applicable.

groups because only species that are native to Europe were assessed and the species that are not endemic to Europe were regionally (not globally) assessed. Further, not all species native to Europe were assessed in each genus – for the legume forages, only species listed in Annex I of the ITPGRFA were assessed and due to the large numbers of species in *Vicia* and *Lathyrus*, only species in GP1b, TG1b, GP2, TG2 (i.e. the closest wild relatives) were assessed. An additional five species of *Vicia* were also assessed for which data were readily available. However, these results provide an indication of the crop gene pools or complexes that are under greatest threat of extinction in Europe.

It is particularly notable that half of the species assessed in the beet gene pool (five species) are threatened – three globally (*Beta patula* and *Patellifolia webbiana* (CR) and *B. nana* (VU)) and two regionally (*B. adanensis* (VU) and *B. macrocarpa* (EN)). The centre of diversity of the beet gene pool is in Europe, with 10 out of the 13 species native to Europe (two of which are single country endemics); therefore, we know that at least 30% of the gene pool (in terms of taxonomic diversity) is threatened with extinction. Beet is a highly important crop for the European economy; the wild relatives have already been used extensively for crop improvement and further genetic diversity may be needed from the wild populations in the future. Therefore, it is clear that urgent attention needs to be paid to the conservation of these species. The brassica complex is also of particular concern as 27% (137) of the species are native to Europe and more than 18% (25) of these are threatened (24 globally), with a further 5% (7) considered to be Near Threatened. The threatened status of the lettuce, wheat and allium gene pools are also of considerable concern because, like beet and brassica crops, these are also highly economically important crops in Europe which have a relatively large proportion of their gene pools native to the region.

We cannot assume that the percentage of threatened species in a gene pool is equivalent to the percentage of threatened genetic diversity; however, in the absence of genetic data to prove otherwise, we have to take the precautionary approach and assume that in percentage terms, the risk of extinction to genetic diversity at least equates to the risk of extinction to taxonomic diversity. In fact, Maxted *et al.* (1997a) and Maxted (2003) pointed out that while it is difficult, if not impossible, to quantify the loss of genetic diversity within CWR species, it must be faster than the loss of species, because there will be some genetic erosion (loss of genetic diversity) from the species that remain extant and complete loss of genetic diversity from those that become extinct, given that both extant and extinct species face the same threats. Therefore, if we assume that genetic diversity is strongly correlated with occurrences of species at particular localities and that some of those occurrences are threatened, then we may validly infer that the percentage of threatened species in a gene pool could signify a greater level of threat to overall genetic diversity in the gene pool than to taxonomic diversity.

Table 28.3 shows the countries with the highest to lowest numbers of regionally and globally threatened or NT species. As would be expected, the highest numbers of species are found in the countries of southern and eastern Europe which are known to have large floras and thus a large number of CWR species. It is notable that many of the threatened and NT species are endemic to the Canary Islands and to the Madeira and Azores archipelagos, as well as to Sicily – this is of course no surprise, since not only do these islands have a high degree of endemism, but many island habitats are highly degraded, fragmented and fragile (Kell *et al.*, 2008).

28.4.2 Least Concern species

It is striking that more than half of the species assessed were evaluated as LC. However, this statistic should be interpreted with great caution as a LC assessment does not necessarily mean that a species or subpopulations of that species do not

Table 28.3. European countries containing regionally and globally threatened or Near Threatened CWR species (out of 591 species assessed).

Country	No. of species	No. endemic to Europe (national endemics)	National endemic species (Red List category)
Spain (including the Balearic and Canary Islands)	33	27 (24)	*Allium melananthum* (NT), *A. pardoi* (VU), *A. pyrenaicum* (VU), *Asparagus arborescens* (VU), *A. fallax* (EN), *A. plocamoides* (VU), *Cicer canariense* (EN), *Crambe arborea* (VU), *C. feuillei* (CR), *C. gomerae* (VU), *C. laevigata* (EN), *C. microcarpa* (EN), *C. pritzelii* (EN), *C. scaberrima* (VU), *C. scoparia* (EN), *C. sventenii* (CR), *C. tamadabensis* (CR), *C. wildpretii* (CR), *Diplotaxis siettiana* (CR), *Lactuca singularis* (VU), *Medicago citrina* (CR), *Patellifolia webbiana* (CR), *Prunus ramburii* (VU), *Rorippa valdes-bermejoi* (CR)
Portugal (including the Azores and Madeira archipelagos)	19	15 (12)	*Beta patula* (CR), *Crambe fruticosa* (NT), *Diplotaxis vicentina* (CR), *Lactuca watsoniana* (EN), *Sinapidendron angustifolium* (CR), *S. frutescens* (EN), *S. gymnocalyx* (NT), *S. rupestre* (CR), *S. sempervivifolium* (EN), *Vicia capreolata* (EN), *V. costae* (CR), *V. ferreirensis* (CR)
Ukraine (including Crimea)	17	7 (5)	*Agropyron cimmericum* (EN), *A. dasyanthum* (EN), *Allium pervestitum* (EN), *Lepidium turczaninowii* (CR), *Medicago saxatilis* (EN)
Greece	12	5 (5)	*Beta nana* (VU), *Cicer graecum* (EN), *Lactuca alpestris* (NT), *Medicago heyniana* (NT), *M. strasseri* (NT)
Italy (including Sardinia and Sicily)	11	6 (5)	*Brassica glabrescens* (VU), *B. macrocarpa* (CR), *B. rupestris* (NT), *B. villosa* (NT), *Lathyrus odoratus* (NT)
Cyprus	11	4 (4)	*Allium exaltatum* (VU), *Brassica hilarionis* (EN), *Lactuca cyprica* (NT), *L. tetrantha* (VU)
France (including Corsica)	6	1 (1)	*Allium corsicum* (CR)
Russian Federation (European part)	5	1 (0)	–
Malta	3	1 (1)	*Allium lojaconoi* (NT)
Serbia	3	1 (0)	–
Montenegro	2	0	–
Romania	2	1 (0)	–
Slovenia	2	1 (0)	–
Turkey (European part)	2	0	–
Croatia	2	1 (0)	–
Bulgaria	2	0	–
The former Yugoslav Republic of Macedonia	2	0	–
Moldova	1	1 (0)	–
Albania	1	0	–
Hungary	1	1 (1)	*Pyrus magyarica* (CR)
Germany	1	0	–
Bosnia and Herzegovina	1	0	–

warrant conservation action. In interpreting this result, there are two important issues that need to be taken into account – the first relates to the application of the Red List Criteria and potential subjectivity of the process, and the second relates to the issue of taxonomic versus genetic diversity assessment, as explained below.

The criteria for assessing a species as threatened (i.e. CR, EN or VU) are rigorous; therefore, when these criteria are not met, an assessor has the choice of assessing the species as NT, LC or DD. According to IUCN guidelines, DD assessments should be avoided when possible; therefore, the assessor is forced to lean towards either a NT or LC assessment. Strong justification is needed to assess a species as NT and where insufficient knowledge has resulted in the threatened criteria not being fulfilled it is also highly likely that an NT assessment could not be justified on the basis of a lack of sufficient knowledge. Therefore, the assessor must either decide to evaluate the species as DD or LC. It is often difficult to make a judgement as to whether there really is insufficient knowledge and the species should be assessed as DD or whether it is in fact an LC species. This decision can be highly subjective depending on the views and attitude of the individual undertaking the assessment – some may be more inclined to take a precautionary approach than others.

Many of the species assessed as LC are relatively widespread in Europe, occurring in several countries; however, some have a relatively narrow distribution and are assessed as LC because despite their restricted range, they do not meet the threatened criteria. The latter group of species is likely to mainly comprise national endemics and may already be included in national conservation plans. However, if they are not already adequately conserved, both *in situ* and *ex situ*, the LC assessment should be carefully interpreted as it does not necessarily mean that the species is not in need of conservation action – at minimum, population monitoring is likely to be needed.

We should also be very careful about interpreting an LC assessment for those species that are relatively widely distributed in Europe – it could be assumed that these species are secure and require no conservation action; however, there are two strong counter arguments. One is that although it is possible to apply the Red List Categories and Criteria (IUCN, 2001) to individual subpopulations,[9] the system does not include genetic diversity within and between subpopulations as a criterion for assessment – it is based on population size and geographic range. As the goal of CWR conservation is to maximize the conservation of genetic diversity, it is vital that sufficient subpopulations are conserved, both *in situ* and *ex situ*, to provide an adequate sample of total genetic diversity. Genetic diversity knowledge is lacking for the majority of species as sampling and analysis is resource intensive; therefore, it is necessary to ensure that as wide a range of ecogeographic diversity is sampled and conserved as possible – ecogeographic diversity being used as a proxy for genetic diversity (see Kell *et al.*, Chapter 2, this volume). This means that conservation of even the most widespread species should be of concern, both at regional and national levels. The second counterargument is that many of the species regionally assessed (or globally assessed if endemic to Europe) as LC are threatened at national level – further analysis is needed to ascertain exactly how many, but based on information documented during the Red Listing process, we estimate that it could be a third or more. Therefore, for the same reason outlined above, these species should be of conservation concern, not only nationally, but also regionally, in order to ensure that the maximum intraspecific genetic diversity is conserved throughout the species' range.

28.4.3 Data Deficient species

The relatively high percentage of species assessed as DD is attributable to two main factors: (i) insufficient knowledge of the species to apply the Red List Criteria; and (ii) resource and time limitations resulting in gaps in data collection and/or application of the criteria. In many cases, knowledge of the species' distribution was available, but there was little, if any information about the population size, structure or trend. General knowledge about the habitats of the species, where known, could often be used to make inferences about threats to the species, but this is not enough to make a reasoned judgement about the threatened status of a species. It is clear that more work needs to be done to improve our knowledge of the threatened and conservation status of these species.

28.5 Threats to European CWR and Population Trends

For 49% (279) of the species assessed, 31 threats were reported, the most frequent being 'livestock farming and ranching', 'tourism and recreation areas' and 'housing and urban areas' (Fig. 28.6). The IUCN threat descriptor, 'livestock farming and ranching' includes the subordinate descriptors 'agro-industry grazing, ranching or farming', 'small-holder grazing, ranching or farming', 'nomadic grazing' and 'scale unknown/unrecorded'. It is important to note that due to the imprecise nature of these descriptors, the significance of the frequency at which this threat was reported should be interpreted with care. It would be erroneous to conclude that farming *per se* is a threat to CWR diversity; in fact, farmed areas (including arable land and pasture) are one of the primary habitats of CWR species. It is unsustainable farming practices, such as severe overgrazing, conversion of land to monocultures and the heavy application of fertilizers, herbicides and pesticides that are the major threats to CWR that grow in agricultural areas. This includes grazing in semi-natural habitats such as Mediterranean maquis.

Of the species assessed, 26% are reported to have no known past or ongoing

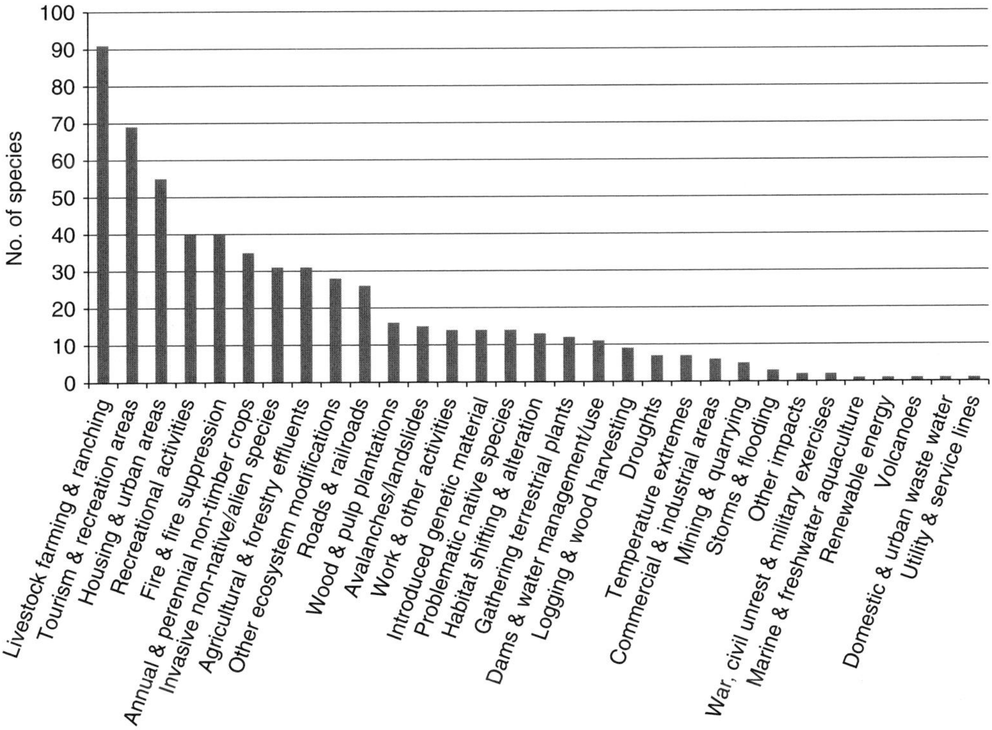

Fig. 28.6. Threats reported to affect 279 of the European CWR species assessed, showing the number of species for which each one was recorded.

threats and are not expected to face any major threats in the foreseeable future. The majority of these were assessed as LC, with only six assessed as DD. For 25% of the species assessed, the threats are unknown – the majority of these species were assessed as DD, but 32 were evaluated as LC.

The majority of threats (nearly 68%) reported were recorded as 'ongoing', but for 27% of the time the threat trend was recorded as 'unknown'. For 15 species, threats were reported as being likely to have an impact in the future, while for the same number of species, the threats were reported to have had an impact in the past and were either thought likely to return (nine species) or unlikely to return (six species).

Figure 28.7 shows the 22 most common habitats in which the species occur. In addition to these, there were another 35 habitat types recorded for nine species or less. The habitat was recorded as unknown for 45 species and 'other' types of habitat were recorded for 19 species. It is difficult to make inferences about threats to particular habitats because threats and habitat types are linked to a species (i.e. the entire European range of the species) and are therefore not directly related. However, it is possible to make some assumptions about the most threatened habitats by looking at the number of threats impacting species found in those habitats (Table 28.4). These data should be interpreted with caution as the greater number of threats shown against the habitat types may be partly attributable to the larger number of species recorded in those habitats.

For 221 of the 571 species assessed, the population trend was reported to be 'stable', for 62 species 'decreasing' and for 13 species 'increasing' – the population trend for 275 species was recorded as 'unknown' (Fig. 28.8). Of the 92 species assessed as threatened or NT, 48 are reported to have a decreasing population trend and 21 are thought to be stable – for 23 of these species, the population trend is unknown (Table 28.5). It is clear that the 48 species assessed as threatened or NT with a decreasing population trend should be flagged up as an urgent priority for conservation action – particularly those endemic to Europe. Those with unknown population trends should have monitoring programmes put in place immediately and the species reported to be stable should also be closely monitored to ensure that potential changes in the trend can be reported.

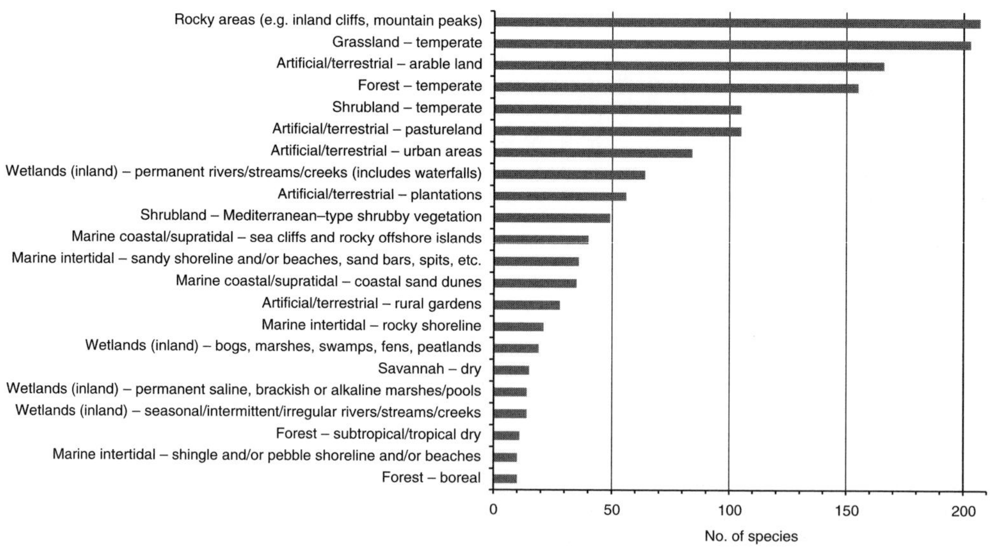

Fig. 28.7. Habitat types recorded for ten species or more.

Table 28.4. The number of threats impacting CWR species found in 45 habitat types recorded for 521 species.

Habitat type	No. of threats
Rocky areas (eg. inland cliffs, mountain peaks)	24
Grassland – Temperate	24
Shrubland – Temperate	23
Wetlands (inland) – Permanent Rivers/Streams/Creeks (includes waterfalls)	22
Forest – Temperate	22
Shrubland – Mediterranean-type Shrubby Vegetation	21
Artificial/Terrestrial – Arable Land	20
Artificial/Terrestrial – Urban Areas	19
Artificial/Terrestrial – Pastureland	19
Marine Coastal/Supratidal – Coastal Sand Dunes	18
Marine Coastal/Supratidal – Sea Cliffs and Rocky Offshore Islands	18
Forest – Subtropical/Tropical Dry	16
Artificial/Terrestrial – Plantations	16
Wetlands (inland) – Permanent Saline, Brackish or Alkaline Marshes/Pools	12
Marine Intertidal – Rocky Shoreline	12
Marine Intertidal – Sandy Shoreline and/or Beaches, Sand Bars, Spits, etc.	11
Forest – Subtropical/Tropical Moist Montane	11
Wetlands (inland) – Bogs, Marshes, Swamps, Fens, Peatlands	11
Shrubland – Subtropical/Tropical Dry	10
Artificial/Terrestrial – Rural Gardens	10
Wetlands (inland) – Seasonal/Intermittent/Irregular Rivers/Streams/Creeks	9
Forest – Boreal	9
Marine Intertidal – Salt Marshes (Emergent Grasses)	9
Shrubland – Boreal	9
Wetlands (inland) – Permanent Freshwater Lakes (over 8 ha)	8
Wetlands (inland) – Seasonal/Intermittent Freshwater Marshes/Pools (under 8 ha)	7
Savanna – Dry	6
Marine Intertidal – Shingle and/or Pebble Shoreline and/or Beaches	6
Marine Intertidal – Mud Flats and Salt Flats	6
Wetlands (inland) – Permanent Freshwater Marshes/Pools (under 8 ha)	5
Marine Coastal/Supratidal – Coastal Brackish/Saline Lagoons/Marine Lakes	5
Wetlands (inland) – Permanent Saline, Brackish or Alkaline Lakes	4
Desert – Temperate	4
Wetlands (inland) – Alpine Wetlands (includes temporary waters from snowmelt)	3
Artificial/Aquatic – Canals and Drainage Channels, Ditches	3
Grassland – Subtropical/Tropical Dry	3
Grassland – Subtropical/Tropical High Altitude	2
Wetlands (inland) – Seasonal/Intermittent Freshwater Lakes (over 8 ha)	2
Shrubland – Subtropical/Tropical High Altitude	2
Wetlands (inland) – Seasonal/Intermittent Saline, Brackish or Alkaline Marshes/Pools	2
Artificial/Aquatic – Ponds (below 8 ha)	2
Wetlands (inland) – Freshwater Springs and Oases	1
Wetlands (inland) – Shrub Dominated Wetlands	1
Artificial/Aquatic – Seasonally Flooded Agricultural Land	1
Artificial/Aquatic – Irrigated Land (includes irrigation channels)	1

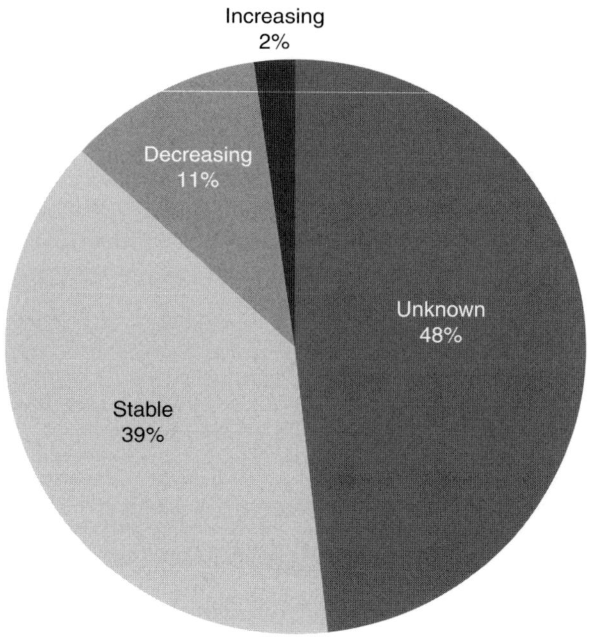

Fig. 28.8. Population trends recorded for 571 species.

Table 28.5. Globally and regionally threatened or Near Threatened CWR species (out of 591 species assessed) and population trends; species endemic to Europe are shown in bold.

Species	Red List status	Population trend[a]
Allium corsicum	CR	↓
Beta patula	CR	?
Brassica macrocarpa	CR	→
Crambe feuillei	CR	↓
Crambe sventenii	CR	↓
Crambe tamadabensis	CR	→
Crambe wildpretii	CR	↓
Diplotaxis siettiana	CR	?
Diplotaxis vicentina	CR	?
Lepidium turczaninowii	CR	↓
Medicago citrina	CR	↓
Medicago fischeriana	CR	↓
Patellifolia webbiana	CR	?
Pyrus magyarica	CR	→
Rorippa valdes-bermejoi	CR	↓
Sinapidendron angustifolium	CR	?
Sinapidendron rupestre	CR	→
Vicia costae	CR	↓
Vicia ferreirensis	CR	↓
Aegilops tauschii	EN	?
Agropyron cimmericum	EN	↓
Agropyron dasyanthum	EN	?
Allium pervestitum	EN	?
Asparagus fallax	EN	↓

Continued

Table 28.5. Continued.

Species	Red List status	Population trend[a]
Asparagus nesiotes	EN	↓
Avena insularis	EN	?
Avena murphyi	EN	↓
Barbarea lepuznica	EN	↓
Beta macrocarpa	EN	↓
Brassica hilarionis	EN	↓
Cicer canariense	EN	↓
Cicer graecum	EN	↓
Crambe laevigata	EN	↓
Crambe microcarpa	EN	↓
Crambe pritzelii	EN	↓
Crambe scoparia	EN	?
Lactuca watsoniana	EN	↓
Lathyrus cassius	EN	↓
Medicago cretacea	EN	↓
Medicago rupestris	EN	↓
Medicago saxatilis	EN	↓
Sinapidendron frutescens	EN	?
Sinapidendron sempervivifolium	EN	?
Vicia capreolata	EN	↓
Aegilops bicornis	VU	↓
Allium exaltatum	VU	↓
Allium pardoi	VU	↓
Allium pyrenaicum	VU	→
Allium schmitzii	VU	↓
Asparagus arborescens	VU	↓
Asparagus pastorianus	VU	↓
Asparagus plocamoides	VU	↓
Beta adanensis	VU	↓
Beta nana	VU	→
Brassica glabrescens	VU	↓
Crambe arborea	VU	→
Crambe aspera	VU	↓
Crambe gomerae	VU	→
Crambe scaberrima	VU	→
Isatis platyloba	VU	↓
Lactuca singularis	VU	↓
Lactuca tetrantha	VU	↓
Medicago glandulosa	VU	↓
Medicago kotovii	VU	↓
Prunus lusitanica	VU	↓
Prunus ramburii	VU	↓
Agropyron tanaiticum	NT	?
Allium albiflorum	NT	?
Allium convallarioides	NT	?
Allium lojaconoi	NT	→
Allium melananthum	NT	?
Brassica insularis	NT	→
Brassica rupestris	NT	→
Brassica villosa	NT	→
Crambe fruticosa	NT	→
Diplotaxis siifolia	NT	?

Continued

Table 28.5. Continued.

Species	Red List status	Population trend[a]
Lactuca alpestris	NT	?
Lactuca cyprica	NT	→
Lactuca triquetra	NT	→
Lathyrus amphicarpos	NT	↓
Lathyrus odoratus	NT	↓
Lathyrus rotundifolius	NT	→
Medicago cancellata	NT	↓
Medicago heyniana	NT	?
Medicago hypogaea	NT	→
Medicago pironae	NT	?
Medicago strasseri	NT	?
Pisum fulvum	NT	→
Rorippa prolifera	NT	?
Sinapidendron gymnocalyx	NT	?
Trifolium argutum	NT	→
Vicia barbazitae	NT	→

[a] ↓ = decreasing; ? = unknown; → = stable

28.6 Conservation Actions and Research Needs

Detailed information about conservation actions in place for each species assessed (primarily the species' *in situ* and *ex situ* conservation status) is recorded in a text field in SIS. There are fields in the database to record whether a species occurs within a protected area (PA) and whether it is conserved *ex situ*, but these data are not reliable and informative enough for analytical purposes. This is because the fact that a species occurs in a PA does not necessarily mean that the population is actively conserved – on the contrary, we know already that many CWR do occur in PAs but that they are only passively conserved as they are not the primary conservation targets of the sites (Maxted *et al.*, 1997b); this means that these populations are not actively managed or monitored and therefore may be in decline, yet this fact is unknown to the PA manager. Further, checking a box to say that a species is conserved *ex situ* tells us nothing about the number of samples conserved, where they were collected (to truly reflect inherent patterns of genetic diversity) and from what source; therefore, it would be misleading to use this data field as an indication of the *ex situ* conservation status of a species. However, using data extracted from EURISCO (2010), it was possible to obtain an indication of the *ex situ* conservation status of the species assessed. Further, it is possible to record conservation and research actions needed in SIS; therefore, we can make inferences from these sources about the adequacy of current conservation measures in place.

Germplasm accessions that are recorded by gene banks as being of wild or weedy origin are reported by EURISCO for 273 of the 571 species assessed (nearly 48%) and these are found in 23 of the 25 crop gene pools/ groups included for assessment. This does not mean that there are not gene bank holdings of the other species because not all gene banks contribute data to EURISCO and not all accessions reported are necessarily tagged as being of wild or weedy origin. Further, germplasm holdings in botanic gardens such as those located in the Canary Islands and Madeira are not reported via EURISCO and these are known to conserve accessions of a number of the CWR species assessed. However, a very high proportion of European gene banks do now provide holdings data to EURISCO; therefore, we can reasonably assume that there are large gaps in the *ex situ* conservation of some of the highest priority CWR in Europe.

Furthermore, most species are represented by very few accessions, are reported by only one gene bank, and have been collected from only a small part of the species' range.

Conservation actions needed were recorded for 483 of the species assessed (Fig. 28.10). The most commonly recorded conservation needs were *ex situ* conservation

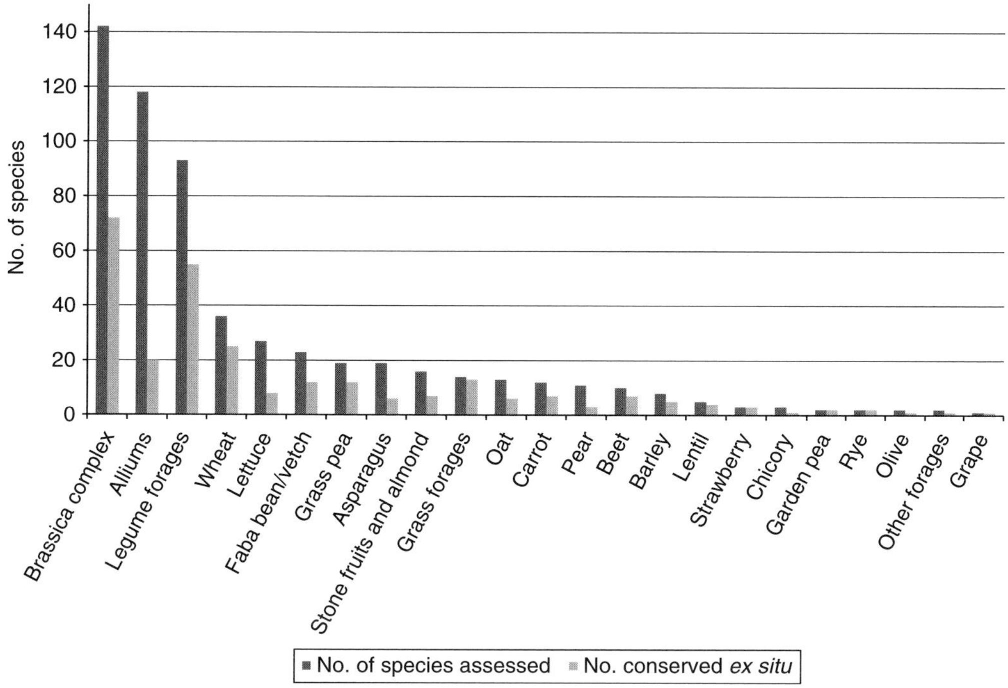

Fig. 28.9. The number of CWR species assessed in each crop gene pool/group compared with the number of species reported by EURISCO (2010) as having wild or weedy accessions in European gene banks.

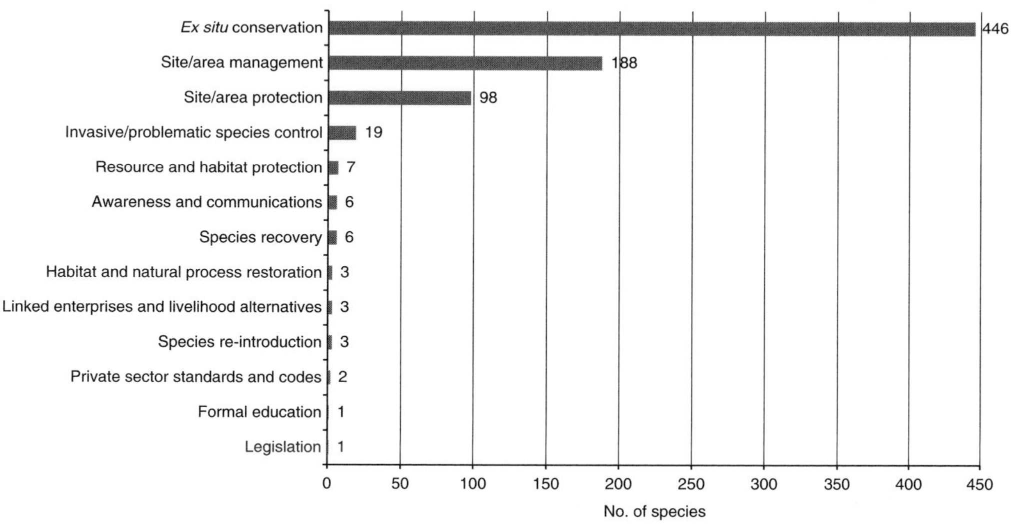

Fig. 28.10. Conservation actions needed for 483 species.

(which was recorded for 446 species – more than 78% of the 571 species assessed), site management (which was noted for 33% of the species) and site protection (17% of species). Research needs recorded included population size, distribution and trends (356 species), threats (163 species), habitat trends (73 species), life history and ecology (69 species) and taxonomy (35 species). It is clear from these results that much needs to be done not only to conserve European CWR, but also to improve our knowledge to enable conservation planning.

28.7 Conclusions and Recommendations

A sample of high priority European CWR species have been regionally assessed using the IUCN Red List Categories and Criteria (IUCN, 2001). These assessments have been published in the first European Red List and those that are endemic to Europe (188 species) have been published in the IUCN Red List of Threatened Species. This is the first time that a concerted effort has been made to carry out Red List assessment specifically for CWR as a group of plants and therefore it represents a significant breakthrough, not only in conservation planning for CWR but also in increasing awareness of their importance and the need for conservation action.

The results of this initiative show that a significant proportion of the species assessed are threatened or are likely to become threatened in the near future and that some crop gene pools or crop groups, such as the cultivated beets, are particularly at risk – these species should be subject to immediate conservation gap analysis and concerted actions. More than half of the species assessed are categorized as Least Concern but many of these are nationally threatened and even for those species that are relatively common and widespread in Europe, there is a need to conserve representative samples from throughout their range (both *in situ* and *ex situ*), to ensure that the widest possible range of genetic diversity is conserved and available for use in crop improvement programmes. A thorough review of the species evaluated as Least Concern should be undertaken to highlight those in most urgent need of conservation attention and those that require monitoring; data recorded on population trends, conservation and research needs and national threatened status can be used to aid priority-setting. Many species were assessed as Data Deficient – re-evaluation of these species is required when resources are available.

An analysis of the threats affecting CWR populations in Europe clearly show that concerted action is needed to alleviate the causal factors, the most commonly reported threats being livestock farming, development for tourism and recreation, and new housing and urban areas; however, with an increasing human population placing pressure on land and resource use, this presents an enormous challenge. An immediate priority should be the establishment of genetic reserves for the highest priority species (see Kell *et al.*, Chapter 2, this volume) with complementary back-up in *ex situ* collections. *In situ* management plans for these species need to address the threats present at the site, such as excessive grazing by livestock. On-farm management may present an option for CWR populations that grow in agro-environments (see Maxted *et al.*, 1997b, 2011).

In addition to the knowledge gained on the threatened status of European CWR species, a positive outcome of this initiative is that a significant quantity of datum has been collated that is not only useful for conservation planning but serves as a baseline for future assessment. Further, a large group of specialists with expert knowledge of wild plant species has received training in IUCN Red Listing and professional collaboration has been fostered within this network. At the same time, undertaking this initiative presented some challenges, including dealing with issues of data quality and consistency, problems associated with information management (data recording and standards), communicating with a network of experts dispersed in many different countries, and the potential subjectivity of the process. It is important to stress that Red Listing depends heavily on the voluntary contributions of experts who have the knowledge and access to information needed to carry out the assessments; however, the demand on their professional time means that they cannot always contribute as much as they might like to Red Listing. Therefore, for future projects of this kind it would be beneficial to allocate funding to acknowledge the contributions of experts (even if it is a nominal amount) because their knowledge is fundamental to the success of such a project.

This initiative should not be viewed as an end in itself but as a springboard for future work in this area. Specifically, we recommend that as a priority, the Crop Wild Relative Specialist Group (CWRSG) of the Species Survival Commission of IUCN coordinates the collation of Red List assessments of national endemic CWR species (both within and outside Europe) for submission to the IUCN Red List of Threatened Species (recognizing that adequate resources will be required). This will have a significant impact by increasing awareness of the importance of CWR, their threatened status and the need for conservation action.

Finally, the application of the IUCN Red List Categories and Criteria to this sample of European CWR species has reinforced the need for the development of an additional means of assessment that takes into account intra-specific genetic diversity. Although the existing system can be used to assess subpopulations of a species (in addition to subspecies and varieties) and has been successfully used in this manner for several mammals (see for example: www.iucnredlist.org/apps/redlist/details/2468/0), the assessment is still based on population size (i.e. the number of mature individuals) and/or geographic range, rather than the genetic diversity (i.e. allelic richness, evenness and/or uniqueness) within and between those subpopulations. It is rarely the case that all subpopulations of a species contain an equal proportion of genetic diversity; therefore, when the aim (as for CWR conservation) is to maximize both inter- and intra-specific genetic diversity to ensure that the widest pool of genes is available for use in crop improvement programmes, the risk of extinction for a species must consider both within and between subpopulation genetic diversity. Indeed, the goal of wild plant species conservation in general should take account of intra-specific genetic diversity as it is the maintenance of this diversity, both within and between subpopulations, that ensures overall population stability. We therefore recommend that genetic diversity is taken into account in the assessment process, either to complement or extend the applicability of the existing system.

Acknowledgements

We would like to acknowledge that this work was carried out under subcontract with IUCN with funding from the European Commission under Service Contract No. 070307/2007/483305/MAR/B2. We are extremely grateful to Cascais Natura which hosted and fully financed the European CWR Red List workshop convened in Cascais, Portugal in April 2010.

We would also like to acknowledge the many people who contributed to this project:

- For Red List workshop fund-raising, organization and facilitation, contributions to assessments, and assessment review and editing: Ana Nieto (IUCN).
- For Red List workshop facilitation and contributions to assessments: Annabelle Cuttelod and Leah Collett (IUCN).
- For provision of substantial data sets: Werner Greuter, Eckhard Raab-Straube and Anton Guentsch (The Botanic Garden Berlin-Dahlem, Germany), Helmut Knüpffer (The Leibniz Institute of Plant Genetics and Crop Plant Research (IPK), Germany), John Wiersema and Stephanie Greene (The United States Department of Agriculture, Agricultural Research Service, USA), Suzanne Sharrock and Meirion Jones

(Botanic Gardens Conservation International), Milko Skofic (Bioversity International), Graham French (Joint Nature Conservation Committee, UK), José M. Iriondo (Universidad Rey Juan Carlos, Spain), Mauricio Parra-Quijano (Centre of Plant Genetic Resources of Spain and Universidad Polytechnica de Madrid, Spain) and Samantha Dobbie (University of Birmingham, UK).

- For production of maps: Joana Magos Brehm and Serene Hargreaves.
- For assistance with data entry and editing: Hannah Fielder, Serene Hargreaves and Joe Osborne.
- Many experts contributed knowledge and gave up their time, without which the assessments could not have been undertaken; they are (in alphabetical order by family name – we apologize if we have inadvertently missed anyone): Carlos Aguiar, Mora Aronsson, Åsmund Asdal, R. Berjano Pérez, G. Blanca, Éva Böhm, Ferdinando Branca, Zofia Bulińska, B. Cabezudo, Miguel Carvalho, Robert Caruana, C.S. Christodoulou, D. Cobo, G.M. Cruz Trujillo, A. Della, Domizia Donnini, David Draper, Maria Cristina Duarte, Ehsan Dulloo, Garifalia Economou, Pavol Eliáš, Brian Ford-Lloyd, Lothar Frese, A. García Gallo, A. Hadjichambis, T. Hadjikyriakou, Vojtech Holubec, José Iriondo, C. Kailis, Andreas Katsiosis, Chris Kik, Helena Korpelainen, Hörður Kristinsson, Angelos Kyratzis, T. Kyriacou, Juozas Labokas, Richard Lansdown, Joana Magos Brehm, Oleg Maslovky, Viktor Melnyk, J. Mota Poveda, J. Naranjo Suárez, P. Navas, Valeria Negri, M. Ortiz-Herrera, Joe Osborne, C. Papasavvas, D. Paraskeva-Hadjichambi, J. Pastor, J. Peñas, A.V. Pérez Latorre, Gert Poulsen, J. Alfredo Reyes Betancort, O. Rodríguez Delgado, I. Santana López, Arnoldo Santos Guerra, Hanno Schaefer, S. Scholz, Tamara Smekalova, Vladimir Stevanović, Silvia Strajeru, S. Talavera Lozano, Maria do Carmo Tavares, Alessandro Tribulato, T. Tsintides, Katya Uzundzhalieva, C. de Vega Durán, Rudolf Vögel, Gabor Vörösváry, Peter Wind.
- Many of the above-named experts are members of the IUCN/SSC Crop Wild Relative Specialist Group, co-chaired by Ehsan Dulloo (Bioversity International) and Nigel Maxted (University of Birmingham).
- We are also grateful to the many people who contributed images for publication with the assessments.

Notes

[1] In February 2011, a major revision of Euro+Med PlantBase was published; the changes made to the families other than Compositae, Poaceae and Roasaceae are not reflected in the list of CWR species selected for inclusion in the European Red List.

[2] A benign introduction is defined as 'an attempt to establish a species, for the purpose of conservation, outside its recorded distribution, but within an appropriate habitat and eco-geographical area. This is a feasible conservation tool only when there is no remaining area left within a species' historic range' (IUCN, 1998).

[3] The field for recording native status primarily contains only one code for each occurrence, but in some cases, multiple codes are used. For example, some records are coded 'DN', 'NE' or 'NS'. Again, this indicates a degree of uncertainty in the data and to take an inclusive approach, all records containing combinations of codes N, S, D and E were included.

[4] Subclasses of food types used in GRIN Taxonomy for Plants (USDA, ARS, National Genetic Resources Program, 2009), which is adapted from the Economic Botany Data Standard (Cook, 1995).

[5] *Brassica* spp. are included in the high priority human food crop list.

[6] *Triticum, Aegilops* and *Secale* spp. are included in the high priority human food crop list.

[7] A species is classified as NA when a very small proportion (usually ca. 1% or less) of its global population occurs in the region of the assessment or because it is not a wild population or not within its natural range in the region, or because it is a vagrant.

[8] A taxon is Near Threatened when it has been evaluated against the criteria but does not qualify for Critically Endangered, Endangered or Vulnerable now, but is close to qualifying for or is likely to qualify for a threatened category in the near future (IUCN, 2001).
[9] The IUCN Red List Categories and Criteria have been applied to a number of mammal species at subpopulation level but no subpopulation assessments of plant species have been published to date.

References

Cook, F.E.M. (1995) Economic Botany Data Collection Standard. Prepared for the International Working Group on Taxonomic Databases for Plant Sciences (TDWG). Royal Botanic Gardens, Kew, 146 pp. Available at: www.kew.org/tdwguses/index.htm (accessed 1 May 2011).

EURISCO Catalogue (2010) Available at: http://eurisco.ecpgr.org (accessed 7 May 2011).

Euro+Med PlantBase (2006) *Euro+Med PlantBase – the Information Resource for Euro-Mediterranean Plant Diversity*, Version 2006. Berlin.

Euro+Med PlantBase Secretariat (2002) *Euro+Med PlantBase – Preparation of the Initial Checklist: Data Standards Version 2.8, July 5th, 2002*. Centre for Plant Diversity and Systematics, School of Plant Sciences, University of Reading, UK.

European Communities (1995–2009) Eurostat. European Commission. Available at: http://epp.eurostat.ec.europa.eu (accessed 7 May 2011).

FAO (2009) FAOSTAT. Food and Agriculture Organization of the United Nations, Rome. Available at: http://faostat.fao.org/ (accessed 7 May 2011).

Hanelt, P. and IPK Gatersleben (eds) (2001) *Mansfeld's Encyclopaedia of Agricultural and Horticultural Crops*. 6 vols. 1st English edn. Springer, Berlin/Heidelberg/New York, 3645 pp.

Harlan, J. and de Wet, J. (1971) Towards a rational classification of cultivated plants. *Taxon* 20, 509–517.

Heywood, V.H. (2009) Foreword. In: Sharrock, S. and Jones, M. (eds) *Conserving Europe's Threatened Plants: Progress Towards Target 8 of the Global Strategy for Plant Conservation*. Botanic Gardens Conservation International, Richmond, UK, p. 3.

IPK Gatersleben (2003) Mansfeld's World Database of Agricultural and Horticultural Crops. Leibniz Institute of Plant Genetics and Crop Plant Research, Gatersleben, Germany. Available at: http://mansfeld.ipk-gatersleben.de/mansfeld/mf-inf_e.htm (accessed 1 May 2011).

IUCN (1998) *Guidelines for Re-introductions*. Prepared by the IUCN/SSC Re-introduction Specialist Group. IUCN, Gland, Switzerland and Cambridge, 10 pp.

IUCN (2001) IUCN Red List Categories and Criteria: Version 3.1. IUCN Species Survival Commission. IUCN, Gland, Switzerland and Cambridge, 30 pp. Available at: www.iucnredlist.org/technical-documents/categories-and-criteria (accessed 1 May 2011).

IUCN (2010) IUCN Red List of Threatened Species. Version 2010.4. Available at: www.iucnredlist.org (accessed 22 April 2011).

Kell, S.P., Knüpffer, H., Jury, S.L., Maxted, N. and Ford-Lloyd, B.V. (2005) Catalogue of Crop Wild Relatives for Europe and the Mediterranean. University of Birmingham, Birmingham, UK. Available online via the Crop Wild Relative Information System (CWRIS – www.pgrforum.org/cwris/cwris.asp) (accessed 1 May 2011) and on CD-ROM.

Kell, S.P., Knüpffer, H., Jury, S.L., Ford-Lloyd, B.V. and Maxted, N. (2008) Crops and wild relatives of the Euro-Mediterranean region: making and using a conservation catalogue. In: Maxted, N., Ford-Lloyd, B.V., Kell, S.P., Iriondo, J., Dulloo, E. and Turok, J. (eds) *Crop Wild Relative Conservation and Use*. CAB International, Wallingford, UK, pp. 69–109.

Kell, S.P., Maxted, N., Ford-Lloyd, B.V. and Magos Brehm, J. (In prep) Broadening the base, narrowing the task: setting priorities for the conservation of crop wild relative diversity.

Mabberley, D.J. (2008) *The Plant-Book. A Portable Dictionary of the Vascular Plants*. Cambridge University Press, Cambridge.

Mathew, B. (1996) *A review of* Allium *section* Allium. Royal Botanic Gardens, Kew.

Maxted, N. (2003) Conserving the genetic resources of crop wild relatives in European protected areas. *Biological Conservation* 113(3), 411–417.

Maxted, N., Hawkes, J.G., Guarino, L. and Sawkins, M. (1997a) The selection of taxa for plant genetic conservation. *Genetic Resources and Crop Evolution* 44, 337–348.

Maxted, N., Hawkes, J.G., Ford-Lloyd, B.V. and Williams, J.T. (1997b) A practical model for *in situ* genetic conservation. In: Maxted, N., Ford-Lloyd, B.V. and Hawkes, J.G. (eds) *Plant Genetic Conservation: the* In Situ *Approach*. Chapman and Hall, London, pp. 339–363.

Maxted, N., Ford-Lloyd, B.V., Jury, S.L., Kell, S.P. and Scholten, M.A. (2006) Towards a definition of a crop wild relative. *Biodiversity and Conservation* 15(8), 2673–2685.

Maxted, N., Kell, S. and Magos Brehm, J. (2011) *Options to Promote Food Security: On-farm Management and* In Situ *Conservation of Plant Genetic Resources for Food and Agriculture*. Food and Agriculture Organization of the United Nations, Rome.

Sharrock, S. and Jones, M. (2009) *Conserving Europe's Threatened Plants: Progress Towards Target 8 of the Global Strategy for Plant Conservation*. Botanic Gardens Conservation International, Richmond, UK.

USDA, ARS, National Genetic Resources Program (2009) Germplasm Resources Information Network – (GRIN) [Online Database]. National Germplasm Resources Laboratory, Beltsville, Maryland. Available at: www.ars-grin.gov/cgi-bin/npgs/html/wep.pl?language=en&chapter=econ1 (accessed 1 May 2011).

29 Documentation Landscape for Plant Genetic Resources – Focus on *In Situ* and On-Farm

S. Harrer, F. Begemann and Th. van Hintum

29.1 Introduction to the Policy Framework

One of the major preconditions for the use of plant genetic resources for food and agriculture (PGRFA) is the knowledge about where these PGRFA are stored *ex situ* or can be found in their natural environment (*in situ*/on-farm) and the knowledge about their traits. The first element is normally described by a set of descriptors called passport data. Information on plant features useful for breeding and agriculture relates either to highly heritable traits or traits with medium to low heritability. Descriptor values of the first category are called characterization data and of the second category evaluation data.

The importance of proper documentation is reflected by the international legal framework in which the PGRFA conservation operates. There are several international agreements as well as European and national legislations and regulations affecting the management and utilization of PGRFA. At the international level the most important ones are the Convention on Biological Diversity (CBD) and the International Treaty on Plant Genetic Resources for Food and Agriculture (Treaty). Countries ratifying the CBD and/or the Treaty agreed upon the need for the conservation and sustainable use of PGRFA. Proper documentation of PGRFA is one of the keys issues. Article 17 (Exchange of Information) of the CBD (CBD, 1992) and Article 17 (The Global Information System on PGRFA) of the Treaty (FAO, 2001) bind contracting parties to develop and strengthen information systems on PGRFA globally to support not only the conservation and sustainable use but also to contribute to the sharing of benefits arising out of their use.

Both treaties are legally binding for the respective contracting parties. Any national or regional information system for PGRFA conserved in gene banks or managed *in situ*/on-farm will have to take into account the aforementioned international policy framework especially with regard to its scope and to the quantity and quality of the data, i.e. all PGRFA regarded as forming part of the national PGRFA system have to be included. Furthermore, a close collaboration between countries seems necessary to ensure a coherent development of any such regional information system.

On the European level, there are several regulations of the European Union (EU) that do have implications on PGFRA conservation management *in situ* and on-farm:

- Council Regulation (EC) 1698/2005 on support for rural development by the

European Agricultural Fund for Rural Development (EAFRD) provides support to preserve plant genetic resources naturally adapted to local and regional conditions and under threat of genetic erosion.

- Commission Directive 2008/62/EC providing for certain derogations for acceptance of agricultural landraces and varieties which are naturally adapted to the local and regional conditions and threatened by genetic erosion and for marketing of seed and seed potatoes of those landraces and varieties regulates the acceptance for inclusion in the national catalogues of varieties of agricultural plant species of landraces and varieties which are naturally adapted to the local and regional conditions and are threatened by genetic erosion (conservation varieties of agricultural crops).
- Commission Directive 2009/145/EC providing for certain derogations, for acceptance of vegetable landraces and varieties which have been traditionally grown in particular localities and regions and are threatened by genetic erosion and of vegetable varieties with no intrinsic value for commercial crop production but developed for growing under particular conditions and for marketing of seed of those landraces and varieties regulates the acceptance for inclusion in the national catalogues of varieties of vegetable plant species (conservation and amateur varieties of vegetable crops).

These regulations and directives have in common that they rely on inventories, i.e. compilations of landraces, former breeder's varieties and similar materials which are naturally adapted to local and regional conditions and are under threat of genetic erosion. Therefore any national or regional system should not only comply with these criteria – they also have a certain legal status as a reference system for the implementation of these regulations.

This paper will give a brief overview on the current state of activities in the field of PGRFA collaboration and documentation in Europe. It also will outline a vision on how an efficient and effective documentation system for PGRFA *in situ* and on-farm could be developed taking into account the aforementioned policy framework. This vision will be restricted to passport data.

29.2 Current State of European PGRFA Collaboration and Documentation

The pan-European platform for collaboration on PGRFA is the European Cooperative Programme for Plant Genetic Resources (ECPGR). ECPGR aims at facilitating the long-term conservation and the increased utilization of plant genetic resources on a cooperative basis in Europe. The programme, which is financed by the participating countries and is coordinated by a Secretariat hosted by Bioversity International, operates through broadly focused Networks dealing with groups of crops or general themes related to plant genetic resources, such as the *In situ* and On-farm Conservation Network and the Documentation and Information Network.

Main infrastructural elements of the ECPGR are the EURISCO web-based catalogue (EURISCO, 2011) and AEGIS (A European Genebank Integrated System) (ECPGR, 2009a). Both elements currently have a clear focus on PRGFA managed *ex situ*. However, the need to also include both *in situ* crop wild relative populations and on-farm managed landraces in the activities is recognized widely, and was also expressed by the external review that the ECPGR underwent recently (ECPGR, 2010).

AEGIS is foreseen to be developed in the next years towards a 'virtual' decentralized and efficient conservation infrastructure for Europe. In this regard the importance of EURISCO as a regional information system will increase in the future since it will play a central role in AEGIS. At present, the web-based catalogue EURISCO provides information about passport data on over 1 million accessions of crop diversity representing 5383 genera and 34,823 species conserved *ex situ* in 41 European countries. Since its launch in September 2003 EURISCO has emerged as the most prominent element in the European *ex situ* information landscape. It can only play that role thanks to the well-developed PGR documentation systems at the institutional and national levels and their strictly organized interaction (for more information see: van Hintum *et al.*, 2010). One of the main elements for the success of EURISCO is its transparent approach with clear responsibilities at the national and European level and standards for data exchange (EURISCO Descriptor List). The organizations in a country maintaining PGRFA provide the National Focal Point (NFP) with their data using the standards for data exchange. The individual data sets form the National Inventories (NIs), which are then sent through the NFP to EURISCO. Such a bottom-up approach allows for a sufficient control of the content and quality of the data at the respective levels: while the data provider at the national, institutional level will have full control and responsibility for the quality of their data, the NFP will have the responsibility to compile in the NI only data on accessions being part of the national PGRFA system. Only these data will then be transferred by the NFP to EURISCO.

29.3 Recent Developments

'*In situ* and on-farm conservation and management' as well as 'documentation and information' are two of the four priorities for the current phase VIII (2009–2013) of ECPGR (ECPGR, 2009b).

Based on the success of EURISCO in the *ex situ* domain, some actors in the *in situ* domain have started to establish a similar infrastructure. Twenty-three countries have officially nominated through their National ECPGR Coordinators 25 NFPs for *in situ*/on-farm NIs a few years ago. These NFPs have the responsibility to compile *in situ*/on-farm data for the respective NI. This includes especially the decision on which landraces and which *in situ* occurrences of PGRFA in a specific county shall be included as well as the competence to officially report this kind of information to others (e.g. to European documentation systems such as EURISCO).

A few years after the establishment of this network the main output is a draft minimum descriptor list for the documentation of on-farm conservation and management activities. The list was developed in 2007 and discussed by the On-farm Conservation and Management Task Force. The purpose of these descriptors (ECPGR, 2007) is to gather information on landraces that are being collected from farmers' fields and to build up a knowledge base. There are 38 descriptors specially developed to describe the landrace(s) as well as the management system (e.g. cropping system, harvest methods, seed processing, storage etc.) and socio-cultural historical as well as traditional uses. Four years after publication of the draft, no further discussion or completion of this list has occurred, and as far as known no adoption or use of the list has taken place.

In the area of *in situ* conservation of PGRFA three important projects have been performed in the past, which have to be taken into account. In the EC-funded project European Crop Wild Relative Assessment and Conservation (PGR Forum, 2011), 23 partners from 21 countries developed a methodological concept for the management of crop wild relative (CWR) data and also established a system called the Crop Wild Relative Information System (CWRIS) between 2002 and 2005 (Kell *et al.*, 2008). Within the EU-funded AEGRO project (An Integrated European *In Situ* Management Work plan: Implementing Genetic Reserves and On-Farm Concepts, 2007–2011) CWRIS was further extended by four independent modules for *Avena*, *Beta*, *Brassica* and *Prunus*, collectively called 'Population Level Information System – PLIS' (see Germeier *et al.*, Chapter 31, this volume) allowing the mapping of all actual and historical occurrences and the identification of occurrences for these species within Natura 2000 protected areas. On the global level a project on '*In situ* conservation of CWR' jointly funded by the United Nations Environment Programme (UNEP) and the Global Environment Facility (GEF) resulted in the establishment of a few National CWR Inventories, based on descriptors developed for CWR. These five NIs (Armenia, Bolivia, Madagascar, Sri Lanka, Uzbekistan) are online searchable directly through the portal created in the framework of this project (CWRGP, 2011).

In conclusion, it is clear that the interest in CWR and on-farm conservation is high and increasing. However the long-term service commitment of the relevant actors to compile and process the relevant *in situ* and on-farm data is in general lacking. As a result, some systems have been developed but sufficient data are in most cases still missing. Furthermore, the previous projects in this field have not resulted in significant products; some of the prototypes and drafts that were developed remained prototypes and drafts.

29.4 Addressing National Documentation Requirements – a German Example

The National Inventory for Plant Genetic Resources for Food and Agriculture in Germany (PGRDEU, 2011) is the central, national documentation of plant genetic resources in Germany. It serves as the national interface for international information systems such as EURISCO and the Global Information System of the Treaty. PGRDEU contains comprehensive information about species and their potential use as well as about the endangerment of species occurring in Germany. PGRDEU includes data of *ex situ* collections of German gene banks and other collections as well as a first data compilation (prototype) of *in situ* occurrences of wild species and landraces. The prototype for *in situ* and on-farm data was developed by a model project 'Establishing a system for reporting and monitoring *in situ* conservation' (see Vögel and Reichling, Chapter 30, this volume) in Brandenburg, Germany. Another element of the German NI is the 'Red List of Endangered Indigenous Domesticated Plants in Germany'. It serves amongst others as official reference list for the national implementation of rural development programmes and for the implementation of the national seed directives for conservation varieties.

29.5 Conclusions and Recommendations

The documentation landscape for CWR and on-farm PGRFA is rather fragmented. There are a few valuable approaches and solutions that can be used to enhance the information exchange and management during the next years. The network of the ECPGR *in situ*/on-farm National Inventory Focal Points can provide the link to the national/institutional level in each country (ECPGR, 2011). The layer of NFP/NI will be the most important one in such a system.

A considerable amount of specific knowledge exists already in the ECPGR Documentation and Information Network. Valuable experiences from the establishment of the system of *ex situ* NIs and EURISCO are also available. As for the establishment of EURISCO as a central information system of *ex situ* holdings, the establishment of an *in situ* information network will depend on the formal commitment of network partners and sufficient funding.

The members of the ECPGR *In situ* and On-farm Conservation Network know the needs for documentation of PGRFA *in situ* and on-farm. However, it will be essential to identify the institutions that are officially committed to compiling the information at the national level. This will require creating ownership of the process and the resulting database among the actors in the community. It can count on the full support of the ECPGR Documentation and Information Network in this process.

There are different systems already available at the national, regional and international levels such as the Crop Wild Relatives Global Portal (CWRGP, 2011) and the Crop Wild Relative Information System (Kell *et al.*, 2008). At the global level data provided by five mainly non-European countries through the Crop Wild Relatives Global Portal are nationally authorized data. At the European level CWRIS provides already comprehensive information about CWR at the taxon level and for four model crops at the occurrence level (see Germeier *et al.*, Chapter 31, this volume). Although data kept in CWRIS are public data they have not run through a formal authorization process similar to CWR Global Portal. EURISCO is the only system designed to provide accession level information, which is of a similar structure as occurrence level data. EURISCO might become relevant for *in situ* and on-farm documentation, too.

Draft standards for exchange of *in situ*/on-farm data already exist, but need to be harmonized with already existing standards such as Descriptors for Farmers' Knowledge of Plants (Bioversity International, 2009). A final agreement at the regional level on minimum standards for data exchange for *in situ* and on-farm data is still missing.

Consequently the following actions are recommended:

1. To develop and agree on one common minimum standard for exchange of *in situ* and on-farm data, based on a draft, jointly prepared by the ECPGR Documentation and Information Network and the *In situ* and On-farm Conservation Network, and involving the relevant institutions which in future may be charged with the task of compiling these data.
2. To seek an agreement on these minimum standards between ECPGR, Bioversity International and FAO.
3. To organize the required support for NFPs to develop NIs for *in situ*/on-farm and to create incentives accordingly.
4. To expand the data structure of EURISCO for inclusion of *in situ*/on-farm data

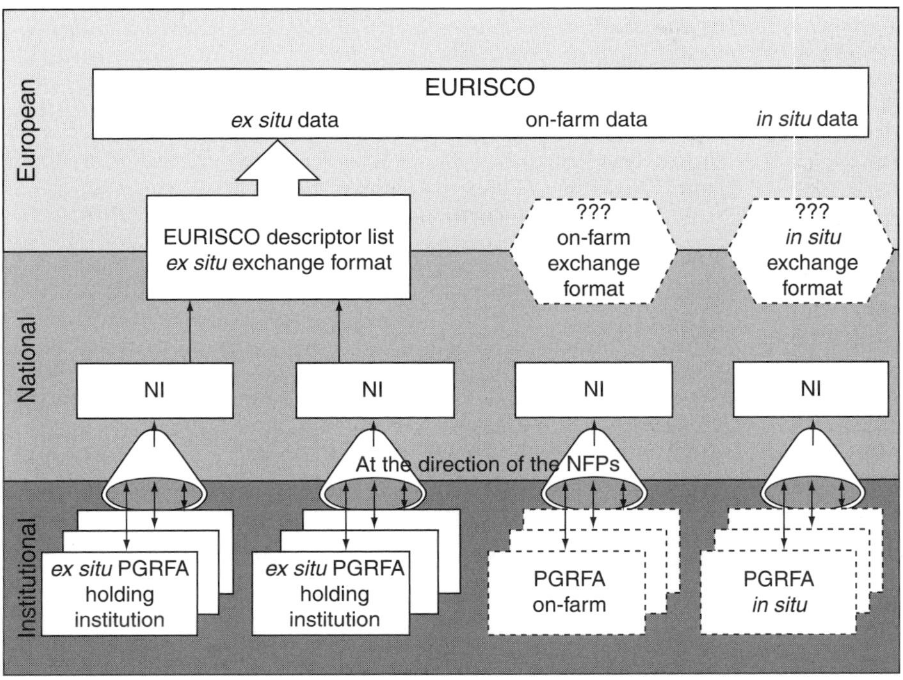

Fig. 29.1. Vision of a future European documentation landscape for PGRFA *ex situ*, *in situ* and on-farm information.

(by Bioversity International on behalf of ECPGR). This step will only be taken if the network of NFPs for *in situ*/on-farm can prove its value to the scope of EURISCO and the partners within the ECPGR system.

5. To develop a transfer mechanism for data from NIs to EURISCO (jointly by Bioversity International on behalf of ECPGR and the network of NFPs).

6. To provide capacity building and training where necessary (by ECPGR and its networks).

Taking into account the experience made during the establishment of EURISCO, it is not at all obvious that a similar approach will be feasible for *in situ* and on-farm data. Inclusion of these data in EURISCO (see Fig. 29.1) will therefore depend on the success of the development of NIs in this domain. Some prerequisites for this success have been met, however, as outlined in the recommendations above there is still a long way to go before an efficient and effective documentation landscape for PGRFA *in situ* and on-farm will be fully established.

References

Bioversity International (2009) Descriptors for farmers' knowledge of plants. Available at: www.bioversityinternational.org/nc/publications/publication/issue/descriptors_for_farmers_knowledge_of_plants.html (accessed 20 January 2011).

CBD (1992) Text of the Convention on Biological Diversity. Available at: www.cbd.int/convention/text/ (accessed 20 January 2011).

CWRGP (2011) Crop Wild Relatives Global Portal. Available at: www.cropwildrelatives.org/national_inventories.html (accessed 26 April 2011).

ECPGR (2007) Minimum descriptor list for the documentation of on-farm conservation and management activities. Available at: www.ecpgr.cgiar.org/Networks/Insitu_onfarm/Docs/OnfarmDescr_DRAFT271107.pdf/ (accessed 20 January 2011).

ECPGR (2009a) *A Strategic Framework for the Implementation of a European Genebank Integrated System (AEGIS)*. A Policy Guide. European Cooperative Programme for Plant Genetic Resources (ECPGR). Bioversity International, Rome, 23 pp.

ECPGR (2009b) About ECPGR. Available at: www.ecpgr.cgiar.org/Introduction/AboutECPGR.htm (accessed 3 February 2011).

ECPGR (2010) ECPGR Review Panel Report. Available at: www.ecpgr.cgiar.org/SteeringCommittee/SC12_Background%20docs/ECPGR%20Panel%20Report%20FINAL%20rev.pdf (accessed 3 February 2011).

ECPGR (2011) ECPGR *in situ*/on-farm National Inventory Focal Points. Available at: www2.bioversityinternational.org/networks/ecpgr/Contacts/ecpgr_PGR_NI_insonfarm_FP.asp/ (accessed 20 January 2011).

EURISCO (2011) The EURISCO Catalogue. Available at: http://eurisco.ecpgr.org/ (accessed 20 January 2011).

FAO (2001) Text of the International Treaty on Plant Genetic Resources for Food and Agriculture. Available at: www.itpgrfa.net/International/content/texts-treaty-official-versions (accessed 20 January 2011).

Kell, S.P., Moore, J.D., Iriondo, J.M., Scholten, M.A., Ford-Lloyd, B.V. and Maxted, N. (2008) CWRIS: an information management system to aid crop wild relative conservation and sustainable use. In: Maxted, N., Ford-Lloyd, B.V., Kell, S.P., Iriondo, J.M., Dulloo, E. and Turok J. (eds) *Crop Wild Relative Conservation and Use*. CAB International, Wallingford, UK, pp. 471–491.

PGRDEU (2011) The National Inventory for Plant Genetic Resources for Food and Agriculture in Germany. Available at: http://pgrdeu.genres.de/ (accessed 26 April 2011).

PGR Forum (2011) European Crop Wild Relative Assessment and Conservation. Available at: www.pgrforum.org (accessed 26 April 2011).

Van Hintum, T., Begemann, F. and Maggioni, L. (2010) The European *ex situ* PGR Information Landscape. In: Maurer, L. and Tochtermann, K. (eds) *Information and Communication Technologies for Biodiversity Conservation and Agriculture*. Shaker Verlag, Aachen, pp. 155–171.

30 Crop Wild Relatives in Brandenburg, Germany: Establishing a System for Reporting and Monitoring for the *In Situ* Conservation of Crop Wild Relatives

R. Vögel and A. Reichling

30.1 Introduction

Environmental agencies and authorities responsible for nature conservation require data on the distribution and abundance of threatened plant and animal species for assessing the species' conservation status. European environmental policy, especially the Flora, Fauna and Habitat (FFH) regulation, calls for regular reports on the state and quality of habitats and species of common European Community interest. Every 6 years the member states are requested to provide data following a standardized procedure which allows the comparison between the reporting years and between EU countries. The procedures are specified in detail by Art. 17 of the Habitats Directive 92/43 EEC. Systematic monitoring of the state habitats and species and the documentation of planned and performed actions is required to stabilize or improve the condition of habitats and target species. Currently, crop wild relatives (CWR) are not addressed as a specific protection target by German or European policy. However, some CWR are classified as endangered and therefore listed in Annex I/II of the EC FFH regulation irrespective of their potential use.

Basically, there is a divergence between the international policy driven by the Convention on Biological Diversity (CBD, 1992) and the International Treaty on Plant Genetic Resources for Food and Agriculture ('Treaty') (FAO, 2001), both strongly emphasizing the need to conserve CWRs more effectively and recommend implementation through national policy. The Treaty should *de facto* accept responsibility for *in situ* conservation and accordingly adapt the traditional environmental policy and acts on nature protection which do not yet foresee the conservation of CWRs as a specific task reaching beyond the rules set out by the aforementioned Art. 17. Until recently, CWR have indeed not attracted sufficient interest of conservation biologists which could, as one of the stakeholder groups, initiate policy changes.

Due to a lack of legal obligations, adequate reporting and monitoring systems for this category of wild plants as well as suitable data sources are largely absent. With the 3-year project 'Reporting and Monitoring System for the *In Situ* Conservation of Crop Wild Relatives' a pilot study was conducted to improve the situation. The project focused on the north-east German state of Brandenburg. Floristic data held in various data sources were identified, analysed, updated and compiled into a regional CWR inventory. The term CWR was used in the project as defined by

Maxted *et al.* (2006). It was implemented by the State Agency for Environment, Health and Consumers Protection Brandenburg (LUGV) in cooperation with the Forest Competence Centre Brandenburg (LFEV) and the University for Applied Sciences, Department Forest and Environment, Eberswalde (HNEE) from 2007 to 2010.

The main objective of the pilot project consisted of the establishment of improved documentation of CWR distribution and conservation status by mobilizing and connecting existing data sources from environmental monitoring projects and the documentation on protected areas and their species inventory formalized by sub-statutory laws issued by each state of the Federal Republic of Germany.

30.2 Sources for CWR Observation Data

Within the context of this paper we define the term 'observation' as the sighting of a species within a 10 × 10 or 7.5 × 7.5 km grid map. If several plants belong to an observation forming a geo-referenced plant group we call it an 'occurrence'. BML (1996) compiled a 'List of Wild Plants with Actual or Potential Value for Food, Agriculture and Forestry' in the first National Report on Plant Genetic Resources for Food and Agriculture (PGRFA) based on a publication of Schlosser *et al.* (1991) and the information system FloraWeb (BfN, 2011) of the Federal Agency for Nature Conservation (BfN). This list was revised and a comprehensive list of German plant species (crops and wild plant species) categorized according to their potential use emerged. The latest issue lists approximately 3600 crops and CWRs in total of which about 2900 species fall into the category CWR distributed in Germany (BLE, 2011). This list is comparable with the main class 'Taxon' of the Crop Wild Relative Information System (CWRIS) (Moore *et al.*, 2008; Kell *et al.*, 2008). This list as well as the CWRIS lack data on the geographic distribution of occurrences, which is an essential requirement for the establishment of a system aiming at: (i) the monitoring of within species variation; and (ii) the reporting on the conservation status of a species. Within the pilot project procedures were investigated and developed using the state of Brandenburg as a model region. The concept definitively scheduled the application of these procedures by the remaining 15 states in Germany. If this would materialize, data on CWR occurrences can be gathered by a network of state level agencies such as the LUGV, transferred to the National Inventory (NI) and passed on via the National Focal Point (NFP) to a central European information system (see Harrer *et al.*, Chapter 29, this volume).

In Germany observation data on wild plant species are kept by a large number of nature conservation agencies, but due to very different formats and technologies applied a compilation of the data in a NI through information technology-based data exchange procedures proved to be difficult. Even within the state of Brandenburg, a broad range of data pools available from environmental monitoring and official protected areas documentation exist of which eight data sources matched the requirements of the project. Most of the records came from biotope mapping, monitoring projects and species inventories linked to nature and environment protection. Additional inventory data were provided by the forest sector.

The data sources differed considerably with respect to data quantity and quality. The data quality correlated with the purpose of data recording, the accuracy of site description and the era the species was observed and the data compiled as well as with the person in charge of the maintenance of the data sets. Against this background, it was not possible to compile the original data and undertake simple analysis. Instead, the data had to be processed and the data sets harmonized. Furthermore, the compiled data sets proved inadequate for monitoring, since most of the information consisted of one observation per geographic unit and species and only very rarely of time series records. In general, the compiled data comprised observation data, taxa and their geographic location. Some data sets also gave information on the abundance, threat factors and the vitality of the observed plant group. In total, out of the 1.7 million investigated data records approximately 700,000 records were processed and used to build the regional inventory of CWR species in Brandenburg. Table 30.1 shows the data sets and the number of records by data sets used to compile the regional inventory.

30.3 Identification of Crop Wild Relatives in the Regional Inventory

Within Germany approximately 300 crop species divided between 190 genera (excluded ornamental plants) are cultivated. Initially, the CWR of these crops were selected from the regional inventory. In a second step only native wild plant species, crop species returned to the wild, or migrant wild plant species were selected resulting in a working catalogue of 215 species divided between 154 genera.

Beets, cabbage and carrot, forage plants, forest plants, fruit crops, grapevine, hop and several pharmaceutical plants and spices are native or are related to species of the German wild flora. The genus *Apium* with *Apium graveolens* and *A. repens* are native examples of Brandenburg. Both species are a potential genetic resource for celery and celeriac breeding. Further remarkable CWR diversity belongs to the group of medicinal plants of economic value such as willows (*Salix* sp.) or Arnica (*Arnica montana*). Wild garlic, bears' garlic (*Allium ursinum*) is an example for a species harvested in the wild which may cause conflicts with nature protection in future. To regulate the harvesting in the wild, a species-specific monitoring system could be established.

30.4 Schema of the Reporting and Monitoring System

A chart showing the data and information flow between the pilot system and the planned national reporting and monitoring system is presented in Fig. 30.1. The NFP and experts at the state level agreed on three standards: (i) a catalogue of PGR taxa; (ii) a CWR catalogue; and (iii) a list of descriptors used for *in situ* management. The national level represented by the Federal Office for Agriculture and Food, Information and Coordination Centre for Biological Diversity (BLE-IBV) provided standardized taxon catalogues and descriptor lists. In addition, federal and state agencies provided geo-data and other geo-services. This service will be regulated within the framework of the national Environment Information Act which implements the Directive 2007/2/EC of the European Parliament and of the Council of 14 March 2007 establishing an

Table 30.1. Data sources, total number of records and number of records used to build the CWR inventory, state of Brandenburg.

Data source	Total number of records on plant species	Number of records on crop wild relatives
Brandenburg biotop mapping BBK	1,300,000	~527,000
Special floristic observations, LUA	598	98
WinArt	811	216
Documentation of genetic resources of elm tree species and black poplar	378	378
Data of forestal gene conservation LFE	775	219
Native, regional trees (GHG-inventory)	2,928	897
OEUB, environmental monitoring of biosphere reserves	9,260	1,801
Florein data	482,819	167,802
In total	~1,700,000	~700,000

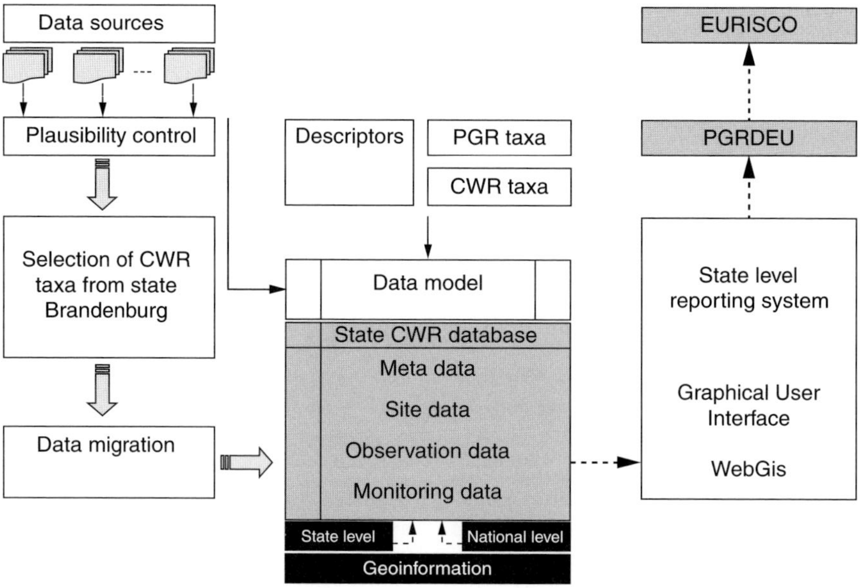

Fig. 30.1. Schema of the development of the state CWR database and the planned data flow to the national and European level. Grey boxes: databases holding CWR data physically today or in future. Black boxes: geo-information. Unshaded boxes: data modelling, data processing or reporting.

Infrastructure for Spatial Information in the European Community (INSPIRE) in the near future. The act will regulate the inter-agency exchange of geo-data by assignment declarations and data release statements.

The state-level pilot information system was designed based on these three standards. It receives and processes data from existing data sets as well as newly recorded data. It has a geographic information system (GIS) interface, which can be used for visualizing the distribution of CWR within Flora-Fauna-Habitat (FFH) sites on maps (Fig. 30.2) or can be applied to report on the state of actively managed CWR occurrences within Brandenburg in future. For such purposes web-based tools were developed and adapted to the pilot information system. It is planned to establish similar state-level information systems within each of the remaining 15 German states. This will facilitate the data flow from the states to the NFP with the objectives of: (i) enabling the NFP to report on the state of *in situ* conservation of CWR to federal authorities more efficiently; and (ii) subsequent transfer of data to the European level.

30.5 Technological Aspects

While developing the relational data structure attention was paid to its facile implementation with different relational database management systems (RDBMS). Using the Open Source Tool SQL Power Architect from the company SQLPower allowed the automatic generation of the generic data structure for systems such as PostgreSQL/PostGIS, MS ACCESS and MySQL by deploying an SQL script. To allow for the integration of the database into a modern geo-data infrastructure in future, the data structure was separated into core data (sites, observations, etc.) and additional data (geo-data, fact data) model. This approach would allow the transfer of external data via web service directly into the system in future without the need to adapt the core data model. The spatial components as well as the topological and geometrical qualifiers of all used spatial components and geometries were implemented compliant with the Simple Feature Specification of the Open Geospatial Consortiums (OGC).

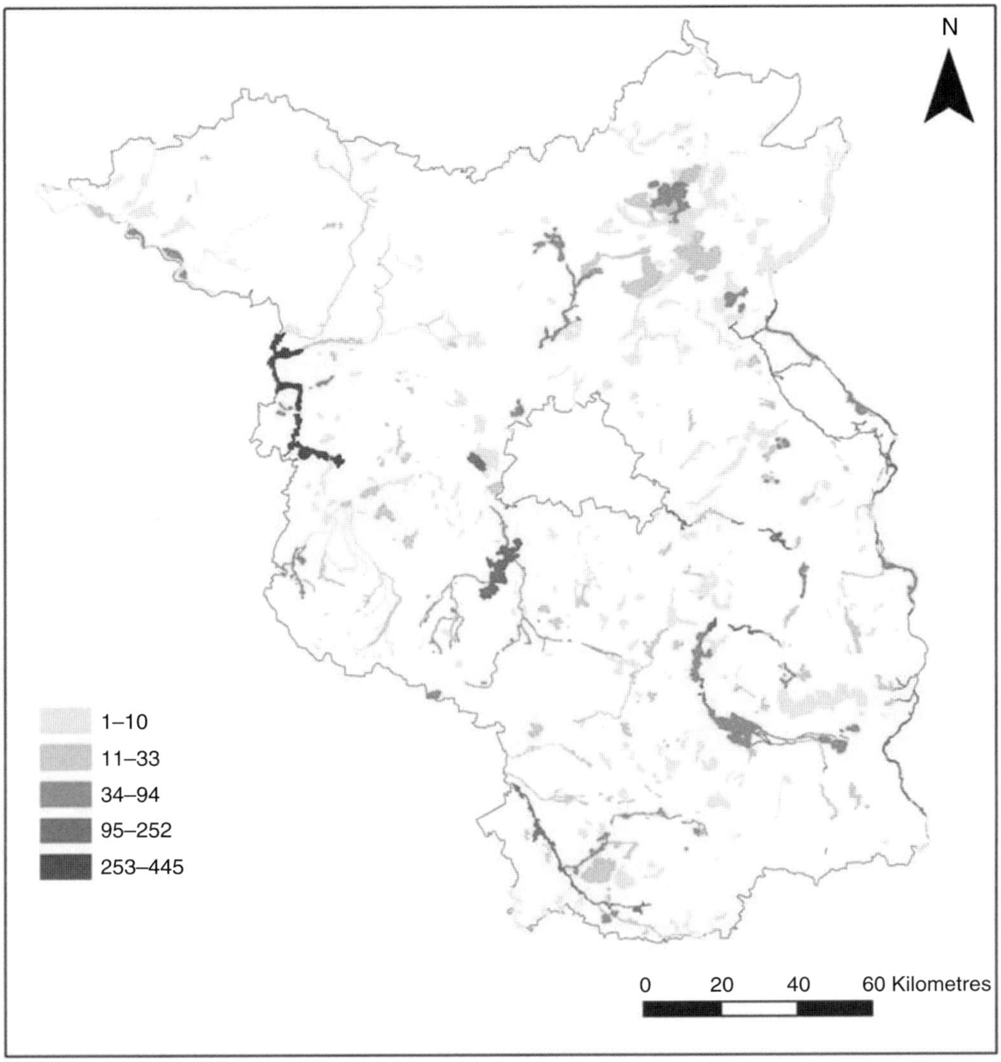

Fig. 30.2. Number of occurrences of selected CWR by FFH site.

Database internal functions were programed to facilitate an XML-based data transfer to other systems. The programs contain only standard functions of the SQL-extension (SQLX/XML) facilitating their adaptation to all RDBMS which support the SQLX-extension. By deploying these functions XML documents can be exported from the State CWR database and processed with XSLT scripts. The software packages XML-Spy and XML-Mapforce from Altova were used to convert the XML documents into KML-files. These files can be directly loaded into the popular Earth-Viewer Google Earth to visualize data on maps.

30.6 Conclusions

Even today no state-level inventory of CWR native to Germany exists. However, an exemplary database for CWR was developed using open source technologies. Open source technologies would allow use of in-house IT skills at no or very low licence costs. The latter is an important aspect if state nature protection agencies are requested to assume additional tasks, which are not regulated by law.

The pilot project improved considerably the information on the distribution of CWR species in Brandenburg. Knowledge required to take informed management actions in the case of endangered CWR can be derived from the information system, whose data content is the result of many years of nature and species protection work of state agencies and volunteers from the civil society who recorded the presence/absence of species in Brandenburg. The information can now be used to improve the *in situ* management of CWR for plant breeding and agricultural production.

In the field of CWR *in situ* conservation there is a clear need for an intensified cooperation between the biodiversity and agrobiodiversity conservation sectors. Without doubt, agriculture has a special responsibility for the management of its own genetic resources and should therefore become more involved in these issues. The project was part of the National Programme for Genetic Resources for Agricultural and Horticultural Crops supervised by the Federal Office for Agriculture and Food (BLE) on behalf of the Federal Ministry for Food, Agriculture and Consumer Protection (BMELV). The establishment of a state-level component of the NI of *in situ* conservation through a programme funded by the BMELV is a first step towards a stronger engagement of the agricultural sector.

The results of the projects can be seen as a contribution to the implementation of the obligation for contracting parties or member states of international agreements such as the Convention on Biological Diversity (CBD, 1992), the Global Plan of Action for Conserving and Sustainable Use of Plant Genetic Resources for Food and Agriculture (BMELF/ZADI-IGR, 1997) and the International Treaty for Plant Genetic Resources for Food and Agriculture (FAO, 2001). It is also a direct contribution to the implementation of the National Programme for Genetic Resources for Agricultural and Horticultural Crops of the German Federal Ministry for Food, Agriculture and Consumer Protection (BMELV, under preparation).

Acknowledgement

We are grateful to L. Frese (JKI) and S. Harrer (BLE-IBV) for critically reviewing our first manuscript and for several valuable contributions to this chapter.

References

BfN (2011) FloraWeb. Available at: www.floraweb.de (accessed 20 January 2011).

BLE (2011) List of Plant Genetic Resources in Germany. Available at: http://pgrdeu.genres.de/index.php?tpl=an_liste/ (accessed 20 January 2011).

BMELF/ZADI-IGR (1997) 4. International Technische Konferenz der FAO über Pflanzengenetische Ressourcen. Konferenzbericht, Leipziger Deklaration, Globaler Aktionsplan und Weltzustandsbericht. *Schriften zu Genetischen Ressourcen. Schriftenreihe des Informationszentrums für Genetische Ressourcen (IGR), Zentralstelle für Agrardokumentation und-information (ZADI)*. ZADI, Bonn, pp. 17–81.

BML (1996) Plant Genetic Resources for Food and Agriculture. *German National Report*. Federal Ministry for Food, Agriculture and Forestry, Bonn.

CBD (1992) Text of the Convention on Biological Diversity. Available at: www.cbd.int/convention/text (accessed 20 January 2011).

FAO (2001) Text of the International Treaty on Plant Genetic Resources for Food and Agriculture. Available at: www.itpgrfa.net/International/content/texts-treaty-official-versions (accessed 20 January 2011).

Kell, S.P., Moore, J.D., Jriondo, J.M., Scholten, M.A., Ford-Lloyd, B.V. and Maxted, N. (2008) CWRIS: an information management system to aid crop wild relative conservation and sustainable use. In: Maxted, N.,

Ford-Lloyd, B.V., Kell, S.P., Iriondo, J.M., Dulloo, E. and Turok, J. (eds) *Crop Wild Relative Conservation and Use*. CAB International, Wallingford, UK, pp. 471–491.

Maxted, N., Ford-Lloyd, B.V., Jury, S.L., Kell, S.P. and Scholten, M.A. (2006) Towards a definition of a crop wild relative. *Biodiversity and Conservation* 15(8), 2673–2685.

Moore, J.D., Kell, S.P., Iriondo, J.M., Ford-Lloyd, B.V. and Maxted, N. (2008) CWRML: representing crop wild relative conservation and use data in XML. *BMC Bioinformatics* 9, 116.

Schlosser, S., Reichhoff, L. and Hanelt, P. (eds) (1991) *Wildpflanzen Mitteleuropas*. Nutzung und Schutz. Deutscher Landwirtschaftsverlag, Berlin.

31 Population Level Information Management for Crop Wild Relatives

C.U. Germeier, J.M. Iriondo, L. Frese, C. Höhne and S.P. Kell

31.1 Introduction

In biology the term 'population' has a wide range of meanings (Menzie *et al.*, 2008). As information systems designed for the *in situ* management of crop wild relatives (CWR) shall support the management of genetic variation distributed between and within populations we have used the definition of Kleinschmit *et al.* (2004), who defined a 'population' as a 'group of individuals which is spatio-genetically different from other groups and forms a reproductively coherent group adapted to the environmental conditions of its growing site'. The genetic reserve conservation technique (Kell *et al.*, Chapter 2, this volume) aims at the management of genetically distinct populations of CWR and schedules by definition the active management of populations and the monitoring of the effects of management measures on populations. As a result of decreasing costs for molecular genetic marker techniques the knowledge on the spatial and temporal distribution pattern of genetic variation within species (e.g. Hirao and Kudo, 2004; Fievet *et al.*, 2007) will increase and allow the delimitation of populations by *inter alia* determining the genetic boundaries between occurrences. Against this background the definition proposed by Kleinschmit *et al.* (2004) for the purpose of forest genetic resources management may gain importance for CWR management in future.

Information on species occurrences, their spatial distribution, genetic structure and the ability to track ongoing and predicted demographic and genetic changes within a species in relation to environmental factors is a precondition for the *in situ* conservation of genetic resources. As demographic and genetic data are not regularly available for CWR as yet, for population level information of CWR, initially two major sub-domains are important: taxonomy and geographic distribution data. Although the Global Biodiversity Information Facility (GBIF) marked a 'quantum leap' forward in biodiversity data accessibility, shortcomings can still be identified in the quality and usefulness of taxonomic and geographic data, in the low levels of sophistication of search mechanisms and the difficulty of linking the data with existing visualization and analysis tools (Guralnick *et al.*, 2007). Mechanisms are therefore needed to report confidence of the data and updated taxonomic and systematic findings and to make textual locality descriptions computer searchable and map compatible (Guralnick *et al.*, 2007).

The reason for insufficient search options and lack of computerized data handling is often the lack of data atomization.

This is frequently the case with biological databases and affects, e.g., taxonomic names and locality descriptions. Concatenation of strings to the lowest level of atomization is the easiest way to bring differently structured data together. Schemas with incomplete data atomization should be considered transient formats for data integration and not biological data standards.

This chapter describes the development of crop-specific population level information systems (CWRIS-PLIS) as an extension of the Crop Wild Relative Information System (Kell *et al.*, 2008a). A great deal of the development work was devoted to improving the level of atomization of taxonomic and geographic information.

31.2 Use Cases for an Information System for Crop Wild Relative Conservation

Maxted *et al.* (1997) proposed the stepwise methodology for the identification of genetic reserve sites for a target crop gene pool. The population level information system has been developed as a device which supports experts in the application of the methodology. The following use cases have been foreseen in order to achieve a prioritization of species and areas, and for *in situ* CWR conservation management and monitoring:

1. Data exploration:
   - (i) Search for occurrences by taxonomic criteria (hierarchical search through taxonomic ranks including synonyms according to different taxonomic views);
   - (ii) Search for occurrences by geographic information (hierarchical search through levels of administrative units or within protected areas);
   - (iii) Combined search by taxonomic and geographic criteria.
2. Data acquisition:
   - (i) Downloading results and displaying them on a map.
3. Data contribution:
   - (i) Editing taxonomic and geographic data for atomization, harmonization and geo-referencing;
   - (ii) Acquisition of population data in the field with portable data assistants and uploading these data to a central database.

The data exploration and data acquisition use cases have been fully implemented in CWRIS-PLIS (http://aegro.jki.bund.de/index.php?id=168), while the data contribution use cases have been only partly implemented.

31.3 Data Sources and Technology Applied

Historical data were analysed to get an overview of the known distribution of CWR and the overlap of their occurrences within protected areas. Different data sources were used depending on the type of data to be retrieved:

1. Occurrence data (including collecting site information for plant genetic resources):
   - (i) The Global Biodiversity Information Facility (GBIF, 2007);
   - (ii) The European Internet Search Catalogue of *ex situ* PGR accessions (EURISCO, 2007);
   - (iii) The NPGS Germplasm Resources Information Network (USDA ARS, 2008);
   - (iv) The European Central Crop Databases for *Avena* (http://eadb.jki.bund.de) and *Beta* (http://idbb.jki.bund.de).
2. Geographic references:
   - (i) The NUTS (Nomenclature des Unités Territoriales Statistiques) nomenclature of territorial units (EUROSTAT, 2003);
   - (ii) The list of local administrative units – LAU (EUROSTAT, 2005);
   - (iii) The database of areas protected under NATURA 2000 regulations (European Environment Agency, 2010).
3. Taxonomic references:
   - (i) Euro+Med PlantBase (2006), which forms the taxonomic core of the Crop Wild Relative Catalogue for Europe and the Mediterranean (Kell *et al.*, 2005, 2008b);

(ii) The taxonomic classification in GRIN Taxonomy for Plants (USDA ARS, 2008). In the case of *Beta* it is essentially based on the revision of *Beta* section *Beta* by Letschert (1993) and *Beta* section *Corollinae* by Buttler (1977);
(iii) The taxonomic system compiled by L. Frese with additional synonyms not represented by the other systems;
(iv) The European *Avena* Database with taxonomic systems by Ladizinsky (1989), Rodionova *et al.* (1994) and Loskutov (2007).

To develop an integrated schema and harmonized data sets, data were downloaded from the mentioned sources, forming a layer of original data. They were redesigned, atomized and adapted in an additional layer of interpreted data. A prototypic application to parse locality strings and identify administrative units was developed in Microsoft Access VBA on a Microsoft Access database. Protected areas were identified for the available geographic coordinates with ESRI software. Finally, the database was migrated to MySQL, and a web application to search, display and download data was designed with JEE 5 Web 2.0 technology (Hibernate, Seam, JSF, Gmaps4Jsf and Google Maps). A prototypic software tool for mobile digital assistants was developed with the CyberTracker development framework (Steventon *et al.*, 2011).

31.4 Modelling Approach

We understand the design of an information system (data model, object model) as the modelling of the scientific domain. Objects and entities should represent domain concepts (domain ontology) in a natural manner and use domain-specific language for naming. Relations should be based on keys meaningful in the domain, so that integrity rules set up by relational database systems help support the logical integrity of the information in the database. To increase usability and support the integrity checking mechanisms, data fields should be atomized, which means that only a single type of information (usually a single figure or word) stands in each field. To facilitate the exchange of data and software, they should be modular; this means they consist of handy components which can work independently of each other and can be used in differently composed information systems. Currently, most biological data structures, even if they consider themselves standards, do not fulfil these principles.

Moore *et al.* (2005–2008, 2008) published a first approach to modelling information for CWR conservation management. In order to implement the mentioned use cases, parts of this model were expanded and modified to further clarify concepts and atomize information. A schematic picture of the modified structure is shown in Fig. 31.1.

The basic unit in *ex situ* collections is the accession, whereas the term occurrence is frequently used for the observation of a taxon *in situ*. In a collecting event one or more accessions are usually retrieved from an occurrence and the material is normally brought to an *ex situ* location where it is stored in a gene bank as a reference sample (accession) and ideally characterized and evaluated. Looking more closely into an occurrence, especially for monitoring and sampling *in situ*, individual plants or patches (agglomerations of individuals) are observed. In experimental monitoring situations, artificial bio-statistical units, plots or transects (not shown in Fig. 31.1) are established. These may contain multiple individuals (cardinality 1:*n*) and patches in full or in part (cardinality *m*:*n*). The mentioned entities are an elaboration of the population structure type described by Moore *et al.* (2008) who neither defined the term population nor arranged explicitly for the documentation of genetic data in the main class 'population'. Following the definition of Kleinschmit *et al.* (2004) the genetic boundaries of a population would have to be determined. For that purpose individual plants need to be geo-referenced and sampled in the natural habitat for genetic analysis. Algorithms based on genetic analysis must then be performed to subordinate individuals or patches to a population.

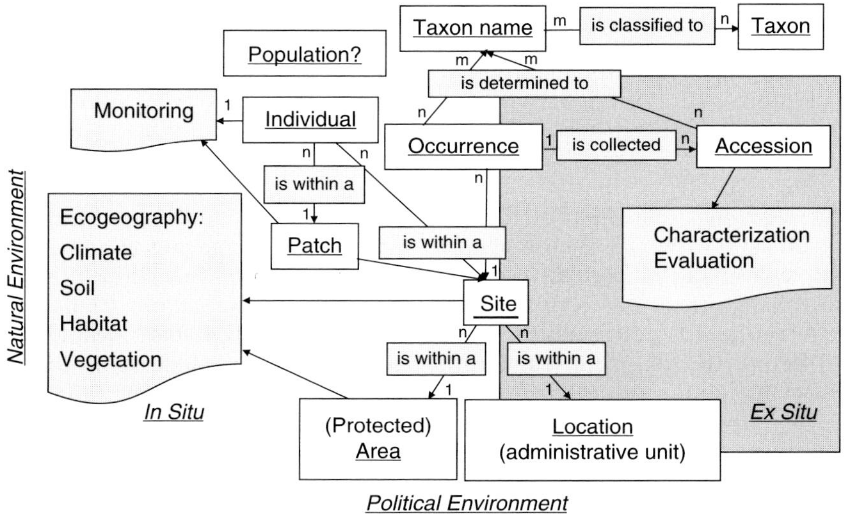

Fig. 31.1. Entities and modules connecting *in situ* and *ex situ* information for protected area selection.

Taxonomic determination is a key issue for the interpretation of occurrence data. The taxonomy module should be able to represent different taxonomic views and differentiate the assignment of names to theoretical concepts (taxonomy) or real world specimens and occurrences (determination). Moore *et al.* (2008) suggest a 'Nomenclature' type for the documentation of taxonomic information. Berendsohn (1995) introduced the concept of a 'potential taxon' as a first approach to separate classification from naming and allow for different taxonomic views. Based on this approach, Berendsohn (1997) published a taxonomic information model for European floristic databases. This model was applied to Euro+Med PlantBase, which forms the taxonomic core of the Crop Wild Relative Catalogue for Europe and the Mediterranean (Kell *et al.*, 2005, 2008b). A similar approach is used for the Taxon Concept Schema (Kennedy *et al.*, 2006), which is used in GBIF. These models grew to high complexity (CATE Project, 2010), making their implementation in working applications and use for data exchange difficult, especially for non-taxonomist communities. Although large and complex class libraries are described, real functionality seen on the web is usually low. Berendsohn *et al.* (1999) consider their model to be a reference model, as it can be modified and simplified for special needs.

A simplification for plant genetic resources and data available in this domain was used here. This was possible by restriction of the model to the part representing the Botanical Code (McNeill *et al.*, 2006), which foresees a limited number of ranks (Art. 4.2). Though assignment of additional ranks is not totally excluded (Art. 4.3), they are rarely seen in digitally available data. Limiting the number of ranks makes modelling of the names table (Taxonname) simpler. Each rank and the original authorship for species and subspecific epithets combinations form an attribute. This atomized information is algorithmically concatenated to the full scientific name, which is used as the primary key for a name object. The 'potential taxon' (Berendsohn, 1995), or 'taxon concept' (Kennedy *et al.*, 2006) is represented by the entity 'Taxon' and exclusively stands for an accepted taxon concept in a taxonomic monograph (system). Classifying names as accepted names or synonyms to taxon concepts within taxonomic systems is the task of 'Taxonomy', an entity modelling the relation 'is classified to' (cardinality *m:n*). A further *m:n* relation 'is determined to' ('TaxonDetermination') represents the taxon identification event (Berendson, 1997).

While in biological taxonomy the existence of a well-formed, codified hierarchical naming concept greatly facilitates data modelling, the situation seems to be different for geographical data. Although the use of data models is widespread in the GIS community, the design process is often *ad hoc* (Glennon, 2010). Classifications can be based on natural geographical features or political administrative boundaries, which may even be interacting (e.g. in the layout of protected areas). We consider a reference point, ideally referenced by a geographical point coordinate as a basic geographical unit, and call it 'Site' in consideration of what has been called collecting or observation site in biological databases. Unfortunately, this term may be somewhat ambiguous, as it is also used in the meaning of area for 'Natura 2000 sites'. We would recommend calling entities representing polygons 'areas' (e.g. protected areas).

For genetic resources conservation, political and administrative issues such as responsibilities for managing *in situ* and *ex situ* resources and the designation and management of protected areas are of importance as well. Community regulation EC 1059/2003 fixed a three-level hierarchical classification of administrative and statistical units (NUTS) for the territory of the EU. Two further levels of local administrative units (LAU) are used and provisions are made for Member States to provide lists of LAUs (Eurostat, 2008). They are still undergoing frequent changes. We combined the three NUTS levels with LAU-2 as attributes of the entity 'Location'.

31.5 Processing Locality Strings

The lack of clear naming concepts in geography is reflected in biological databases, where full text descriptions, usually called 'locality strings', have dominated the documentation of collecting sites and *in situ* occurrences. Processing this information to make it usable for computerized search and geo-referencing is a considerable challenge, which has been undertaken in the BioGeomancer project (BioGeomancer Working Group, 2007). As we found no way to batch process locality strings with the BioGeomancer workbench available at that time, we implemented two approaches for parsing the locality strings in a prototypic MS Access application: (i) parsing along keywords and marks, which frequently occur as attribute references (e.g. 'region', 'district', 'county' etc.) or as logical operators. Lists of these terms and their meanings were used to automatically parse the strings; and (ii) searching NUTS/LAU entries as sub-strings in the locality strings to link the information to a NUTS/ LAU region. While the first approach proved very difficult and obtained poor parsing results, the Eurostat reference information greatly improved the chance of extracting administrative unit information from the locality strings. Obviously this is only applicable for European countries contributing to the NUTS/LAU system.

31.6 Applications for Field Work and Online Search

A CyberTracker tool developed for the recording of monitoring data or the documentation of sampling work covers the following use cases:

- User and survey identification;
- Documentation of administrative unit and protected area where the survey is carried out;
- Documentation of site details (name of a farm, field, natural site, distance and direction to next village or town, GPS reference and elevation, landscape, soil texture);
- Definition and geo-referencing a monitoring plot;
- Counting of individuals species-wise in plots or transects;
- Description of patches by species, their shape and density, and geo-referencing

a fixed reference point or outlining the margins by GPS;

- Performing a demographic census by counting cohorts of different developmental stages (seedling, juvenile, flowering, senescent) in a predefined plot, transect or patch; and
- Geo-referencing and description of individuals (sex, developmental stage, health condition) during sampling.

Data from the CyberTracker tool can be exported into Excel format and imported into databases.

Furthermore, a web application has been developed for the online search of historical occurrence data by taxon, administrative unit and protected area for four genera (AEGRO, 2011). Result sets can be displayed in GoogleMaps and downloaded to Excel format.

31.7 Perspective and Conclusion

A main target in the AEGRO project was to identify resources (data, tools) already available and needs for services to improve online usability of data relevant to the *in situ* conservation of CWR. Figure 31.2 portrays a data flow and data harmonization services architecture considered necessary for CWR work. GBIF offers web services for acquiring occurrence records (http://data.gbif.org/ws/rest/occurrence). Alternatively, data could be accessed with the Biological Collection Access Service for Europe (bioCASE). GBIF intends to have sources such as EURISCO, GRIN or institutional collections as data providers. Technically, this is implemented with some shortcomings (e.g. gene bank accession identifiers are lost in the communication between EURISCO and GBIF, duplicate checks are not performed, and only part of the genetic resources information is currently available in GBIF). Several data types in the source formats (ABCD, MCPDv2) still need improvement. This especially applies to taxonomic and geographic information, which is often incomplete or has concatenated strings. The transformers and interpreters indicated in Fig. 31.2 have been prototyped as semi-automatic interactive applications in Microsoft Access using downloaded data. It would be desirable to implement data flow and interpreter functionality as machine to machine communication and web services for translating synonyms to

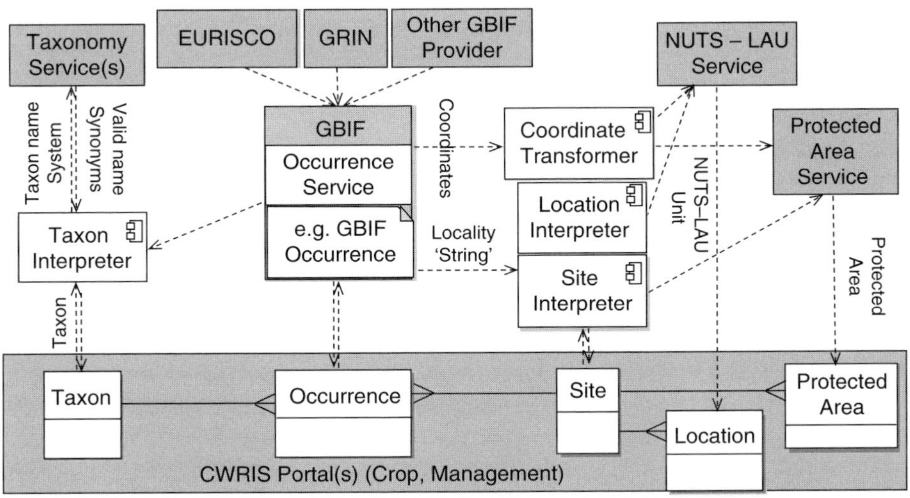

Fig. 31.2. Data flow architecture with transformers and interpreters needed for a portal on historic data relevant for prospection of genetic reserves as designed in AEGRO CWRIS-PLIS.

valid taxa or for providing administrative units or protected areas in response to locality strings or geographic coordinates.

CWRIS-PLIS is an extension of CWRIS aimed at developing concepts and partly implementing tools mainly for practical work in the field or with databases of historical species occurrences. To really cover the population level as defined by Kleinschmit *et al.* (2004), an integration of this information with molecular genetic data and tools will be required.

References

AEGRO (2011) CWRIS-AEGRO PLIS. Available at: http://aegro.jki.bund.de/aegro/index.php?id=168/ (accessed June 2011).

Berendsohn, W.G. (1995) The concept of 'potential taxa' in databases. *Taxon* 44, 207–212.

Berendsohn, W.G. (1997) A taxonomic information model for botanical databases: the IOPI model. *Taxon* 46, 283–309.

Berendsohn, W.G., Anagnostopoulos, A., Hagedorn, G., Jakupovic, J., Nimis, P.L., Valdés, B., Güntsch, A., Pankhurst, R.J. and White, R.J. (1999) A comprehensive reference model for biological collections and surveys. *Taxon* 48, 511–562.

BioGeomancer Working Group (2007) BioGeomancer Workbench. Available at: www.biogeomancer.org/workbench.html (accessed 10 October 2007).

Buttler, K. (1977) Revision von *Beta* Sektion *Corollinae* (Chenopodiacea). I. Selbststerile Basisarten. *Mitteilung Botanik München* 13, 255–336.

CATE Project (2010) About CATE Data Model. Available at: www.cate-project.org/cate-model/hbm2doc/entities/index.html (accessed 1 September 2010).

EURISCO (2007) Quick Search. Available at: http://eurisco.ecpgr.org/search/quick_search_step1_taxonomy.php/ (accessed 10 October 2007).

Euro+Med PlantBase (2006) *Euro+Med PlantBase – the information resource for Euro-Mediterranean Plant Diversity*. Version 2006. Berlin.

European Environment Agency (2010) Natura 2000 data – the European network of protected sites. Available at: www.eea.europa.eu/data-and-maps/data/natura-2000 (accessed 1 September 2010).

Eurostat (2003) *Administrative units/Statistical units*. http://epp.eurostat.ec.europa.eu/cache/GISCO/geodatafiles/NUTS_RG_03M_2003_0.zip (accessed 4 May 2008).

Eurostat (2005) *Local Administrative Units (LAU)*. http://epp.eurostat.ec.europa.eu/portal/page/portal/nuts_nomenclature/local_administrative_units (accessed 4 May 2008).

Eurostat (2008) *European Regional and Urban Statistics – Reference Guide, 2008 edn*. Office for Official Publications of the European Communities, Luxembourg.

Fievet, V., Touzet, P., Arnaud, J.-F. and Cuguen, J. (2007) Spatial analysis of nuclear and cytoplasmic DNA diversity in wild sea beet (*Beta vulgaris* ssp. *maritima*) populations: do marine currents shape the genetic structure? *Molecular Ecology* 16, 1847–1864.

GBIF (2007) The Global Biodiversity Information Facility, Occurrence search. Available at: http://data.gbif.org/occurrences/ (accessed 10 October 2007).

Glennon, A. (2010) Creating and validating object-oriented geographic data models: modelling flow within GIS. *Transactions in GIS* 14, 23–42.

Guralnick, R.P., Hill, A.W. and Lane, M. (2007) Towards a collaborative, global infrastructure for biodiversity assessment. *Ecology Letters* 10, 663–672.

Hirao, A.S. and Kudo, G. (2004) Landscape genetics of alpine-snowbed plants: comparison along geographic and snowmelt gradients. *Heredity* 93, 290–298.

Kell, S.P., Knüpffer, H., Jury, S.L., Maxted, N. and Ford-Lloyd, B.V. (2005) Catalogue of Crop Wild Relatives for Europe and the Mediterranean. University of Birmingham, Birmingham, UK. Available via the Crop Wild Relative Information System (CWRIS) at: www.pgrforum.org/cwris/cwris.asp (accessed 1 May 2011) and on CD-ROM.

Kell, S.P., Knüpffer, H., Jury, S.L., Ford-Lloyd, B.V. and Maxted, N. (2008a) Crops and wild relatives of the Euro-Mediterranean region: making and using a conservation catalogue. In: Maxted, N., Ford-Lloyd, B.V., Kell, S.P., Iriondo, J.M., Dulloo, E. and Turok J. (eds) *Crop Wild Relative Conservation and Use*. CAB International, Wallingford, UK, pp. 69–109.

Kell, S.P., Moore, J.D., Iriondo, J.M., Scholten, M.A., Ford-Lloyd, B.V. and Maxted, N. (2008b) CWRIS: an information management system to aid crop wild relative conservation and sustainable use. In: Maxted, N., Ford-Lloyd, B.V., Kell, S.P., Iriondo, J.M., Dulloo, E. and Turok J. (eds) *Crop Wild Relative Conservation and Use*. CAB International, Wallingford, UK, pp. 471–491.

Kennedy, J., Hyam, R., Kukla, R. and Paterson, T. (2006). Standard data model representation for taxonomic information. *Omics* 10, 220–230.

Kleinschmit, J.R.G., Kownatzki, D. and Gregorius, H.-R. (2004) Adaptational characteristics of autochthonous populations – consequences for provenance delineation. *Forest Ecology and Management* 197, 213–224.

Ladizinsky, G. (1989) Biological species and wild genetic resources in *Avena*. In: *Report of a working group on Avena (third meeting)*. IBPGR, Rome.

Letschert, J.P.W. (1993) *Beta* section *Beta*: biogeographical patterns of variation and taxonomy. *Wageningen Agricultural University Papers* 93, 1.

Loskutov, I.G. (2007) Oves (*Avena* L.): rasprostranenie, sistematika, evoljucija i selekcionnaja Cennost' [Oat (*Avena* L.), distribution, taxonomy, evolution and breeding value], russ., Rossijskaja Akad. Sel'skochozjajstvennych Nauk, Saint-Petersburg.

Maxted, N., Hawkes, J.G., Ford-Lloyd, B.V. and Williams, J.T. (1997) A practical model for *in situ* genetic conservation. In: Maxted, N., Ford-Lloyd, B.V. and Hawkes, J.G. (eds) *Plant Genetic Conservation: the* in situ *approach*. Chapman & Hall, London, pp. 545–592.

McNeill, J., Barrie, F.R., Burdet, H.M., Demoulin, V., Hawksworth, D.L., Marhold, K., Nicolson, D.H., Prado, J., Silva P.C., Skog, J.E., Wiersema, J.H. and Turland, N.J. (2006) International Code of Botanical Nomenclature (Vienna code), *Regnum Vegetabile 146*. A.R.G. Gantner Verlag KG. Available at: www.ibot.sav.sk/icbn/main.htm (accessed 15 January 2011).

Menzie, C., Bettinger, N., Fritz, A., Kapustka, L., Regan, H., Moller, V. and Noel, H. (2008) Population protection goals. In: Barnthouse, L.W., Munns, W.R. Jr and Sorensen, M.T. (eds) *Population-level Ecological Risk Assessment*. SETAC press, Taylor and Francis, Florida, pp. 41–68.

Moore, J.D., Kell, S.P., Iriondo, J.M., Ford-Lloyd, B.V. and Maxted, N. (2008) CWRML: representing crop wild relative conservation and use data in XML. *BMC Bioinformatics* 9, 116.

Moore, J.D., Kell, S.P., Maxted, N., Iriondo, J.M., Ford-Lloyd, B.V. and contributors (2005–2008) International Crop Wild Relative Information Schema Release 2.2. University of Birmingham, UK. Available at: www.pgrforum.org/CWRML.xsd (accessed on 1 May 2011).

Rodionova, N.A., Soldatov, V.N., Merezhko, V.E., Yarosh, N.P. and Kobylyansky, V.D. (1994) *Kulturnaya flora SSSR*. Oves. (Cultivated flora of the USSR), Vol. 2, Part. 3, pp. 1–367. Kolos, M. (In Russian).

Steventon, J., Liebenberg, L. and Pepler, C. (2011) CyberTracker – Towards a worldwide environmental monitoring network. Available at: www.cybertracker.co.za (accessed 1 May 2011).

USDA ARS (2008) National Plant Germplasm System, pcGRIN. Available at: www.ars-grin.gov/npgs/pcgrin (accessed 4 May 2008).

32 Development of an *In Situ* Database Inventory – On the Way to a Swiss Solution for Forages

B. Boller, B. Schierscher-Viret and C. Koehler

32.1 Introduction

The Swiss National Plan of Action for the conservation and sustainable use of plant genetic resources (NAP) was established in 1997 after the adoption of the Global Plan of Action in Leipzig 1996. Since 1999, the Swiss Federal Office of Agriculture has financed 'NAP-Projects'. While the first NAP projects focused on inventory of vegetative propagated perennials especially of fruit trees and regeneration of gene bank material, it soon became apparent that forages play an important role for such a programme. Even though there were only a limited number of traditional varieties of just very few species, such as the persistent 'Mattenklee' type of red clover (Kölliker *et al.*, 2003), it was recognized that the high diversity of permanent, semi-natural grassland in Switzerland deserves particular attention as a reservoir of potentially useful genetic diversity of forage plants.

As part of the NAP programme, Italian ryegrass (*Lolium multiflorum* Lam. ssp. *italicum* Vokart ex Schinz et Keller) and meadow fescue (*Festuca pratensis* Huds.) were chosen to investigate the relationship between the diversity of permanent grassland sites and the diversity of forage plant populations originating from such sites. Molecular studies revealed more than 90% of genetic diversity to be due to within-site variance (Peter-Schmid *et al.*, 2008a) but still, habitat and agricultural management affected genetic structure of the meadow fescue populations. This was more evident and true for both species when morpho-physiological characteristics were taken into account (Peter-Schmid *et al.*, 2008b). For example, meadow fescue populations from high altitudes were more susceptible to crown rust, were shorter, had a more prostrate growth habit and headed later than those from the lowlands. Italian ryegrass accessions from intensively managed sites headed later, grew taller and were more susceptible to crown rust than those from sites with a more relaxed management. These spaced plant results were corroborated by plot experiments showing large differences in agronomic performance among Swiss ecotype populations of Italian ryegrass (Boller *et al.*, 2009). This study showed a clear relationship between the performance of the populations and the botanical composition of the grassland they were collected from. The abundance of species typical for *Arrhenatheretum* alliances was negatively correlated with yield, vigour, snow mould and bacterial wilt resistance of the Italian ryegrass population collected from such grasslands. It was concluded that current efforts to protect only extensively managed, species-rich grasslands were insufficient to preserve agronomically useful variation within grassland species considered important as genetic resources for food and agriculture.

32.2 Swiss Concept for *In Situ* Conservation of Forage Plants

Based on the results of the first NAP projects, a concept for the *in situ* conservation of forage plants was developed by Weyermann (2007) (see Fig. 32.1). The concept aims at maintaining the largest possible variation in plant traits by considering as wide a range of habitats as possible. Habitat diversity is assumed to be captured by the diversity of plant alliances and by considering habitats in different biogeographic regions (see Fig. 32.2). The concept recognizes that agricultural management, together with climatic and soil factors, plays an important role.

The idea behind the concept is represented in Fig. 32.1. Environment and agricultural management are the driving forces of natural selection, which act on both a pool of locally occurring grassland species and a pool of plant genotypes within each species. It is assumed that the process of natural selection will result in a typical association of grassland species which can be described by a certain plant alliance, and at the same time will result in a population of plant genotypes of a given forage plant species which is typical for that alliance. Therefore, including a range of plant alliances in a conservation programme will generate a range of combinations of environment and management enhancing the genetic diversity of plant populations within grassland species.

Plant alliances relevant to the conservation of forage plant species in Switzerland (Table 32.1) were defined on the basis of Delarze *et al.* (1999) and Dietl and Jorquera (2003). The latter source allows for a more detailed consideration of plant alliances typical for more intensively utilized grassland. This was considered appropriate because the studies with Italian ryegrass and meadow fescue ecotypes (Peter-Schmid *et al.*, 2008a, b; Boller *et al.*, 2009) had shown a differentiation between populations originating from sites with moderate or high management intensity.

The concept proposes to create an inventory of five to nine sites per biogeographic region (see Fig. 32.2) for each of the 17 plant alliances listed in Table 32.1. The farmers are to be involved in the choice of the sites. Each site should have a defined agricultural management, which is to be determined using a farmer questionnaire. The feasibility of the concept was tested by

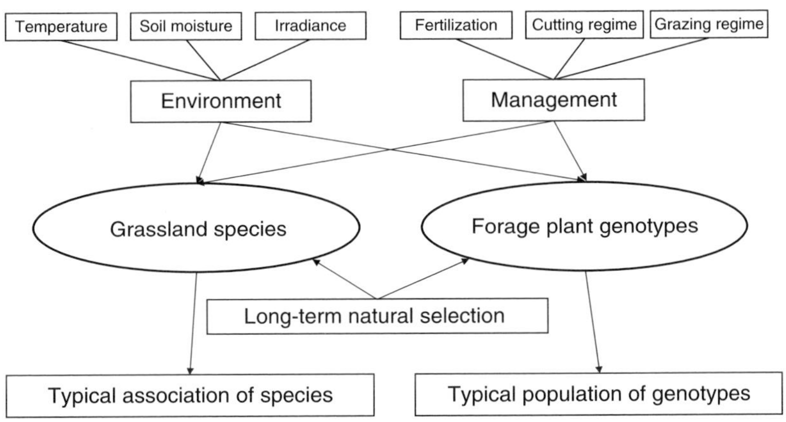

Fig. 32.1. Action of environmental and management factors as driving forces of selection of plant species and of genotypes within a species.

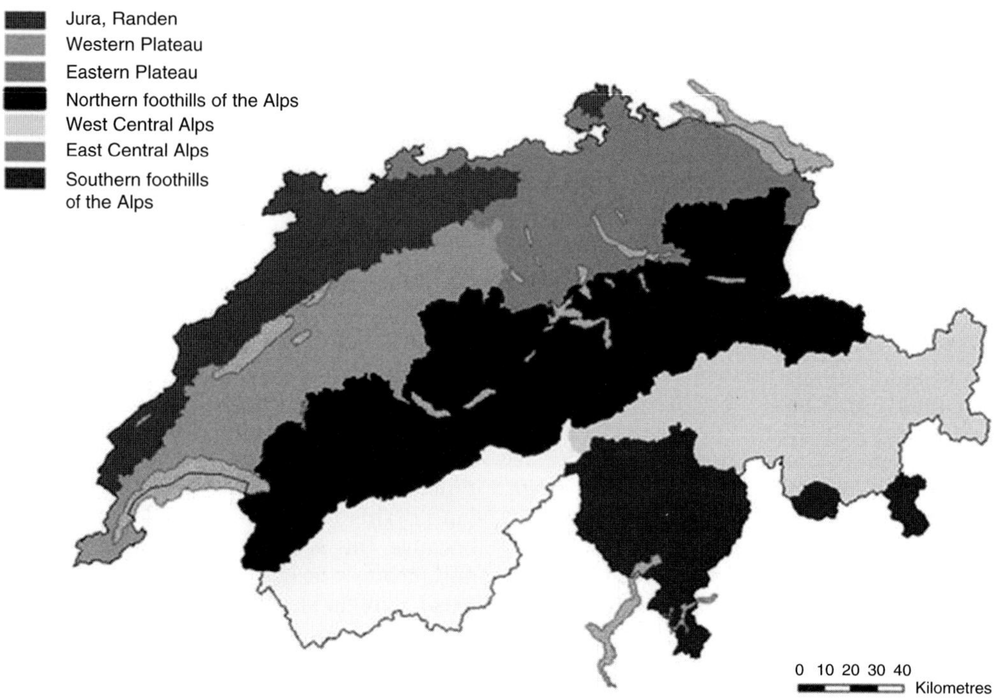

Fig. 32.2. Biogeographic regions of Switzerland according to Gonseth *et al.* (2001).

Table 32.1. Plant alliances relevant to the conservation of forage plant species (from Weyermann, 2007, with supplement from Bosshard *et al.*, 2009).

Alliances/associations	Type of habitat	Altitudinal range (m)
Caricion fuscae/davallianae	Moor	200–2000
Calthion/Filipendulion	Wetland meadow	200–1500
Molinion	Wetland meadow	200–1400
Arrhenatherion	Fertilized meadow, rather extensive	200–800
Taraxacum-Apiaceae meadow[a]	Fertilized meadow, rather intensive	200–1200
Lolietum multiflori[a]	Multi-cut meadow, intensive	200–700
Trifolio-Alopecuretum[a]	Multi-cut meadow, intensive	200–1400
Poo pratensis-Lolietum perennis[a]	Grazing/cutting grassland, intensive	200–1400
Poo trivialis-Ranunculetum repentis[a]	Multi-cut meadow, very intensive	200–1400
Polyono-Trisetion	Fertilized meadow, moderately intensive	900–2000
Cynosurion	Fertilized pasture	200–1600
Poion alpinae	Fertilized pasture	1400–2500
Seslerion	Non-fertilized meadow or pasture	1000–2500
Festuco-Agrostion[b]	Non-fertilized meadow	200–2000
Nardion	Non-fertilized pasture	800–2200
Caricion ferrugineae	Non-fertilized meadow or pasture	200–2200
Mesobromion, Xerobromion	Non-fertilized, dry meadow	200–1400

[a] according to Dietl and Jorquera (2003)
[b] supplement suggested by Bosshard *et al.* (2009)

a pilot study by Bosshard *et al.* (2009) covering the biogeographic region 'Northern foothills of the Alps'. The authors identified 119 sites to which they were able to attribute one of the units listed in Table 32.1, so that each unit was represented by approximately seven sites.

32.3 Representation of *In Situ* Data in the National Database

An *in situ* database was created to implement the Swiss concept for the *in situ* conservation of forages and was developed under the auspices of the national 'Working Group on forages' of SKEK (Swiss Commission for the Conservation of Cultivated Plants). It was then made fit to allocate the data collected in the pilot study of Bosshard *et al.* (2009). The database was integrated into the public Swiss national database for the conservation and sustainable use of plant genetic resources for food and agriculture (NDB-PGRFA), www.bdn.ch (SKEK-CPC, 2011a).

Each site represented in the database is identified by a unique number (COLLNUMB). This number is intended to serve as a bridge to eventual accessions of individual forage plant species collected on the same surface and maintained *ex situ*. This is why a descriptor name was chosen that is also used by EURISCO for *ex situ* conserved material. The sites are essentially characterized by two lists connected by this unique number. The first list with a total of 109 descriptors refers to the identification and description of the site in terms of geographic location, agricultural management and details of the data collection. The second list results from the detailed botanical survey and contains botanical names and abundance scores (Braun-Blanquet, 1928) for all plant species present. Examples of *in situ* descriptors are given in Table 32.2. The complete list of descriptors can be downloaded from www.bdn.ch/descriptors/insitu (SKEK-CPC, 2011b).

Table 32.2. Examples of descriptors for *in situ* data for forage plants in the NDB-PGRFA (Swiss national database for the conservation and sustainable use of plant genetic resources).

Descriptor name	Details
Species list descriptors	
FAMILY	
GENUS	EURISCO-5
SPECIES	EURISCO-6
SUBSPECIES	EURISCO-8
ISFS_ID	Code from Index Synonymique de la Flore Suisse (www.crsf.ch)
ABUNDANCE	Abundance scores according to Braun-Blanquet (1928), +, 1, 2a, 2b, 3, 4, 5
Site descriptors: Collection of vegetation data	
NPANUMB	Project number NAP
INSTCODE	Institute code (FAO Code)
COLLNAME	Collector's name (name of person responsible for botanical survey)
COLLNUMB	Collecting number (identifies site)
COLLDATE	Collecting date of botanical survey
Site descriptors: Identification (passport data)	
MUNICIPALITY	Municipality
FIELDNAME	Name of field (under which it is locally known)
ELEVATION	Elevation (m above sea level)
EXPO	Aspect (S, N, W, E)
SLOPE	Slope
BIOGEOREGION	Biogeographic region (see Fig. 32.2)

Continued

Table 32.2. Continued.

Descriptor name	Details
Site descriptors: Agricultural management	
PARCELSIZE	Plot size (refers to management unit)
PARCELUSE	Main type of utilization (mowing/grazing/mowing–grazing)
ACCESS	Accessibility
CUTTINGFORM	Cutting equipment
SLICEDATE	Usual date of first utilization in spring
SLICENUMB	Number of cuts
GRASSCONSERV	Type of use of mown grass (zero grazing/hay/silage)
FERTILIZINGTYPE	Type of farmyard fertilizer (manure/slurry/compost etc.)
Site descriptors: Characterization and classification of vegetation	
GRAMCOVER	Coverage by Gramineae
LEGCOVER	Coverage by Leguminosae
ALLIANCECLASSIFICATION	Alliance according to *in situ* concept (see Table 32.1)
HABITATHOMOGENEITY	Homogeneity of target alliance
HABITATPORTION	Proportion of target alliance (% of plot size)

32.4 Conclusions and Perspectives

A Swiss concept for *in situ* conservation of forage plants has been developed based on previous project results. Grassland plant alliance and biogeographic region are basic criteria for the choice of sites to become part of the *in situ* inventory. A first biogeographic region, the 'Northern Foothills of the Alps', was investigated in a pilot project. A database was developed to allocate vegetation and site characterization data. Data of the pilot project have been made publicly available online. Future projects are to complement the database with *in situ* data from the Italian ryegrass and meadow fescue studies, and to carry out characterization and evaluation of *Festuca rubra* accessions sampled from sites of the 'Northern Foothills of the Alps' inventory.

Acknowledgement

The NAP Projects with forage grasses and the development of the *in situ* database for forages were financed by the Swiss Federal Office for Agriculture. The *in situ* database for forages was implemented by Claude Paroz, a freelance computer scientist (2XLibre) working under a mandate of the Swiss Commission for the Conservation of Cultivated Plants.

References

Boller, B., Peter-Schmid, M., Tresch, E., Tanner, P. and Schubiger, F.X. (2009) Ecotypes of Italian ryegrass from Swiss permanent grassland outperform current recommended cultivars. *Euphytica* 170, 53–65.

Bosshard, A., Eggenberg, S., Huber, C. and Volkart, G. (2009) In-situ-Erhaltung von Futterpflanzen – Pilotprojekt Alpennordflanke. Bericht im Rahmen des Nationalen Aktionsplanes zur Erhaltung und nachhaltigen Nutzung von pflanzengenetischen Ressourcen (NAP), NAP 03-113. Available at: www.cpc-skek.ch/fileadmin/pdf/NAP-PhaseIII/NAP_03-113_JB08.pdf (accessed 22 February 2011).

Braun-Blanquet, J. (1928) *Pflanzensoziologie: Grundzüge der Vegetationskunde*. Biologische Studienbücher, Julius Springer Verlag, Berlin.

Delarze, R., Gonseth, Y. and Galland, P. (1999) *Lebensräume der Schweiz*. Ott Verlag, Thun.

Dietl, W. and Jorquera, M. (2003) *Wiesen- und Alpenpflanzen: Erkennen an den Blättern, Freuen an den Blüten*. Österreichischer Agrarverlag, Leopoldsdorf.

Gonseth, Y., Wohlgemuth, T., Sasonnens, B. and Buttler, A. (2001) Die biogeographischen Regionen der Schweiz. Bundesamt für Umwelt, Bern. Available at: www.bafu.admin.ch/publikationen/publikation/00207/index.html (accessed 22 February 2011).

Kölliker, R., Herrmann, D., Boller, B. and Widmer, F. (2003) Swiss Mattenklee landraces, a distinct and diverse genetic resource of red clover (*Trifolium pratense* L.). *Theoretical and Applied Genetics* 107, 306–315.

Peter-Schmid, M.K.I., Boller, B. and Kölliker, R. (2008a) Habitat and management affect genetic structure of *Festuca pratensis* but not *Lolium multiflorum* ecotype populations. *Plant Breeding* 127, 510–517.

Peter-Schmid, M.K.I., Kölliker, R. and Boller, B. (2008b) Value of permanent grassland habitats as reservoirs of *Festuca pratensis* Huds. and *Lolium multiflorum* Lam. populations for breeding and conservation. *Euphytica* 164, 239–253.

SKEK-CPC (2011a) Conservation of plant genetic resources – Swiss National Database. Available at: www.bdn.ch (accessed 22 February 2011).

SKEK-CPC (2011b Conservation *in situ* des plantes fourragères. Available at : www.bdn.ch/pages/culture/futterpflanzen/in-situ (accessed 22 February 2011).

Weyermann, I. (2007) Konzept zur in situ Erhaltung von Futterpflanzen. Bundesamt für Landwirtschaft, Bern. Available at: www.cpc-skek.ch/pdf/ConceptFutterpflanzen_Vprovisoire.pdf (accessed 22 February 2011).

33 The Role of EURISCO in Promoting Use of Agricultural Biodiversity

S. Dias, M.E. Dulloo and E. Arnaud

33.1 Introduction

The conservation of agricultural biodiversity in gene banks and the promotion of its sustainable use are fundamental in addressing the challenges posed by climate change and crop adaptation, to ensure food security and quality worldwide. In order to achieve these goals, the provision of easily accessible, good quality and comprehensive data on conserved accessions is a pre-condition to facilitating access to suitable germplasm. FAO (2010) estimates that more than 7.4 million accessions of plant genetic resources for food and agriculture (PGRFA) are conserved worldwide, of which 1,725,315 accessions are maintained in European gene banks, representing about one quarter of the global PGRFA conserved *ex situ*, thus entrusting Europe with a major responsibility in maintaining these resources.

33.2 EURISCO Network – Infrastructure, Roles and Responsibilities

EURISCO, the European Plant Genetic Resources Search Catalogue, was established in 2003 as a European *ex situ* information portal on plant genetic resources (PGR) (Fig. 33.1) with two objectives: (i) to provide technical support to allow countries to develop and build their national inventories, and consequently allow a better understanding of the in-country collections, the conserved material, the diversity conservation coverage and their status; and (ii) to develop and maintain a European PGR Search Catalogue with passport data on *ex situ* collections maintained in Europe. As such, EURISCO has encouraged and enabled countries to make their collections visible to the world. It has been a vehicle to assist countries in meeting national obligations and international commitments, namely to: the International Treaty on Plant Genetic Resources for Food and Agriculture (ITPGRFA) (Art. 5, 12, 13, 14, 16 and 17); the Global Plan of Action (GPA) of the United Nation's Food and Agriculture Organization (GPA Priority Activity 17); and the Convention on Biological Diversity (CBD) (Art. 7; 9; 17), among others.

33.3 What EURISCO Contains

EURISCO currently consists of an online website publishing 41 National Inventories (NI) of *ex situ* collections maintained in Europe (see Fig. 33.2). It holds passport information on almost 1.1 million accessions from 313 European gene banks

Fig. 33.1. EURISCO Home page (http://eurisco.ecpgr.org/).

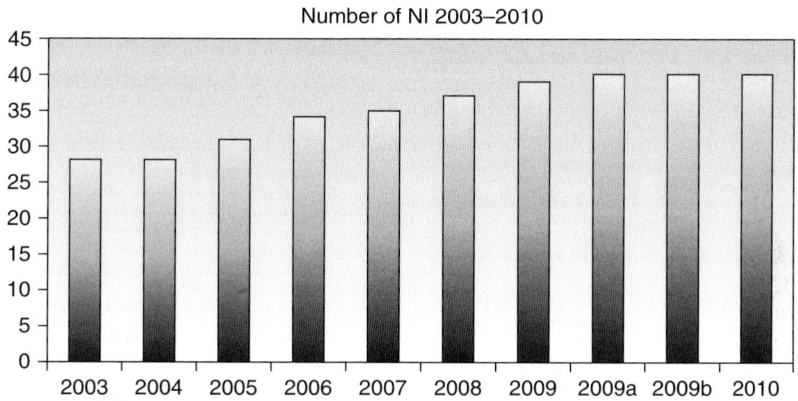

Fig. 33.2. Availability of NIs from the period 2003–2010.

(individual data providers) covering more than 5400 genera and 35,000 plant species, thus representing more than half of the estimated *ex situ* accessions maintained in Europe and roughly 15% of total worldwide holdings (EURISCO, 2010). EURISCO currently provides information on passport data, i.e. unique information about an accession such as identification number, species name, family name, date of collection and georeference data, among others. These data sets are available in diverse collection groups, including food crop genetic resources, forages, wild and weedy species (including crop wild relatives – CWR), medicinal and ornamental plants. EURISCO does not include forest genetic resources, which has its own catalogue – EUFGIS (www.eufgis.cgiar.org). EURISCO covers data on material collected from 217 countries, of which 9% is wild material, 29% breeding/research material, 25% traditional cultivars/landraces, 19% improved cultivars and 18% not known (Fig. 33.3).

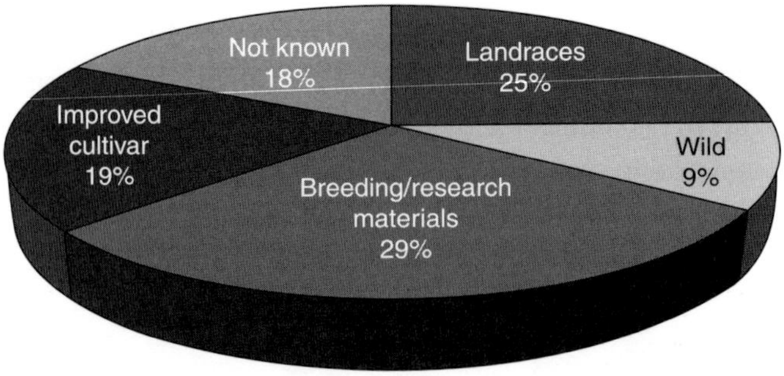

Fig. 33.3. Percentage of data type groups.

33.4 EURISCO Roles and Responsibilities

The NI data available through EURISCO is provided through voluntary contributions from European countries (see Fig. 33.4). However, the data type, standards, content and process of these contributions have been agreed by the network through a Data Sharing Agreement (DSA). Thus the provision and sharing of data is legally implemented within EURISCO. The DSA clearly defines the roles and responsibilities between the data providers and the central aggregation point or institutions holding that role (Gaiji *et al.*, 2008).

The sharing of data has two levels.

1. The national level where countries meet their commitments, where, as a first step, countries appoint a National Focal Point (NFP) that is a EURISCO country representative who is the sole point of contact between the NI and Bioversity International. Additionally, the NFP manages the NI and ensures it is uploaded to EURISCO on a regular basis (at least once a year).

2. At the central level, EURISCO is hosted and maintained by Bioversity International on behalf of the Secretariat of the European Cooperative Programme for Plant Genetic Resources (ECPGR). Bioversity is responsible for coordinating and sustaining the network, ensuring that the existing tools, such as the upload system – intranet – is functioning, the helpdesk is active, quality reports are running and data are published and publicly available through the EURISCO website. One important dissemination action is to promote the portal and encourage users to acknowledge EURISCO as the source of the data accessed through the EURISCO website (EURISCO, 2009).

33.5 Data Exchange Mechanism

EURISCO has evolved a data sharing mechanism that includes three key components.

33.5.1 Data standards

EURISCO members agreed to adopt the Multi-Crop Passport Descriptors (MCPD) for data exchange and also include an extended set of specific information (the first descriptor (0) and the last seven (numbered 29–35) are additional, and specific) to EURISCO. Only the four fields identifying the accession are mandatory, all other fields are highly recommended. The mandatory fields are NICODE (0), INSTCODE (1), ACCENUMB (2) and GENUS (5), the combination of these fields has to be unique) for identification of the NIs and other users' needs, e.g. it has been necessary to accommodate a flag for reporting to the Multilateral System (MLS) set by the ITPGRFA (http://eurisco.ecpgr.org/documents/MCPD%20_EURISCO_Descriptors_111.pdf/). The combination of

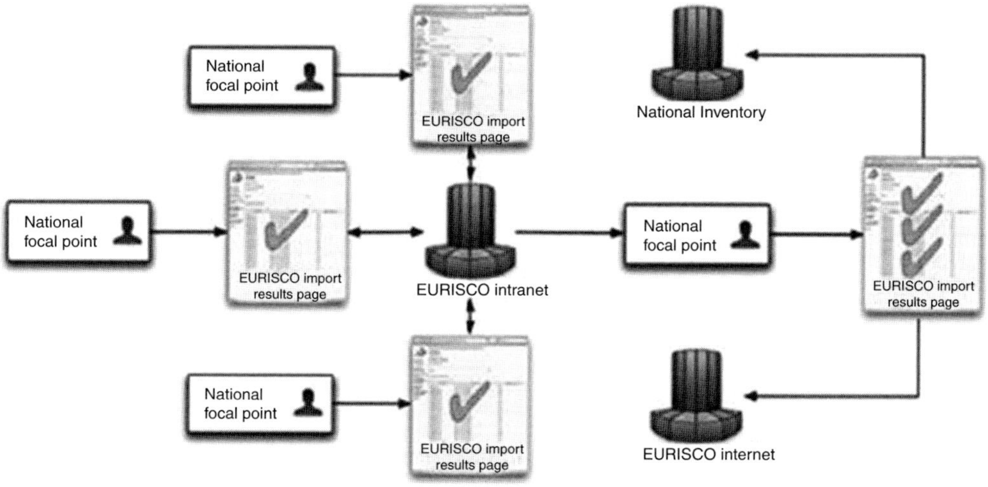

Fig. 33.4. NI collation and upload process.

the MCPD and EURISCO standards enables the countries' NIs to be compiled and shared effectively through EURISCO.

33.5.2 Data upload

An upload system was created to facilitate submission of data to EURISCO (see Fig. 33.5). This tool is a pyramidal and centralized system to which the data is sent by the NFP to be published online, after the NFP receives the relevant data from the gene banks. However, before it goes public, the NFPs are encouraged to follow the following steps:

1. Upload (import) the tab-delimited NI data file.
2. View the transfer reports file to ascertain the status (acceptance/rejection) of the file.
3. View the import reports file for data quality.
4. View the data before uploading them to EURISCO.
5. Upload to EURISCO, when the data are ready to go online.
6. View EURISCO update reports for the success/rejection of the online NI.

From this point on users have free and full access to the information provided.

33.5.3 Data quality checks

The upload system is accessed through an intranet page with an individual password per country NFP. After the NI is uploaded on the system, the central database automatically starts the quality control to check the data files for essential descriptors, dates and type formats, ISO country codes, FAO institute codes, latitude/longitude coordinates and taxonomy. A data quality report is then generated and is available to the NFP on the individual intranet country page to assist the NFPs in deciding on whether to continue the upload of data or abort the session and resubmit the data after the necessary corrections have been made. The step by step upload system allows the NFP to interrupt the process at any stage and restart it whenever the data are considered ready to be published. Only then are the data made universally available to users through the EURISCO website. This process has greatly enhanced the information quality of the collections and assisted in their management through a data quality checking process that would otherwise be a burdensome and costly process for gene banks. With this process EURISCO data are constantly evolving.

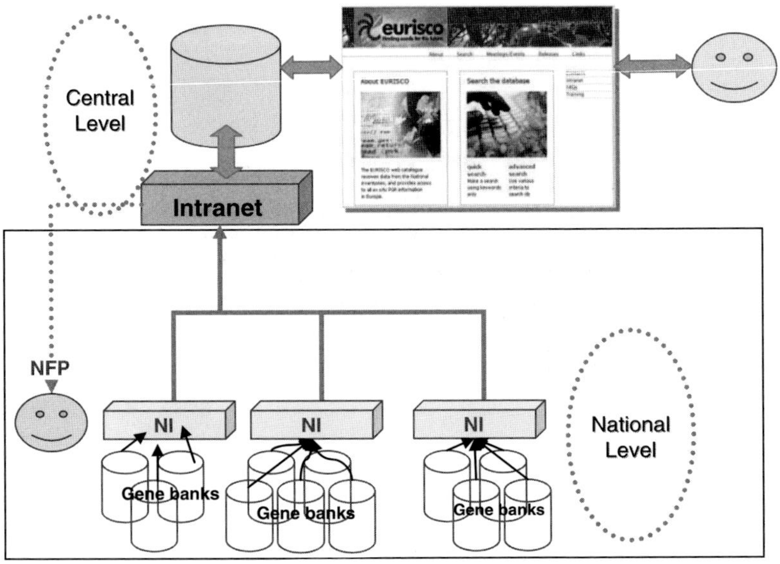

Fig. 33.5. Data upload mechanism and levels of roles and responsibilities.

33.6 EURISCO Use and Users

The EURISCO website has a very flexible and dynamic search functionality allowing the user to answer many PGR research questions, such as 'Who is conserving *Helianthus* from xyz country of origin?', 'At which latitude and longitudes have *Lupinus* been collected from 1975 to 1995?', 'List all the wild material being conserved in European gene banks', view the data and download it. The search capacity ranges from simple to advanced/combined searches, using all 35 data fields simultaneously in a simple, user-friendly format. The portal is widely accessed by scientists, breeders, students and policy makers, among others. Scientists use the information available to carry out gap analysis, diversity studies, climatic modelling and crop adaptability. Breeders look for genetic material of actual or potential interest containing useful traits for breeding.

EURISCO supports the European policy makers in meeting country commitments and national, regional and international obligations such as CBD, GPA and ITPGRFA. More than 200,000 accessions are registered as being part of the MLS of the ITPGRFA (see Fig. 33.6), amounting to approximately 20% of the total material searchable in EURISCO, and covering around 66% of the genera from Annex I of the ITPGRFA (Dias, 2010). EURISCO also allows data analysis on the status of PGR conservation strategies for conservation managers.

Monitoring use of the EURISCO portal during 2010 revealed 8650 visitors, of which half are frequently returning users. It also shows a slight increase for log-on origin of the EURISCO website visitors traffic (109 countries in 2010 compared to 105 countries in 2009). Users searching for data directly from EURISCO have increased by 31.15% since 2009, proving that representation at national and international fora, and networking and public awareness efforts are having a direct and positive impact (Dias, 2010). The addition of characterization and evaluation data is likely to exponentially increase the number of visitors in the coming years.

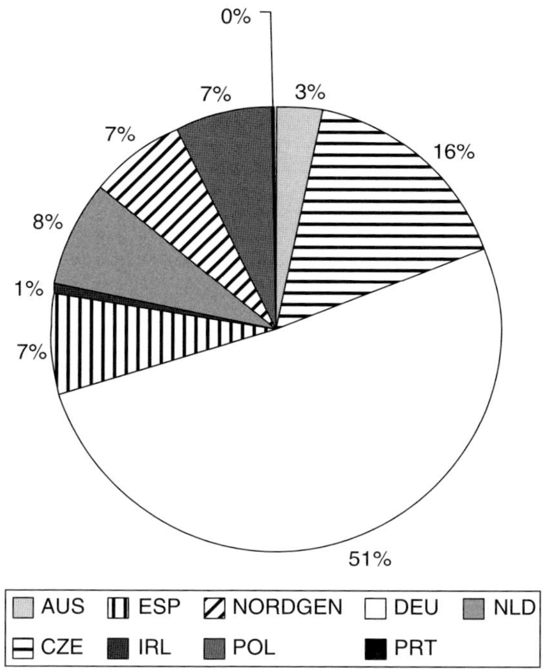

Fig. 33.6. Current NIs that have flagged within EURISCO accessions in the MLS (EURISCO, 2010).

33.7 EURISCO in Other Global Portals – GBIF and GENESYS

In 2006, a collaborative agreement was signed between Bioversity and the Global Biodiversity Information Facility (GBIF) to publish data from the PGR information systems that Bioversity manages and maintains (GBIF, 2010). Therefore, EURISCO serves as a GBIF thematic regional network node, providing the data of the European PGR collections and supporting the deployment of the data-sharing tools.

EURISCO, along with the System-Wide Information Network on Genetic Resources (SINGER) and GRIN-Global, the Genebank Information System on Genetic resources by the Agricultural Research Service of the US Department of Agriculture (USDA-ARS) are the major source of data for GENESYS (see Fig. 33.7), the largest online global accession-level plant information gateway. GENESYS currently holds 2,333,733 records, offering the ability to query records from these three sources simultaneously. GENESYS also aims to be the entry point for users to mine genetic variation by using combinations of data on the characteristics, environments and other aspects of genetic diversity in order to identify accessions of interest and order them online. From this perspective, GENESYS offers access to first sets of characterization and evaluation data.

To facilitate access to the samples, GENESYS will be equipped with an online germplasm request system, based on the SINGER model that is compliant with the recommendations of the ITPGRFA regarding the Standard Material Transfer Agreement (SMTA). GENESYS has been developed and supported by Bioversity International, the ITPGRFA Secretariat and The Global Crop Diversity Trust to be the global information system, as called for by Article 17 of the ITPGRFA. The ECPGR community is a key partner in the further development of this global system and related data standards (Arnaud *et al.*, 2010).

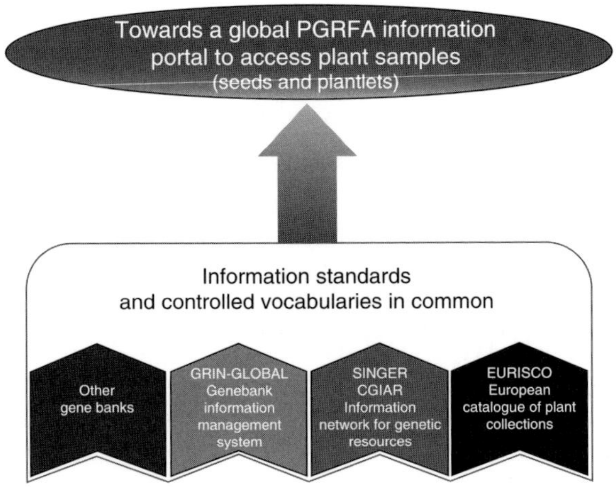

Fig. 33.7. Contributors to the PGRFA Global information portal – GENESYS.

EURISCO partners are also contributing to the development of GRIN-Global, in collaboration with Bioversity and the Global Crop Diversity Trust. GRIN-Global is composed of scalable data management software suitable for any gene bank with a multilingual design to facilitate adoption by gene banks throughout the world (initially supporting Arabic, English, French, Russian and Spanish). Bioversity assists in deploying the software, in gene banks that request it, through which the gene banks will produce formatted data sets that can be exchanged and published online on any platform, including the global accession-level information portal GENESYS.

33.8 EURISCO Future Scope

Currently EURISCO contains passport information on almost 1.1 million accessions. It is envisaged that the scope will be widened to include characterization and evaluation data, further enabling users to identify useful diversity in EURISCO to solve problems such as tolerance to salinity, production, resistance, and adaptation to climate change. Additionally, the recent ECPGR review (ECPGR, 2010) recommended to 'expand EURISCO's structure, in order to include relevant data for the management of the *in situ* and on-farm components'. It has also been suggested to provide information on CWR and landraces traits, as well as the national inventories for diversity present *in situ* and on farm. In this context, a recently approved EU-funded project, PGR Secure, which started in March 2011, aims at developing and establishing *in situ* and on-farm national inventories and linking them to *ex situ* data. These will include molecular, phenotypic, transcriptomics, characterization and evaluation data of *in situ* and on-farm diversity of four main crops of importance to Europe (*Beta*, *Brassica*, *Medicago* and oats). Within this framework there will be research using novel GIS techniques to predict which CWR and landrace (LR) accessions are likely to contain desirable traits. The information derived from these activities will be organized and made available to users through a Trait Information Portal (TIP). The ultimate objective is to integrate TIP with existing genomic (e.g. EMBL Nucleotide Sequence Database) and PGRFA information systems (e.g. Crop Wild Relatives Information System (CWRIS), EURISCO, European Central Crop Databases (ECCDB)) for development of European CWR and LR conservation strategies.

As more information about characterization and evaluation data is requested at trait and population level, it is envisaged that data providers will also expand from the traditional gene bank managers to conservationists, farmers and managers of protected areas; in turn, the user community will expand to new types of users including environmentalists, landscape managers and ecologists. With regard to database development and infrastructure, there will be a major shift focusing more on interoperability options and taking into consideration the development of ontologies, web services, multidimensional databases and diversification of providers. These developments will not only be important in providing support to the network of data providers, but also ensure data availability to users. The EURISCO infrastructure and network has proven to be an easy and effective model which works well as a ready to go tool/mechanism for the deployment of information in other regions and countries, and can even be used at a thematic and global level.

In conclusion, our vision is that EURISCO will provide a wider service for national, regional and international initiatives on PGR. In particular EURISCO could become an instrumental tool to the European Commission Directive 2008/62/EG on Conservation Varieties and help countries to enact specific policies such as 'biodiversity indicators' and the 'Adaptation to Climate Change in the Agricultural Sector', and possibly contribute to the implementation of the EC Community Biodiversity Strategy, The Biodiversity Action Plan for Agriculture (COM/2001/0162) (Dias *et al.*, 2010). EURISCO is a knowledge hub on the plant diversity maintained in European countries, providing data and information on PGR to help countries meet the multiple reporting commitments for the various legally binding and voluntary international instruments, in addition to helping raise general awareness of the need to value and conserve CWR and LR for present and future generations.

References

Arnaud, E., Dias, S., Mackay, M., Cyr, P., Gardner, C., Bretting, P., Kinard, G., Guarino, L. and Louafi, S. (2010) A portal enabling worldwide access to information on conservation and use of biodiversity for food and agriculture. In: Maurer, L. and Tochtermann, K. (eds) *ICT-SECURE: Information and Communication Technologies for Biodiversity Conservation and Agriculture*. Shaker Verlag, Aachen, Germany, pp. 173–186.

Dias, S. (2010) EURISCO E-Bulletin, December 2010. Available at: http://eurisco.ecpgr,org/releases/ebulletin/december2010.php/.

Dias, S., Arnaud, E and Dulloo, M.E (2010) *Info for food - EURISCO and promoting agrobiodiversity*. Oral Presentation at Session 5 – *In situ* PGR conservation: policy implications and implementation. In Symposium 'CWRs & Landraces in Europe' book of abstracts. University of Madeira, Funchal 13–16 September 2010. Portugal.

ECPGR (2010) Report of the Independent External Review Panel. July 2010. Rome. Available at: www.ecpgr.cgiar.org/Evaluation_Report_ECPGR_final.pdf (accessed in December 2010).

EURISCO (2009) *EURISCO data Sharing Agreement*. ECPGR Secretariat, Rome.

EURISCO (2010) EURISCO home page. Available at: http://eurisco.ecpgr.org (accessed 30 December 2010).

FAO (2010) *Second Report on the State of the World's Plant Genetic Resources for Food and Agriculture*. Food and Agriculture Organization of the UN, Rome.

Gaiji, S., Dias, S., Dag, T.F.E. and Franco, T. (2008) Desarrollo de un sistema global de información a nivel de accesiones en apoyo al Tratado Internacional sobre los Recursos Fitogenéticos para la Alimentación y la Agricultura 1. *Recursos Naturales y Ambiente* 53, 126–135.

GBIF (2010) Global Biodiversity Information Facility. Available at: www.gbif.org (accessed 30 December 2010).

34 SWOT Analysis of the German Seed Savers Sector

J. Efken

34.1 Introduction

In Germany, as in all industrialized countries, farmers do not need to maintain and improve their own seed materials. Task sharing between farmers producing goods for the market and breeders producing seeds for the farmers is far advanced. Thus, plant genetic resources management activities by farmers have almost completely disappeared (Frese and Efken, 2002). But this development was attended by a concentration of agricultural production on few crops vice versa the disappearance of the majority of species and landraces formerly cultivated (Porceddu *et al.*, 1988). Reasons are the change of society due to the industrialization and market orientation based on division of labour. Consequently consumer needs changed too. Products had to be marketable and in conformity with market demand since subsistence farming was no longer important or possible for the majority of the population (Swanson, 1995; Swift *et al.*, 2004).

In Germany the problem of loss of diversity is barely apparent in daily life due to a broad range of products with similar ingredients, or products imported from all over the world. Consequently this topic engenders neither broad attention nor engagement (Maggioni and Lipman, 2009). As a result there is the phenomenon of neglected species or rare breeds and moreover, only a few people are involved in the on-farm management (OFM) of these formerly used and now disappearing species, landraces and breeds. In Germany knowledge about the existing private activities to conserve on-farm genetic resources (who maintains what, how many and how) was limited to case studies (Becker *et al.*, 2003) and to a very few publicly known actors, companies and organizations. Detailed information on the attitudes and form of involvement could be gained through an online survey. The questionnaire was made available to approximately 4000 households via newsletters. More than 1200 persons clicked on the questionnaire and a total of 485 questionnaires flowed into the analysis.

These data, in conjunction with results from further studies about agrobiodiversity and OFM in Germany, as well as an international literature review, built the basis for the following 'Strengths, Weaknesses, Opportunities and Threats' analysis (SWOT). SWOT analysis is a tool to evaluate a market, sector or branch by classifying the findings. It is a starting point for generating strategies and implications for organizations, public bodies, etc. Here, rather than review the extensive literature regarding OFM, I focus explicitly on factors with significant influence on the German OFM sector. However, the results presented are not restricted to the German situation specifically, because agrobiodiversity and in consequence OFM is influenced not only by local and specific factors but also by more general issues.

34.2 Strengths

34.2.1 Content

In the questionnaire mentioned 380 persons, or 80% of the participants, used the opportunity to explain their reasons for conserving rare plants or animals in their own words. Besides the dominant topic 'conservation of diversity' (nearly 100%) the topics 'self-realization' and 'enjoyment' played a dominant role. In other studies people connect biodiversity with attitudes such as balance, naturalness, uniqueness (Fischer and van der Wal, 2007). In other words: diversity of agricultural genetic resources and of food are more than an obligation of society and every individual; diversity is a way to improve public and even personal well-being. These aspects should be considered because public awareness is a basis for sustainable strategies for the conservation of biodiversity (FAO, 2010, p. 116). To point out the potential and advantages, one should compare these statements with controversial themes such as pesticide use or genetic engineering. It is obvious that these themes will not have such broad and unanimous support.

34.2.2 Knowledge base

In Germany there is a strong network of more than ten scientific institutes in universities and further research centres such as the Max Planck Gesellschaft and the Fraunhofer Gesellschaft dealing with plant breeding issues. Beside other special gene banks, the Leibniz Institute of Plant Genetics and Crop Plant Research (IPK) in Gatersleben hosts the central German *ex situ* gene bank, the stock currently totalling approximately 147,000 accessions. Specific to Germany there are a high number of independent small-scale breeding companies that are involved in agricultural, horticultural and ornamental breeding (BDP, 2010) (one central information platform about genetic resources in Germany on the web is available at: www.genres.de/en).

34.2.3 Activists

The results of the questionnaire indicate a strong involvement and commitment of those who are active in conservation of agrobiodiversity in Germany. On average they invest more than 10 h/week, and 60% more than 5 h/week, in conservation activities. The majority are involved directly in on-farm activities, specifically on-garden work, but people are also active in public awareness, organizations and marketing. Putting these results together with information about other movements such as 'slow food' or the success of 'pro specie rara' in Switzerland, positive developments are possible. In particular the rise of the mainly private funded 'pro specie rara' to a respectable actor in Swiss agrobiodiversity policy and to a recognized supplier of niche products on the basis of rare plant genetic resources in formal marketing chains shows what is possible in this sector (Hörler and Pro Specie Rara, 2007; Andersen and Winge, 2008).

34.3 Weaknesses

34.3.1 Content

In a study prepared on behalf of the German Federal Ministry of Food, Agriculture and Consumer Protection (BMELV) on communication strategies for agricultural biodiversity, the authors impressively highlighted the difficulties tied to this topic:

> Massive problems do not only occur in the semantic comprehension of the meaning of

> the term, but rather also in relation to the perception and acceptance as a socially relevant topic, as a national and supra-national problem, affecting not only policy and agricultural economics, but also private consumption.
>
> (Kleinhückelkotten *et al.*, 2006)

This outcome is in line with results of an Euro-barometer survey where 65% of European (EU 27) respondents have heard of the term 'biodiversity' but only 35% have 'heard of it and know what it means' (EU Commission, 2007). These findings are not surprising if one thinks about the different dimensions, the complexity and the sometimes ambiguous research findings regarding the valuation of biodiversity and even more of agrobiodiversity (Bulte and van Kooten, 2000; Martin-Lopez *et al.*, 2008): (i) there are the three levels: habitat, inter-species and intra-species diversity one has to deal with (Rubenstein *et al.*, 2005); (ii) there is the public good characteristic and, hence, the difficult integration of agrobiodiversity into private but also public, respectively general, decision-making; (iii) and bound with the latter, the – until now – missing direct effects in daily life in industrialized countries lead to an undervaluation of the issue; independent of the existing level; and (iv) if we talk about the role of agrobiodiversity it is connected with some kind of forecasting future needs and market demand. Prognoses incorporate, of course, uncertainty as the following citation exemplarily shows:

> biodiversity is a source of resources and services that are central to our economies; ... albeit difficult to quantify. Even if this is not immediately apparent, the truth is that the loss of biodiversity damages human interests at large and economic interests in particular.
>
> (Rosa, 2008)

34.3.2 Know-how and resources

Participants of the online survey were asked about the areas in which they possess adequate abilities or in which they desire support (see Fig. 34.1). The results for rare cultivated plants were surprising: only 20% of the active conservers had adequate competence in the breeding conservation of rare plants according to their own information. Here, a significant deficit became apparent, since this ability is the core of the OFM. Consequently, support is desired in terms of maintenance activities. The results indicate a lack of cooperation between scientific and educational institutions and OFM. Furthermore a minor part of the participants rate their own skills in marketing and networking as sufficient. Similar results were found by Kleinhückelkotten *et al.* (2006, p. 154) in their study: 'The actor landscape is

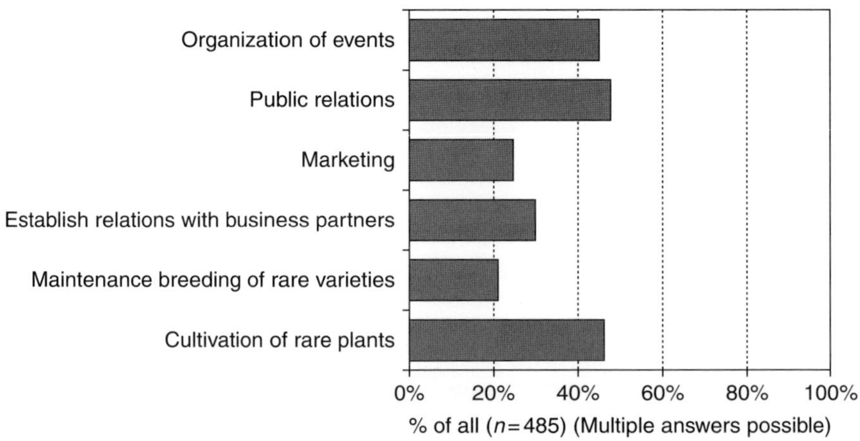

Fig. 34.1. Areas in which adequate competencies are available (Source: Efken, 2010).

shattered, cooperation between the different breeding and maintenance association is rare'. Thus, cooperation within the branch and between OFM and further public and private partners needs to be established or intensified.

OFM is not a field predominantly led by farmers: 70% of the survey participants do not own a farm, but still are involved in the maintenance of rare animals or plants. This is of high importance in two respects. First, OFM and home garden management requires resources of land, labour and technical equipment. It soon reaches its limits if it is not a farm-based activity. Second, since agricultural support schemes (national as well as EU) are almost entirely related to farmers, a majority of actors in this field are excluded from existing public support possibilities. Here, governments and public bodies have to think about new ways of support in order not to miss the core groups of OFM and thus, not to miss objectives of agrobiodiversity at all.

34.4 Opportunities

34.4.1 Social and behavioural trends

According to Maslow's theory (see Fig. 34.2) additional personal needs emerge from increasing income with respect to welfare (Maslow, 1943). Thus, in high income countries with a high level of physical and social infrastructure such as Germany, consumer preferences derive from esteem and self-actualization needs since biological, safety and belongingness needs are mainly fulfilled. This goes in line with social and consumer trends that market research observes nowadays: 'the majority of Europeans associate food and eating with enjoyment' (EU Commission, 2010a). In consequence, consumers like to connect with products and meals more than being satisfied in biological terms. Although price sensibility and financial restrictions always play a decisive role, consumers wish for 'add-ons' like health and wellness, convenience, environmentally friendly, regional, fair trade, beauty, etc. More general trends are individualism, experimentalism and change, differentiation, pleasure, etc. (Toops, 2008; Food channel, 2009; Nestlé, 2009). Most of these issues could be summarized by the so-called consumer group 'Lifestyle of Health and Sustainability' (LOHAS). Although the described developments may represent main trends, they do not stand for typical mass markets of single 'bulk' products but for product differentiation amongst others through niche markets and an increasing market share of niche and speciality products. The main message in this case is that these trends correspond to a high degree with characteristics like uniqueness, exclusiveness, naturalness and environmental friendliness, regionalism, tradition etc. of products deriving from rare plants. Thus, it is a matter of marketing and resources to capitalize on OFM products. An OECD working group on economic aspects of biodiversity (WGEAB) sees information as the primary factor in order to create markets for biodiversity products (OECD, 2003, 2004). In other words: public awareness and marketing tools such as product-, price-, distribution- and communication-policy are key factors (for examples see Secretariat of the CBD, 2008). Of particular importance is the subject health. The health issue is directly connected with

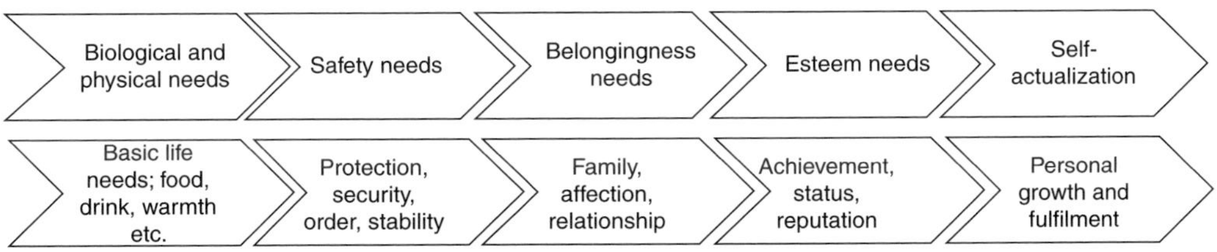

Fig. 34.2. Maslow's theorem of the hierarchical composition of human needs.

agrobiodiversity through medicinal and aromatic plants but more generally linked with agrobiodiversity because of the fact that 'International food guidelines consistently recommend eating diversely because humans need to consume more than 40 essential nutrients which can only be obtained from a wide range of food items,' and that 'Apart from direct health effects, scientists suppose that it affects consumer satisfaction as it increases the enjoyment of eating, encourages personal choices, and enlarges consumer opportunities for food selection and thereby satisfies various taste preferences' (Drescher, 2007, pp. 25, 28). These citations from a PhD study sound just like an advertising slogan – 'There is nothing to add.'

34.4.2 New and modified marketing channels

Since the 1990s there has been a renaissance of alternative food channels and consumer–producer networks (Marsden *et al.*, 2000). Consumers want to judge for themselves and with their own capacities what they eat. Highly standardized and treated/processed food represents anonymity, complexity and opacity (Murdoch and Miele, 1999). Beside the product price, consumer judgements are based on other aspects, like their cultural roots, health issues, nature and environment, etc. (Nygård and Storstad, 1998). This leads to new and modified marketing channels in which participants try to connect new consumer attitudes with production, processing and trade (Murdoch and Miele, 1999). Thus, producers of products from rare plants have to evolve by themselves local networks or go into existing networks and use distribution mixes in order to meet demand (Spiller, 2005). Keeping in mind the limited resources of local producers and traders in contrast to the potential of the established food industry and food retailers, government and public bodies ought to seek support possibilities, not only in the field of marketing, but also in the preceding activities of product creation.

34.5 Threats

34.5.1 Marketing trends

Although marketing trends offer opportunities, there are at the same time trends that hamper agrobiodiversity. As a starting point one has to bear in mind that a product based on a rare plant usually can be offered to a limited amount only and will probably not be produced in a low cost way compared to standard agricultural products. Then, all marketing channels which require large amounts of highly standardized and fast moving products at a low price cannot be used. But these are just the mainstream requirements of the vast majority of products and of the connected retailers in which discounters depict the extreme case of low cost retailing (EU Commission, 2010b; Nielsen, 2010). In Germany market share of discounters is the highest within Europe and perhaps worldwide. But also in other European countries, the market share of discounters is increasing. According to Nielsen (2010) consumers wish to do their shopping quickly and have it standardized, albeit they like to compare to some extent. In conclusion, all these findings do not support marketing of diverse agrobiodiversity products. Competition matters in a market economy and will be the driver of changes. Here, the important question will be whether and how speciality products and niche market products survive or even gain importance.

34.5.2 Handling of agrobiodiversity

It is commonly accepted that broad and diverse genetic resources for cultivated plants are needed for successful plant breeding in order to increase output and to avoid serious damage by pathogens (Heal *et al.*, 2004). Major challenges are the identification, integration and use of genetic resources in commercial crops. Various studies underscore the benefit of genetic diversity through adding value and increased social welfare by improved crop varieties (Rubenstein *et al.*, 2005). This stands in contrast to the discussion regarding agrobiodiversity on the species level. There is no intensive discussion and research strategy or proposal, whether and how to enhance diversity of the used and consumed agricultural and horticultural plants (FAO, 2010; p. 112). This applies even more to industrialized countries. Concentration on a few main crops leads to fewer investments in breeding and product development of the other plants and consequently they become less important respectively and lose competitiveness (FAO, 2010, p. 103).

34.5.3 Missing link to the consumer

OFM links agricultural and socio cultural aspects because both local knowledge about cultivated plants or animals and all possible forms of use are obtained (Virchow, 1999). An important component of OFM besides the immediate maintenance of rarer varieties is the production for consumption. Thus, only when the reintroduction of rare cultivated varieties succeeds, does a conservation strategy show a complete and sustainable picture. In consequence, as long as the discourse about strategies and possible implementation areas is limited to the scientific community and public agents in ministries and offices, strategies were threatening to fail due to a missing valuable link to the general public (Nordbeck and Kvarda, 2006). It is more and more common sense that public involvement supports successful solutions through better understanding, acceptance and participation (Fischer and van der Wal, 2007). Kleinhückelkotten *et al.* (2006, p. 71) propose the development of a concept that appeals to the basic values, needs and desires of the consumer and the general public in order to raise public awareness, as well as to increase direct demand for diverse and, consequently, healthy nutrition.

References

Andersen, R. and Winge, T. (2008) Success Stories from the Realization of Farmers' Rights Related to Plant Genetic Resources for Food and Agriculture. Background Study 7, FNI Report 4/2008, Lysaker. Available at: www.fni.no/doc&pdf/FNI-R0408.pdf (accessed 22 October 2010).

Becker, H.C., Bergmann, H., Jantsch, P. and Marggraf, R. (2003) Darstellung und Analyse von Konzepten des On-farm-Managements pflanzengenetischer Ressourcen unter besonderer Berücksichtigung der ökonomischen Rahmenbedingungen in Deutschland, Studie für das Bundesministerium für Verbraucherschutz, Ernährung und Landwirtschaft. Staatliche Universitätsbibliothek, Göttingen. Available at: http://webdoc.sub.gwdg.de/ebook/y/2003/becker/on-farm-management.pdf (accessed 20 October 2010).

Bulte, E. and van Kooten, G.C. (2000) Economic Science, Endangered Species, and Biodiversity Loss. *Conservation Biology* 14 (1), 113–119.

Bund Deutscher Pflanzenzüchter (BDP) (2010) Members of the German Plant Breeders Association. Available at: www.bdp-online.de/en/Ueber_uns/Mitglieder (accessed 22 October 2010).

Drescher, L.S. (2007) *Healthy Food Diversity as a Concept of Dietary Quality*. Cuvillier Verlag, Göttingen.

Efken, J. (2010) Online-survey of keepers of rare animals or plants about their activities and attitudes in Germany. *Landbauforschung* 60(1), 35–44.

EU Commission (2007) Attitudes of Europeans towards the issue of biodiversity. Flash Eurobarometer Series 219. Available at: http://ec.europa.eu/public_opinion/flash/fl_219_en.pdf (accessed 20 October 2010).

EU Commission (2010a) Food-related risks. Special Eurobarometer 354. Available at: www.efsa.europa.eu/en/factsheet/docs/reporten.pdf (accessed 20 October 2010).

European Commission (2010b) Commission Staff Working Document, On Retail Services in the Internal Market. Available at: http://ec.europa.eu/internal_market/retail/docs/working_document_en.pdf (accessed 02 November 2010).

FAO (2010) *The Second Report on the State of the World's Plant Genetic Resources for Food and Agriculture*. FAO, Rome.

Fischer, A. and van der Wal, R. (2007) Invasive plant suppresses charismatic seabird – the construction of attitudes towards biodiversity management options. *Biological Conservation* 135(2), 256–267.

Food Channel (2009) The Food Channel's Top Ten Trends for 2009. Available at: www.foodchannel.com/articles/article/the-food-channels-top-ten-trends-for-2009 (accessed 20 October 2010).

Frese, L. and Efken, J. (2002) A case of community-based, traditional on-farm germplasm management in Germany. *Plant Genetic Resources Newsletter* 130, 73–76.

Heal, G.M., Walker, B., Levin, S., Arrow, K.J., Dasgupta, P., Daily, G., Ehrlich, P., Maler, K.-G., Lubchenco, J., Kautsky, N., Starrett, D.A. and Schneider, S.H. (2004) Genetic diversity and interdependent crop choices in agriculture. *Resource and Energy Economics* 26(2), 175–184.

Hörler, H. and Pro Specie Rara (2007) ProSpecieRara – 25 Jahre für die Vielfalt. LID Dossier No 422, 26 April 2007. Available at: www.prospecierara.ch/uploads/LID-Dossier_ProSpecieRara_2007.pdf (accessed 20 October 2010).

Kleinhückelkotten, S., Wippermann, C., Behrendt, D., Fiedrich, G., Schürzer de Magalhaes, I., Klär, K. and Wippermann, K. (2006) Kommunikation zur Agro-Biodiversität: Voraussetzungen für und Anforderungen an eine integrierte Kommunikationsstrategie zu biologischer Vielfalt und genetischen Ressourcen in der Land-, Forst-, Fischerei- und Ernährungswirtschaft (einschließlich Gartenbau). Available at: www.bmelv.de/cln_163/SharedDocs/Downloads/Broschueren/Agrobiodiversitaet_Kommunikation.pdf?__blob=publicationFile (accessed 09 August 2010).

Maggioni, L. and Lipman, E. (2009) ECPGR Report of a Task Force on On-farm Conservation and Management, Third Meeting, 2–3 October 2007, Ljubljana, Slovenia. Available at: www.bioversityinternational.org/fileadmin/bioversity/publications/pdfs/1320_Report%20of%20a%20task%20force%20on%20on-farm%20conservation%20and%20management.pdf?cache=1290438161 (accessed 9 August 2010).

Marsden, T., Banks, J. and Bristow, G. (2000) Food supply chain approaches: exploring their role in rural development. *Sociologia Ruralis* 40, 424–438.

Martin-Lopez, B., Montes, C. and Benayas, J. (2008) Economic valuation of biodiversity conservation: the meaning of numbers. *Conservation Biology* 22(3), 624–635.

Maslow, A.H. (1943) A Theory of Human Motivation. Originally Published in *Psychological Review*, 50, 370–396. In: Green, C.D. (ed.) Classics in the History of Psychology. Available at: http://psychclassics.yorku.ca/Maslow/motivation.htm (accessed 20 October 2010).

Murdoch, J. and Miele, M. (1999) 'Back to nature': changing 'worlds of production' in the food sector. *Sociologia Ruralis* 39, 465–483.

Nestlé (ed.) (2009) *So is(s)t Deutschland, Ein Spiegel der Gesellschaft*, Frankfurt a.M.

Nielsen, (2010) Universen 2010 Deutschland, Handel Verbraucher Werbung. Available at: http://de.nielsen.com/pubs/documents/Universen_2010_2.pdf (accessed 2 November 2010).

Nordbeck, R. and Kvarda, E. (2006) New Modes of Governance for Sustainability in Austria: Comparing the Legitimacy and Effectiveness of the Austrian Biodiversity Strategy and the Austrian Forest Dialogue. In: Berlin Conference on the Human Dimensions of Global Environmental Change – Resource Policy: Effectiveness, Efficiency, and Equity. Available at: http://userpage.fu-berlin.de/ffu/akumwelt/bc2006/download.htm (accessed 2 September 2010).

Nygård, B. and Storstad, O. (1998) De-globalization of food markets? Consumer perceptions of safe food: the case of Norway. *Sociologia Ruralis* 38, 35–53.

OECD (ed.) (2003) *Harnessing Markets for Biodiversity, Towards Conservation and Sustainable Use*. OECD Publishing, Paris.

OECD (ed.) (2004) *Handbook of Market Creation for Biodiversity, Issues in Implementation*. OECD Publishing, Paris.

Porceddu, E., Ceoloni, C., Lafiandra, D., Tanzarella, O.A. and Scarascia Mugnozza, G.T. (1988) Genetic resources and plant breeding: problems and prospects. In: Miller, T.E. and Koebner, R.M.D. (eds) *Proceedings of the Seventh International Wheat Genetics Symposium*. Institute of Plant Science Research, Cambridge, pp. 7–22.

Rosa, H.D. (2008) Business: a key player in coping with the biodiversity challenge. *CBD Business Newsletter* 3(3), 4.

Rubenstein, K.D., Heisey, P., Shoemaker, R., Sullivan, J. and Frisvold, G. (2005) Crop genetic resources: an economic appraisal. *Economic Information Bulletin* 2, 1–41.

Secretariat of the Convention on Biological Diversity (Secretariat of CBD) (2008) Business 2010. *CBD Business Newsletter* 3(2).

Spiller, A. (2005) Vermarktungsoptionen für Biodiversitätsprodukte. In: Efken, J. (ed.) *Vermarktungsstrategien für innovative Produkte und Verfahren auf der Basis genetischer Ressourcen für Ernährung und Landwirtschaft,* Bonn, *Schriften zu Genetischen Ressourcen* 25, 13–22.

Swanson, T.M. (1995) Why does biodiversity decline? The analysis of forces for global change. In: Swanson, T.M. (ed) *The Economics and Ecology of Biodiversity Decline: The Forces Driving Global Change*. Cambridge University Press, Cambridge, pp. 1–9.

Swift, M.J., Izac, A.-M.N. and van Noordwijk, M. (2004) Biodiversity and ecosystem services in agricultural landscapes – are we asking the right questions? *Agriculture, Ecosystems & Environment* 104(1), 113–134.

Toops, D. (2008) News and Trends Editor Consumer Trends for 2009, With the economy affecting shopping, consumers grapple with prices and desires. Available at: www.foodprocessing.com/articles/2008/ConsumerTrends2009.html?page=1 (accessed 20 October 2010).

Virchow, D. (1999) *Conservation of Genetic Resources*. Springer-Verlag, Bonn.

35 Achievements of the European Native Seed Conservation Network – ENSCONET

R.J. Eastwood and J.V. Müller

35.1 Introduction

'Europe is very fortunate having a well organised network for the *ex situ* conservation of seeds of plant species – ENSCONET' (Heywood, 2009). The European Native Seed Conservation Network (ENSCONET) project for the first time unified all key facilities for European native seeds. Much has changed in the *ex situ* conservation of native plants in Europe over the last 5 years. Today, European native seed banks speak with one voice, have defined high level, common working standards, exchange staff regularly, and most share their holding data in a freely accessible web-based database, to list only a few of the transformations.

35.2 ENSCONET

ENSCONET is part of the wider Millennium Seed Bank Partnership (MSBP), a global *ex situ* conservation initiative, which, having collected 10% of the world's seed-bearing flora, has set its next target at 25%. Biodiversity in Europe is threatened in the same way as biodiversity globally for example by habitat degradation and loss, invasive species and climate change. Europe's threatened plant diversity includes landraces, crop wild relatives (CWR) and wild species. ENSCONET was initiated to facilitate the *ex situ* conservation of the native European seed-bearing flora.

The ENSCONET project was implemented as a Coordination Action, from November 2004 to October 2009 and received funding through the European Union's (EU) 6th framework programme (FP6) instrument. The Royal Botanic Gardens, Kew (RBG Kew) initiated and coordinated the ENSCONET project from the Millennium Seed Bank. The project has now grown into the ENSCONET Consortium with a Secretariat provided by RBG Kew.

Seed banking is one of the most powerful *ex situ* conservation tools available to combat the loss of biodiversity. The aim of seed banking wild species is not to compensate for *in situ* conservation but to work in concert with it. *Ex situ* conservation is also used to:

- Provide high quality and well identified seed material (for reintroduction programmes and reinforcement of endangered populations at their natural site).
- Conserve genetic resources (for forestry and plant breeding programmes).
- Provide and control access to seed material for other types of research.

Seed banking takes advantage of the natural properties of seed-bearing plants (spermatophytes), namely the production of storage capsules – seeds – and high incidence of out-cross breeding. The consequence of out-crossing is that most genetic diversity resides within (rather than between) populations. Consequently, in most cases, seed samples taken from relatively few populations will contain a high proportion of the alleles found within the species (Neel and Cummings, 2003). Furthermore, seeds occupy a small volume. Therefore, a large amount of diversity can be stored in a small space. Compared with other methods of plant *ex situ* conservation (such as field gene banks, botanical garden plant collections, *in vitro* material, pollen banks and DNA banks), seed banking offers many advantages including potential storage for centuries, implementation of simple technology, relatively low costs and maintenance of pure lineages.

ENSCONET was designed to bring together isolated activities and institutions to maximize effectiveness in achieving the common goal of successful *ex situ* conservation of the European native flora. Only a few years ago, before the launch of ENSCONET in November 2004, little coordination between the native seed banks in Europe occurred. Seed banks largely acted alone adopting different working standards. Today this has changed: European native seed banks speak with one voice, have defined high level, common working standards, exchange staff regularly, and most share their holding data in a freely accessible web-based database.

The purposes of ENSCONET are to:

- Improve quality, coordination and integration of European seed conservation practice, policy and research for native plant species.
- Assist EU conservation policy and its obligations to the CBD and its Global Strategy for Plant Conservation (GSPC).

ENSCONET achieves these by providing a platform for discussion and the exchange of experience through annual general meetings, workshops and exchange visits. It builds relationships between staff, scientists and horticulturalists, from seed banks across Europe, who in the past rarely met but who share similar goals. It is these people and their enthusiasm that are vital to the success of the network.

ENSCONET was initiated in 2004 with 19 members. Membership grew over time and at the end of the EU-funded phase there were 24 members and 7 associates from 18 countries across Europe. Partners were distributed from Portugal to Cyprus and from Norway to the Canary Islands. They ranged from well-established institutes such as the seed banks in Madrid and Warsaw to much smaller, newer facilities such as the botanic garden seed banks in Brussels and Helsinki. A full list of members can be found on the ENSCONET website (www.ensconet.eu).

ENSCONET covers nine of the biogeographical regions of Europe (European Environment Agency, 2005). Ensuring adequate bioregional coverage in the network is more important than representing each political unit because species distributions have more affinity with environmental conditions than country boundaries (MacArthur, 1984). Some countries are only part of one biogeographical region, e.g. the UK (Atlantic) and Cyprus (Mediterranean). Other countries span several biogeographical regions, e.g. Spain (Atlantic, Mediterranean, Alpine, Macaronesian).

ENSCONET is a lead organization for the European Strategy for Plant Conservation (ESPC) 2008–2014 (Planta Europa, 2008), Europe's response to the GSPC (Secretariat of the Convention on Biological Diversity, 2002), which was born from the Convention of Biodiversity. The project's work focused on GSPC targets 3 (protocols for plant conservation) and 8 (*ex situ* conservation) but also contributed to targets 14 (education, public awareness) and 16 (networks).

The success of ENSCONET has been widely acknowledged. It was noted as a key success in the ESPC (Planta Europa, 2008). The Secretariat of the Convention on Biological Diversity (2009) commented 'the *ENSCONET* network has linked a powerful group and diversity of institutions to target seed conservation needs for European wild plants for the first time'.

35.3 ENSCONET Activity Areas

It is through the activity areas that the project has been achieving its aims. The ENSCONET project comprised four activity areas: Collecting, Curation, Data Management and Dissemination. Management formed a separate action. The major achievements in each activity area are documented below.

35.3.1 Collecting

For the first time it is possible to find out easily whether a European wild plant species is held as an *ex situ* seed collection. This is due to the unified holdings list compiled by ENSCONET. Using this list, gaps in holdings have been identified. Collecting plans have been prepared at the European, national, bioregional and institutional levels using an innovative collaborative approach that focuses on biogeographic plant distributions. The Collecting Plan Report (Eastwood, 2009) documents this process and provides case studies of species included. The advantages of planning collaboratively include avoidance of duplication, improved geographic representation, wider genetic sampling representation and strengthened working relationships between institutes across country borders. The development of members' collecting plans has allowed institutes to prepare for the future and have a basis for planning different future funding scenarios. The approach itself has potential wider application as it could be implemented at any scale to coordinate governance or conservation of a region of shared or overlapping responsibility, e.g. waterways, nature reserves or marine areas.

The largest achievement of this activity area has been the production of the ENSCONET Collecting Manual for wild species (ENSCONET, 2009a). In contrast to crop species where there is ample literature, there is little published on the collection and storage of wild plant species. Seeds of wild species are diverse in many aspects, for example in their maturation time, ripening requirements and storage needs. The ENSCONET Collecting Manual aims to promote high standards across all aspects of seed collecting and maximize the capture of genetic diversity. It documents best practice for collecting Europe's native plant species and covers all stages of collecting from planning expeditions to caring for collections post harvest. Importantly, a data passport form is included, which helps to ensure accurate and detailed data are collected in the field.

The manual was developed through targeted workshops and extensive testing in the field. It has also been internationally peer reviewed. Anne Cochrane from the Threatened Flora Seed Centre in Western Australia commented 'The ENSCONET Collecting Manual provides the collector with comprehensive up-to-date knowledge about the correct protocols required to sample diversity for *ex situ* conservation, and the data required to document these collections.' Adherence to these standards should result in high quality and long-lived collections. To facilitate wide coverage and easy accessibility, pdf files of the manual can be downloaded free from the ENSCONET website (www.ensconet.eu/downloads) in nine European languages.

35.3.2 Curation

Under this activity area an inventory of European seed conservation facilities and resources for wild species was carried out, allowing the network as a whole to identify its strengths. This has been particularly useful to target exchange visits for specific training or study (including seed bank design advice). Additionally, ENSCONET identified research necessary to further enhance the value of seed banking in the conservation, restoration and sustainable use of the native European seed-bearing flora.

Preparation of the ENSCONET Curation Protocols and Recommendations (ENSCONET, 2009b) documenting best practices for native seed bank curation filled the gap in written documentation for processing wild species. Furthermore, it defined standard curatorial practices. It was developed through a number of workshops discussing practical techniques and methods. Topics covered include seed cleaning, drying, long-term storage and seed regeneration. Like the ENSCONET Collecting Manual, the Curation Protocols and Recommendations publication has been peer reviewed and is available online in eight European languages.

35.3.3 Data management

Through the Data Management activity area, the ENSCONET partners developed a database schema (including standards) to facilitate sharing seed bank data. This schema includes a protocol for recording germination data. ENSCONET has ensured that each partner has a data management system that is compatible with the ENSCONET database (http://enscobase.maich.gr/). This central database, ENSCOBASE, provides a location to share information and increase accessibility to collections. Most partners, and some non-ENSCONET seed banks, contributed their seed bank holdings and associated data to ENSCOBASE. Collections, to date, come from 40 countries across 11 biogeographic regions and are held in 29 institutes. ENSCOBASE holds data on over 42,400 seed accessions from 9294 taxa. ENSCONET members have not specially targeted CWR, yet 62% of the taxa in ENSCOBASE are CWR (as identified by Maxted *et al.*, 2006). Data stored include conditions and results of nearly 21,000 germination tests. Users can make use of simple pre-defined or complex customized searches to view data. The database shows seed availability along with contact details for all holding seed banks.

ENSCOBASE has been fundamental in monitoring conservation progress. The database shows that 44% of EC Habitats Directive plant species and 27% of the European threatened plant list are held in seed banks across Europe. ENSCOBASE provided an invaluable contribution to the review of progress towards GSPC target 8 (Sharrock and Jones, 2009), which demonstrated that at least 42% of European threatened plant taxa are in accessible *ex situ* collections within their region of origin.

The network members developed ENSCOTOOL, a Virtual Seed Bank Data Manager, to allow partners to manage their content in ENSCOBASE. They can amend or update existing data and add new records. To ensure the data are useful and of high quality, standards are implemented in the tool including a set of mandatory fields, which include taxon name, accession ID, availability status, collecting date and number of seeds. ENSCOTOOL allows the utility of ENSCOBASE to seamlessly extend beyond the initial project.

To facilitate field data collection, a prototype system for utilizing portable computers with GPS in the field was developed. This system was demonstrated and trialled during a joint meeting in the Italian Alps in summer 2009. Further development of this system is in discussion.

35.3.4 Dissemination

Communicating with the public and the wider scientific community is very important for ENSCONET. The main portal for this is the website, which partners created and maintained in English, French and Spanish. During the project period an internal members' only section of the website allowed internal communication including document exchange. Additionally an e-forum was started, stimulating exchange and discussion with the public and within the project. Using this for e-meetings also reduced travel requirements. Production and circulation of the annual bulletin, ENSCONEWS (www.ensconet.eu/News_Bulletin.htm) documenting project activities to decision makers, members of national and European parliaments and international organizations disseminates the work and results of ENSCONET further. Advertising through news coverage including radio and television brought the ENSCONET project to the public whereas the scientific community was reached through journal articles and representation at more than 90 conferences.

One of the major products of the dissemination activity area is the seed bank educational tool, the Virtual Tour, aimed at users aged 9 and upwards. The Virtual Tour is available, free, online (www.ensconet.eu/Tour.tml) in nine languages. A DVD version has also been distributed to educational establishments, including primary schools, within ENSCONET member countries. The Virtual Tour has three parts. During the first two parts an animated cartoon seed character explains why *ex situ* conservation is needed and follows the seed banking process through a virtual seed bank. The third section is interactive allowing the visitor to explore and view photos, videos and descriptions of processes and equipment in a seed bank for native plants.

35.4 Outlook

After the end of the FP6 funded phase of the project, the network members agreed to continue the momentum of ENSCONET. They formed the ENSCONET Consortium with 30 partners by December 2010.

The initial aim is to sustain the dialogue, impetus and the collaborative spirit generated through ENSCONET and alongside this to maintain the website and continue managing the database. However, the ambition is to obtain further funded projects to expand the activities of ENSCONET. Additional activities may include facilitating access to seed material and removing constraints to seed germination and storage.

A priority for the future is increased data sharing and linkages, e.g. with CWRIS, ECPGR European Central Crop databases and EURISCO. It is clear that there is scope for native seed banks and crop genetic resource banks/reserves to work more closely together. There are differences in philosophy and practicality to conserving crop and wild species though in reality there is a continuum between the two. Both communities share common interests and alongside these face similar threats (habitat fragmentation, changes to farming and forestry practice, invasive aliens and climate change), challenges (sustainable development, biofuels) and problems (funding, networking, building and maintaining partnerships). Each can offer complementary activities and knowledge. Collaborative areas could include germination problem solving, storage behaviour investigations, knowledge exchange, research into alternative storage methods, comparison of standards and curation methods, provision of material for pre-breeding and seed storage and duplication.

35.5 Conclusion

Only a few years ago, before the launch of ENSCONET, there was little harmonization between the native seed banks in Europe. Seed banks acted in an un-coordinated way at local, regional and national levels, adopting different working standards. The accessions that existed in Europe had been collected over many years in an *ad hoc* fashion with varied quality and effectiveness in storing methods. Although individual seed banks offered access to seed material, there was no integrated database listing available material, making it very difficult for users to locate and request seeds.

Through a spirit of collaboration, ENSCONET has for the first time unified all key facilities in a network for European native seed conservation, which has during the last five years changed significantly the way the native seed conservation community is organized. Today, European native seed banks speak with one voice, have defined common high-level working standards, exchange staff regularly and most share their holding data in a freely accessible web-based database, to list only a few of the achievements.

The EU member states and the European Commission have identified the need to continue the support for native seed banks and seed conservation on a European scale in order not to lose the significant dialogue and momentum gained. This will allow seed conservation activities to be expanded and implemented more widely. Further native seed banking projects will be initiated and some will provide the opportunity for wider cooperation.

References

Eastwood, R.J. (2009) ENSCONET European Collecting Plan Report. Unpublished report. Millennium Seed Bank Partnership, Ardingly, UK. Available at: www.ensconet.eu/download.htm (accessed November 2010).

ENSCONET (2009a) ENSCONET seed collecting manual for wild species. Millennium Seed Bank Partnership, Ardingly, UK. Available at: www.ensconet.eu/download.htm (accessed November 2010).

ENSCONET (2009b) ENSCONET curation protocols and recommendations. Millennium Seed Bank Partnership, Ardingly, UK. Available at: www.ensconet.eu/download.htm (accessed November 2010).

European Environment Agency (2005) Biogeographical regions of Europe. Available at: www.eea.europa.eu/data-and-maps/data/biogeographical-regions-europe-2005 (accessed November 2010).

Heywood, V.H. (2009) The impacts of climate change on plant species in Europe. Standing Committee on the Convention on the Conservational European Wildlife and Natural Habitats.

MacArthur, R.H. (1984) *Geographical Ecology: Patterns in the distribution of species*. Princeton University Press, New Jersey.

Maxted, N., Ford-Lloyd, B.V., Jury, S., Kell, S. and Scholten, M. (2006) Towards a definition of a crop wild relative. *Biodiversity and Conservation* 3, 333–345.

Neel, M.C. and Cummings, M.P. (2003) Effectiveness of conservation targets in capturing genetic diversity. *Conservation Genetics* 17, 219–229.

Planta Europa (2008) *A Sustainable Future for Europe; the European Strategy for Plant Conservation 2008–2014*. Plantlife International (Salisbury, UK) & the Council of Europe (Strasbourg, France).

Secretariat of the Convention on Biological Diversity (2002) *Global Strategy for Plant Conservation*. Secretariat of the Convention on Biological Diversity, Montreal, Quebec, Canada.

Sharrock, S. and Jones, M. (2009) Conserving Europe's threatened Plants: Progress towards Target 8 of the Global Strategy for Plant Conservation Richmond, UK. Botanical Gardens Conservation International.

Secretariat of the Convention on Biological Diversity (2009) The Convention on Biological Diversity Plant Conservation Report: A Review of Progress in Implementing the Global Strategy of Plant Conservation (GSPC), 48 pages.

36 Medicinal and Aromatic Plants ECPGR Working Group: Objectives and Achievements

A.M. Barata and Å. Asdal

36.1 Introduction

Medicinal and aromatic plants (MAP) have been an important resource for human health care. According to the World Health Organization, the majority of the world's human population, especially in developing countries, depends on traditional medicine based on MAP (WHO, 2002). Between 50,000 and 70,000 plant species are known to be used in traditional and modern medicinal systems throughout the world (Schippmann *et al.*, 2006). In Europe, at least 2000 MAP species are used on a commercial basis, of which 1200–1300 are native to Europe. The increased global interest in the use of MAP and the increasing demand on raw materials by various processing industries (pharmaceutical, food, cosmetic, perfume, etc.) have resulted in an expanding market. Increasing demand for MAP is placing pressure on natural resources, since most species used are still collected in the wild. Uncontrolled over-exploitation of wild plants, their habitat loss and alteration are the main reasons why medicinal plants, their study, evaluation, utilization and conservation have become essential parts of the programmes of international organizations, such as WHO, IUCN and WWF.

Recognizing the importance of these species/genera, the Steering Committee of the European Cooperative Programme for Plant Genetic Resources (ECPGR), during its eighth meeting, in October 2001, agreed on the establishment of the Medicinal and Aromatic Plants Working Group (MAPWG). The Working Group was initially included in the Vegetables Network and at the moment is a part of the Sugar, Starch and Fibre Network. The first meeting was held in Gozd Martuljek, Slovenia, in September 2002. This meeting brought together 18 members representing 20 European countries and 12 observers, including several Slovenian officials and scientists and representatives from the World Wildlife Fund for Nature (WWF-UK and WWF-TRAFFIC Europe) (Baricevic *et al.*, 2004). The main objectives of MAPWG were to develop conservation strategies for Europe by inventory of MAP genetic resources, *ex situ* and *in situ* conservation, characterization/evaluation, development of general and crop-specific descriptors, documentation of *ex situ* collections and *in situ* populations. From 2002 until 2010 three more MAP Working Group meetings have taken place: in 2004 in Strumica, Macedonia FYR, in 2007 in Olomouc, Czech Republic together with the Vegetable Network and in 2009 in Izmir, Turkey. During the Fourth Meeting held in Izmir, the MAPWG members were able to review the proposed tasks and their achievement and agreed on those to be carried out until the end of the current phase of ECPGR (2007–2013):

- Crop-specific characterization and evaluation descriptors, for ten priority species.
- Documentation of *ex situ* collections, following the EURISCO procedures.
- Documentation of *in situ* populations.
- Development of a common Project: 'Conservation and characterization of oregano (*Origanum vulgare* L.) wild populations in Europe'.

The project's main objectives are: the inventory, survey and characterization of habitats of native populations of *Origanum vulgare* L. in the partner countries; collecting herbarium specimens; study of genetic and chemical variability of oregano populations; and the documentation of characterization data. The data collected within this project will also contribute to the selection of oregano accessions to be included in a European Collection according to the initiative for 'A European Genebank Integrated System' (AEGIS).

36.2 Selection of a Priority List of MAP Species

After consultation the Working Group members agreed a priority list of ten model species/genera for further study using the following criteria:

- The species must be important from the point of view of both biological and chemical diversity.
- The species must be important from the point of view of medicinal plant production (either from cultivation and/or collection).
- There must be some common interest in the gene bank conservation of the chosen species for different European regions (at least a third of countries for each).
- The conservation and characterization/evaluation results achieved could be used as a model for the conservation of other MAP species.

The final 10 species/genera proposed for study are: *Mentha piperita* and *M. spicata*; *Thymus vulgaris* and *T. serpyllum*; *Origanum* spp. *Carum carvi*; *Melissa officinalis*; *Achillea millefolium* agg. *Salvia officinalis*; *Artemisia absinthium*; *Gentiana lutea*; and *Hypericum perforatum*. The methodology developed for these priority species/genera will serve as a model for other species and will encourage appropriate activities to be carried out on MAP at national level. Further, the MAPWG proposes that the original 'priority list' of MAP species should in time be expanded. The expanded priority list will be used to produce a document showing the level of threat and the Red List status for MAP species in Europe.

36.3 General and Crop-specific Descriptors

The WG developed a proposal for a MAP general descriptor list, which include the following categories: passport; management; environment and site. These descriptors are now available in the MAPWG webpage (ECPGR MAP WG page, 2010). The main purpose of creating crop-specific descriptors for the priority species is to be able to:

- Characterize *in situ* populations and/or gene bank accessions *ex situ* for their taxonomic and chemotaxonomic status.
- Define morphological, chemical and genetic characteristics of gene bank accessions for selection, breeding work and cultivar development.
- Find potential traits for food, industry and trade.

The final version of the ten target species descriptors will then be uploaded on to the MAPWG page.

36.3.1 MAP collections in Europe

The status of MAP collections in 19 European countries (Austria, Czech Republic, Estonia, Finland, Iceland, Hungary, Israel, Italy,

Latvia, Lithuania, Norway, Poland, Portugal, Romania, Serbia, Slovenia, Spain, Sweden and Turkey) and from NordGen (representing the five Nordic countries) was generated from the country reports presented at the last working group (WG) meeting in 2009. A synthesis of the information provided by WG members responding to a questionnaire about the status of their MAP collections prior to the meeting also gave very valuable information. The country reports gave comprehensive information about collections in all reporting countries, including which species and subspecies are conserved and also how many accessions are conserved in each collection (Barata *et al.*, 2010).

Based on the compilation of data provided, some key statistics of MAP collections are: (i) there are 19,500 European MAP collections, and out of these 15,277 accessions are collected from domestic flora; and (ii) there are approximately 14,000 *ex situ* seed or 5000 field gene bank accessions and a further 226 accessions are conserved *in situ*. The summaries of the reports provided interesting data about the size of the national species collections. The size of the collections, i.e. the number of accessions/species, shows that many of the collections are small and thus conserve only small portions of genetic diversity. Only 482 out of the total number of 2168 collections have five or more accessions (see Table 36.1); the MAP species with the largest number of collections is shown in Table 36.2. For each of the ten priority species, Table 36.3 shows the: number of countries holding collections, total number of accessions, number of indigenous accessions, number of accessions conserved *ex situ* (seed bank/field) and number of accessions preserved *in situ*.

Table 36.1. Relation between the size and number of collections.

No. of accessions	No. of collections
≥ 5	482
≥ 10	286
≥ 50	73
≥ 100	35*

*includes some crops other than MAP.

Table 36.2. Table with the largest MAP collections and in which countries.

Country	Species	No. of accessions
Hungary	*Papaver somniferum*	828
Israel	*Sinapis alba*	363
Spain	*Rosmarinus officinalis* L.	335
Turkey	*Salvia* spp.	332
Spain	*Lavandula latifolia* Medik.	277
Turkey	*Origanum* spp.	219
Turkey	*Sideritis* spp.	206
Austria	*Papaver somniferum*	196
Portugal	*Humulus lupulus*	162
Spain	*Salvia lavandulifolia* Vahl.	151
	Total	3069

Morphological characterization and evaluation data of the conserved accessions was provided and has been carried out in 3446 of the total of 19,500 accessions. Chemical evaluation has been carried out in 1645 accessions and molecular research has been carried out in only 308 accessions.

36.3.2 Documentation of *ex situ* collections and *in situ* populations

The European Internet Search Catalogue EURISCO has been used to upload information from the MAP collections registered in the national databases. A total number of 14,287 accessions are documented in national databases and out of these 6863 MAP accessions have been uploaded in EURISCO. Evaluation data from Latvia and Hungary have been uploaded in national databases. Efforts have been made to be able to document MAP *in situ* populations, using existing databases used for instance to document *in situ* populations of crop wild relatives.

36.3.3 AEGIS initiative

The task-sharing initiative which currently has the highest priority in ECPGR is AEGIS.

Table 36.3. Summary by priority number of countries holding collections, total number of accessions and preservation status.

Taxa	Number of countries with collections	Number of accessions		Conserved		
				Ex situ		
		Total	Indigenous	Seed	Field	*In situ*
Achillea millefolium	14	185	159	111	55	6
Artemisia absinthum	13	108	91	44	67	5
Carum carvi	15	622	449	441	243	0
Gentiana lutea and *Gentiana* sp.	8	61	50	49	8	0
Hypericum perforatum	15	493	390	425	99	6
Melissa officinalis incl. *Melissa* sp.	16	166	108	127	43	0
Mentha spp. incl. 13 species and hybrids	18	558	480	329	345	0
Origanum vulgare, incl. *Origanum* sp. and *O. vulgare* subsp.	19	745	568	463	298	5
Salvia officinalis incl. *Salvia* sp. and subsp.	17	805	482	650	154	0
Thymus spp.	21	881	650	580	278	36

The MAP WG during its last meeting discussed how their activities could contribute to the overall implementation of the AEGIS concept for MAP species. The WG believes that MAP *ex situ* accessions can be offered to be designated as European Collection accessions and agreed that the AEGIS quality standards for conservation of accessions in *ex situ* gene banks will be adopted as far as possible in the national collections and gene banks. The WG will be involved in identifying Most Appropriate Accessions for MAP species. The WG has agreed that relevant criteria for choosing species for inclusion into AEGIS could be adequacy of knowledge about the accessions, e.g. documentation about landraces and varieties.

36.4 Project 'Conservation and characterization of oregano (*Origanum vulgare* L.) Wild populations in Europe'

During the ECPGR Steering Committee meeting in June 2008 the project 'Conservation and characterization of oregano (*Origanum vulgare* L.) Wild populations in Europe' was submitted and approved by the MAP WG, but with a substantial budget reduction. Due to the financial reduction the project tasks had to be revised and a new project plan was elaborated. The project will be carried out in the period from the 1 June 2010 to 31 March 2011. The project will have the following countries as partners: Slovenia, Portugal, Albania, Czech Republic, Bulgaria, Finland, Italy, Israel, Latvia, Macedonia FYR, Norway, Slovakia, Turkey, Croatia, Serbia, Spain, Greece, Hungary and Lithuania.

The main objectives of the project are to make an inventory of, and to survey native populations of wild oregano (*Origanum vulgare* L.), to characterize their genetic and chemical variability, and to find out the distribution pattern of taxonomically defined populations in European countries. The data on characterization of plant material will be documented and will be available for exchange between all partners. This project aims at providing preliminary results and the background information for the establishment of future regional collection(s) (according to the AEGIS concept) of *Origanum vulgare* L. in Europe. The specific objectives of this project are:

- Inventory, survey and characterization of habitats and of native populations of *Origanum vulgare* L. in European countries, members of ECPGR;
- Collecting of herbarium specimens;
- Study of genetic and chemical variability of oregano populations;
- Documentation; and
- Exchange of characterization data among partners.

The partners were also encouraged, according to available resources and funds, to collect seeds for gene bank conservation and to characterize the populations, using the general descriptors and the descriptors for *Origanum* that have been developed by the MAP WG.

The study of genetic and chemical variability of oregano populations, will be performed by Dr Johannes Novak, at the Institute for Applied Botany and Pharmacognosy in the University of Veterinary Medicine in Vienna. The main outcome of the project is data on the distribution and properties of native populations of taxonomically defined oregano (*Origanum vulgare* L.) in the ECPGR MAPWG member countries that are partners of the project. The material will be characterized genetically and chemically. The data on characterization of plant material and of habitats will be documented and distributed among partners. The results of the project will be used in further studies and activities of the ECPGR MAPWG. Knowledge about the valuable characters of the plants will be used in the planning of future collection and implementation of conservation measures using concepts similar to AEGIS. Results of this project are to be presented, in a report format, before the next mid-term ECPGR SC meeting, planned in the second half of 2012.

36.5 Definition of Tasks and Responsibilities for Phase VIII of ECPGR

- *Documentation of ex situ collections* should continue to follow EURISCO. The data providers should thus organize and forward their passport data, following the EURISCO descriptors format, to their respective National Focal Point (NFP). The information needs to be continuously updated and therefore this should be a permanently ongoing activity.
- *Documentation of* in situ *populations* should be undertaken by each member country, continuing to use their own tools until a common database can be developed and used.
- *Characterization and evaluation descriptors* of the ten priority species are to be finalized and will be made available through the MAPWG web page.

The ten target species will remain as the highest priority, but a new expanded list will show priorities for a broader range of species. This updated MAP list should be the basis for the development of a document showing the level of threat and the Red List status for MAP species in Europe.

36.6 Priorities to be Undertaken in the Future

- *European Red List for MAP species.* Many MAP species are threatened in the wild flora and have already been red listed, and it is important to address this fact in future work. The possibilities for making a European Red List for MAP species has been discussed, and it was decided to carry out such an exercise, alone as a working group or if possible integrated in the ongoing initiative for making a European Red List for crop wild relatives (CWR).
- *In situ conservation of MAP species* would need increased focus in the coming years and experiences made in some of the countries will be available for other partners.
- *MAP species* in vitro *or by cryopreservation.* Safety duplication of MAP collections is insufficient and the WG

believes that *in vitro* and cryopreservation could be used to improve this situation. Projects with the aim to propagate rare and threatened species (not exclusively traditional MAP species) by micro-propagation methods will be useful for conservation and reintroduction of species into former habitats, and also for commercial production of plants and MAP products.

References

Barata, A.M., Asdal, Å. and Lipman, E. (2010) *Report of a Working Group on Medicinal and Aromatic Plants*. Fourth Meeting, 29 September–1 October 2009, Kuşadası, Turkey. Bioversity International, Rome.

Baricevic, D., Bernàth, J., Maggioni, L. and Lipman, E. (compilers) (2004) *Report of a Working Group on Medicinal and Aromatic Plants*. First Meeting, 12–14 September 2002, Gozd Martuljek, Slovenia. IPGRI, Rome.

ECPGR MAP WG page (2010) Medicinal and Aromatic Plants Working Group. Available at: www.ecpgr.cgiar.org/Workgroups/Med_aromatic/med_aromatic.htm (accessed on 20 September 2010).

Schippmann, U., Leaman, D.J. and Cunningham, A.B. (2006) Cultivation and wild collection of medicinal and aromatic plants under sustainability aspects. In: Bogers, R.J., Craker, L.E. and Lange, D. (eds) *Medicinal and Aromatic Plants*. UR Frontis Series no. 17. Springer, Dordrecht, the Netherlands.

WHO (2002) *WHO Traditional Medicine Strategy 2002–2005*. World Health Organization, Geneva, Switzerland.

37 A Community Divided: Lessons from the Conservation of Crop Wild Relatives Around the World

D. Hunter, L. Guarino, C. Khoury and H. Dempewolf

37.1 Introduction

Global agreements such as the Convention on Biological Diversity (CBD) and the International Treaty on Plant Genetic Resources for Food and Agriculture (ITPGRFA) have brought considerable attention to bear on the conservation of biodiversity, and increasingly on agricultural biodiversity. However, despite the strategies, action plans and projects that have ensued, and indeed proliferated, the species and genetic diversity represented by crop wild relatives (CWR) remains under-conserved, and under-utilized as a resource for crop improvement. Estimates of the proportion of *ex situ* holdings worldwide made up by CWR accessions range from 2% (Astley, 1991) to 2–6% (Maxted and Kell, 2009), to 15% (Hammer *et al.*, 2003) and to 18% (FAO, 2010). By any standard this is inadequate, given the large amount of genetic diversity which these species represent, certainly relative to their cultivated cousins (Petersen *et al.*, 1994; Vollbrecht and Sigmon, 2005; Damiana, 2008) and their actual contribution to agricultural output (Phillips and Meilleur, 1998). This contribution should only increase, as breeders are better equipped through improving biotechnologies to utilize novel variation in order to produce adequate food, under more challenging conditions, with less energy, on at most the same amount of land, in a more ecologically sustainable manner. The *in situ* conservation of CWR is equally inadequate, with very few examples of effective conservation actions inside protected areas, and even fewer outside these areas (FAO, 2010). It seems both strategies are failing. Why?

37.2 The Challenge of Conserving CWR in Gene Banks

Large gaps in species and genetic coverage remain to be filled in *ex situ* collections. Maxted and Kell (2009) estimate that 94% of European CWR species are entirely missing from *ex situ* collections. In a recent study of the state of conservation of CWR of the *Phaseolus* (bean) genus, Ramírez-Villegas *et al.* (2010) found that of the 85 taxa in the genus, 48 (57%) are either severely under-represented or not present in gene banks, and only five taxa were assessed as well represented in *ex situ* collections. Collecting of CWR germplasm may have been neglected in the past due to a view that wild species are less threatened in the field in comparison to other types of germplasm, such as landraces, as well as being less easy to use and more complicated to maintain. In fact, many of these species are threatened in the field, in particular by land use and climate change, invasive species and over-exploitation. Collecting may also have been constrained by a lack of data on where exactly to collect. But recent advances in spatial data management and analysis, and in the quantification of genetic diversity, should considerably facilitate targeting of specific areas for maximum diversity return on investment (Ramírez-Villegas *et al.*, 2010).

The world's gene banks lack an efficient, internationally coordinated system to maintain high standards, and many gene banks face insecure funding; therefore it should not be taken for granted that the CWR diversity currently held in gene banks is cared for adequately. Vital gene bank activities, such as the periodic regeneration of accessions, are constrained by lack of funding, expertise and research, leading to regeneration backlogs and the subsequent loss of unique genetic diversity (Engels and Rao, 1998; Schoen *et al.*, 1998; Fowler and Hodgkin, 2004; Dulloo *et al.*, 2009; FAO, 2010; Khoury *et al.*, 2010). In some cases, we simply do not know how best to conserve the seeds of wild relatives, let alone how to regenerate them, as the basic seed longevity and storage behaviour research has not been carried out. Gene banks are in any case not exempt from the dangers posed by natural disasters, as experienced by the national gene bank in the Philippines during cyclone Xangsane in September 2006, nor by war and civil strife, as the gene banks in Iraq, Afghanistan and Egypt have seen in recent years. Or indeed the whims of decision makers, as exemplified by recent examples of the gene banks at Pavlosk and Brogdale. For most CWR accessions, the reduction of such risks that safety duplication could bring about is still no more than a fond hope.

Even in crops with substantial collections of CWR germplasm, the genetic diversity within the accessions themselves may be limited, as has been noted for wild maize (Khoury *et al.*, 2010). In a number of cases, the variability present has been shown to be reduced over time due to gene bank procedures, particularly regeneration (Gómez *et al.*, 2005; Negri and Tiranti, 2010). And, of course, even if diversity is properly stored, access to it may be constrained by plant health considerations, or indeed political imperatives. Although this issue has been addressed by the ITPGRFA, facilitated access under the Treaty is only currently available for the species listed in its Annex 1, which specifically excludes, for example, the wild relatives of cassava and maize, as well as important crops such as groundnut and soybean in their entirety.

37.3 The Challenge of Conserving CWR *In Situ*

With the goal of conserving species within a context that supports their continued evolution, and given political considerations and the challenges of adequately conserving CWR in gene banks, protected areas have been an increasing focus of CWR conservation efforts in the last few decades. Maxted and Kell (2009) have suggested the establishment of a worldwide network of *in situ* genetic reserves that is independent of national political borders, which could parallel the global system of gene banks. At present, however, there are few examples of protected area management plans which specifically incorporate the genetic conservation of CWR diversity (FAO, 2010; Hunter and Heywood, 2010). Further, there has been little information published or documented that provides guidance in working with protected area authorities and managers or other relevant actors. For example, no mention is made of CWR, genetic reserves or genetic resource management in Lockwood *et al.* (2006), a global guide to managing protected areas (V. Heywood, England, 2010, personal communication). It has too often been assumed that affording CWR protection in protected areas is relatively easy or can be achieved with a minimal amount of intervention, and that this task can be left up to the protected area manager to formulate through the necessary adjustment of the management plan with the corresponding management actions. This fails to take account of the many other demands on the resources and time available to the protected area manager, or indeed the fact that in the majority of protected areas, management plans do not even exist (Hunter *et al.*, in press).

More importantly, reliance on the continued existence of protected areas in their current location is a risky strategy for CWR conservation in the face of global, especially climatic, change. Jarvis *et al.* (2003), in a bioclimatic modelling study of wild groundnuts (*Arachis* spp.) in South America found that only about 2% of the geo-referenced observations occurred in a protected area. And that within the foreseeable future hotspots of wild groundnut species diversity would change drastically in composition, strength and location over time due to climate change. Similarly, Lira *et al.* (2009) found that, under a drastic climatic change scenario, the eight wild Cucurbitaceae species closely related to the cultivated taxa are likely to encounter suitable habitat in just 29 out of the 69 natural protected areas where they currently occur.

Given that protected areas and other conservation areas cover only 12–13% of the earth's surface, it is clear that they cannot by themselves ensure the survival of all species and ecological communities, even without the impacts of accelerated global change. It is crucial, therefore, that lands outside the protected areas be managed in ways that allow as much biodiversity as possible to be maintained. The *in situ* conservation of species outside protected areas, where the majority of species distributions occur, is a seriously neglected aspect of biodiversity conservation and is in need of greater attention from conservation agencies. It may be possible to take actions to ensure that areas outside formal protection, whether on public or private lands, can provide a sufficient degree of protection to target species so as to ensure maintenance of viable populations, through some form of agreement with the landowner (Hunter and Heywood, 2010). Opportunities for such actions surely exist, given the increasing interest in ecosystem and eco-agriculture landscape approaches (Scherr and McNeely, 2007). Landscape initiatives such as FAO's Globally Important Agricultural Heritage Systems (GIAHS), the recently launched *Satoyama* Initiative and the expanding network of Indigenous and Community Conserved Areas (ICCAs) appear to provide entry points for integration of CWR concerns and actions with the additional opportunity to work with local communities and alternative institutional partnerships to explore novel community-led approaches to sustainable management of CWR and their genetic interactions with crops. Taking this to its logical next step could see CWR conservation concerns being part of matrix approaches to better link agroecosystems to deal with increasing fragmentation (Perfecto *et al.*, 2009). However, we have very little experience so far of how to safeguard CWR in such a context.

37.4 The Need for Information

There are some glimmers of hope. The information baseline for CWR conservation today is certainly much improved compared to 10 years ago (Thormann *et al.*, 1999; Meilleur and Hodgkin, 2004). Most recently, a UNEP/GEF CWR project helped establish national CWR information systems in Armenia, Bolivia, Madagascar, Sri Lanka and Uzbekistan linked through a CWR Global Portal (Thormann *et al.*, 2010). But information still remains fragmented, incomplete and poorly managed. National efforts need to be linked up and up-scaled. Much more needs to be done in order to develop a truly integrated global portal accessing and making available both *in situ* and *ex situ* information including the accession level data, including trait information, which can support national conservation efforts and at the same time meet the needs of plant breeders and other users. This may be starting to be addressed by the Genesys portal (www.genesys-pgr.org), which brings together accession-level information on gene bank holdings worldwide, currently including the international collections maintained by the CGIAR Centres, the USDA's NPGS and the gene banks contributing to the EURISCO information network. To date, Genesys reports basic passport information on about 250,000 wild accessions, but little characterization and evaluation data. Users of CWR must do a better job of feeding back information on what has worked and what has not.

37.5 The Imperative of Use

Wild relatives have become a well-established source of genes for crop improvement, particularly for pest and disease resistance and tolerance to abiotic stresses, for crops such as banana, barley, beans, cassava, chickpea, lettuce, maize, oats, pearl millet, potatoes, rice, sugarcane, sunflower, tomato and wheat (Tanksley and McCouch, 1997; Phillips and Meilleur, 1998; Hoisington *et al.*, 1999; Gur and Zamir, 2004; Hajjar and Hodgkin, 2007; McCouch *et al.*, 2007; Damiana, 2008; Maxted and Kell, 2009; FAO, 2010). This has largely been driven by the efforts of individual breeders, rather than by gene banks making their material more available and easier to use, by pull rather than push. The challenge is that the true value of any single CWR accession for crop improvement often only becomes apparent once initial crosses with cultivated varieties have been performed and the results evaluated (Tanksley and McCouch, 1997; McCouch, 2004). This of course poses a significant challenge to the curation of CWRs in *ex situ* collections, since conventional evaluation of the material itself, as commonly done in landraces, cannot be depended upon for a prediction of future value. Modern molecular techniques can, and indeed are, playing an important role here. Where the phenotype of a CWR accession can be misleading to a potential user, its genotype cannot. And the genotype is becoming increasingly easy, fast and cheap to read, analyse, and explore for useful traits. However, even though genotypic information may aid in the selection of accessions of CWR for pre-breeding efforts, it still is a considerable challenge to draw predictive conclusions from such genotypic information with regards to the phenotype of the progeny from a cross with improved material. It has become conventional to call for closer links between gene banks and breeding programmes. However, we believe that there is a more active role for gene banks to play, by getting directly involved in pre-breeding efforts, conducted in close collaboration with, and as a service to, breeding efforts.

Of course, crop improvement is not the only possible use that can be made of *ex situ* conserved material. Its role in ecosystem recovery has only recently begun to be explored, for example by Kew's Millennium Seed Bank, and we may also need to consider reintroduction and translocation in the future (Malcolm *et al.*, 2002). Here will be ample scope for gene banks and *in situ* conservation programmes to work hand in hand.

In order to raise the awareness of the value of CWR with the public and with policy makers, more data is certainly needed that shows how the use of CWR as a genetic resource positively impacts crop production and livelihoods. Research in this area requires collaborative data sharing across fields, between collectors, gene banks, breeders, distributors and economists. Certainly, unless CWR are used in breeding and restoration, and are seen to be thus used, there will be minimal support – and resources – for their conservation.

37.6 The Need for Integrated Strategies

How do we move forward? Both *ex situ* and *in situ* conservation of CWR have largely been applied with little reference to the other. A dedicated Global Strategy for CWR Conservation and Use (Heywood *et al.*, 2008) and a Global Plan of Action for PGRFA (FAO, 1996) promoting both *in situ* and *ex situ* conservation exist, but the institutions and practitioners of both approaches have tended to work in isolation, if not, occasionally, in opposition. Few attempts have been made to elaborate the collaborative frameworks that are necessary to guide activities to support truly complementary conservation decision making. The agencies that need to work together to effectively undertake complementary CWR conservation – in the agriculture, forestry and environment sectors – often have no linkages or tradition of collaboration, and seldom is there any. CWR *in situ* conservation often falls between two stools – agricultural agencies think of these species as the province of environmental agencies, and vice versa. And as for *ex situ* conservation, collecting of CWR from important regions of diversity has been constrained by limited cooperation at the international level, as well as limited resources at the national level. Most countries' national biodiversity strategies and action plans do not specifically refer, let alone emphasize, CWR. Even the more specifically agriculture-focused country reports to the Global Plan of Action on PGRFA reveal few examples of specific strategies at the national or regional level for the conservation of CWR.

Given the complexity and multidisciplinary nature of CWR conservation and the range of roles and responsibilities across sectors, a comprehensive national CWR strategy may be vital to providing a coherent, coordinated and complementary approach to their conservation and utilization. This long-term focus will avoid the deficiencies inherent in the short-term, project-based actions that have been the normal mode of operation to date. Given that the distribution of wild species rarely matches political borders, these national strategies will necessarily require international collaboration, and for a truly adequate conservation of large gene pools, will be necessary to fit within overarching international strategies. This may be the only way to translate global commitments into effective regional, national and local action.

37.7 Conclusions

We have suggested that the keys to the sustainable conservation and use of CWR lie in integrated strategies and support for their actualization, more and better data, long-term thinking and more effective use. Underlying all these is the need for better communication among actors. Divisions have dogged this field – between *ex situ* and *in situ*, between conservation and use, and between laboratory biologists and field practitioners. To some extent, this is a problem of biodiversity conservation in general, but it is particularly acute, and damaging, in CWR conservation. As the challenges of climate change become more severe, the need for agriculture to develop solutions with lower ecological impact increases, and our interdependence for genetic resources becomes clearer, there may indeed be hope for traditional divisions in the field to be abandoned in favour of approaches that can truly conserve these invaluable resources for current and future use.

References

Astley, D. (1991) Exploration: methods and problems of exploration and field collecting. In: Hawkes, J.G. (ed.) *Genetic Conservation of World Crop Plants*. Academic Press, London, pp. 11–22.

Damiana, A. (2008) History, achievements, and current status of genetic resources conservation. *Journal of Agronomy* 100(1), 9–21.

Dulloo, M.E., Ebert, A.W., Dussert, S., Gotor, E., Astorga, C., Vasquez, N., Rakotomalala, J.J., Rabemiafara, A., Eira, M., Bellachew, B., Omondi, C., Engelmann, F., Anthony, F., Watts, J., Qamar, Z. and Snook, L. (2009) Cost efficiency of cryopreservation as a long-term conservation method for coffee genetic resources. *Crop Science* 49(6), 2123–2138.

Engels, J.M.M. and Rao, V.R. (1998) *Regeneration of seed crops and their wild relatives: proceedings of a consultation meeting, 4–7 December 1995*. ICRISAT, Hyderabad, India. IPGRI, Rome.

FAO (1996) *Global Plan of Action for the Conservation and Sustainable Utilization of Plant Genetic Resources for Food and Agriculture*. Food and Agriculture Organization of the UN, Rome.

FAO (2010) *The Second Report on the State of the World's Plant Genetic Resources for Food and Agriculture, Rome*. Food and Agriculture Organization of the UN, Rome.

Fowler, C. and Hodgkin, T. (2004) Plant genetic resources for food and agriculture: assessing global availability. *Annual Review of Environment and Resources* 29(11), 143–179.

Gómez, O.J., Blair, M.W., Frankow-Lindberg, B.E. and Gullberg, U. (2005) Comparative study of common bean (*Phaseolus vulgaris* L.) landraces conserved *ex situ* in genebanks and *in situ* by farmers. *Genetic Resources and Crop Evolution* 52(6), 371–380.

Gur, A. and Zamir, D. (2004) Unused natural variation can lift yield barriers in plant breeding. *PLoS Biology* 2(10), 1610–1615.

Hajjar, R. and Hodgkin, T. (2007) The use of wild relatives in crop improvement: a survey of developments over the last 20 years. *Euphytica* 156, 1–13.

Hammer, K., Arrowsmith, N. and Gladis, T. (2003) Agrobiodiversity with emphasis on plant genetic resources. *Naturwissenschaften* 90, 241–250.

Heywood, V.H., Kell, S.P. and Maxted, N. (2008) Towards a global strategy for the conservation and use of crop wild relatives. In: Maxted, N., Ford-Lloyd, B.V., Kell, S.P., Iriondo, J., Dulloo, E. and Turok, J. (eds) *Crop Wild Relative Conservation and Use*. CAB International, Wallingford, UK.

Hoisington, D., Khairallah, M., Reeves, T., Ribaut, J.-M., Skovmand, B., Taba, S. and Warburton, M. (1999) Plant genetic resources: what can they contribute toward increased crop productivity? *Proceedings of the National Academy of Sciences of the United States of America* 96(11), 5937–5943.

Hunter, D. and Heywood, V.H. (2010) *Crop Wild Relatives: a manual of* in situ *conservation*. Earthscan, London.

Hunter, D., Maxted, N., Heywood, V., Kell, S. and Borelli, T. (in press) Protected areas and the challenge of conserving crop wild relatives. PARKS

Jarvis, A., Ferguson, M.E., Williams, D.E., Guarino, L., Jones, P.G., Stalker, H.T., Valls, J.F.M., Pittman, R.N., Simpson, C.E. and Bramel, P. (2003) Biogeography of wild *Arachis*: assessing conservation status and setting future priorities. *Crop Science* 43(3), 1100–1108.

Khoury, C., Laliberté, B. and Guarino, L. (2010) Trends in *ex situ* conservation of plant genetic resources: a review of global crop and regional conservation strategies. *Genetic Resources and Crop Evolution* 57(4), 625–639.

Lira, R., Tellez, O. and Davila, P. (2009) The effects of climate change on the geographic distribution of Mexican wild relatives of domesticated Cucurbitaceae. *Genetic Resources and Crop Evolution* 56 (5), 691–703.

Lockwood, M., Worboys, G.L. and Kothari, A. (2006) *Managing Protected Areas: A Global Guide*. Earthscan, London.

Malcolm, J.R., Markham, A., Neilson, R.P. and Garaci, M. (2002) Estimated migration rates under scenarios of global climate change. *Journal of Biogeography* 29(7), 835–849.

Maxted, N. and Kell, S.P. (2009) Establishment of a global network for the *in situ* conservation of crop wild relatives: status and needs. Unpublished consultancy report. Food and Agriculture Organization of the UN, Rome, pp. 1–265.

McCouch, S. (2004) Diversifying selection in plant breeding. *PLoS Biology* 2(10), e347. doi:10.1371/journal.pbio.0020347.

McCouch, S., Sweeney, M., Li, J., Jiang, H., Thomson, M., Septiningsih, E., Edwards, J., Moncada, P., Xiao, J., Garris, A., Tai, T., Martinez, C., Tohme, J., Sugiono, M., McClung, A., Yuan, L. and Ahn, S. (2007) Through the genetic bottleneck: *O. rufipogon* as a source of trait enhancing alleles for *O. sativa*. *Euphytica* 154, 317–339.

Meilleur, B.A. and Hodgkin, T. (2004) *In situ* conservation of crop wild relatives. *Biodiversity Conservation* 13, 663–684.

Negri, V. and Tiranti, B. (2010) Effectiveness of *in situ* and *ex situ* conservation of crop diversity. What a *Phaseolus vulgaris* L. landrace case study can tell us. *Genetica* 138(9) (October), 985–998.

Perfecto, I., Vandermeer, J. and Wright, A. (2009) *Nature's Matrix: linking agriculture, conservation and food sovereignty*. Earthscan, London.

Petersen, L., Østergard, H. and Giese, H. (1994) Genetic diversity among wild and cultivated barley as revealed by RFLP. *Theoretical Applied Genetics* 89(6), 676–681.

Phillips, O.L. and Meilleur, B.A. (1998) Usefulness and economic potential of the rare plants of the United States: a statistical survey. *Economic Botany* 52(1), 57–67.

Ramírez-Villegas, J., Khoury, C., Jarvis, A., Debouck, D.G. and Guarino, L. (2010) A gap analysis methodology for collecting crop genepools: a case study with *Phaseolus* beans. *PLoS ONE* 5(10), e13497. doi:10.1371/journal.pone.0013497.

Scherr, S.J. and McNeely, J.A. (2007) *Farming with Nature: the science and practice of ecoagriculture*. Island Press. Washington, DC.

Schoen, D.J., David, J.L. and Bataillon, T.M. (1998) Deleterious mutation accumulation and the regeneration of genetic resources. *Proceedings of the National Academy of Sciences of the United States of America* 95(1), 394–399.

Tanksley, S.D. and McCouch, S.R. (1997) Seed banks and molecular maps: unlocking genetic potential from the wild. *Science* 277, 1063–1066.

Thormann, I., Jarvis, D., Dearing, J. and Hodgkin, T. (1999) International available information sources for the development of *in situ* conservation strategies for wild species useful for food and agriculture. *Plant Genetic Resources Newsletter* 118, 38–50.

Thormann, I., Hunter, D., Danielyan, A., Djataev, S., Ramelison, J., Wijesekara, A. and Zapata Ferrufino, B. (2010) Meeting the information challenge for crop wild relatives *in situ* conservation: A global portal hosting national and international data on CWR. *IAALD XIIIth World Congress*, 26–29 April 2010 Montpellier, France.

Vollbrecht, E. and Sigmon, B. (2005) Amazing grass: developmental genetics of maize domestication. *Biochemical Society Transactions* 33(6), 1502–1506.

38 Interactive Characterization of Scottish *Avena strigosa* Schreb. Landraces on the Outer Hebrides

M. Scholten, B. Spoor, N. Green, S. Carter and N. MacPherson

38.1 Introduction

In recent years Scottish field crop landraces have been surveyed, collected and a start has been made with their characterization (Scholten *et al.*, 2009). In addition, a scheme was set up to provide maintainers of landraces with *ex situ* conservation support (Green *et al.*, 2009). Characterization carried out so far was conducted 'on station' rather than 'on location'. However, a planned characterization study of Scottish landraces of bristle oat (*Avena strigosa* Schreb.) was redesigned and became an 'on farm' project in the largest area of its cultivation, on the remote Outer Hebrides. The aim of the characterization was to assess *A. strigosa* diversity between and within Scotland and to position the Scottish gene pool within the wider region.

Avena strigosa has survived in Scotland on three island groups: Shetland and Orkney in the north, and the Hebrides in the west (Scholten *et al.*, 2009). However, only on the southern Outer Hebrides – the Uists – is it still grown on a substantial scale, often mixed with a rye landrace and *bere*, an ancient barley landrace. It is used for winter feed for cattle and sheep, which are the most important crofting products of the islands.

The survival of landraces in Scotland is strongly associated with 'crofting', a form of regulated land tenure of agricultural (small) holdings or 'crofts', unique to mainly the Highlands and Islands of Scotland (Hunter, 1976). Crofting areas, especially in the Outer Hebrides, have retained the highest proportion of native Gaelic-speakers in Scotland and Gaelic is still very much the vernacular language on the Uists. Of the many challenges facing crofting, the ageing of the crofting population is a major one (Western Isles Fact File, 2010). This demographic pattern has been observed elsewhere (Negri, 2009), but is rarely targeted in landrace research.

Scotland has defined a biodiversity strategy with indicators (Mackey and Mudge, 2010) but has no official policy for landraces. The Outer Hebrides have rich marine and terrestrial diversity and carry several conservation designations. Landraces are cultivated on the coastal grasslands, a Priority Habitat in Scotland listed in Annex 1 under the EC Habitats Directive. Although biodiversity conservation brings in tourism and subsidies, sometimes this has locally been received with resentment or resistance (Mitchell, 2004). The strong presence of conservation designations is sometimes seen as hampering an area described by the OECD (2008) as one of the weakest economies and most fragile human populations in Scotland.

Although the importance of cereal cultivation has been acknowledged in local biodiversity conservation plans, there was little awareness among farmers and conservationists of the significance of these local varieties as landraces and rare historical crops. Similar lack of awareness among maintainers of landraces had been observed in Italy (Negri, 2009) and the importance to raise awareness about landraces was emphasized in a recent overview of on-farm conservation in Europe (Veteläinen *et al.*, 2009).

An opportunity to combine an 'on farm' characterization study of landraces of *A. strigosa* with raising awareness arose at the end of 2008 with the start of a new school module about crofting at the Community Secondary School in Lionacleit on the isle of Benbecula. Teacher and crofter-instructor, a maintainer of a barley landrace himself, agreed to include the characterization trial into the course and to make land available for a trial. The characterization study was then turned into an 'on croft' experiment, run over two seasons, aiming at raising awareness about landraces. The outreach element will be presented here in a chronological summary.

38.2 Publicity

Almekinders' (2001) suggestions for local publicity as a starting point were followed up with articles in the Uist community newspaper *Am Pàipear*. In 2007 a series explained the importance of landraces, summarizing results of a recent barley landrace diversity study and explaining *ex situ* conservation in Scotland (Scholten, 2007; Scholten and Green, 2007; Scholten and Southworth, 2007). In the following year (2008) germplasm collection was announced (Scholten, 2008a) and in 2009–2010 the trial's first results as well as visitors and wider impact were very regularly reported on (Scholten, 2008b, 2009a, b, c, d, 2010a, b, c; Scholten and Carter, 2009; Scholten and MacLellan, 2009). The threshing of the oats by students was filmed and shown on national television in *Monty Halls Great Escape*. In 2010 BBC Nan Gaidheal reportage about the trial was shown on Gaelic-medium television.

38.3 Research Site

The research site was situated approximately 300 m from the Atlantic Ocean and a few metres above sea level. It was located at the harsher end of Hebridean growing conditions: highly exposed to winds, with light soils with low organic matter, prone to droughts; highly alkaline with high pH and potential micronutrient deficiencies. Although manganese deficiency is known to occur in common oat in this region, the local *A. strigosa* is known to perform well (Darling, 1955). The site itself demonstrated a 'genetic history of a crop's adaptation to the local place' (Simmonds, 1979).

The on-farm research design was laid out following Mutsaers *et al.* (1997) as a one replicate, single treatment study, to maximize the demonstration effect. In the second year, a three-replicate design was used. The site was managed by the crofter-teacher on whose land the experiment was located. Seeds were sown as collected in the previous year from the Outer Hebridean islands of Lewis, North Uist, Berneray, Benbecula, South Uist, Tiree and from Orkney; Shetland oat seed was received directly from growers on the islands of Burra and Yell in Shetland. Welsh, Irish, Finnish and German *A. strigosa* accessions were obtained from gene banks and formed reference material. Some common oat (*A. sativa* L.) cultivars were included to test for local adaptation, thus introducing a variety-testing component which was thought to make the trial more interesting to crofters.

38.4 Education

The field experiment formed a teaching facility for the 15–16-year-old students of Community Secondary School to learn about landraces, *ex situ* conservation and crop research as well as performing practical activities such as sowing, weeding and fencing (as protection against rabbits), harvesting and threshing. In the second year the students also learned to identify oat species, score lodging tendencies and evaluate oat species and varieties for silage quality. One outcome of the educational experiment was the uptake of landraces as a topic in a Scotland-wide school project. Diversity kits, i.e. seed packs of landraces, were sent to 25 schools in 2010 as the 'Living gene bank' project within the Crofting Connections project (www.croftingconnections.com).

38.5 Outreach on Site

The site was open to the general public and guided visits were held every week in both years. It attracted many visitors including crofters, tourists, parents of students, journalists, local general public, conservationists and farmers. The plots were signposted and an explanatory brochure was available. In it, the significance of the Uists as one of the last remaining areas of cultivation of *A. strigosa* in north-western Europe was further enhanced with information on the (pre)history of cereal landraces on the islands. The story of *A. strigosa* was told as ongoing crop evolution and contrasted with gene bank conservation. The survival of *A. strigosa* on the Outer Hebrides was set against the cessation of plant breeding and seed production for marginal environments (Wright *et al.*, 2002). The role of crofters as maintainers was brought in and set against the background of a lack of a Scottish national policy on landraces. The ecological benefits of landraces, i.e. as low-input agriculture in semi-natural habitats (High Nature Value farming) (Bignal and McCracken, 1996; EEA, 2004), were explained as well.

The experiment was not intended to be 'participatory' because it was unclear at the start what form of participation could be expected or how feasible, given the large distances between villages, or how acceptable this would be. None the less, the crofter-teacher was an active participant-observer during the 2 years of the experiment and is now planning his own experiment next year. Variety performances were evaluated by visiting crofters and compared with their own growing practices and preferences. Some crofters visiting the site showed interest in new oat cultivars and have started their own experiments.

38.6 Outreach at Local Events

Outreach activities were not confined to the research site but also comprised participation in local events. At the annual local cattle and sheep shows, organized by the two local agricultural societies, plants and seeds of *A. strigosa* were displayed and the research explained in a brochure. In 2009 this was done at the annual sheep show of the Hebridean Sheep Society and at the agricultural shows at Hosta, North Uist, and at Iochdar, South Uist. A talk about the oat trial was given at *An Fheill-Micheil*, St Michaelmas Open Day on 26 September.

In 2010 the International Year of Biodiversity was celebrated with a special stand; agrobiodiversity, i.e. landrace plants and seeds, were on display on Lewis and South Harris as well as the Uists. One immediate impact of the seed displays on the former growing islands of Lewis and Harris was the realization by crofters and conservationists that seed of the landraces was still available and that reintroduction was feasible.

Local Gaelic names were used in publications and in displays. Participating in a Gaelic-cultural event led to an invitation to run a special workshop about landraces at a Gaelic-cultural summer school Ceolas in 2010. Participants represented experienced 'baking ladies' from four different islands of the Uists. In the workshop the meaning and importance of landraces was explained and a diversity of *Coirce beag* (*A. strigosa*) seeds shown. The use of *Coirce beag* as breakfast porridge, reported in the diary of a parish priest on the tiny island of Eriskay in the late 19th century, was discussed but not recollected by any of the participants. Traditional recipes of barley and oat were discussed. As a practical activity, traditional barley flour (from the landrace *bere*) was used to produce *Aran eorna*, the traditional barley bread of the Outer Hebrides.

38.7 Informing Stakeholder Organizations

Prior to the trial, contacts had been made with various stakeholders organizations such as island authorities, local agricultural extension service and statutory (Scottish Natural Heritage – SNH) and NGO conservation agencies. The latter are important players on the islands because of the importance of wildlife conservation.

The trial site functioned as a demonstration of oat genetic diversity for technical working parties of these organizations. *Avena strigosa* was shown as distinct species and rare crop; the importance of the Uists as one of the last remaining areas of cultivation in Europe was flagged up; the importance of conserving multiple seed sources was stressed as well as good seed quality in order to maintain genetic diversity.

A special technical visit was organized in 2009 for a newly established working group on landraces and traditional varieties within the Scottish Biodiversity Forum. This is an umbrella working partnership of Scottish Government, its agencies, such as SNH, local authorities, voluntary bodies, farmers, business and scientists. At the trial site, national representatives of Scottish Government and SNH were introduced to *A. strigosa*, its regional diversity and local growing conditions. Fields of crofters were visited as well so that crofters could explain their own practices and problems. This visit was reported nationally in the *Scottish Farmer* and in regional newspapers.

Contact with the other major stakeholder group, the Scottish Crofting Federation (SCF), the organization representing crofters, had been made at the first visit to the Uists by the first author in 2004. At the time of the trial collaboration was strengthened and at the regional agricultural show in Dingwall in 2010, landrace seeds and plants, brought in from the trial, were on display at the SCF stand. Further collaboration is planned to lobby Scottish politicians for recognition of landraces and their maintainers; to promote the use of landraces in an ongoing project to stimulate and support crofting produce.

38.8 Evaluation

Interactive research aims at innovation (Röling, 1996; Mutsaers *et al.*, 1997). In this respect this experiment led to a new teaching module, a new theme at shows and events and agrobiodiversity as a theme in celebrations of the International Year of Biodiversity. The issue of landraces was picked up in two Scotland-wide projects both dealing with (the future of) crofting. As an outreach experiment, it appears to have been successful as throughout the Uists many crofters were aware of the experiment (S. Huband, Benbecula, 2010, personal communication).

Impact at local level contrasted with, and may have been hampered by, the lack of policy at national Scottish level. The publicity campaign about landraces may have run the risk of raising expectations too high, a risk that has been associated with interactive research (Chambers, 2008). In addition, national and international publicity around the trial raised the issue of ownership of landraces and led to shelving an oat diversity register.

Some of the trade-offs of an 'on-farm' experiment may have been blessings in disguise. Lack of technical support and other resources meant a lot of manual work and time-consuming public transport and cycling. This however, contributed to the visibility of the experiment, lowered the threshold between researcher and public, led to many brief interactions with passing crofters with opportunities for cross-checking observations or other information. The facilitating role of the researcher in interactive research (Röling, 1996) sometimes meant assisting in the logistics of sourcing seed, but leads to a better understanding of the seed system and to developing new contacts and promotion of the reintroduction of landraces in previous growing areas.

Follow-up work is planned to present the results of the genetic diversity study to crofters and crofters' organizations and to explore how this genetic research can support landraces and their maintainers.

Acknowledgements

The research on Scottish landraces was supported through a Wingate Scholarship. The trial on the Outer Hebrides was financially supported by an SAC Trust Fund Grant in 2009 and 2010 and received EM Cruickshank funding in 2009. Crofters and crofting students are acknowledged for their interest, assistance and advice. Carolina Tania Camacho Villa is acknowledged for drawing our attention to 'action' research.

References

Almekinders, C.J.M. (2001) *Management of Crop Genetic Diversity at Community Level.* Deutsche Gesellschaft für Technische Zusammenarbeit (GTZ) GmbH.

Bignal, E.M. and McCracken, D.I. (1996). Low-intensity farming systems in the conservation of the countryside. *Journal of Applied Ecology* 33, 413–424.

Chambers, R. (2008) *Revolutions in Development Inquiry*. Earthscan London.

Darling, F. (ed.) (1955) *West Highland Survey. An Essay in Human Ecology*. Oxford University Press, Oxford.

EEA (2004) *High Nature Value Farmland. Characteristics, trends and policy challenges*. European Environmental Agency/United Nations Environment Programme Denmark.

Green, N., Campbell, G., Tulloch, R. and Scholten, M. (2009) Scottish Landrace Protection Scheme. In: Veteläinen, M., Negri, V. and Maxted, N. (eds) *European Landraces: On-farm Conservation, Management and Use*. Bioversity Technical Bulletin No. 15. Bioversity International, Rome, pp. 233–243.

Hunter, J. (1976) *The Making of the Crofting Community*. John Donald Publishers Ltd., Edinburgh.

Mackey, E.C. and Mudge, G. (2010) *Scotland's Wildlife: An assessment of biodiversity in 2010*. Scottish Natural Heritage, Inverness.

Mitchell, I. (2004) *Isles of the West*. Birlin Press, Edinburgh.

Mutsaers, H.J.W., Weber, G.K., Walker, P. and Fisher, N.M. (1997) *A Field Guide for On-Farm Experimentation*. International Institute of Tropical Agriculture, Ibadan, Nigeria.

Negri, V. (2009) 'Fagliolina' (*Vigna unguiculata* subsp. *unguiculata* (L.) Walp.) from Trasimeno Lake (Umbria Region, Italy). In: Veteläinen, M., Negri, V. and Maxted, N. (eds) *European Landraces: On-farm Conservation, Management and Use*. Bioversity Technical Bulletin No. 15, Bioversity International, Rome, pp. 177–182.

OECD (2008) *OECD Rural Policy Reviews: Scotland, UK Assessment and Recommendations*. Produced on behalf of OECD for the Scottish Government by RR Donnelley B54749 02/08.

Röling, N. (1996) Towards an interactive agricultural science. *European Journal of Agricultural Education and Extension* 3(4), 35–48.

Scholten, M. (2007) Uist cereals attract research interest. Part 1: Small oat. *Am Pàipear*, March 2007.

Scholten, M. (2008a) Canadian Genebank and Scottish Heritage Collection request Uist Seed. *Am Pàipear*, May 2008.

Scholten, M. (2008b) A wee note on coirce beag – with a riddle about stooks. *West Highland Free Press*, 5 December 2008.

Scholten, M. (2009a) Benbecula Small oat trial: the start. *Am Pàipear*, July 2009.

Scholten, M. (2009b) Sowing tamed oats in Benbecula and what it says about breeding. *The Shetland Times*, July 2009.

Scholten, M. (2009c) Small oat trial: mid-July favourites, mid-August drop-outs. *Am Pàipear*, September 2009.

Scholten, M. (2009d) Brussels insider about Uist corn: keep the seeds on the islands! *Am Pàipear*, December 2009/January 2010.

Scholten, M. (2010a) Widespread attention for small oat project. *Am Pàipear*, April 2010.

Scholten, M. (2010b) Oats: the Scottish year of oat diversity. *Am Pàipear*, May 2010.

Scholten, M. (2010c) An update on coirce beag with news from Holland. *Am Pàipear*, August 2010.

Scholten, M. and Carter, S. (2009) Genebank on the machair. Scotland National Rural Network. Available at: www.ruralgateway.org.uk/en/node/876 (accessed 29 June 2009).

Scholten, M. and Green, N. (2007) Conservation of local varieties in seedbanks. *Am Pàipear*, May 2007.

Scholten, M. and MacLellan, A. (2009) Working group on landraces and traditional varieties visits Uist. *Am Pàipear*, October 2010.

Scholten, M. and Southworth, C. (2007) Tìr an Eòrna/land of barley – land of diversity. *Am Pàipear*, April 2007.

Scholten, M., Green, N., Campbell, G., Maxted, N., Ford-Lloyd, B., Ambrose, M. and Spoor, W. (2009) Landrace inventory of the UK. In: Veteläinen, M., Negri, V. and Maxted, N. (eds) *European Landraces: On-farm Conservation, Management and Use*. Bioversity Technical Bulletin No. 15. Bioversity International, Rome, Italy, pp. 161–170.

Simmonds, N.W. (1979) *Principles of Crop Improvement*. Longman, London and New York.

Veteläinen, M., Maxted, N. and Negri, V. (2009) A European strategic approach to conserving crop landraces. In: Veteläinen, M., Negri, V. and Maxted, N. (eds) (2009) *European Landraces: On-farm Conservation, Management and Use*. Bioversity Technical Bulletin No. 15, Bioversity International, Rome, pp. 305–325.

Western Isles Fact File (2010) Available at: www.cne-siar.gov.uk/factfile/index.asp (accessed 23 November 2010).

Wright, I.A., Dalziel, A.J.I., Ellis, R.P. and Hall, S.J.G. (2002) *The Status of Scottish Rare Breeds and Plant Varieties*. Macauley Institute and SCRI, Dundee.

39 Modern Landrace Crops: The *De Novo* Synthesis of New Crop Landraces

M.J. Ambrose

39.1 Introduction

The conservation of important agrobiodiversity in the form of landrace crops and their on-farm management is an area of active methodological development in the plant genetic resources community. When the term landrace is used, most picture materials that pre-date the onset of modern plant breeding, that have been maintained and developed over extended periods of time by traditional maintainers and are well adapted to local climatic and edaphic factors. Key aspects in the definition of landrace crops relates to their historical origin, their heterogeneous nature and their dynamic nature, which may change over time in response to changes in their market use, consumer preferences or changes in the environment or climate (Camacho Villa *et al.*, 2009; Osman and Chable, 2009). A less appreciated aspect of landraces study is that new landraces are continually being created.

This paper details a case study of the *de novo* synthesis of new landrace crops, the drivers for their development in terms of market need and how they are traded and exchanged. These activities represent an example of on-farm informal plant breeding with defined objectives, selection and evaluation strategies and the continual active management of materials that are distinct in form and function. These processes coexist in parallel but are separate from the formal plant breeding and the associated legal and policy framework.

39.2 Drivers of *De Novo* Landrace Synthesis

The key driver to the continued use of any landrace or development of new landraces is utility. Material is only developed and maintained to meet a demand or niche in a particular market. One of the drivers to new landraces is that the variation within existing landraces and available germplasm is insufficient or fails to meet specific user criteria and thus new better suited landraces are required. There may also be an element of serendipity and certainly there is scope for unexpected or chance forms/variants to arise and for advantage to be recognized for particular traits or qualities that will result in a form creating its own distinct market demand.

The use of long straw wheat as a roofing material for use as thatch is well documented throughout Europe since medieval times and in the UK remains an important rural industry in the south-west, the

Midlands and eastern England (Ambrose and Letch, 2009). The majority of long straw wheats actively grown in the UK today are heritage varieties from the 1880s to the 1960s (Scholten *et al.*, 2003; Ambrose and Letch, 2009). This is wholly attributable to the incorporation of reduced height (*Rht*) genes that increased the partitioning coefficient and yield, thus they became the cornerstone of modern wheat breeding programmes. Virtually no new long straw types have been made available since the 1960s so thatches have been restricted in the materials they have available to them.

39.3 Breeding of *De Novo* Landraces

The development of new landrace crops is an ongoing activity within thatcher/growers in the East Anglian region (Ambrose, in prep.), a user community that show a high degree of organization, planning and resource allocation. The majority of thatcher growers are self-employed small businesses and this investment of scarce resources only serves to underline the vision and potential they see in their ability to bring about improvements in their thatching landraces. The activities these individuals engaged in match the characteristics of traditional farmers described by Wood and Lenné (1997).

This first involves the ongoing search by farmers to sources of novel variation. This is met by thatching landrace farmers sharing experiences and seeds with others from different regions of the UK, and with contacts and materials from UK *ex situ* collections. The second is the ability to experiment and use this variation, which is illustrated by their readiness to compare and contrast with other material in their own trials (with respect to growing and the quality of the end product), and to actively combine different lines to produce new mixtures for trialling. A summary of their 'breeding' programme is provided in Table 39.1.

During interactions with these thatchers it is clear that they are running small informal on-farm breeding programmes and plant out trial areas of newly sourced materials before going on to readily experiment with combining seed lots of different lines to generate new composite mixtures for trialling against their existing best materials.

Table 39.1. Informal breeding of *de novo* landrace long straw wheats.

Objectives	Improved straw quality with emphasis on: • improved consistency of year to year performance • improved standing ability for easier crop management • grades of straw thickness for different end use
Selection and Evaluation	• dedicated areas of land given over to growing of trial plots • attention to the straw is focused at heading, towards harvesting • standing ability • ease of cutting and harvesting • evaluation of straw during the growing, thrashing and handling of the straw during the thatching process • spread of maturity times to spread the harvesting workload
Active management	Thatching lines may be highly heterogeneous. Use spike characteristics and maturity times between crops to help keep separate material in the field.
Utility	Quality straw that can be stored and used for thatching on a diverse range of buildings. Thatching employs a range of straws in different areas (roof ridges versus main areas).
Trading and marketing	Traded by the cart/trailer By appearance of finished work After harvest, a stook of straw may be presented as sample prior to negotiation over price

The commodity underpinning the market is the quality of the straw and how it handles in the field at harvest or after wetting on the bed immediately prior to use and its longevity. It is clearly the case that the best samples of straw produced by some thatchers are the product of highly heterogeneous mixtures, which was clearly recorded by Biffen and Engledow in 1926 and is still in evidence today and can be viewed as a low-intensity recurrent selection system form of dynamic management as outlined by Goldringer *et al.* (2001). The end result is the last characteristic referred to by Wood and Lenné (1997) in their analysis of actual farmer's activities and how they contribute to on-farm conservation, which is the management of a dynamic portfolio of varieties which may include modern commercial as well as landrace and heritage varieties.

39.4 A Role for Collection Managers

As has been suggested by Wood and Lenné (1997), the most practical way of fostering a dynamic informal sector is access to diversity. Experience as a germplasm curator has shown that the best method of achieving this is by providing opportunities to view demonstrations of growing materials, preferably in environments that growers can readily relate to and extrapolate from. Such a programme of demonstrations has been in progress at the John Innes Centre since 2004 with small-scale autumn-sown plots in the field in line with farmers practice (JIC GRU, 2004). Viewings of the material are organized close to the time when the material is ripening and would be ready to cut and people come and evaluate the material in their own way and make notes of particular lines, which they can then order and grow on their own land and evaluate for themselves. Over the intervening years, feedback from discussions with the growers and thatchers has fed directly into the planning of what is grown over in the next season. Word of this initiative has spread as has the source of older heritage material, all of which can be demonstrated to increase the level of and uptake of materials.

39.5 The Role of Landrace Growers

There is clear evidence that individual farmers, while experimenting with novel material, are in contact and show other trusted growers what they are doing. Small numbers of growers might even share material and extend the evaluation of new forms on to other sites, with this collaboration often extending into other activities of mutual benefit such as sharing of equipment and resources. This activity can nevertheless be characterized as self-generated and is supported exclusively by the 'community of interest'. This is in contrast to the more direct public-sector support characterized variously as participatory plant breeding, base broadening or dynamic management, of which there are numerous examples (Henry *et al.*, 1991; Witcombe, 1996, 2001; Ceccarelli *et al.*, 2000). These bring with them increases in scale both in terms of the number of sites and the diversity of the materials which are made available for individual farmers to sample and perform their own selection programmes (decentralization). These collaborative initiatives are also characterized by significant investment in assessments of diversity and the inclusion of multiple genetic crosses to broaden the gene pool so that through the processes of introgression and recombination there will be much greater range of genetic variation that the farmers can select from. The value of farmer selection to improving local adaptation has been quantified for a range of agronomic and morphological as well as allelic diversity as part of a long standing wheat experiment in France initiated in 1984 (Goldringer *et al.*, 2001; Dawson *et al.*, 2010; Enjalbert *et al.*, 2011).

39.6 Concluding Remarks

In summary, it is clear that farmers are still actively engaged in the dynamic management and on-farm breeding of certain crops where, for a variety of reasons, their requirements are not matched by modern elite varieties. By default these crops are focused on niche markets and offer material and often monetary advantages to the grower as their investment of time and resources must be met through the quality and utility of the product. Contact with these growers shows them to have a very open and fluid view of the material that they exchange and work with and are extremely open to viewing new material that is targeted at their needs. In their hands, landraces are far from fixed entities and I fully expect other cases of *de novo* synthesis of modern landraces to come to light with in-depth investigations into other niche products based on attributes of heritage crops.

References

Ambrose, M. and Letch, S. (2009) Thatching with long-straw wheat in relation to on-farm conservation in England. In: Vetelainen, M., Negri, V. and Maxted, N. (eds) *European Landraces: On-farm conservation, management and use*. Bioversity Technical Bulletin No.15, pp. 197–202.

Biffen, R.H. and Engledow, F.L. (1926) Wheat-breeding investigations at the Plant Breeding Institute, Cambridge. *Research Monograph No.* 4, Ministry of Agriculture and Fisheries, London.

Camacho Villa, T.C., Maxted, N., Scholten, M. and Ford-Lloyd, B. (2009) Defining and identifying crop landraces. *Plant Genetic Resources* 3, 373–384.

Ceccarelli, S., Grando, S., Tutwiler, R., Baha, J., Martini, A.M., Salahieh, H., Goodchild, A. and Michael, M. (2000) A methodological study on participatory barley breeding I. Selection phase. *Euphytica* 111, 91–104.

Dawson, J.C., Rivière, P., Galic, N., Pin, S., Serpolay, E., Mercier, F. and Goldringer, I. (2010) On-farm conservation and farmer selection as a strategy for varietal development in organic agricultural systems. In: Goldringer, I., Dawson, J.C., Chable, V., Lammerts van Bueren, E., Finckh, L. and Barot S. (eds) *Breeding for Resilience: a strategy for organic and low-input farming systems?* Proceedings of EUCARPIA 2nd Conference of the 'Organic and Low-Input Agriculture'. Le Comptoir General, Paris, France, 1–3 December 2010, pp. 123–126.

Enjalbert, J., Dawson, J.C., Paillard, S., Rhone, B., Rouselle, Y., Thomas, M. and Goldringer, I. (2011) Dynamic management of crop diversity: from an experimental approach to on-farm conservation. *Comptes Rendus Biologies*, (in press).

Goldringer, I., Enjalbert, J., David, J., Paillard, S., Pham, J.L. and Brabant, P. (2001) Dynamic management of genetic resources: a 13-year experiment on wheat. In: Cooper, H.D., Spillane, C. and Hodgkin, T. (eds) *Broadening the Genetic Base of Crop Production*. CAB International, Wallingford, UK, pp. 245–260.

Henry, J.P., Pontis, C., David, J.L. and Gouyon, P.H. (1991) An experiment on dynamic conservation of genetic resources with metapopulations. In: Seitz, A. and Loeschcke, V. (eds) *Species Conservation: a population biological approach*. Birkhaüser Verlag, Basel, pp. 185–198.

JIC GRU (2004) Growing demonstrations of UK cereal landraces, landrace selections and early cultivars from the BBSRC Small Grain Cereal Collections. Available at: www.jic.ac.uk/GERMPLAS/Cereal%20landraces.htm (accessed 5 May 2011).

Osman, A. and Chable, V. (2009) Inventory of initiatives on seeds of landraces in Europe. *Journal of Agriculture and Environment for International Development* 103, 95–130.

Scholten, M., Maxted, N. and Ford-Lloyd, B. (2003) UK National Inventory of plant genetic resources for food and agriculture. Unpublished Report, University of Birmingham, Birmingham, UK.

Wood, D. and Lenné, J.M. (1997) The conservation of agrobiodiversity on-farm: questioning the emerging paradigm. *Biodiversity Conservation* 6, 109–129.

Witcombe, J.R. (1996) Participatory approaches to plant breeding and selection. *Biotechnology and Development Monitor* 29, 26.

Witcombe, J.R. (2001) The impact of decentralisation and participatory plant breeding on the genetic base of crops. In: Cooper, H.D., Spillane, C. and Hodgkin, T. (eds) *Broadening the Genetic Base of Crop Production*. CAB International, Wallingford, UK, pp. 407–418.

40 Ecological Strategies for Crop Plants and Their Wild Relatives

R. Milla

40.1 Crop Yields, Plant Breeding and Ecological Theory

Crop yields experienced extraordinary growth during the Green Revolution, both due to plant breeding and to agronomic development. Yet, crop improvement has not seen comparable advances for increased productivity in the last decades (Denison *et al.*, 2003). Artificial selection over pre-existing natural variation might have reached a dead end, and the environmental impacts of intensified agronomy are unsustainable in the long term (Hoisington *et al.*, 1999; Miflin, 2000). Thus, plant breeding may have to look for novel strategies to ensure food security and environmental sustainability for a steadily increasing human world population (Miflin, 2000).

Some advocate that new biotechnological tools may offer qualitative advancement in breeding for higher productivity. Doebley *et al.* (2006), for example, state that, once we know which are the genes important to domestication and their expression dynamics and protein functions, we will be able to produce 'super-domesticates', for example, transferral of C_4 photosynthetic machinery to C_3 plants (Vaughan *et al.*, 2007). Other sources propose alternative approaches. Miflin (2000) says that technological transfer from genetic approaches is too slow to provide achievements soon enough, and advocates for turning to a phenocentric approach. Denison *et al.* (2003) point that focusing on traits that were under stabilizing selection in the wild for millions of years (e.g. photosynthesis) will be unfruitful. Traits that were less important in the wild would probably be less fixed, and working on improving those traits, many of which have great relevance to agriculture, may prove more profitable.

In this context of change in breeding strategies and understanding of domestication, ecological theory may provide a significant contribution from diverse angles. Here, I focus on one of those angles, and propose that a fair amount of the evolutionary changes that wild plants have experienced upon domestication could be framed in the theory of plant ecological strategies. The application of ecological strategies to plant domestication has been scarce and can bear promising advances in understanding and guidance for breeding programmes. I will briefly comment on what is commonly understood as the pool of traits that have been generally modified through artificial selection by humans (i.e. the domestication syndrome). Then, I will present plant strategies and multi-trait syndromes of wild plants as proposed in ecological theory. Multi-trait syndromes of wild plants typical of a given habitat or of a given level of resources are classically used in ecology to understand plant adaptation in the wild and provide guidelines for wilderness conservation actions or restoration strategies. This is analogous to the usage of the domestication syndrome to understand artificial selection and to supply bases for crop improvement. Finally, I will provide theoretical reasons to expect that traits related to plant strategies in the wild should have changed with domestication, and some examples supporting this view.

40.2 The Domestication Syndrome

The domestication syndrome is the set of traits that distinguishes the majority of crop species from their wild ancestors (Harlan *et al.*, 1973; Hammer, 1984). In itself, the domestication syndrome is a generalization, and its defining traits cannot be common to events as different as the domestication of wheat for grain yield or the selection of clonally propagated cassava for tuber production. However, most grain, fruit and legume herbaceous crops tend to share a common suite of traits when compared to their closest wild relatives: larger seeds or fruits (albeit fewer per plant), increased apical dominance and low branching frequencies, earlier and more synchronous reproductive phenology, augmented reproductive allocation, loss of diaspore dispersal and seed dormancy mechanisms, decreased levels of secondary metabolites and different amino acid profiles, etc. (Doebley *et al.*, 2006). Geneticists have extensively researched the domestication syndrome, and provided interesting findings. For example, domestication traits tend to be regulated by few genes, which facilitate fast evolution. Alleles are generally recessive, and their loci are usually clustered in nearby regions within the genome. In this way, direct selection for a given trait may easily drag the inheritance of other traits of interest through selective sweeps (see Hancock, 2004 for a review on genetic trends accompanying domestication).

The domestication syndrome chiefly relates to traits that were selected, either consciously or unconsciously, for their agronomic benefit to humans (Doebley *et al.*, 2006). However, domestication may have also impacted other traits that were not of interest to humans. There are evolutionary and ecological mechanisms that may theoretically account for such changes to occur (see below).

40.3 Character Syndromes of Wild Plants and Plant Ecological Strategies

According to Craine (2009), plant strategies are sets of interlinked adaptations that arose as a consequence of natural selection and that promote Darwinian fitness in a given environment. In essence, plant strategies' schemes pursue three main objectives:

- To define groups of species that function similarly under different ecological scenarios of resource supply, disturbance and stress in the wild;
- To identify the suites of traits that are typical to each ecological setting (trait syndromes) and infer which are current and past selection agents that may have shaped those phenotypes; and
- To investigate how variation in those key traits relates to variation in Darwinian fitness or plant performance in particular environments (Craine, 2009).

To date, plant strategies theory has focused on how plants use resources, especially mineral nutrients, for vegetative growth and performance, and much less effort has been invested on the identification of reproductive strategies. Also, debates on plant strategies have been profuse in the ecological literature since the late 1970s and a set of competing theories have been proposed (see review in Craine, 2009). Relevant discussion in this area focuses on which is the role of competition in environments with different resource supply, on whether plants respond similarly to different sources of stress or resource scarcity, or on which are the key traits that define plant strategies (Grubb, 1998; Reich *et al.*, 2003; Craine, 2009).

One of the seminal papers on plant strategies theory is Terry Chapin's *The Mineral Nutrition of Wild Plants* (Chapin, 1980; see Aerts and Chapin, 2000 for an updated version). In his earlier paper Chapin highlighted that traits related to resource acquisition and use are remarkably different among plant species typical of low versus high nutrient supply habitats, including crops and related species in the latter category. He proposed that low nutrient species should generally show low nutrient absorption rates, slow leaf, root and whole plant growth rates, slow organ turnover rates, high efficiency in nutrient use, linked to small nutrient losses through leaf and root senescence, and slow pace in biochemical processes such as photosynthesis or respiration, and vice versa for species from richer habitats (Chapin, 1980). Chapin called those two character syndromes as the competitive/ruderal strategy and the nutrient-stress tolerant strategy (similarly to the leading plant strategy scheme at the time: Grime, 1979). Today the idea remains the same in the core, although the terminology is somewhat different, i.e. resource-conservation versus resource-acquisition strategies, or low-resource versus high-resource strategies; and attention is paid to resources other than nutrients, with low-resource strategies being sometimes split into several categories depending on the resource in focus (e.g. Reich *et al.*, 2003; Craine, 2009). An important aspect of this scheme is how plant defence relates to the above plant strategies. Generally, resource-conservation is linked to increased investment in constitutive defence against pathogens and herbivores at the expense of reducing investment in growth, and the opposite for resource-acquisition (Herms and Mattson, 1992). The growth–defence trade-off is critical to the usefulness of bringing into play ecological strategies in the context of plant domestication (see below).

Chapin made his case by opposing mineral nutrition traits of wild plants from unfertile sites to those of plants typical of fertile sites and of crops, clumping together cultivated species and their wild ancestors in the high resource supply category. However, we know that the evolution from wild ancestors to their cultivated counterparts has profoundly changed plant phenotypes, and selective pressures have changed remarkably for the species involved. A shift from resource scarcity, unpredictable supply, and frequent herbivorism–parasitism in the wild, to higher resource availability, predictability and crop protection in agricultural environments has occurred (McKey *et al.*, in press). A parallel shift in plant phenotypes, if demonstrated, may have important implications for designing efficient crops. As Chapin (1980) highlighted in his concluding paragraph:

> At a time when the agricultural trend toward use of high-yield grains with high fertilizer requirements is threatened by energy shortages and rising costs (and in many countries by scarcity of fertilizer), it may be necessary to breed crops with somewhat lower growth and reproductive rates but with lower nutrient requirements.

40.4 Did Domestication Favour the Evolution of Resource-Acquiring Strategies?

There are reasons to expect that changes in traits other than the domestication syndrome, and in plant strategies, have occurred during domestication. There are, at least, two ecological-evolutionary mechanisms that may have led to a shift in non-domestication-syndrome traits and plant strategies between wild ancestors and their domesticated derivatives.

1. *Relaxed selection*: artificial environments have changed the main environmental pressures that plants had to overcome to successfully reproduce and persist. In agricultural lands water and nutrients are provided in higher and more predictable supply (Chapin, 1980), probably relaxing natural selection towards efficient nutrient and water use mechanisms and thus favouring resource-acquisition strategies (McKey *et al.*, in press). Also, in artificial environments, crop protection is a common management action whereby pests and pathogens are controlled through inputs that bear no metabolic cost to the individual plant. Therefore, it is likely that selection for genotypes with efficient, but costly, anti-herbivory and anti-pathogen machineries has been relaxed (Mondolot *et al.*, 2008).

2. *Phenotypic integration*: plants are phenotypically integrated organisms, and individual traits are functionally interdependent. It is very unlikely that the advent of gigantism in one organ through artificial selection (e.g. fruit size) can arise independently of correlated changes in other functions of the plant (e.g. hydraulic conductivity, vegetative growth rate, source–sink interactions, etc.). First, genes are not randomly located across genomes, and direct selection for a given trait may easily drag the inheritance of other traits through selective sweeps (Fay and Wu, 2000), which has been demonstrated for several domestication syndrome traits (e.g. Chapman *et al.*, 2008). Alleles with pleiotropic effects over phenotypic expression may also contribute to coordinated evolution of interlinked functions. Gene linkage, pleiotropy, genetic 'hitchhiking' and selective sweeps may have provoked the indirect selection of traits other than those directly related to the domestication syndrome (Hancock, 2004). Moreover, irrespective of genome architecture and of whether a gene affects single or multiple functions, phenotypes have to be functionally fully viable. This (almost) surely impedes the evolution of, for example, fruit gigantism and resource-conserving hydraulic systems in the same individual.

There is not much empirical evidence backing or rejecting the tenet that domestication has generally favoured the evolution of resource-acquisition strategies, and a good deal of systematic and extensive work should be done in this regard. However, a few species-specific studies point that this may be the case. Perhaps the best documented case on a probable shift of strategy during domestication is that of cassava (*Manihot esculenta* Crantz). Domesticated cassava invests less in tannins and cell wall compounds related to defence than its wild relatives (Mondolot *et al.*, 2008). Also, during domestication, cassava evolved epigeal germination and photosynthetic cotyledons, both of which confer high seedling productivity but higher attractiveness to herbivores (Pujol *et al.*, 2005). At the organ level, leaves of domesticated cassava deploy more leaf area per unit of investment in dry mass (i.e. higher Specific Leaf Area, SLA) and show higher carbon fixation rates than its wild counterparts (Pujol *et al.*, 2008). McKey *et al.* (in press) has studied other interesting aspects of cassava domestication, such as the mechanisms of artificial selection, or molecular dynamics, not so directly related to the issues discussed in this chapter (see references in McKey *et al.*, in press). Making a great simplification of their work on cassava, McKey *et al.* have demonstrated that the domestication of this species implied a shift to a more resource-acquiring plant strategy.

Another interesting example comes from a detailed herbivory study on maize and teosinte (Rosenthal and Dirzo, 1997). Rosenthal and Dirzo (1997) found that growth rate and reproductive output traits were higher the more advanced the domestication stage of their accessions, whereas susceptibility to herbivory was lowest among the wild and early domesticate lines. This is also consistent with a shift in plant strategy during maize domestication and further selection.

There are several worked case studies in the literature that document phenotypic changes in response to the relaxation of natural selection over traits of key importance to achieve fitness in the wild, but of lesser relevance in agricultural environments. Two examples are provided here. One frequently advocated argument, but with scarce empirical support, is that selective pressures over maintaining root mutualisms with mycorrhizal fungi and N-fixing bacteria should have been relaxed in fertile agricultural environments, disrupting pre-existing symbioses (An *et al.*, 2010). Kiers *et al.* (2007) showed that this was the case in soya, *Glycine max* (L.) Merr., where mechanisms for association with beneficial strains of *Rhizobium* were ineffective in highly improved soya lines. A completely different example of relaxed selection comes from research in apple trees. Using observational and experimental data, Archetti (2009) showed that:

- The beautiful range of autumnal leaf colours, that leaves of wild apples exhibit during leaf fall, are costly honest signals that warn aphids about the strength of parasite resistance of the host tree.
- Modern varieties have lost this mechanism, concurrently with the evolution of domestication traits such as big fruit sizes, which is compatible with the hypothesis of relaxed selection over defence traits in artificial environments.

Other examples of disruptions of species interactions following domestication, many of them caused by relaxed selection, are reviewed by Macfadyen and Bohan (2010).

40.5 Conclusion

In synthesis, domestication has been extensively researched by molecular biologists and agronomists, but comparatively little by evolutionary ecologists (McKey *et al.*, in press). It may be enriching to attract more ecological and evolutionary research to this subject. Combining new approaches to breeding, making use of crop wild relatives and their likely more resource-conserving strategies (Hancock, 2004), and taking into account other proposals based on evolutionary theory and community and ecosystem ecology (e.g. Denison *et al.*, 2003; Macfadyen and Bohan, 2010) may prove useful to reshuffle plant breeding strategies in the face of current demographic and environmental challenges.

References

Aerts, R. and Chapin, F.S. (2000) The mineral nutrition of wild plants revisited: a re-evaluation of processes and patterns. *Advances in Ecological Research* 30, 1–67.

An, G.H., Kobayashi, S., Enoki, H., Sonobe, K., Muraki, M., Karasawa, T. and Ezawa, T. (2010) How does arbuscular mycorrhizal colonization vary with host plant genotype? An example based on maize (*Zea mays*) germplasms. *Plant and Soil* 327, 441–453.

Archetti, M. (2009) Evidence from the domestication of apple for the maintenance of autumn colours by coevolution. *Proceedings of the Royal Society B: Biological Sciences* 276, 2575–2580.

Chapin, F.S. (1980) The mineral nutrition of wild plants. *Annual Review of Ecology and Systematics* 11, 233–260.

Chapman, M.A., Pashley, C.H., Wenzler, J., Hvala, J., Tang, S., Knapp, S.J. and Burke, J.M. (2008) A genomic scan for selection reveals candidates for genes involved in the evolution of cultivated sunflower (*Helianthus annuus*). *Plant Cell* 20, 2931–2945.

Craine, J.M. (2009) *Resource Strategies of Wild Plants.* Princeton University Press, Oxford.

Denison, R.F., Kiers, E.T. and West, S.A. (2003) Darwinian agriculture: when can humans find solutions beyond the reach of natural selection? *The Quarterly Review of Biology* 78, 145–167.

Doebley, J.F., Gaut, B.S. and Smith, B.D. (2006) The molecular genetics of crop domestication. *Cell* 127, 1309–1321.

Fay, J.C. and Wu, C.I. (2000) Hitchhiking under positive Darwinian selection. *Genetics* 155, 1405–1413.

Grime, J.P. (1979) *Plant Strategies and Vegetation Processes.* Wiley, UK.

Grubb, P.J. (1998) A reassessment of the strategies of plants which cope with shortages of resources. *Perspectives in Plant Ecology, Evolution and Systematics* 1, 3–31.

Hammer, K. (1984) Das Domestikationssyndrom. *Kulturpflanze* 32, 11–34.

Hancock, J.F. (2004) *Plant Evolution and the Origin of Crop Species*, 2nd edn. CAB International, Wallingford, UK.

Harlan, J.R., de Wet, J.M.J. and Price, E.G. (1973) Comparative evolution of cereals. *Evolution* 27, 311–325.

Herms, D.A. and Mattson, W.J. (1992) The dilemma of plants: to grow or to defend. *The Quarterly Review of Biology* 67, 283–335.

Hoisington, D., Khairallah, M., Reeves, T., Ribaut, J.V., Skovmad, B., Taba, S. and Warburton, M. (1999) Plant genetic resources: what can they contribute toward increased crop productivity? *Proceedings of the National Academy of Sciences USA* 96, 5937–5943.

Kiers, E.T., Hutton, M.G. and Denison, R.F. (2007) Human selection and the relaxation of legume defences against ineffective rhizobia. *Proceedings of the Royal Society B: Biological Sciences* 274, 3119–3126.

Macfadyen, S. and Bohan, D.A. (2010) Crop domestication and the disruption of species interactions. *Basic and Applied Ecology* 11, 116–125.

McKey, D., Elias, M., Pujol, B. and Duputié, A. In press. Ecological approaches to crop domestication. In: Gepts, P., Bettinger, R., Brush, S.B., Famula, T., McGuire, P.E. and Qualset, C.O. (eds) *Biodiversity in Agriculture: Domestication, Evolution and Sustainability*. Cambridge University Press, Cambridge.

Miflin, B. (2000) Crop improvement in the 21st century. *Journal of Experimental Botany* 51, 1–8.

Mondolot, L., Marlas, A., Barbeau, D., Gargadennec, A., Pujol, B. and McKey, D. (2008) Domestication and defence: foliar tannins and C/N ratios in cassava and a close wild relative. *Acta Oecologica* 34, 147–154.

Pujol, B., Mühlen, G.S., Garwood, N., Horoszowski, Y., Douzery, E.J.P. and McKey, D. (2005) Evolution under domestication: contrasting functional morphology of seedlings in domesticated cassava and its closest wild relatives. *New Phytologist* 166, 305–318.

Pujol, B., Salager, J.L., Beltran, M., Bousquet, S. and McKey, D. (2008) Photosynthesis and leaf structure in domesticated cassava (Euphorbiaceae) and a close wild relative: have leaf photosynthetic parameters evolved under domestication? *Biotropica* 40, 305–312.

Reich, P.B., Wright, I.J., Cavender-Bares, J., Craine, J.M., Oleksyn, J., Westoby, M. and Walters, M.B. (2003) The evolution of plant functional variation: traits, spectra, and strategies. *International Journal of Plant Sciences* 164, 143–164.

Rosenthal, J. and Dirzo, R. (1997) Effects of life history, domestication and agronomic selection on plant defence against insects: evidence from maizes and wild relatives. *Evolutionary Ecology* 11, 337–355.

Vaughan, D.A., Balazs, E. and Heslop-Harrison, J.S. (2007) From crop domestication to super-domestication. *Annals of Botany* 100, 893–901.

41 AEGIS: A Regionally Based Approach to PGR Conservation

J.M.M. Engels and L. Maggioni

41.1 Introduction

Globally, there are about 1800 gene banks and/or germplasm collections, holding approximately 7.4 million accessions of which an estimated 2 million are unique (FAO, 2010). Of the 3.8 million accessions with status information available, about 700,000 (18%) are crop wild relatives (CWR) and 1,700,000 (44%) landrace accessions, held in 724 and 901 gene banks, respectively. In Europe, approximately 625 gene banks and/or germplasm collections have been reported, holding more than 2 million accessions of which 30–40% have been estimated to be unique (FAO, 2010). Further, approximately 84,500 accessions of the 1.1 million accessions included in EURISCO are wild species (including CWR) and 268,840 landrace accessions maintained in 120 and 223 gene banks, respectively (EURISCO, 2011).

In the late 1990s, the European Cooperative Programme for Plant Genetic Resources (ECPGR) reported difficulties with the maintenance of plant genetic resources, in particular a lack of long-term conservation facilities, insufficient safety-duplication, regeneration backlogs and of inhomogeneous quality of conserved germplasm. Since 1998 the Steering Committee of ECPGR has been discussing options for the sharing of conservation responsibilities between European countries and decided in 2003 to initiate a feasibility study on a European Genebank Integrated System (AEGIS), using four model crops (*Allium*, *Avena*, *Brassica* and *Prunus* spp.), representing different biology- and policy-related scenarios. In 2006 the Steering Committee decided to formally establish AEGIS as a programme component on the basis of the outcome of the aforementioned study (ECPGR, 2006).

41.2 AEGIS

The goal of AEGIS is to create a European Genebank Integrated System for plant genetic resources for food and agriculture, aimed at conserving the genetically unique and important accessions for Europe and making them readily available for breeding and research purposes. The identified material, constituting the European Collection, will be safely conserved under well-defined storage or maintenance conditions that will ensure the genetic integrity and viability for the long term. According to the ECPGR Strategic Framework Policy Guide (ECPGR, 2009), the *ex situ* conservation of germplasm will be carried out according to common, agreed quality standards, wherever the germplasm is physically located, and will be carried out in such a way that it will facilitate close linkages with *in situ* conservation, the use of and research into the conserved germplasm.

AEGIS focuses primarily on the *ex situ* conservation and use of genetically unique and important accessions of PGRFA (following the definition of the International Treaty, as well as medicinal and ornamental species) and their wild relatives of European origin. It also includes other important species introduced and maintained in gene banks of the AEGIS member countries. It is intended to develop AEGIS within the existing legal framework of the International Treaty and, where necessary, to extend its scope according to the spirit and intentions of the Treaty, thereby contributing to its effective implementation (ECPGR, 2009).

AEGIS will be building on existing organizational bodies of the ECPGR Programme, in particular that of its Steering Committee (for the oversight and provision of funds), the Crop Working Groups (as the principal coordinating technical bodies) and the National Coordinators (as the focal point for all AEGIS-related activities in a given country). It is anticipated that the existing European information management systems, in particular EURISCO and the Central Crop Databases, will play a key role in orchestrating the information management at the national and regional level and as such to contribute to global information management systems such as Genesys (www.genesys-pgr.org).

41.3 Anticipated Benefits of Establishing AEGIS

AEGIS has been perceived to provide a number of benefits to participating countries and institutions. However, the establishment of the European Collection and all its supporting components and procedures will take time and only when reaching a critical threshold will it be possible to note or measure these benefits. The perceived benefits include:

- Improved security of germplasm through long-term commitment and systematic safety-duplication.
- Facilitated access to and availability of high quality germplasm and related information.
- Improved and widely agreed minimum quality standards of conserved material.
- Cost-efficient conservation activities.
- Reduced duplication of germplasm material within and across collections.
- Improved sharing of knowledge and information.
- Shared responsibilities among European countries and their institutions.
- A direct contribution to the implementation of the International Treaty at the national as well as at the regional level.
- An established European reference system for the effective and efficient conservation and the facilitated use in Europe. This should become attractive for longer-term financial commitments to maintain the system and/or for using the system in jointly agreed projects on priority areas.

41.4 Key Components of AEGIS

The establishment of AEGIS requires a number of key components. These include:

1. *A Strategic Framework* for the Implementation of AEGIS – A Policy Guide. This is the document that sets out the wider strategy and policies of establishing and operating AEGIS and as such it is the 'constitutional' backbone of the integrated system. The document was endorsed by the ECPGR Steering Committee (ECPGR, 2009).

2. *Memorandum of Understanding* (MoU). This is a formal agreement that ECPGR concludes with countries (Members) and institutions (Associate Members) that hold the germplasm or provide services to AEGIS. Membership in AEGIS is open to all the countries of the European Region and all the institutions of an AEGIS member country, i.e. also organizations that can provide conservation and use-related activities that complement what gene banks traditionally offer. Full participation of a country in the ECPGR is necessary prior to joining AEGIS, since AEGIS is based on the ECPGR structures. The signing of the MoU is the expression of the willingness and interest of a given country to become a member of AEGIS and to accept the commitment of long-term conservation of genetically unique and important accessions maintained in its gene bank(s) and to make these accessions available under the terms and conditions of the Standard Material Transfer Agreement (SMTA) of the International Treaty. By doing so, the designated accessions of a given country become part of the European Collection. By the end of February 2011, 26 countries had signed the MoU (http://aegis.cgiar.org/membership.html/).

3. *The European Collection.* This decentralized collection is composed of the selected 'European Accessions' that the countries accept to conserve according to agreed standards and to make readily available. AEGIS will focus primarily on the conservation and use of genetically unique and agronomically and/or historically/culturally important accessions for Europe. The Steering Committee has agreed on so-called selection requirements for the identification of the accessions that eventually will make up the European Collection (see Box 41.1).

Where duplicates or quasi duplicates are identified among accessions, ECPGR Working Groups have the task to select the Most Appropriate Accession (MAA) among them based on crop gene-pool specific selection criteria. These criteria will include aspects such as the comprehensiveness of existing passport data, the number of regeneration cycles, the health status, the existence of characterization and evaluation data, whether the accession is maintained in the country where it was collected or originated, and others. Whereas the general requirements are intended to discriminate between accessions for inclusion in the European Collection, the selection criteria are meant to facilitate the decision-making process which accession of a group of duplicates to accept for inclusion. Further details on the European Collection and the selection process details can be found on the AEGIS website (http://aegis.cgiar.org/european_collection.html/).

4. Quality System for AEGIS (AQUAS). The framework for the quality system is

Box 41.1. Selection Requirements for European Accessions

The following requirements have been agreed by the ECPGR Steering Committee during its Eleventh Meeting in Sarajevo, Bosnia and Herzegovina, September 2008. All European Accessions will need to comply with all requirements:

1. Material under the management and control of the governments of member countries and their associate members, in the public domain and offered by the associate members for inclusion into AEGIS.
2. Genetically unique within AEGIS, to the best available knowledge (i.e. genetically distinct accessions; assessment based on available data and/or on the recorded history of the accession).
3. Plant genetic resources for food and agriculture as defined in the International Treaty as well as medicinal and ornamental species.
4. European origin or introduced germplasm that is of actual or potential importance to Europe (for breeding, research, education or for historical and cultural reasons).

formulated in a Discussion Paper, where a number of underpinning principles have been recognized:

(i) Quality assurance is based on the principle that you:
(a) Plan – say what you do;
(b) Do – do what you say;
(c) Check – let an independent body check that you do what you say;
(d) Act – correct and improve what you say you do.

(ii) AQUAS is based on the principle of consensus.

(iii) With respect to the technical standards, agreement has to be reached through a well-defined process on what the 'lowest' acceptable standards are, i.e. standards that will ensure long-term and secure conservation, genetic integrity, identity and availability of the accessions. Such standards have been coined by the Steering Committee as the 'agreed minimum standards'.

(iv) Capacity building is a central activity in building and operating the virtual European gene bank system at an appropriate level of quality management and thus establishing and operating AQUAS. Capacity building efforts, in particular with regard to training, possibly both from within the gene bank or country as well as from outside, will be required to ensure the establishment of widely acceptable standards in all the gene banks hosting European Accessions.

(v) The AQUAS should be as little bureaucratic as possible, pragmatic rather than doctrinaire, and it should be recognized that different participating collections can achieve agreed minimum standards in different ways. Furthermore, the general principles are more important than over-prescriptive protocols.

(vi) A monitoring system should encourage participants to improve the performance, and thus to strengthen the capacity, rather than to feel they are being policed. Therefore, an 'effective guiding and advisory approach at the AEGIS level' will need to be developed to facilitate compliance of the partners with the collectively agreed management and minimum standards. A system of 'record keeping' of the performed activities will have to be developed in a participatory manner. The performance monitoring itself should be conducted by an ECPGR or a completely independent body.

Besides the above mentioned principles, there are the 'generic operational standards' (which were prepared in close collaboration with FAO through the revision process of the Genebank Standards), the 'operational genebank manual' (for which a template has been designed that will guide the Associate Members in preparing their respective gene bank manual) and the 'agreed minimum standards (by crop or crop group; they are defined in a process that has just been initiated by the respective Working Groups)'.

5. EURISCO (http://eurisco.ecpgr.org/) is the information portal for the European Collection. Accessions belonging to the European Collection are marked and become clearly identifiable through the catalogue.

6. A dedicated AEGIS website (see http://aegis.cgiar.org/about_aegis.html/) has been recently revised and is intended to provide easy access to the various documents, guidelines and other relevant information by the PGRFA community in Europe.

41.5 Achievements

Following the completion of the feasibility study (2004–2008), the results of which are available online (http://aegis.cgiar.org/documents/crop_specific_documents.html/), the AEGIS coordinator was appointed in 2008 as part of the ECPGR Secretariat, based at Bioversity International. An AEGIS Advisory Committee (AC) was also appointed to guide the process of AEGIS implementation (http://aegis.cgiar.org/structure/governance/avisory_committee.html/).

A Strategic Framework Policy Guide was compiled and endorsed in 2008 by the ECPGR Steering Committee (SC). On the basis of this document, the Memorandum of Understanding (MoU) text was developed, endorsed by the SC and sent for signature to all ECPGR member countries in April 2009. AEGIS entered into force in July 2009, after the signature of the tenth country, and by February 2011 the membership was raised to 26 countries with 19 Associate Member institutions. The principles of a quality system for AEGIS (AQUAS) have been endorsed and published in the form of a discussion paper. A template for the preparation of the gene bank operational manual was also endorsed in September 2010 by the AEGIS Advisory Committee. All Associate Members are expected to fill in this template and to publish online their operational manuals. Generic gene bank standards are in the process of being endorsed at the international level in early 2011, in collaboration with FAO. Working Groups need to define crop-specific gene bank standards. A few groups have already defined minimum standards that are at different levels of completion (http://aegis.cgiar.org/documents/crop_specific_documents.htm/).

A system of record keeping, reporting and monitoring still needs to be agreed and implemented across Working Groups, before a fully functional quality system can be considered operational. The Working Groups are defining the criteria for selecting the European Accessions. Draft lists are available for a few crops. The first accessions designated as part of the European Collection are expected in the course of 2011.

A Competitive Small Grant Scheme was launched in 2009 with the intention of facilitating the establishment and operation process of AEGIS; three proposals were awarded in 2009 and these should facilitate the assessment of the European collection of umbellifer crops and of pea, as well as to improve the quality of cryopreservation of garlic. More projects will be awarded in 2011 as a result of a second call.

41.6 AEGIS and *In Situ*/On-Farm Conservation

ECPGR has had a historic focus on *ex situ* conservation, but its objective also includes 'long-term *in situ* conservation of PGR in Europe'. The ECPGR *In situ* and On-farm Conservation Network was established in 2000 and a number of outstanding achievements have been realized by the Network. But so far, *in situ*/on-farm conserved germplasm has not been integrated into AEGIS. The recent ECPGR External Review recommended 'to fully integrate *in situ* and on-farm activities in AEGIS, expanding in a second step the Genebanks' *ex situ* coverage to both *in situ* crop relatives' populations and on-farm managed landraces; this will imply an expansion of EURISCO's structure, in order to include relevant data for the management of the *in situ* and on-farm components'. Consequently, the assessment of how this could be implemented has to be undertaken and a number of questions and issues have been identified and are listed in Box 41.2.

In conclusion, the *In situ* and On-farm Network should aim to aid the AEGIS Steering Committee to include *in situ* germplasm within AEGIS and there should be formal agreement of the AEGIS Steering Committee to include *in situ* germplasm (based on supporting documentation, including the benefits and feasibility of implementation of such decision) within its remit. Similar developments in establishing a European Forest Genetic Resources information system, EUFGIS, have been undertaken by the European Forest Genetic Resources Network (EUFORGEN) with respect to the management of identified forest populations and accessions (http://portal.eufgis.org/).

Box 41.2. Questions and Issues for *In Situ*/On-Farm Conservation

1. What kind of aspects should *in situ* and on-farm 'accessions' have to facilitate inclusion into AEGIS?

- **(i)** To be identified as 'genetically unique and important' population/landrace (including definition of an *in situ*/on-farm 'accession'!);
- **(ii)** To be placed by the respective country into the public domain;
- **(iii)** Countries to accept long-term conservation responsibility for *in situ* material;
- **(iv)** To be available for distribution (together with relevant information);
- **(v)** To be managed in accordance with to-be-established standards.

2. What kind of adjustments need to be undertaken in AEGIS?

- **(i)** A Steering Committee decision on the inclusion of *in situ* and on-farm material is required;
- **(ii)** Scope of the AEGIS MoU will need to be expanded to include *in situ* and on-farm managed germplasm;
- **(iii)** Adjustments of MoU text will be required;
- **(iv)** *In situ* and On-farm Conservation Network will have to assume technical and coordinating responsibilities;
- **(v)** Adjustments in EURISCO will be required to allow relevant information to be included.

3. What kind of tools and procedures need to be developed in order to allow quality management?

- **(i)** Protocol and/or manual of currently followed management procedures by Associate Member Institutions;
- **(ii)** Generic technical management practices/standards for *in situ*/on-farm material;
- **(iii)** Species-specific technical standards or requirements, e.g. minimum population size; specific management practices, etc.; and
- **(iv)** Reporting and monitoring procedures.

References

ECPGR (2006) *Report of the Steering Committee. Tenth Meeting, 5–8 September 2006, Jurmala, Riga, Latvia.* Bioversity International, Rome.

ECPGR (2009) *A Strategic Framework for the Implementation of a European Genebank Integrated System (AEGIS). A Policy Guide.* European Cooperative Programme for Plant Genetic Resources (ECPGR). Bioversity International, Rome.

EURISCO (2011) EURISCO: finding seeds for the future. Available at: http://eurisco.ecpgr.org (accessed 12 February 2011).

FAO (2010) *The Second Report on the State of the World's Plant Genetic Resources for Food and Agriculture.* Commission on Genetic Resources for Food and Agriculture. Food and Agricultural Organization of the United Nations, Rome.

42 European On-Farm Conservation Activities: An Update from Six Countries

V. Negri, D. Fasoula, M. Heinonen, V. Holubec, M. Musayev, G. Spataro, M. Veteläinen and R. Vögel

42.1 Introduction

There has been a growing interest in on-farm conservation in Europe in recent years and the membership to the On-farm Conservation and Management Working Group of the ECPGR (European Co-operative Programme for Plant Genetic Resources) has increased as a consequence. The Working Group presently has 31 state members and one representative from NGOs. The aim of the Working Group is to promote the conservation and use of landrace diversity on-farm across Europe. A number of activities and tools were prepared by the group in the last couple of years to raise awareness of the on-farm conservation community. Two publications (Bailey *et al.*, 2009; Veteläinen *et al.*, 2009) reviewed the status and problems related to on-farm and home garden conservation in Europe, a website was launched to facilitate the sharing of information (www.sharinginformation.eu) and an on-farm data recording scheme was established and put online (see www.ecpgr.cgiar.org/Networks/Insitu_onfarm/Docs/OnfarmDescr_DRAFT271107.pdf) to help recording information on-farm. Finally, contacts were established with several stakeholders (especially farmers, farmer organizations and NGOs), European funded projects (AEGRO, FSO, PGR secure and SOLIBAM) and other ECPGR Working Groups.

A survey on the status of the national plans for preserving genetic resources was carried out during the first joint meeting of the 'Wild species in genetic reserves' and 'On-farm conservation and management' Working Groups, of ECPGR *In-situ* and On-farm Conservation Network, held in Funchal, Madeira, in September 2010. It showed that most countries have initiated the inventory of their landraces and old cultivars, and one (Switzerland) has completed its own (B. Schierscher-Viret, Madeira, 2010, personal communication). The survey revealed the following constraints in fulfilling inventory tasks:

- Lack of resources;
- Lack of proper methodology; and
- Lack of purposely developed national policies and local difficulties in applying the present EU seed legislation.

It seems that there is a long way ahead before complete inventories will be available for each country. This severely hampers possible conservation, rescue and further dynamic use of European landraces and old cultivars.

Following the 2008/62/EC, 2009/145/EC and 2010/60/EU Commission Directives on seed commercialization of conservation

 Agrobiodiversity Conservation: Securing the Diversity of Crop Wild Relatives and Landraces (eds N. Maxted *et al.*)

varieties, 110 of them were registered in the European common catalogue (Table 42.1). Most of them are open-field crop (bread and spelt wheat, barley, oat, maize, potato, swede and *Vicia* spp.) varieties coming from Austria, Estonia, Finland, Germany, Italy, Romania, Slovenia, Spain, Sweden and the UK (Table 42.1). Only a few horticultural crop conservation varieties have been registered to date, probably because the relative Directive has not been received at each National level yet. These (including cardoon, pepper, leek, celery, common bean, pea and tomato) conservation varieties come from Italy and Spain (Table 42.1). It should also be mentioned that Finland also has landraces (two potato, two white clover and four timothy) registered as varieties. It is yet to be seen whether, overall, these Directives will benefit landrace and on-farm conservation. They do not specifically address conservation *per se* but only seed production and marketing and it is unlikely that it will be possible for all the surviving landraces to be commercialized.

42.2 Updated Reports from Individual Countries

42.2.1 Azerbaijan

Agriculture appeared in Azerbaijan several millennia BC, as evidenced by the excavations

Table 42.1. Landraces registered as conservation varieties (CV) by each single country and in total in January, 2011.

Latin name	Member state	CVs	Latin name	Member state	CVs
×*Triticosecale* Wittm. ex A. Camus	RO	2	*Pisum sativum* L.	IT	1
Total		2	Total		1
Allium porrum L.	IT	1	*Secale cereale* L.	FI	6
Total		1		DE	1
Apium graveolens L.	IT	1	Total		7
Total		1	*Solanum tuberosum* L.	EE	1
				ES	2
Avena sativa L. (including *A.*	RO	2		SE	14
byzantina K. Koch)	SE	5		DE	4
Total		7	Total		21
Brassica napus var. *napobrassica* (L.) Rchb.	UK	4	*Sorghum bicolor* (L.) Moench	AT	1
Total		4	Total		1
Capsicum annuum L.	ES	1	*Triticum aestivum* L.	AT	2
	ES	2		DE	2
	IT	3		RO	9
Total		6		SE	14
Cynara cardunculus L.	IT	1	Total		27
Total		1	*Triticum spelta* L.	AT	1
Hordeum vulgare L.	FI	1	Total		1
	RO	2	*Vicia faba minor* L.	DE	1
	SE	7	Total		1
	UK	1	*Vicia sativa* L.	RO	1
Total		11	Total		1
			Zea mays L.	ES	5
Lycopersicon esculentum Mill.	ES	1		IT	8
Total		1		SI	1
Phaseolus vulgaris L.	ES	1	Total		14
	IT	1			
Total		2	Overall total		110

in Chalagantapa, Aghdash region, and Misharchay Jalilabad region, which show evidence of settlements sowing grain-crops 5000–6000 years ago. Azerbaijan also has quite different pedoclimatic conditions, resulting in the presence of more than 75% of the higher plants of the Caucasus (i.e. 4500 species).

A wide diversity of modern fruit crops' wild relatives is present with more than 149 species of fruit crops belonging to 39 genera and 15 families. For example, Azerbaijan sea-buckthorn *Hippophae rhamnoides* L. has over 80 forms. This species is widely used as a medicinal plant, as food and feed, as well as an ornamental and as protective belts and fences for the prevention of erosion and for soil reclamation. Three cultivars of the Azerbaijan sea-buckthorn were bred and released by the Genetic Resources Institute using the natural diversity. They showed high yield (18–25 t/ha), big fruit (fruit weight, 50–60 g) and weak or absent thorns (Musayev, 2008).

In addition, each region of the country is famous for different fruits: for example, Shirvan for its quince and pomegranate, Nakhchivans for its apricots and peaches, Zakatala for its nuts and walnuts, Apsheron for its figs and grapes, Guba for its apples and pears. There is an uncountable number of ancient varieties for each species and each of them is peculiar for the taste and the quality of the fruit. Exploration missions carried out by the Genetics Resources Institute recorded a high number of species present, but also documented that many wild species (including wild relatives) and landraces are at risk. This instigated the creation of germplasm collections for their conservation and use.

42.2.2 Cyprus

Ex situ conservation activities began in the late 1970s (Della, 2002) and led to the current National Gene Bank, which hosts 26 barley and 58 durum wheat, 19 lentil, 28 chickpea, 15 bitter vetch, 12 ochrus vetch, 19 grass pea, 29 lucerne and 6 pea landrace accessions. In the framework of the EU-funded RUBIA project endemic and rare plants, medicinal and aromatic plants, pistachio, almonds and carob trees (Della *et al.*, 2006) and tomato, melon, and bean landraces were also collected as part of the INTERREG IIIC Programme (see http://farvaldi.maich.gr/home/). These actions provided the necessary framework upon which present on-farm conservation activities of barley and vegetable species' landraces have been initiated (Fasoula and Kyratzis, 2010; Fasoula, 2011) between the Agricultural Research Institute in Nicosia and the Ministry of Agriculture, Natural Resources and the Environment. Local farmers have been preserving landraces with various degrees of persistence and success, depending on the region and their individual capacity. However, many of them have already disappeared, mostly due to the ageing of farmers and the spread of new cultivars. At present, the Agricultural Research Institute is trying to identify, register and locate the surviving landraces and serve as a focal point for their conservation, to monitor their most efficient propagation and evolution in the field, and to make available the seeds to interested farmers. There is a growing interest among younger farmers in their use. It is hoped that this approach will contribute to the revival of some of the most valuable landraces, along the lines of other previously successful cases in the EU, e.g. cowpea landraces in Italy (Polegri and Negri, 2010).

42.2.3 Czech Republic

After political changes in the Czech Republic, there were important changes in agricultural priorities. The Czech market has opened up for useful traditional materials especially in connection with organic farming, due to the rise in demands for both vegetarian and health foods and for the diversification of species and crops. In this context it would be useful to use crops that are unique for the traditional use for which they were bred, cultivated and valued, as in the case of the Carpathian emmer wheat, *Triticum dicoccon* Schrank, that was used for making peeled hulled grain for blood sausages and for soups. Several attempts have been made to set up on-farm conservation activities that involve national parks, regional organizations, museums (addressing educational and demonstration needs) and private citizens (addressing landrace production), all based on reintroduction from material preserved in gene banks. While on-farm conservation of fruit trees presently seems to be very successful for the future, long-term on-farm conservation of herbaceous plants is usually temporary and not as certain.

Several typical landraces, from various Czech regions, have been chosen for on-farm conservation trials (Holubec *et al.*, 2010), among them landraces of emmer and einkorn wheat from the Czech-Slovak border, perennial tufty rye, grass pea, common bean, medicinal and condiment plants, shallot onion, red cabbage, lettuce and several fruit trees (apples, pears, cherries, plums and rowan). As for fruit trees in particular, based on the plant determination and description in the different regions of the Czech Republic, landraces were selected for the establishment of on-farm conservation fields (Paprštein *et al.*, 2010) that were established in Vrchlabí, Krkonoše National Park, in Neratov, Orlické Mts, in Znojmo, Podyjí National Park and in the Šumava National Park (Table 42.2). Several more places are presently being proposed and discussed among stakeholders as conservation sites.

42.2.4 Finland

The multidisciplinary project 'On farm conservation in Finland' (2006–2008), carried out by MTT Agrifood Research, studied the on-farm management and the social and cultural aspects and values anchored to landraces that motivate farmers to grow cereal landraces at the present time and in the future. In total 34 farms were contacted, 47 notifications of cereal landraces or old cultivars in cultivation were received and 14 thematic interviews were carried out.

There is a wide range of reasons for landrace cultivation. Aged farmers valued strongly the use value of their landraces in traditional cooking (e.g. baking bread). They also underlined the good cultivation properties (e.g. yield reliability). Young farmers considered it as a family heritage and had very personal, close and intimate connection to the landrace itself. The more market-oriented landrace farmers saw landraces' potential to niche markets because of their rarity, taste and history. In contrast, hobby farmers highlight the cultural and historical as well as genetic diversity values and their own searching for specialities and flavour experiences as motivation to cultivate landraces.

A broad and versatile range of actors are needed to keep cereal landraces in cultivation (Heinonen, 2009). The recent national support system for on-farm maintenance of some crops is targeted to active farmers, but it was not seen suitable or useful by most of the landrace farmers. However, the national implementation of the EU directive 2008/62/

Table 42.2. On-farm conservation of fruit trees in the Czech Republic (Paprštein *et al.*, 2010).

	No. of landraces per region			
Fruit species	Vrchlabí, Krkonoše NP 2002	Neratov, Orlické Mts 2004	Znojmo, NP Podyjí 2005	NP Šumava - 2008
Apple	25	29	5	14
Pear	4	21	8	3
Sweet cherry	10			
Sour cherry	4			
Plum	4	2		
Total	47	52	13	17

EC which led to register several conservation varieties in the national list (Table 42.1) has raised somewhat the awareness of landrace cultivation or even functioned as an incentive. This can be seen by the increased number of contacts from farmers and private persons to the National Programme for PGR so that there is evidence of a slight positive trend. However, it needs to be stressed that new activities, such as product development of landrace-based products and cultural activities, will promote the continuity of the cultivation of landraces. Local museum gardens have a potential to demonstrate and to promote especially heritage plants. A comprehensive national inventory on landraces, especially on horticultural crops, is still needed for further promoting landrace conservation.

42.2.5 Germany

The Federal Ministry of Agriculture (BMELV) has launched programmes for conservation and use of plant genetic resources since 2002. Up to now 58 projects have been granted, 39 of them covering plant genetic resources. Objectives and themes are widespread, including orchards and fruit trees, regional inventories of cherries, *in situ* measures for wild apple, database establishment of regional crop wild relatives, managing and testing of regional lettuce collections for small scale and commercial use and the establishing of a gene bank for selected crop wild relatives under national responsibility.

EC agro-environmental programme (ELER-Directive 1698/2005) is used by Nord-Rhein-Westfalen and Brandenburg federal member states, offering advice and subsidies for the cultivation and use of rye, wheat, barley and oat cultivars and landraces of historic importance. As a result, a network of farmers dedicated to old cultivars' cultivation and to food technology for new products, has been established involving about 500 ha, 60 farmers and more than 50 varieties.

A national list of available and described landraces and old cultivars was brought up as a draft (BMELV/BLE 2010) and is recently under discussion. For the purpose of the common agro-environmental measures (ELER-Dir. 1698) a regional inventory of described, available and tested grains has been published (www.isip.de/coremedia/generator/isip/Start, documentId=103084.html). Further activities are running under the EC scheme of LEADER+- and INTERREG programmes and under a national umbrella of programmes for renewable resources. The national implementation of the EC-directives for conservation varieties actually sees eight varieties listed (Table 42.1). Although a growing community of gardeners and farmers has become engaged to plant for genetic resources conservation in the past, commercialization, industrial techniques in agri- and horticulture and the present restrictive behaviour of European seed legislation still limit their use and conservation.

42.2.6 Italy

Following the publication of the National Plan for Agricultural Biodiversity and the implementation of the above mentioned Commission Directives on seed marketing of conservation varieties at the national level, the Italian Ministry of Agriculture has funded a work aimed to promote and facilitate the implementation of the plan and of the Directives at the national level. The first step of this work is to provide the Italian Regions, which have jurisdiction for preserving plant genetic resources, with a set of operative instruments. Prior to the EU regulations, Italy and some Italian Regional Governments passed and implemented laws to safeguard genetic resources and to implement the provisions of the FAO Treaty (Lorenzetti *et al.*, 2009; Porfiri *et al.*, 2009). These led to regional inventories of genetic resources. In order to help the Regions to inventory landraces, a manual on *in situ* (on-farm) and *ex situ* conservation problems, methods and possible outcomes is being developed (Marino,

2010) by a group including experts, NGO representative persons, officers of the Variety Registration office, gene bank manager, professors and researchers of several Universities, the National Research Council and the Agricultural Research Council. The manual would also help in identifying those landraces that are compliant with the requirements foreseen by EU Directives on marketing of seed of conservation varieties and be useful to other countries to develop an appropriate approach to *in situ* on-farm conservation.

42.3 Conclusions

This short review of significant updates shows an increased awareness of landrace importance serving farmer needs. It is hoped that it will help in further enhancing on-farm conservation activities.

References

Anonymous (2010) Finnish Plant Variety Journal. *EVIRA* 2010, 2, 9.5.2010.

Bailey, A., Eyzaguirre, P. and Maggioni, L. (eds) (2009) *Crop Genetic Resources in European Home Gardens*. Proceedings of the ECPGR *In Situ* and On-farm Workshop, Ljubljana, Slovenia, October 3–4, 2007. Bioversity International, Rome, pp. 72–80.

Della, A. (2002) Germplasm in Cyprus National Plant Genetic Resources Inventory. Available at: www.ecpgr.cgiar.org/EPGRIS/intranet/CountryReports/CountryReports2002/Cyprus2002.doc (accessed 14 February 2011).

Della, A., Paraskeva-Hadjichambi, D. and Hadjichambis, A. (2006) An ethnobotanical survey of wild edible plants of Paphos and Larnaca countryside of Cyprus. *Journal of Ethnobiology and Ethnomedicine* 2, 34–43.

Fasoula, D.A. (2011) An overlooked cause of seed degradation and its implications in the efficient exploitation of plant genetic resources. *Plant Genetic Resources: characterization and utilization*, DOI: 10.1017/S1479262111000219.

Fasoula, D.A. and Kyratzis, A. (2010) Linking on farm and *ex situ* activities for conservation of vegetable landraces in Cyprus. *First meeting of the ECPGR Wild Species Conservation in Genetic Reserves Working Group and On-farm Conservation and management Working Group*, 13–16 September 2010, Funchal, Madeira, Portugal (abstract).

Heinonen, M. (ed.) (2009) Maatiaiskasvien ylläpitoviljely Suomessa. [On-farm conservation of cereal landraces in Finland] *Maa- ja elintarviketalous* 144, 102 s. Available at: www.mtt.fi/met/pdf/met144.pdf [in Finnish, abstract in English].

Holubec, V., Vymyslický, T. and Paprštein, F. (2010) Possibilities and reality of on farm conservation. *Czech Journal of Genetics and Plant Breeding* 46, S60–S64.

Lorenzetti, F., Lorenzetti, S. and Negri, V. (2009) The Italian law on conservation varieties and the National implementation of the EU Directive 2008/62 EC. In: Vetelainen, M., Negri, V. and Maxted, N. (eds) *European Landraces: On-farm Conservation, Management and Use*. Bioversity Technical Bulletin No. 15, Bioversity International, Rome, pp. 300–304.

Marino, M. (2010) Implementazione del Piano Nazionale per la Biodiversità in Agricoltura. *Dal Seme* 4, 37–41.

Musayev, M. (2008) Originating center and domesticating history of sea-buckthorn (*Hippophae rhamnoides* L.) in the Azerbaijan Republic. Global Facilitation Unit for Underutilized Species. October 2007. Available at: www.underutilized-species.org/feature-hippophae-aze.pdf. *FAO-Hippophae Countri Compas Non-Wood News* 17 July 2008.

Paprštein, F., Sedlák, J. and Holubec, V. (2010) On farm orchards of fruit trees. *Czech Journal of Genetics and Plant Breeding* 46, S65–S69.

Polegri, L. and Negri, V. (2010) Molecular markers for promoting agro-biodiversity conservation: a case study from Italy. How cowpea landraces were saved from extinction. *Genetic Resources and Crop Evolution* 57, 867–880.

Porfiri, O., Costanza, M.T. and Negri, V. (2009) Landrace inventories in Italy and the Lazio Region case study. In: Vetelainen, M., Negri, V. and Maxted, N. (eds) *European Landraces: On-farm Conservation, Management and Use*. Bioversity Technical Bulletin No. 15, Bioversity International, Rome, pp. 117–123.

Veteläinen, M., Maxted, N. and Negri, V. (eds) (2009) *European Landraces: On-farm Conservation, Management and Use*. Bioversity Technical Bulletin No. 15, Bioversity International, Rome, 305 pp.

43 Current and Future Threats and Opportunities Facing European Crop Wild Relative and Landrace Diversity

N. Maxted, Z.I. Akparov, M. Aronsson, Å. Asdal, A. Avagyan, B. Bartha, D. Benediková, T. Berishvili, R. Bocci, Z. Bullinska-Radomska, J. Cop, T. Curtis, K. Daugstad, S. Dias, M.C. Duarte, S. Dzmitryeva, J. Engels, D.A. Fasoula, N. Ferant, L. Frese, P. Freudenthaler, R. Hadas, L. Holly, A. Ibraliu, J.M. Iriondo, S. Ivanovska, T. Jinjikhadze, G. Kamari, S.P. Kell, C. Kik, L. Koop, H. Korpelainen, K. Kristiansen, A. Kyratzis, J. Labokas, L. Maggioni, J. Magos Brehm, E. Maloupa, J.J.R. Martinez, P.M.R. Mendes Moreira, M. Musayev, M. Radun, P. Ralli, D. Sandru, K. Sarikyan, B. Schierscher-Viret, T. Smekalova, Z. Stehno, T. Stoilova, S. Strajeru, A. Tan, M. Veteläinen, R. Vögel, G. Vorosvary and V. Negri

43.1 Introduction

Conservation budgets are increasingly under pressure from alternative demands for funding and, as conservation is largely funded from the State and there are many competing funding demands, there is an imperative to maximize the efficiency of conservation expenditure. Any activity that helps target expenditure is thus a priority. Addressing this issue is as important for those conserving agrobiodiversity as it is for those with a more generic biodiversity conservation mandate.

Identifying both short and longer term priorities were discussed and prioritized during the first joint meeting of the 'Wild species in genetic reserves' and 'On-farm conservation and management' Working Groups, of the European Cooperative Programme for Plant Genetic Resources (ECPGR) *In situ* and On-farm Conservation Network. This meeting, held in Funchal, Madeira, on 16 September 2010, was held at the end of the symposium 'Towards the establishment of genetic reserves for crop wild relatives and landraces in Europe', which was the final dissemination meeting of the EC AGRI GENRES 057 project An Integrated European In Situ Management Workplan: Implementing Genetic Reserve and On-farm Concepts (AEGRO). For the *In situ* and On-farm Conservation Network meeting 41 delegates from 31 countries were present. The aim of the Network is to promote the conservation and use of landraces and crop wild relative diversity on-farm and in genetic reserves across Europe. As such the short and longer term priorities discussed and prioritized relate to these aims.

The particular relevance of horizon scanning as a participatory approach to establishing future priorities is increasingly recognized by governments (King and Thomas, 2007), commercial organizations and conservation agencies

(Sutherland *et al.*, 2008, 2010). It is now routinely making a contribution to strategic planning, risk management, research priorities and policy making. For policy makers and practitioners to make informed resource allocation decisions, they require an evidence base and a comparative assessment of the potential options for conservation action. This evidence base needs to cover all relevant policy aspects: political, social and economic, as well as environmental and scientific (Sutherland *et al.*, 2010). Sutherland and Woodroof (2009) suggested that horizon scanning could identify both potential new threats to biological diversity (in terms of structure, composition and function) and new opportunities for its conservation. Information and evidence to support policy choices concerning agrobiodiversity conservation may not be readily available at the right time, e.g. who would have foreseen 25 years ago the current imperative of climate change modelling, assessment of resilience and mitigation? However, many agrobiodiversity conservation challenges are the result of technical developments, evolving consumer demands and market changes or agroenvironmental change. Many changes will have known or suspected impacts on agrobiodiversity, while others result from an acceleration of current patterns of change or new legislation. Examples of previous issues in European plant genetic resources include the impact upon agrobiodiversity conservation targets of genetic pollution, lack of appreciation of the significance of crop landraces, or disconnect between the agrobiodiversity and biodiversity communities in Europe. On the other hand conservation, particularly genetic, and digital technology has been advancing so rapidly that we can now answer questions that previously were not even considered; for instance, historically, ecogeography was routinely used as a proxy for genetic distance when planning collecting, but now genetic markers are used routinely to review population differentiation and genetic distance in currently conserved germplasm accessions, so that the decision can be made whether additional collecting is warranted and if it is, in which geographic locations the collecting should be targeted.

Sometimes policy makers and conservationists can spot a problem in advance, but fail to act because of competing short-term interests or the problem identified may not be communicated well enough or there may be insufficient collaboration with other specialists when an interdisciplinary approach is necessary. We argue that these problems can be addressed by finding more effective ways for agrobiodiversity conservationists to work across disciplines (including social sciences) and for scientists and policy makers to communicate with one another across the boundaries between disciplines about future problems. In this paper we present the results of such an exercise in reviewing short-term priorities and longer term priorities through consensual horizon scanning. The aim is to identify the major issues that currently are challenging agrobiodiversity conservation and also future issues that might challenge conservation of agrobiodiversity (Crop Wild Relatives – CWR and Landraces – LR) conservation in Europe in the next 20–30 years. Our approach was to use the collaborative expertise of the ECPGR *In situ* and On-farm Conservation Network to identify and prioritize relevant issues.

43.2 Methods

To help identify the short-term agrobiodiversity conservation priorities a list of actions associated with effective CWR and LR was agreed and each country, via its ECPGR representative, was asked to comment on their country's status as regard to each action. For CWR diversity the actions were whether there existed: (i) a National Action Plan for CWR survey, monitoring and conservation; (ii) a National CWR inventory; (iii) a prioritization list of CWR species; (iv) a CWR information system; (v) systematic gap analysis had been used to aid CWR conservation;

(vi) *in situ* genetic reserves for CWR conservation; (vii) *ex situ* germplasm holdings of CWR diversity; (viii) threat assessment using IUCN Red List Criteria of CWR diversity; (ix) routine national utilization of CWR diversity; (x) public awareness of CWR value; and (xi) legislative/policy framework to enhance CWR conservation. For LR diversity the actions were whether there was a complete, partial or no national LR inventory, and if not complete what was the limiting factor. The data were collected for 32 European countries for the CWR conservation actions and 22 countries for LR conservation actions. The results of this data collation exercise would help identify the immediate priorities for CWR and LR conservation action both Europe-wide and also nationally.

To establish the longer term agrobiodiversity conservation priorities horizon scanning was used, amended from the approach taken by Sutherland *et al.* (2008, 2010). All 83 delegates at the symposium, whether members of the ECPGR *In Situ* and On-Farm Network, were asked to identify emergent issues that they felt were of European importance or may have a local effect on CWR and LR diversity in Europe in the future 20–30 years. For each issue they submitted they were asked to outline the Strengths, Weaknesses, Opportunities, Responsibilities and Threats associated with the issue for European diversity. This identified a set of 15 issues for CWR concerns and 13 issues for LR concerns; these issues were raised by 12 people from 10 European countries for CWR issues and 11 people from 9 European countries for LR concerns. The table of CWR and LR issues were written on white boards and open for amendment/discussion. At the end of the symposium all delegates were given five points and allowed to attribute their points to the CWR and LR issues they regarded as the highest European conservation priorities. The points were then added up to identify the priority CWR and LR conservation issues in the longer term between 2020 and 2035.

43.3 Results and Discussion

43.3.1 Short-term issues (2010–2020)

CWR-related issues

The results of the survey of short-term CWR-related conservation issues in 33 European countries are summarized in Table 43.1. The results indicate that nearly all countries have some *ex situ* conservation of CWR diversity, and most countries have some form of national CWR inventory and national CWR threat assessment, although IUCN Red List threat of CWR has been undertaken as part of overall national threat assessment and was not focused on CWR species. The EC-funded FP5 PGR Forum project generated national CWR inventories that were sent to all ECPGR National Coordinators, therefore all countries should be aware of their existence; however, the responses indicate this may not be the case. The recently EC-funded FP7 PGR Secure project should address this requirement as it will hold a workshop for national CWR focal points nominated by the ECPGR National Coordinators to develop and promote the use of national CWR inventories.

Most countries have some form of CWR information systems, commonly with a more generalized information system or in association with the EURSICO national inventory. Most countries also have some, though not systematic, use of CWR diversity in breeding and have some national public awareness of the value of CWR diversity. However, few countries have a prioritized list of national CWR species, have developed national CWR action plans, or undertaken genetic gap analysis for even their most important CWR species and it is rare to have specific mention of CWR conservation or protection in national conservation legislation. Perhaps most surprising, given the increasing publicity given to *in situ* CWR conservation in the last 20 years, is the minimal progress in establishing working genetic reserves to conserve CWR diversity, perhaps with the notable exception of Armenia, Azerbaijan, Israel, Turkey and

Table 43.1. Summary of short-term issues that related to CWR conservation in Europe.

	Components of CWR national strategies											
						In situ conservation of CWR						
Country	National action plan in place	National inventory	List of priority species	CWR information systems	Gap analysis completed	Within genetic reserves	Within on-farm systems	*Ex situ* conservation	Threat assessment	National utilization of CWR	Public awareness of CWR value	Legislative/ policy framework
Albania	First draft	No	No	No	No	No	Partial	No	Yes	Yes	Partial	No
Armenia	CWR Conservation Action plan drafted	Yes	List of CWR priority species is prepared	www.cwr.am, EURISCO, National inventory, C&E data –limited	Partial	State reserve for CWR, Management plan is drafted	No	On average about 50% of CWR species require more seed collections	Armenian Red Data Book (2010)	Partial	Booklets, leaflets, round tables, workshops	Biodiversity Conservation Action Plan, Protected Area Strategy, and Sustainable Agricultural development strategy
Azerbaijan	Preparing as part of National strategy on conservation and sustainable utilization of PGR	Partial	On going	EURISCO	No	Yes, within 15 national parks	No	We have seed and field gene banks	A new Red Book is being prepared	Partial	Some	President of Azerbaijan Republic adopted 2006 'National Strategy and Action Plan for Conservation and Sustainable Utilization of the biodiversity'
Belorussia	No	No	No	No	No	Some	No	Partial	No	No	No	No
Bosnia	?	No	No	No	No	No	No	Yes	No	No	No	National Assembly of Republika Srpska adopted Programme for PGRFA protection

Cyprus	No	No	No	No	No	No	No	Yes	Yes	No	Some	No
Czech Republic	No	No	No	No	No	Partial	No	Yes	No	No	Some	Yes
Denmark	Partial	Yes	Draft	No	No	No	No	Partial	Partial	Partial	No	No
Finland	Partial	Yes	No	Partial	No	No	No	Yes	Yes	Partial	Partial	No
France	–	–	–	–	–	–	–	–	–	–	–	–
Germany	Part of the national PGR programme	Yes	No	Yes	No	No	No	Yes	Yes	Partial	Partial	Strategy for Agrobio-diversity in place
Greece	Partial	Yes, but requires updating	Priority species are identified by research bodies, most important species are agriculture and natives	Yes (but needs amending)	Partial	Some actions commenced by Greek gene bank and other institutes	Yes actions com-menced by Greek gene bank and other institutes	Yes	Greek Red Data Book (2009)	Partial	Partial	PD 80/90 / National Strategy for Biodiversity conservation in progress / New draft of the PD 80/90 in progress
Hungary	Within LIFE+ Pannon Seed Bank project covering CWR and other wild species of the Pannonian flora	Yes	Done	EURISCO, National Inventory	Partial	Partial, some actions started in collaboration of National Parks	No	Yes	Hungarian Red Data Book	Partial	Started as part of the PR actions related to the Pannon Seed Bank project	National Strategy for Agrobio-diversity is under development
Ireland	Yes	Done	Done	No	No	No	No	Yes	Yes	Partial	Limited	Not CWR specific
Israel	Partial	Yes	Yes	Yes, integrated in the IGB information system	Partial	Yes	No	Partial	Yes (Red Data Dook vol. 1 2007)	Partial	Yes	Partial
Italy	No	Yes	No	?CWR list published in PGR forum	No	No	Some actions	Partial	Some are highly threatened but no specific assess-ment	Partial	Limited	Not CWR specific

Continued

Table 43.1. Continued.

Country	Components of CWR national strategies										
					In situ conservation of CWR						
	National action plan in place	National inventory	List of priority species	CWR information systems	Gap analysis completed	Within genetic reserves	Within on-farm systems	*Ex situ* conservation	Threat assessment	National utilization of CWR	Public awareness of CWR value	Legislative/ policy framework
Lithuania	No	No	Partial	Yes	No	Partial	No	Yes	No	Partial	Partial	Not CWR Specific
Montenegro	?	No	No	No	No	No	No	Yes	No	No	Partial	Not CWR Specific
Netherlands	No	Yes	No	No	No	No	Partial	Yes	No	Partial	Partial	Not CWR Specific
Norway	Partial	Yes	No	No	No	No	No	Partial	Yes	Yes	Partial	Biodiversity Act implemented 2009
Poland	No	Yes	No	No	No	No	No	Partial	No	Yes	Partial	Partial
Portugal	No	Yes (inventory exists, but not approved by national authority)	Yes (list exists, but not approved by national authority)	Partial (unofficial CWR list available at www.jb.ul.pt)	Partial (only for a few priority species)	Passive conservation of some CWR in Protected Areas	No	Partial (few CWR are conserved *ex situ*; need for systematic *ex situ* conservation recognized)	Partial (threat assess-ment exists for a few CWR but not validated by IUCN or national authority)	Partial	No	Not CWR specific
Romania	No	No	No	*Ex situ* information only	No	Passive conservation of some CWR species	No	Partial	No	Partial	Limited	Not CWR specific
Russia	No	Yes	Yes	No	No	No	No	Partial	Yes	Partial	No	Not CWR specific

Serbia	No	No	No	No	No	Passive conservation of some CWR species	Partial	Partial	No	Partial	No	Mentioned in National Biodiversity strategy
Slovakia	Yes	Yes	No	Yes	Partial	No	No	Yes	Yes	Yes	Yes	Not CWR specific
Slovenia	No	No	No	*Ex situ* information only	No	No	No	Partial	Red Book, but not complete for CWR	No	Partial	Not CWR specific
Spain	No	No	No	No	No	No	No	Yes	Yes	No	No	Not CWR specific
Sweden	No	Yes	No	Yes	Yes	No	No	Partial	Yes	No	No	Not CWR specific
Switzerland	No	Yes	Yes	yes (www.bdn.ch)	No	No	No	No	Yes	No	Partial	Not CWR specific
Turkey	Yes	Yes	Yes	Yes	Yes (for some)	Yes	Yes	Yes	Red Data Book (2000) being revised and Important Plant Areas (2003) identified	Yes (many of the national breeding pro-grammes utilize CWR)	Yes	Yes various legislations and policy framework, incl. Plan of Action for *in situ* conservation and National Strategy and Action Plan
Ukraine	No	Yes	No	EURISCO	Partial	Yes	No	Yes	Red Book	Partial	Partial	Yes
UK	No	Yes	Yes	Yes	Partial	No reserves, one being developed	No	Yes Two popula-tions per species	Yes	Partial	Partial	Not CWR specific

Ukraine. Also the potential to conserve CWR alongside on-farm conservation programmes has only been investigated in Portugal and Turkey, which would be likely to prove beneficial without significant additional expenditure. European countries vary in their effectiveness in approaching CWR conservation with the most systematic approach by far being taken in Turkey and Armenia, though it should be added that both countries benefited from Global Environment Facility CWR-related projects (see for Turkey Zencirci *et al.*, 1998; and for Armenia see Hunter and Heywood, 2010). No information was available for CWR conservation activities in France.

On the basis of the process of collecting the data and its analysis, the five leading short-term priorities for CWR conservation to be achieved by 2020 are:

1. Systematic genetic reserve implementation to conserve *in situ* CWR diversity at both individual national and European scales.
2. Integration of *in situ* CWR conservation into on-farm initiative, so undertaking complementary CWR and LR *in situ* conservation.
3. National genetic gap analysis of CWR species that compare intrinsic CWR diversity with that sample of diversity that is currently actively conserved, as a basis for the establishment of national CWR conservation action plans.
4. Incorporation of legislative protection for CWR species and genetic diversity at both national and European levels.
5. Although many countries have existing use of CWR diversity in crop breeding, CWR use is not spread evenly across all crops, there is significant scope to enhanced utilization, as utilization is likely to hold the key to sustainability of conservation, breeder's use of CWR diversity should be promoted.

If these actions are implemented they will contribute significantly to ensuring the longer term food security of Europe.

LR-related issues

The results of the survey of short-term LR-related conservation issues in 33 European countries are summarized in Table 43.2. The analysis of the LR-related issues focused on whether European countries had a National LR Conservation Strategy in place and if not what constraints were impeding progress with its establishment. Only one country, Switzerland, has a National LR Conservation Strategy in place, 23 countries have partially complete inventories and related conservation strategies, and eight countries have thus far not commenced formulating a national LR inventory or a National LR Conservation Strategy. No information was available for LR conservation activities in France.

The major constraints on the establishment of a National LR Conservation Strategy are the lack of tried and tested methodologies, perverse national incentives and differential national application of EU seed legislation. These constraints are perhaps surprising given the relative ease with which plant breeders can exploit LR diversity. LR ecogeographic surveys, threat assessment and use potential, genetic gap analysis, actual or virtual characterization of LR diversity and generation of national and European LR conservation strategies are all priorities for action. Conversely, it can be argued that in countries with a well-developed breeding industry LR have in the past been extensively used by breeders to select improved varieties (around 1900) and today breeders will only use LR if a trait is lacking in the breeding pool. Further, the interest in the creation of LR inventories is led by public demand, policy makers and EU legislation, not by a strong demand from breeders. The simple counter for this argument is that breeders today do not know what adaptive traits they will require in which crops in 20–30 let alone 50 years' time, they are assuming that in 20–50 years' time the same or sufficient LR diversity in all crops will be available for them to mine traits. They have not taken into account the various threats facing LR diversity, particularly widespread LR replacement and the largely still unknown impact of climate change. Unless we conserve now breeders will have a reduced LR gene pool to select from for the yet unknown desired traits.

Table 43.2. Summary of short-term issues that related to LR conservation in Europe.

Country	National LR conservation strategy in place	Constraints reported
Albania	Not attempted	Lack of funding
Armenia	Not attempted	Lack of funding
Azerbaijan	Partially complete	Lack of funding
Belorussia	Not attempted	Lack of funding
Bosnia	Not attempted	Lack of funding
Cyprus	Partially complete	–
Czech Republic	Partially complete	–
Denmark	Not attempted	Lack of funding
Finland	Partially complete	Resources/methodology
France	–	–
Germany	Partially complete	–
Greece	Partially complete	–
Hungary	Partially complete	Lack of resources/EU seed legislation and national policies
Ireland	Partially complete (complete for apples and Brassicas)	Ageing farmers
Israel	Partially complete	Lack of funding
Italy	Partially complete	–
Lithuania	Not attempted	Lack of funding
Montenegro	Partially complete	Lack of funding
Netherlands	Partially complete	Lack of funding
Norway	Partially complete	–
Poland	Partially complete	Lack of resources/methodology/ people/national policies
Portugal	Partially complete	Lack of funding
Romania	Partially complete	Lack of funding
Russia	Not attempted	Lack of funding
Serbia	Partially complete (complete for fruit trees)	Lack of funding
Slovakia	Partially complete	Policy
Slovenia	Partially complete	Lack of resources and methodology
Spain	Not attempted	Lack of funding
Sweden	Partially Complete	Lack of resources and methodology
Switzerland	Complete	N/A
Turkey	Partially complete	Lack of funding
Ukraine	Partially complete	Lack of legislation
UK	Partially complete	Lack of funding

43.3.2 Long-term issues (2020–2035)

CWR-related issues

The results of the horizon scanning exercise to identify long-term CWR-related conservation issues in European countries are summarized in Table 43.3; the table is ranked with the highest priority issues identified by ECPGR *In situ* and On-farm Conservation Network members and other conference delegates at the top of the table. Although the establishment of national and European genetic reserves has been previously identified as a short-term priority, the establishment of a European network of CWR genetic reserves that systematically conserves the highest priority CWR diversity, particularly that associated with crop gene pool 1b or 2, or taxon group 1b, 2 or 3 (see discussion of CWR prioritization concepts in Maxted *et al.*, 2006) will

Table 43.3. Summary of longer-term issues that related to CWR conservation in Europe.

Votes	Priority action	Strengths	Weaknesses	Opportunities	Threats	Responsibilities
24	European network of CWR genetic reserves	Systematic *in situ* CWR conservation	Weak link between PA and PGRFA community	Establish closer PGRFA/ biodiversity links	PA community do not see value of CWR diversity	ECPGR + PlantaEuropa + other European Agencies
14	Improve IUCN Red List criteria to take account of intra-specific genetic diversity	Improve IUCN Red List criteria to take account of intra-specific genetic diversity	Time taken to develop and test new criteria; lack of suitable data to apply new criteria	Highlight the need to conserve specific populations/full range of ecogeographic and genetic diversity of CWR species	IUCN may not support the proposal and publication of a parallel system may be ineffective without IUCN's support	CWRSG
14	To develop participatory management and monitoring models for CWRs conservation	Significant experience is available of public participation and it engenders strong support from public	Insecure financing and requires professional oversight	Increased emphasis on holistic approach to conservation strategies and methodologies, and integration of PGR conservation into mainstream biodiversity conservation	In some countries lack of motivation to move forward and for conservationists to work through local communities or NGOs	ECPGR, individual countries
13	Improved national CWR inventories and prioritization of inventory on economic value, breeding demands and threat and biogeographic responsibility	Greater comparability of CWR data, allowing systematic transparent planning *of in situ/ ex situ* conservation actions	Availability of data, accuracy and reliability of data, problem of imprecise or agreed taxonomy	A significant step towards a European biodiversity information system with a focus on PGR (crops and trees)	Lack of staff, resources and agreement among national agencies	National level implementation, but introduced and promoted by ECPGR
13	Pilot national networks of CWR genetic reserves in test countries as model for European network of CWR genetic reserves	Systematic *in situ* CWR conservation	Weak link between PA and PGRFA community	Establish closer PGRFA/ biodiversity links	PA community do not generally see value of CWR diversity so does not prioritize CWR conservation	National PGR programmes and national PA communities

10	Improve methodologies for CWR conservation: inventorying; basic investigation (not only molecular); conservation outside PAs	Scientific basis for conservation measures and for monitoring	Each country employs different approaches to prioritization etc., limited discussion between scientists and authorities and lack of national inventories	Scientifically justified and objective methodologies	Lack of funds, lack of integration, lack of unified criteria	ECPGR, Bioversity International, FAO, Global Crop Diversity Trust, NGOs
10	Promotion of biodiversity friendly agriculture systems	Positive trends are visible, programmes are available	Economy often seems to work against biodiversity conservation, i.e. perverse incentives	Some consumers recognize value of biodiversity and are willing to pay for high quality or diversity products. NGOs are improving (organizational structures and knowledge)	Agricultural business in Europe tends to limit species number. However, CWR species diversity is critical to food security	Agricultural ministries, public relation offices, political parties, NGOs
9	Establishment of CWR genetic reserves in secondary ecosystems (human disturbed, e.g. roadside and railroad banks)	1. Provides additional value to human disturbed areas; 2. Provides corridors to promote gene flow between natural populations	1. Lack of interest by those in charge of managing these habitats; 2. Difficulties in setting up a sustainable network	1. Shift toward sustainable development includes more stringent infrastructure development requirements; 2. It can be seen as a way of mitigating other biodiversity losses due to infrastructure development	The fate of these genetic reserves might be vulnerable to political shifts as they are not located in protected areas	National programmes and ministries of infrastructures
9	Web-accessible central European database providing a comprehensive inventory of CWR genes or alleles used in crops	Information availability is essential to support long-term CWR conservation activities	Information may be difficult to retrieve especially if it involves plant breeding companies and sensitive information	1. It would enhance the value of CWR in the eyes of the society; 2. It may motivate plant breeders to make a greater use of CWR germplasm	It may be a product difficult to fund	ECPGR, Global Crop Diversity Trust, individual researchers, plant breeding companies
8	Submit IUCN Red List assessments of national endemic CWR to the global Red List	CWR included in the global Red List	Lack of availability of population data for threat assessment; lack of expertise and time	Increase recognition of CWR and threats to CWR in the nature conservation community	Funding; limitations of IUCN criteria in terms of assessing threat to genetic diversity	National PGR programmes, CWRSG

Continued

Table 43.3. Continued.

Votes	Priority action	Strengths	Weaknesses	Opportunities	Threats	Responsibilities
7	Ensure that ⩾70% of European CWR are conserved reliably (Target 9 of GSPC)	Better safeguarding crop diversity and sustainability	Different countries employ different systems which lack European coordination	Strengthening collaboraion between European and national research and decision-making bodies	Lack of will for European and national biodiversity and agrobiodiversity communities to work in a coordinated manner	Requires link between European and national biodiversity and agrobiodversity communities
7	Establishment of information network for CWR	Availability of data to prevent species losses and facilitate CWR use	Different structures of CWR responsible agencies in different European countries	Links between *ex situ* and *in situ* conservation; early warning systems and rapid response actions	Lack of funds and national information exchange mechanisms on CWR	ECPGR, Bioversity International, gene banks, national authorities
6	Intensive use of genomic information: detailed knowledge of agronomic genes	Systematic targeted breeding actions using CWR diversity	Expensive and high technical requirement	Improved cultivars and greater food security, coordination and delivery of knowledge base and products for Europe and individual countries	Too often a narrow range of characteristics are considered	ECPGR, plant breeding companies
5	*In situ* and on-farm National Focal Point network linked to *ex situ* NFPs	Better country representation and 'joined-up' conservation	Need for recognition and accountability on commitments and engagements	Strengthen communication; share experiences/tools; improved European PGR conservation and use; reduce gap between system and user	Funding; sustainability	Countries, ECPGR, Bioversity, EU
5	Maintenance of taxonomic expertise	Wealth of knowledge still available in botanic gardens, botanic institutes and gene banks	Trends towards replacing classical taxonomic knowledge by molecular genetic approaches alone	PGRFA advisory boards, ECPGR working groups have identified the problem. Need to launch programmes to attract younger scientists interested in classical botany/taxonomy	Lack of funding for botanic gardens; low impact factors for classical taxonomic publications	Bioversity International, European Commission, Science Ministries Universities with botanic gardens
2	CWR communication platform within networks	Oriented combined knowledge of CWR *in situ*/*ex situ* conservation	Weak link between research community and farmers' cooperatives or groups	Create a framework programme from the existing data and knowledge	Funding; sustainability	ECPGR, Planta Europa, BGCI, other stakeholders users

necessarily be a longer-term action because of the current lack of priority given to CWR conservation by the broad biodiversity conservation community. The establishment of closer links between the agrobiodiversity and biodiversity communities will raise awareness of the need for both CWR species and their intra-species diversity conservation, which ultimately should lead to the systematic CWR *in situ* conservation within existing protected areas. The reasons for locating reserves in existing areas are that: (i) these sites already have an associated long-term conservation ethos and are less prone to hasty management changes associated with private land or roadside where conservation value and sustainability is not a consideration; (ii) it is relatively easy to amend the existing site management to facilitate genetic conservation of CWR species; and (iii) it means creating novel conservation sites can be avoided so avoiding the possibly prohibitive cost of acquiring previously non-conservation managed land (Maxted *et al.*, 2008). In Europe logically this should be linked to the existing Natura 2000 network of protected areas (see www.natura.org) to avoid the wasteful resource expenditure on establishing a parallel protected area system. As such the fuller integration and the establishment of the European Network of CWR Genetic Reserves is a high priority longer-term goal.

The most commonly applied means of assessing threats to wild taxa is the application of the IUCN Red List criteria (IUCN, 2001), but CWR are now being specifically prioritized for assessment in Europe through an EC-funded project (Kell and Maxted, in prep.). This project has undertaken threat assessment of 586 European high priority CWR species, but this initial project has confirmed three important features of Red List assessment: (i) that Red List assessment is a very useful tool for aiding conservation prioritization, the greater the threat the higher the conservation priority; but (ii) Red List assessment does require a significant level of knowledge about the threats faced by species and time series data for population level, which is currently not widely available for most CWR species; and (iii) Red List assessment does not take account of the threat to or quantify changes in the genetic diversity within CWR species over time. While (ii) should not be seen as a criticism of Red List assessment, the level of accuracy is required to provide objectivity to the Red List assessment, but in relation to (iii) there is a need to work with IUCN to develop an 'extension' to the current IUCN Red List criteria that does take into account the need to threat assess genetic as well as species diversity. Thus, the longer term priority is to systematically Red List-assess European CWR species and also to work with IUCN to extend current IUCN Red List criteria to permit genetic diversity threat assessment.

Perhaps too often CWR conservation is seen by conservationists in purely objective, scientific terms and insufficient attention is paid to the potential of taking a more participatory approach. There is a need for rethinking the management and monitoring models for CWR conservation to take greater account of the role of local communities in genetic reserve conservation, so that genetic reserve conservation is more a partnership between the local communities and the conservationist. There are already good models of how this might be achieved (see Maxted *et al.*, 1997; Zencirci *et al.*, 1998; Iriondo *et al.*, 2008; Hunter and Heywood, 2010) and the evidence from these experiences shows that involving local communities directly benefits the project in meeting its scientific goals but also raises public awareness of the value of agrobiodiversity conservation and may also help integrate PGR conservation into mainstream biodiversity conservation.

The ultimate raison d'être for having a focus on CWR conservation is the contribution these CWR species can make through adaptive trait transfer to crops, therefore it is logical that any shorter or longer term priorities should address the need to facilitate the flow of traits from CWR to crops. Steps towards achieving this goal are making sure countries have national CWR inventories, that the inventory is prioritized on the potential of CWR species to contribute economically valuable traits, that as a CWR

Table 43.4. Summary of longer-term issues that related to LR conservation in Europe.

Votes	Priority action	Strengths	Weaknesses	Opportunities	Threats	Responsibilities
21	Promote European legislation in favour of LR conservation on-farm	Enhancing diversity maintenance in the field in the face of climate change	Low awareness of value of on-farm diversity among politicians	Maintenance/creation of new variation/ adaptation for breeding (incl. Participatory Plant Breeding), new farming systems and new products	Bureaucratic impasse, lack of effective lobbying actions	European Commission, ECPGR, individual country
19	Research level of unique crop diversity held in European home gardens and ensure its adequate conservation	Initial research shows home gardens contain significant unique diversity not held on-farm or *ex situ*	Difficult to identify unique home garden diversity as currently based on growers names not genetic distinction	Empower ordinary gardeners in conservation action, raise awareness of agrobiodiversity conservation and ensure unique crop diversity is conserved	Garden based germplasm is subject to severe periodic genetic erosion as individual gardeners die and material is not taken by younger gardeners	ECPGR and *In situ*/ On-farm Network to lobby EC to fund this research, European Commission
18	Establish European on-farm inventory of landraces, elder cultivars and farm saved seeds with maintainer details and basic descriptors	First overview and baseline of European on-farm diversity; baseline for monitoring system about LR, EC, FSS diversity	Availability of LR, EC, FSS information, individual maintainer data protection	Availability is a precondition for breeding and use actions, material exchange and network activities	Funding; sustainability, perverse governmental, breeders and commercial incentives	ECPGR and *In situ*/ On-farm Network, national PGR networks, NGOs
14	To develop participatory management and monitoring models for LR conservation	Large experiences and knowledge available, would aid public support of agrobiodiversity conservation	Insecure financing and requires professional oversight	Increased emphasis on holistic approach to conservation strategies and methodologies, and integration of PGR conservation into mainstream biodiversity conservation	In some countries lack of motivation to move forward and for conservationists to work through local communities or NGOs	ECPGR, individual countries

14	Promotion of biodiversity friendly agriculture systems	Positive trends are visible, existing programmes are available	Economy often seems to work against biodiversity conservation, i.e. perverse incentives	Some consumers recognize value of biodiversity and are willing to pay for high quality or diversity products. NGO are improving (organizational structures and knowledge)	Agricultural business in Europe tends to limit species number. However, CWR species diversity is critical to food security	Agricultural ministries, public relation offices, political parties, NGOs
12	Improve methodologies for LR conservation: inventorying; basic investigation (not only molecular)	Scientific basis for conservation measures and for monitoring	Each country employ different approaches to prioritization etc. limited discussion between scientists and authorities and lack of national inventories	Scientifically justified and objective methodologies applied	Lack of funds, lack of integration, lack of unified criteria	ECPGR, Bioversity International, FAO, Global Crop Diversity Trust, NGOs
12	Involve farmer networks in policies and decisions	Change perspective: 1. From conservation to sustainable use; 2. Detailed analysis of existing seed systems	Conservation is still widely viewed as being in opposition to agriculture and vice versa	Diversify farming systems and improve conservation of diversity	Lack of market for more diverse products	Farmers based NGOs, national PGR agencies, politicians
12	Network of LR maintainers (individual farmers or institutions) that are maintaining variable LR populations long-term	Enhancing diversity maintenance in the field in the face of climate change	If maintainer incentive were required it could prove costly	Maintenance/creation of new variation/ adaptation for breeding (including Participatory Plant Breeding), new farming systems and new products	Possibly limited by requirement for long-term funding	Farmers, ECPGR and *In situ*/ On-farm Network
8	*In situ* and on farm National Focal Point network linked to *ex situ* NFPs	Better country representation and 'joined-up' conservation	Need for recognition and accountability on commitments and engagements	Strengthen communication; share experiences/tools; improved European PGR conservation and use; reduce gap between system and user	Funding; sustainability	Countries, ECPGR; Bioversity; EU

Continued

Table 43.4. Continued.

Votes	Priority action	Strengths	Weaknesses	Opportunities	Threats	Responsibilities
8	Pre-breeding and participatory efforts to encourage farmers to maintain LR diversity	Better local, national and European understanding of LR potential and value	Time consuming and costly	Link maintenance, conservation and potential new products. Awareness of biodiversity and its use by public in general	Only achievable with long-term funding	ECPGR, farmers associations, national PGR agencies
5	Intensive use of genomic information: detailed knowledge of agronomically important genes	Systematic targeted breeding actions using LR diversity	Expensive and high technical requirement	Improved cultivars and greater food security, coordination and delivery of knowledge base and products for Europe and individual countries	Too often a narrow range of characteristics are considered	ECPGR, plant breeding companies
2	LR Communication platform within networks	Oriented combined knowledge of CWR *in situ*/*ex situ* conservation	Weak link between research community and farmers' cooperatives or groups	Create a framework programme from the existing data and knowledge	Funding; sustainability	ECPGR, Planta Europa, BGCI, other stakeholders users

community we need to engage in a dialogue with the breeders to understand their changing demands. Given that CWR conservation has a real cost and the fact that CWR diversity is neither conserved *ex situ* or *in situ* systematically or effectively currently, it can be argued that more efficient utilization, or the option value associated with more efficient utilization, is the primary justification for the resources required to systematically and effectively conserve CWR diversity.

The highest long-term priority is the establishment of a European network of CWR genetic reserves but, as was recognized above, achieving this goal is complex and has yet to be supported by the European protected area community. Part of persuading protected area managers of the value of *in situ* CWR conservation may be the establishment of pilot National networks of CWR genetic reserves in test countries as a model for a European network of CWR genetic reserves. Once the methodology is tested in Europe and the value of CWR genetic reserves to both the breeder and protected area communities is demonstrated, then these pilot National networks of CWR genetic reserves would form the foundation of the European network of CWR genetic reserves that in turn will ensure the systematic *in situ* conservation of the highest priority CWR diversity in Europe.

These top five long-term priority issues for European CWR conservation, together with the other priority issues suggested by the ECPGR *In situ* and On-farm Conservation Network members and other conference delegates, provide a lead for European CWR conservation action until 2030. It is worth noting that one of the long-term issues raised was the need to ensure that ≥70% of European CWR are conserved reliably, which is obviously linked to Target 9 of the Global Strategy for Plant Conservation (CBD, 2002), which calls for '70 per cent of the genetic diversity of crops and other socio-economically valuable plant species conserved, while respecting, preserving and maintaining associated indigenous and local knowledge'. More explicitly within the European context, the European Strategy for Plant Conservation (Planta Europa, 2008) calls for the establishment of 25 CWR genetic reserves in Europe, along with undertaking gap analysis of current *ex situ* CWR holdings, followed by the systematic filling of diversity gaps. Drawing on the experience of the ECPGR *In situ* and On-farm Conservation Network members the necessary expertise and methodologies to achieve these goals are available, the likely constraint to implementation over the next 20 years is available resources.

LR-related issues

The results of the horizon scanning exercise to identify long-term LR-related conservation issues in European countries are summarized in Table 43.4. The table is ranked with the highest priority issues identified by ECPGR *In situ* and On-farm Conservation Network members and other conference delegates at the top of the table.

It is widely believed that European and subsequent national seed legislation intended to standardize crop names and protect both consumers and breeders has had the unintended consequence of drastically reducing the numbers of traditional varieties or LR grown as there is a cost to DUS testing and impinging on the ability of farmers to grow older varieties or landraces not present on the list (Louwaars, 2007; Negri *et al.*, 2009, 2010). Therefore, it is not surprising that the highest longer-term priority issue for LR conservation is both to promote European legislation in favour of LR and on-farm conservation. By implication there is also an equal adverse priority to identify and nullify legislation that threatens LR and on-farm conservation. The recent legislative developments at the European level (i.e. Commission Directives 2008/62/EC, 2009/145/EC and 2010/60/EU) on Conservation Varieties seed production and marketing have opened a new way to safeguard biodiversity of interest for agriculture, however the legislation is not being implemented uniformly across Europe and it is yet to be seen whether overall it benefits LR and on-farm conservation. In fact, it could be argued that these Directives do not specifically address conservation per se and it is unlikely that it will be possible for all the surviving LRs to be commercialized. The European Commission also makes significant funding available each year for various forms of agroenvironmental schemes but again the national application of these schemes varies significantly among individual EU member states, some prioritize landrace inclusion and conservation, while in other such as the UK, landraces are excluded even though the animal equivalent, traditional animal breeds, are included. It is clear that European and national legislation could make a very significant contribution to LR and on-farm conservation but the proof of concept will be in the legislations implementation.

Recent surveys of home garden diversity in Europe (Stickland, 2001; Bailey *et al.*, 2009, Vincent and Maxted, 2009) have re-confirmed the wealth of traditional LR and heirloom varieties maintained in home gardens. Although the hypothesis that home gardens are a persistent reservoir of LR diversity has yet to be comprehensively tested, it does appear that LR lost from even small scale commercial production and many others that may not even have reached the market may be retained within home gardens. Further experience has shown that new LR is continuously being created in home gardens (Galluzzi *et al.*, 2010).

All of which reinforces the need to research the unique nature of crop diversity held in European home gardens and ensure its adequate conservation. Though one possible limitation of such research is the difficulty in identifying unique home garden diversity, do growers' names correlate with genetic distinction? Further, in a recent survey of crop varieties grown in Birmingham city allotments, although 97% of varieties were seed saved and planted annually, the percentage of growers that knew the variety name for the crop they grew varied from 10% for fruit trees to 90% for chilli peppers and many were sketchy concerning the varieties' origin (Dobbie and Maxted, 2010). A similar picture is observed in Portugal (see www.portau.org) – does this picture hold true elsewhere in Europe? Whether or not it does, it is an interesting hypothesis worth testing and the research is likely to empower ordinary gardeners in conservation action and raise awareness of agrobiodiversity conservation; it may also encourage younger gardeners to take up home garden production.

In recent years there has been an initiative led by the ECPGR *In situ* and On-farm Conservation Network to promote the inventory of national LR diversity and although none except Switzerland are comprehensive, there are national reports of LR inventories in Veteläinen *et al.* (2009) for Bulgaria, Denmark, Finland, Germany, Greece, Hungary, Italy, Portugal, Rumania, Russia, Sweden and the UK. The question can be asked how can we hope to use and conserve LR unless we have access to an inventory and know what we have available? So although a start has been made on national inventories, their completion to cover LR, former breeders' varieties and farm-saved seeds along with maintainer details and basic descriptors is critical. The latter information could then be used to facilitate the establishment of national Networks of LR maintainers (individual farmers or institutions) that maintain variable LR populations long term, with perhaps an institutional base like the Scottish Landrace Protection Scheme (see Green *et al.*, 2009) to provide a safety backup if original LR samples are lost on-farm.

Perhaps even more critically than for CWR diversity, the participatory involvement of local communities is central to LR conservation; after all here it is the farmer that does the conservation not the conservationist. The ideal model is one where the conservation involves the local communities and conservationists working in partnership; farmer networks are equally involved in policies and decisions. Although there is a growing literature on LR and on-farm conservation there is perhaps more scope for researching how best the conservationist might promote the conservation goals while at the same time not impinging on the local communities' development aspirations.

Among the other longer-term LR issues identified there are those that mirror priority CWR issues, such as: improved methodologies for inventorying and basic studies, encouraging pre-breeding and closer links with breeders, greater use of genomic information to locate adaptively important genes, promotion of participatory approaches to encourage farmers to maintain diversity, promotion of more biodiversity friendly agriculture systems and establishment of a communication platform within networks. Also specifically within the context of ECPGR, the Documentation and Information Network has via the national PGR coordinator nominated *ex situ* National Focal Points that among other activities collate gene bank holding information to generate national *ex situ* holdings, that are in turn combined at a European level to form the basis of the EURISCO (European Internet Search Catalogue of Ex Situ PGR Accessions – http://eurisco.ecpgr.org/). The Documentation and Information Network has also more recently promoted the nomination of *in situ* and on-farm National Focal Points and linking more closely the two sets of NFPs would enhance country representation and result in more 'joined-up' conservation.

43.4 Conclusions

The approach of this priority CWR and LR conservation exercise was to use the skills and knowledge of 52 ECPGR *In situ* and On-farm Conservation Network members and other conference delegates, representing a broad European range of PGR-related academic, research and NGO organizations, to identify the short- and longer-term issues that are/will impact on European CWR and LR conservation in the next 30 years. The experts identified issues and constraints on 12 CWR- and two LR-related short-term issues and using horizon scanning techniques (Sutherland *et al.*, 2006) identified a further 16 CWR and 12 LR issues. Furthermore, the issues generated from the horizon scanning were prioritized.

Notable among the results of the short-term CWR priority setting were the requirements for systematic genetic reserve implementation to conserve *in situ* CWR diversity at both individual national and European scales, wherever possible the integration of *in situ* CWR conservation into on-farm initiatives, the undertaking of national genetic gap analysis of CWR species as a basis for the establishment of national CWR conservation action plans, the incorporation of legislative protection for CWR species and genetic diversity at both national and European levels and promotion of closer conservationist/breeder links and through that link enhance CWR diversity utilization. For short-term LR priority the production of national LR inventories, and the generation and implementation of a National LR Conservation Strategy is of the highest priority. It is recognized that in the short term, CWR/LR use by breeders is likely to be channelled through *ex situ* collection because of the convenience for breeders, but long-term breeders may find it just as convenient to access germplasm from *in situ* sites, that is providing it is adequately characterized. In terms of the priorities between 2020 and 2035 identified by horizon scanning the likely highest priority CWR issues are implementing a European network of CWR genetic reserves (most likely in association with the existing Natura 2000 network of protected areas), systematic IUCN Red Listing of European CWR species and complementing the current criteria to take into account threat assessment of intra-specific genetic diversity, promotion of more participatory management and monitoring models for CWRs conservation, and improving national CWR inventories and prioritization of inventory on economic value to better address breeders' demands. The longer term priority issues for LR were identified as improved European and national legislation to promote LR on-farm conservation, research into the value of unique crop diversity held within European home gardens, the establishment of European on-farm inventory of LR, former breeders' varieties and farm-saved seeds, the development of participatory management and monitoring models for LR conservation and the promotion of biodiversity friendly agriculture systems. It was noted that the ECPGR *In situ* and On-farm Conservation Network was strategically well placed to achieve many of these goals through Network activities or through the newly EC-funded PGR Secure project.

We anticipate that the results presented in this paper will be used in four ways. First, that policy makers will examine how the issues identified here might impact upon their interests and then decide whether any warrant action and on what time scale. Second, we expect that this exercise will help the PGR research community engage in the likely issues of the future. The authors hope researchers, funders and those working on policy and regulation will use the outcome from this exercise when considering the future direction of strategic CWR and LR research. Third, the approach was used in this specific case to identify issues related to CWR and LR conservation but we hope the success of the exercise will be used in other areas of PGR research and priority-setting inside and outside Europe. As this action was linked to the ECPGR *In situ* and On-farm Conservation Network it is suggested that this exercise might prove useful for ECPGR as a whole and possibly be repeated for each successive 5-year Phase of ECPGR. Finally, this exercise may encourage further consideration and debate about the issues that are on the horizon and the ways in which scientists and decision makers can best communicate about them. The exercise involved the full range of stakeholders from 28 European countries and the European Union, as such the resultant issues identified and the priorities formulated are authorative and will have buy-in from the European PGR community. Strategic targeting of research efforts as implied here may lead to a more timely appreciation of the conservation threats and provide a basis for action before further CWR and LR diversity is lost and European food security is further undermined.

Acknowledgements

We acknowledge the financial support of the EC (Council Regulation no. 870/2004, Agreement no. 057 AGRI GEN RES, Contract no. AGRI-2006-0396) and ECPGR (European Cooperative Programme for Plant Genetic Resources) for funding this exercise. We also thank the many people from across the participating organizations that made suggestions and contributions.

References

Bailey, A., Eyzaguirre, P. and Maggioni, L. (eds) (2009) *Crop Genetic Resources in European Home Gardens*. Proceedings of a Workshop, 3–4 October 2007, Ljubljana, Slovenia. Bioversity International, Rome.

CBD (2002) *Global Strategy for Plant Conservation*. Secretariat of the Convention on Biological Diversity, Montreal, Quebec, Canada.

Dobbie, S. and Maxted, N. (2010) Sowing New Seeds: a vegetable survey of Birmingham allotments. Unpublished report, University of Birmingham.

Galluzzi, G., Eyzaguirre, P. and Negri, V. (2010) Home gardens: neglected hotspots of agro-biodiversity and cultural diversity. *Biodiversity and Conservation* 19, 3635–3654.

Green, N., Campbell, G., Tulloch, R. and Scholten, M. (2009) Scottish Landrace Protection Scheme. In: Veteläinen, M., Negri, V. and Maxted, N. (eds) *European Landraces: On-farm Conservation, Management and Use*. Bioversity Technical Bulletin 15. Bioversity International, Rome, pp. 233–243.

Hunter, D. and Heywood V.H. (eds) (2010) *Crop Wild Relatives: A Manual of in situ Conservation*. Earthscan, London.

Iriondo, J.M., Maxted, N. and Dulloo, E. (eds) (2008) *Conserving Plant Genetic Diversity in Protected Areas: Population Management of Crop Wild Relatives*. CAB International, Wallingford, UK.

IUCN (2001) IUCN Red List Categories and Criteria: Version 3.1. IUCN Species Survival Commission. IUCN, Gland, Switzerland and Cambridge. Available at: www.iucnredlist.org/static/categories_criteria_3_1 (accessed 21 October 2008).

King, D.A. and Thomas, S.M. (2007) Taking science out of the box – foresight recast. *Science* 316, 1701–1702.

Louwaars, N. (2007) Seeds of confusion – The impact of policies in seed systems. PhD thesis, University of Wageningen, the Netherlands.

Maxted, N., Ford-Lloyd, B.V. and Hawkes, J.G. (eds) (1997) *Plant Genetic Conservation: the* in situ *approach*. Chapman and Hall, London, 451 pp.

Maxted, N., Ford-Lloyd, B.V., Jury, S.L., Kell, S.P. and Scholten, M.A. (2006) Towards a definition of a crop wild relative. *Biodiversity and Conservation* 15(8), 2673–2685.

Maxted, N., Iriondo, J., Dulloo, E. and Lane, A. (2008) Introduction: the integration of PGR conservation with protected area management. In: Iriondo, J.M., Maxted, N. and Dulloo, E. (eds) *Plant Genetic Population Management*. CAB International, Wallingford, UK,.pp. 1–22.

Negri, V., Maxted, N. and Veteläinen, M. (2009) European Landrace Conservation: an Introduction. In: Veteläinen, M., Negri, V. and Maxted, N. (eds) *European Landraces: On-farm Conservation, Management and Use*. Bioversity Technical Bulletin 15. Bioversity International, Rome, pp. 1–22.

Negri, V., Barocco, R., Pacicco, L., Veronesi, F. and Venanzoni, R. (2010) Landraces in Europe: an approach towards identifying landrace rich areas to be protected with priority. In: *'CWRs and Landraces in Europe. Towards the establishment of genetic reserves for crop wild relatives and landraces in Europe' and the joint meeting of the ECPGR* In situ *and On-farm Conservation Network and the final meeting of the EU project AGRI GENRES 057 – AEGRO*. University of Madeira, Funchal, 40 pp.

Planta Europa (2008) *A Sustainable Future for Europe; the European Strategy for Plant Conservation 2008–2014*. Plantlife International (Salisbury, UK) and the Council of Europe (Strasbourg, France).

Stickland, S. (2001) *Back Garden Seed Saving. Keeping our vegetable heritage alive*. Eco-logic Books, Bristol.

Sutherland, W.J. and Woodroof, H.J. (2009) The need for environmental horizon scanning. *Trends in Ecology and Evolution* 24, 523–527.

Sutherland, W.J., Armstrong-Brown, S., Armsworth, P.R., Brereton, T., Brickland, J., Campbell, C.D., Chamberlain, D.E., Cooke, A.I., Dulvy, N.K., Dusic, N.R., Fitton, M., Freckleton, R.P., Godfray, H.C., Grout, N., Harvey, H.J., Hedley, C., Hopkins, J.J., Kift, N.B., Kirby, J., Kunin, W.E., MacDonald, D.W., Markee, B., Naura, M., Neale, A.R., Oliver, T., Osborn, D., Pullin, A.S., Shardlow, M.E.A., Showler, D.A., Smith, P.L., Smithers, R.J., Solandt, J.-L., Spencer, J., Spray, C.J., Thomas, C.D., Thompson, J., Webb, S.E., Yalden, D.W. and Watkinson, A.R. (2006). The identification of one hundred ecological questions of high policy relevance in the UK. *Journal of Applied Ecology* 43, 617–627.

Sutherland, W.J., Bailey, M.J., Bainbridge, I.P., Brereton, T., Dick, J.T.A., Drewitt, J., Dulvy, N.K., Dusic, N.R., Freckleton, R.P., Gaston, K.J., Gilder, P.M., Green, R.E., Heathwaite, A.L., Johnson, S.M., Macdonald, D.W., Mitchell, R., Osborn, D., Owen, R.P., Pretty, J., Prior, S.V., Prosser, H., Pullin, A.S., Rose, P., Stott, A., Tew, T., Thomas, C.D., Thompson, D.B.A., Vickery, J.A., Walker, M., Walmsley, C., Warrington, S., Watkinson, A.R., Williams, R.J., Woodroffe, R. and Woodroof, H.J. (2008) Future novel threats and opportunities facing UK biodiversity identified by horizon scanning. *Journal of Applied Ecology* 45(3), 1365–2664.

Sutherland, W.J., Clout, M., Côté, I.M., Daszak, P., Depledge, M.H., Fellman, E., Fleishman, E., Garthwaite, R., Gibbons, D.W., De Lurio, J., Impey, A.J., Lickorish, F., Lindenmayer, D., Madgwick, J., Margerison, C., Maynard, T., Peck, L.S., Pretty, J., Prior, S., Redford, K.H., Scharlemann, J.P.W., Spalding, M. and Watkinson, A.R. (2010) A horizon scan of global conservation issues for 2010. *Trends in Ecology and Evolution* 1198, 1–7.

Veteläinen, M., Negri, V. and Maxted, N. (eds) (2009) *European Landraces: On-farm Conservation, Management and Use*. Bioversity Technical Bulletin 15. Bioversity International, Rome, pp. 1–359.

Vincent, H. and Maxted, N. (2009) A Vegetable Survey of Worcestershire Allotments. Unpublished report, University of Birmingham.

Zencirci, N., Kaya, Z., Anikster, Y. and Adams, W.T. (eds) (1998) *Proceedings of an International Symposium on* in situ *Conservation of Plant Diversity*. Central Research Institute for Field Crops, Ankara, Turkey.

Index